D0138573

PULSATION, ROTATION AND MASS LOSS IN EARLY-TYPE STARS

INTERNATIONAL ASTRONOMICAL UNION

UNION ASTRONOMIQUE INTERNATIONALE

PULSATION, ROTATION AND MASS LOSS IN EARLY-TYPE STARS

PROCEEDINGS OF THE 162ND SYMPOSIUM OF THE
INTERNATIONAL ASTRONOMICAL UNION,
HELD IN ANTIBES-JUAN-LES-PINS, FRANCE, OCTOBER 5–8, 1993

EDITED BY

LUIS A. BALONA

South African Astronomical Observatory, Cape Town, South Africa

HUIB F. HENRICHS

Astronomical Institute, University of Amsterdam, Amsterdam, The Netherlands

and

JEAN MICHEL LE CONTEL

Observatoire de la Côte d'Azur, Nice, France

KLUWER ACADEMIC PUBLISHERS

DORDRECHT / BOSTON / LONDON

A C.I.P. Catalogue record for this book is available from the Library of Congress.

ISBN 0-7923-3044-7

Published on behalf of
the International Astronomical Union
by
Kluwer Academic Publishers, P.O. Box 17, 3300 AA Dordrecht, The Netherlands.

Kluwer Academic Publishers incorporates
the publishing programmes of
D. Reidel, Martinus Nijhoff, Dr W. Junk and MTP Press.

Sold and distributed in the U.S.A. and Canada
by Kluwer Academic Publishers,
101 Philip Drive, Norwell, MA 02061, U.S.A.

In all other countries, sold and distributed
by Kluwer Academic Publishers Group,
P.O. Box 322, 3300 AH Dordrecht, The Netherlands.

Printed on acid-free paper

Printed in the Netherlands

TABLE OF CONTENTS

1. PULSATION AND ROTATION

2. MAGNETIC FIELDS

3. X-RAY OBSERVATIONS

4. POLARIZATION

5. Be STARS: SPECTROSCOPY AND PHOTOMETRY

6. Be STARS: CIRCUMSTELLAR ENVIRONMENT

7. OB STELLAR WINDS

Scientific Organizing Committee

D. Baade (Germany)
L.A. Balona (South Africa), Chair
J.P. Cassinelli (USA)
H.F. Henrichs (The Netherlands)
A.M. Hubert (France)
M. Jerzykiewicz (Poland)
S.P. Owocki (USA)
J.R. Percy (Canada)
H. Saio (Japan)
M.A. Smith (USA)
F. Vakili (France)

Local Organizing Committee

N. Berruyer
D. Le Contel
J.M. Le Contel
J.E. Lefèvre
F. Mignard
D. Mourard
F. Vakili, Chair

PREFACE

The last IAU Symposium on Be Stars was held in München during April 1981. This was followed by IAU Colloquium 92 on the *Physics of Be Stars* in Boulder during August 1986. At both these meetings, the emphasis was on the circumstellar envelope rather than on the source of the mass loss—the underlying star. Extension of observations into the infrared, ultraviolet and X-ray region has greatly increased our knowledge and stimulated further research into the circumstellar material and the stellar wind as is evident from these Proceedings. Very recently, it has even become possible to resolve the circumstellar disks around some Be stars. The presentation of these results at the Symposium was a major highlight.

The development of high-resolution digital spectroscopy in the late 1970's led to the discovery of new types of pulsating variables among the B stars (53 Per, ζ Oph). This was soon followed by the discovery of periodic line-profile and light variations in Be stars. These discoveries have focussed attention on nonradial pulsation, in combination with rapid rotation as a possible mechanism for enhanced mass loss in Be Stars. The interest in nonradial pulsation in Be and other early-type stars motivated the Workshops on *The connection between nonradial pulsations and stellar winds in massive stars* in Boulder (April 1985) and on *Rapid Variability of OB-Stars* in Garching (October 1990). Since that time, the long-standing problem of the mechanism which drives pulsation in β Cephei stars has been resolved by the revision in metal opacities. This led to the possibility of understanding the pulsations in Be stars, if indeed it is pulsations that are responsible for the periodic line-profile and light variations (see the debate in these Proceedings). Recently, evidence has been found which suggests that the photospheres of Be stars are very active indeed. If we are to understand the complexities that abound in Be stars, it is necessary to exchange information about the effects of pulsation, rotation, magnetic fields, stellar winds, etc. in other early-type stars. This was one of the main motivating factors for the Symposium.

The proposal that a Symposium of this nature should be organized originated among the committee members of the *Working Group on Be Stars* under the chairmanship of D. Baade in 1988. We are grateful to Dietrich Baade, Vera Doazan, Tony Hearn, Mike Marlborough, Joachim Dachs, John Percy and Gerrie Peters for taking the first steps in getting the Symposium started. By 1991, it became clear that the Observatoire de la Côte d'Azur, in particular the Optical Interferometry Group, was interested in hosting the meeting. This is particularly appropriate as it was this Group who pioneered the technique of interferometric resolution of Be-star disks. The excellence of the work done at the Observatoire de Nice on pulsating B stars is also well known by the international community. Later that year the LOC under the chairmanship of Farrokh Vakili was established. Nominations for the SOC and its chairperson were solicited from the committee of the *Working Group on Be Stars*, a vote was taken, and the SOC was established.

The meeting was held at the beautiful coastal resort city of Juan-les-Pins just outside Nice from 5–8 October 1993 at the Palais des Congrès. The attractions of Juan-les-Pins and the calibre of the invited speakers assured a very good attendance of young and established scientists from all over the world. The format of the meeting was somewhat unusual in that there were no contributed talks. Instead, 15 minutes were devoted to discussion after each invited talk. The well-lit top floor of the Palais des Congrès was used for poster presentations: one hour was set aside each afternoon for viewing posters.

Through the support of the International Astronomical Union, and the Local Organizing Committee, many young astronomers were able to attend the Symposium. We are very grateful for this support and wish to express our thanks to the Executive of the IAU and the LOC.

We are very grateful to IAU Commission 27 (Variable Stars) who acted as sponsor for IAU support. In addition, we would like to thank the chairpersons and members of the following commissions for co-sponsorship: Commission 29 (Stellar Spectra), Commission 30 (Radial Velocities), Commission 35 (Stellar Constitution), Commission 36 (Theory of Stellar Atmospheres) and Commission 45 (Stellar Classification).

Without the financial assistance of many institutions the Symposium could not have taken place. We are very grateful indeed to the following groups who contributed so generously:

The International Science Foundation,
Observatoire de la Côte d'Azur (OCA),
OCA, Département A. Fresnel, URA-CNRS 1361,
OCA, Département CERGA, URA-CNRS 1360,
Collège de France, LAO.
The CNRS Research Groups:
Milieux Circumstellaires,
Structure Interne,
Astrophysique et Méthodes Interférométriques,
CNRS, Section Système Solaire et Univers Lointain.

Finally, the editors thank Erica Veenhof for her help with the preparation of the manuscript of these proceedings, which consist of 20 invited talks and 130 poster contributions, including 283 authors.

June 1994   L.A. Balona, H.F. Henrichs, J.M. Le Contel.

LIST OF PARTICIPANTS

AARON SIGUT, T.	Toronto, Canada	sigut@vela.astro.utoronto.ca
AERTS, Conny	Leuven, Belgium	kgafa01@cc2.kuleuven.ac.be
ALBERTS, Frank	Amsterdam, Netherlands	frank@astro.uva.nl
ANANDARAO, B.G.	Ahmedabad, India	anandarao@prl.ernet.in
ARNOLD, Luc	Caussols, France	ocar01::arnold
ARSENIJEVIĆ, Jelisaveta	Belgrade, Yugoslavia	
ARTRU, Marie-Christine	Lyon, France	artru@physique.ens-lyon.fr
BAADE, Dietrich	Garching, Germany	dbaade@eso.org
BABEL, Jacques	Gif-sur-Yvette, France	jbabel@ariane.saclay.cea.fr
BAGLIN, Annie	Meudon, France	baglin@frmeu51
BALLEREAU, Dominique	Meudon, France	mesioa::ballereau
BALONA, Luis	Cape Town, South Africa	lab@saao.ac.za
BARRERA, Luis	Bochum, Germany	(see J. Dachs)
BERGHÖFER, Thomas	Garching, Germany	thb@mpe-garching.mpg.de
BERNABEU, Guillermo	Alicante, Spain	bernabeu@ealiun11.bitnet
BERRUYER, Nicole	Nice, France	ocan01::nicole
BERTHOMIEU, Gabrielle	Nice, France	ocan01::bertho
BESKROVNAYA, Nina	St. Petersburg, Russia	beskr@gaoran.spb.su
BJORKMAN, Jon	Madison, WI, USA	bjorkman@madraf.decnet
BJORKMAN, Karen	Madison, WI, USA	kbjorkman@madraf.decnet
BLAZIT, Alain	Caussols, France	ocar01::blazit
BLOMME, Ronny	Brussels, Belgium	ronny@astro.oma.be
BOHLENDER, David	Kamuela, HI, USA	david@cfht.hawaii.edu
BOLTON, Tom	Richmont Hill, Canada	bolton@struve.astro.utoronto.ca
BONIFACIO, Piercarlo	Trieste, Italy	38439::bonifacio
BONNEAU, Daniel	Caussols, France	ocar01::bonneau
BOURGUINE, Michail	Moscou, Russia	mburgin@esoc1.bitnet
BRIOT, Danielle	Paris, France	iapobs::briot
CASSINELLI, Joseph	Madison, WI, USA	cassinelli@madraf.astro.wisc.edu
CHALABAEV, Almas	Grenoble, France	chalabaev@gag.observ-gr.fr
CHAUVILLE, Jacques	Meudon, France	mesioa::chauville
CHEN, Haiqi	Alberta, Canada	haiqi@bear.ras.ucalgary.ca
CIDALE, Lydia	La Plata, Argentina	lydia@fcaglp.edu.ar
CLARK, J.Simon	Southampton, England	jsc@uk.ac.soton
CLEMENT, Maurice	Toronto, Canada	mclement@apus.astro.utoronto.ca
COTE, Jacqueline	Groningen, Netherlands	jcote@sron.rug.nl
CRANMER, Steven	Newark, DE, USA	cranmer@bartol.udel.edu
CRUZALEBES, Pierre	Caussols, France	mpei2::cruzalebes
CUGIER, Henryk	Wroclaw, Poland	hcugier@plwrum11
CUYPERS, Jan	Brussel, Belgium	jan@astro.oma.be
DACHS, Joachim	Bochum, Germany	astrorub@
		ruba.rz.ruhr-uni-bochum.dbp.de
DE ARAÚJO, Francisco	Nice, France	ocan01::francisco
DEMEY, Katrien,	Leuven, Belgium	fgafa27@cc1.kuleuven.ac.be

xix

DENISSENKOV, Pavel	Garching, Germany	dpa@cvxastro. mpa.ipp-garching.mpg.de
DENOYELLE, Jozef	Brussels, Belgium	jozef@astro.oma.be
DOAZAN, Vera	Paris, France	iapobs::doazan
DOUGHERTY, Sean	Liverpool, England	smd@starlink. liverpool-john-moores.ac.uk
DUDOROV, Alexander	Chelyabinsky, Russia	dudorov@tphys.cgu.chel.su
DZIEMBOWSKI, Wojciech	Warsaw, Poland	wd@camk.edu.pl
EVANS, Nancy	North York, Canada	evans@nereid.sal.ists.ca
EVERALL, Chris	Southampton, England	ce@uk.ac.soton.phaster
EVERSBERG, Thomas	Bochum, Germany	(see J. Dachs)
FABREGAT, Juan	Valencia, Spain	16444::fabregat
FLIEGNER, Jens	Garching, Germany	jfl@ibm-1.mpa-garching.mpg.de
FLOQUET, Michèle	Meudon, France	mesioa::floquet
FRANDSEN, Soren	Aarhus, Denmark	srf@obs.aau.dk
FREITAS PACHECO, José	Sao Paolo, Brazil	
FRIEDJUNG, Michael	Paris, France	iapobs::friedjung
FRISCH, Hélène	Nice, France	
FULLERTON, Alex	Newark, DE, USA	fullerton@bartol.udel.edu
GARRIDO, Rafael	Granada, Spain	16488::garrido
GHOSH, K.K.	Alangayam, India	kkg@iiap.ernet.in
GIES, Douglas	Atlanta, GA, USA	gies@chara.gsu.edu
GLATZEL, Wolfgang	Göttingen, Germany	wglatzel@dgogwdg1.bitnet
GONZÁLEZ BEDOLLA, S.	Mexico, Mexico	salgb@astroscu.unam.mx
GONZALEZ, Jean-François	Lyon, France	jfgonzalez@physique.ens-lyon.fr
GOODE, Philip	Newark, NJ, USA	pgoode@solar.stanford.edu
GUERRERO, Gianantonio	Merate (Como), Italy	39217::Guerrero
HAHULA, Michael	ATLANTA, GA, USA	hahula@chara.gsu.edu
HANUSCHIK, Reinhard	Bochum, Germany	(see J. Dachs)
HAO, Jinxin,	Beijing, China	bmabao@ica.beijing.canet.cn
HARMANEC, Petr	Ondrejov, Czech Republic	hec@sunstel.asu.cas.cz
HASHIMOTO, Masaaki	Fukuoka, Japan	e76051a@kyu-cc.cc.kyushu-u.ac.jp
HEAP, Sara	Greenbelt, MD, USA	hrsheap@stars.gsfc.nasa.gov
HEARN, Tony	Utrecht, Netherlands	ahearn@fys.ruu.nl
HENRICHS, Huib	Amsterdam, Netherlands	huib@astro.uva.nl
HIESBERGER, Felicitas	Wien, Austria	feli@tycho.ast.univie.ac.at
HIRATA, Ryuko	Kyoto, Japan	hirata@kusastro.kyoto-u.ac.jp
HOLBERG, Jay	Tucson, AZ, USA	holberg@looney.lpl.arizona.edu
HORN, Jiri	Ondrejov, Czech Republic	horn@sunstel.asu.cas.cz
HUBERT, Anne Marie	Meudon, France	mesioa::hubert
HUBRIG, Swetlana	Potsdam, Germany	shubrig@aip.de
HUMMEL, Wolfgang	Brussel, Belgium	whummel@vub.ac.be
ILIEV, Lubomir	Sofia, Bulgaria	liliev@bgearn
JAENSCH, Mario	Potsdam, Germany	jaensch@betty.astro.uni-jena.de
JANOT-PACHECO, E.	Sao Paulo, Brazil	47556::janot
JEFFERY, C.Simon	St. Andrews, Scotland	c.s.j.@st-and.ac.uk
JERZYKIEWICZ, M.	Wroclaw, Poland	mjerz@plwruw11.bitnet

JIANG, Yiman	Vancouver, Canada	jiang@geop.ubc.ca
KAMBE, Eiji	Vancouver, Canada	kambe@geop.ubc.ca
KAPER, Lex	Amsterdam, Netherlands	lex@astro.uva.nl
KAUFER, Andreas	Heidelberg, Germany	t98@vm.urz.uni-heidelberg.de
KENNELLY, Ted	Vancouver, Canada	kennelly@geop.ubc.ca
KHOLTYGIN, Alexander	St. Petersburg, Russia	vayak@astro.lgo.spb.su
KILAMBI, Gopal	Hyderabad, India	
KIRBIYIK, Halil	Ankara, Turkey	hhes@trmetu
KIRIAKIDIS, Mihail	Göttingen, Germany	mkiriak@dgogwdg1.bitnet
KJELDSEN, Hans	Garching, Germany	hkjeldse@hg.eso.org
KOGURE, Tomokasu	Kyoto, Japan	
KOLB, Manfred	Garching, Germany	mkolb@eso.org
KOLKA, Indrek	Toravere, Estonia	indrek@aai.ee
KOUBSKÝ, Pavel	Ondrejov, Czech Republic	koubsky@sunstel.asu.cas.cz
KUPKA, Friedrich	Vienna, Austria	kupka@avia.una.ac.at
KUSCHNIG, Rainer	Vienna, Austria	kuschnig@avia.uva.ac.at
LABEYRIE, Antoine	Caussols, France	ocar01::labeyriea
LAFON, Jean Pierre	Meudon, France	lafon@frmeu51
LAVAL, Annie	Marseille, France	laval@obmara.cnrs-mrs.fr
LE CONTEL, Danielle	Nice, France	ocan01::danielle
LE CONTEL, Jean Michel	Nice, France	ocan01::lecontel
LEISTER, Nelson	Sao Paolo, Brazil	47556::leister
LOPEZ, Bruno	Nice, France	ocan01::lopez
LORENZ MARTINS, Silvia	Nice, France	ocan01::silvia
MACFARLANE, Joseph	Madison, WI, USA	jjm@vms3.macc.wisc.edu
MARLBOROUGH, Mike	London, Canada	marlboro@phobos.astro.uwo.ca
MATHYS, Gautier	Santiago, Chile	gmathys@eso.org
MAZZALI, Paolo	Trieste, Italy	mazzali@astrts.astro.it
McDAVID, David	Pipe Creek, TX, USA	mcdavid@pinet.aip.org
MEGESSIER, Claude	Meudon, France	mesioa::megessier
MEKARNIA, Djamel	Nice, France	mekarnia@rameau.obs-nice.fr
MIGNARD, François	Nice, France	ocan01::mignard
MIROCHNITCHENKO, A.	St. Petersburg, Russia	anat@gaoran.spb.su
MORAND, Frédéric	Caussols, France	ocar01::morand
MOSKALIK, Pawel	Warsaw, Poland	pam@camk.edu.pl.
MOSS, David	Manchester, England	mbbgsdm@cms.mcc.ac.uk
MOURARD, Denis	Caussols, France	ocar01::mourard
NORTH, Pierre	Lausanne, Switzerland	north@scsun.unige.ch
NORTON, Andrew	Milton Keynes, England	a.j.norton@open.ac.uk
OKAZAKI, Atsuo	Sapporo, Japan	okazaki@ astro1.phys.hokudai.ac.jp
OWOCKI, Stan	Newark, DE, USA	owocki@bartol.udel.edu
PAMYATNYKH, Aleksey	Moscou, Russia	alesza@camk.edu.pl
PAVLOVSKI, Kresimir	Zagreb, Croatia	pavlovski@uni-zg.ac.mail.yu
PEÑA, H. José	Mexico, Mexico	
PENICHE, Rosario	Mexico, Mexico	rpeniche@astroscu. unam.mx
PERCHERON, Isabelle	Caussols, France	ocar01::perchero

PERCY, John	Toronto, Canada	percy@utorphys.bitnet
PETERS, Geraldine	Los Angeles, CA USA	hyades::peters
PETROV, Romain	Nice, France	ocan01::petrov
PIGULSKI, Andrzej	Wroclaw, Poland	pigulski@plwruw11
PITERS, Ankie	Amsterdam, Netherlands	ankie@astro.uva.nl
POGODINE, Mikhail	St. Petersburg, Russia	pogodin@gaoran.spb.su
POLOSUKHINA, N.	Crimea Nauchny, Ukraine	postmaster%crao.crimea.ua@ relay.ussr.eu.net
PRINJA, Raman	London, England	rkp@starlink.ucl.ac.uk
PROVOST, Jeanine	Nice, France	ocan01::provost
QUIRRENBACH, Andreas	Washington, DC, USA	quirren@rira.nrl.navy.mil
RABBIA, Yves	Grasse, France	earn::yvrabb@fron,51
REID, Andy	London, England	19752::ahnr
REIG, Pablo	Valencia, Spain	16526::pablo
ROBBE, Sylvie	Caussols, France	ocar01::robbe
ROMANOV, Yuri S.	Odessa, Ukraine	root@astro.odessa.na
RONS, Nadine	Brussels, Belgium	nrons@vnet3.vub.ac.be
ROUNTREE, Janet	Tucson, AZ, USA	rountree@nssdc.gsfc.nasa.gov
RUŽIĆ, Željko	Zagreb, Croatia	zeljko.ruzic@uni-zg.ac.mail-yu
SAIO, Hideyuki	Sendai, Japan	saio@astroa.astr.tohoku.ac.jp
SAPAR, Arved	Toravere, Estonia	sapar@aai.ee
SARASWAT, Priyamvada	Bombay, India	bisht@tifrvax.bitnet
SAREYAN, Jean Pierre	Nice, France	
SCHNEIDER, Hartmut	Göttingen, Germany	hschnei@ibm.gwdg.de
SCHUMACHER, Gèrard	Caussols, France	
SCIGOCKI, David	Caussols, France	ocar01::scigocki
SHIMADA, Michihiro	Kyoto, Japan	shimada4@kusartrou.ac.jp
SIROUSSE ZIA, Haydeh	Paris, France	
SMITH, Myron	Lanham-Seabrook, MD, USA	msmith@iuegtc.dnet.nasa.gov
SOFIA, Sabatino	New Haven, CI, USA	sofia@astroo.astro.yale.edu
SOLANO MARQUEZ, E.	Mardrid, Spain	28843::esm
STEE, Philippe	Caussols, France	ocar01::stee
ŠTEFL, Stanislav	Ondrejov, Czech Republic	sstefl@sunstel.asu.cas.cz
TALLON BOSC, Isabelle	Caussols, France	bosc@rameau.obs-nice.fr
TALON, Suzanne	Paris, France	talon@iap.fr
TELTING, John	Amsterdam, Netherlands	john@astro.uva.nl
TORREJÓN, Jose Miguel	Valencia, Spain	jmt@evastr.mtapl.uv.es
VAKILI, Farrokh	Caussols, France	vakili@ocar01.span.nasa.gov
VALTIER, Jean Claude	Nice, France	ocan01::valtier
VERDUGO, Eva	Madrid, Spain	28847::ev
WAELKENS, Christoffel	Leuven, Belgium	fgafa01@cc1.kuleuven.ac.be
WALKER, Gordon	Vancouver, Canada	walker@astro.ubc.ca
WATERS, Rens	Amsterdam, Netherlands	rens@sron.rug.nl
WEISS, W.Werner	Vienna, Austria	weiss@avia.una.ac.at
WOOD, Kenneth	Glasgow, Scotland	gapv58@uk.ac.gla.cms
ZAPPALA, Rosario Aldo	Italy	aldo@astrct.ct.astro.tt
ZOREC, Jean	Paris, France	iapobs::zorec

1. PULSATION AND ROTATION

OBSERVATIONS OF PULSATING OB STARS

MIKOŁAJ JERZYKIEWICZ

Wrocław University Observatory, ul. Kopernika 11, PL-51-622 Wrocław, Poland

Abstract. Using examples from the literature, I try to show how astrophysical parameters of pulsating stars are derived and how certain (or, occasionally, uncertain) they are. The examples are arranged in the order of increasing input of *a priori* knowledge. Thus, I start with pulsation periods, because in most cases they can be obtained directly from observations. Then I consider the two stars for which pulsation constants have been determined without recourse to photometric calibrations. Next I show how multicolour photometry can help identify pulsation modes. Finally, I discuss the photometric T_{eff}, $\log g$, and pulsation constants. This leads to conclusions which contain some recommendations for the observers, including a list of promising targets for asteroseismology.

1. Introduction

Observations of pulsating stars can be used to derive information about their (1) equilibrium state, (2) eigenvalues, and (3) eigenfunctions. The equilibrium state is characterized by such parameters as mass, age, chemical composition, effective temperature and surface gravity, the latter two averaged over the pulsation cycle. The eigenvalues are the period (or frequency) and the pulsation constant (or the dimensionless frequency). The eigenfunction determines whether the mode of pulsation is radial or nonradial, fundamental or overtone, pressure or gravity, etc.

2. Pulsation Periods

They can be derived by induction or by deduction. In the present context, induction consists in deriving a period from observations, and then proving that it can only be accounted for by pulsations. Deduction, on the other hand, starts with assuming that pulsation in a specific mode is the cause of the observed variations. A phase in the pulsation cycle can then be found for each observation by a comparison with the assumed model. Finally, the phases yield the period.

2.1. AN EXAMPLE OF DERIVING PULSATION PERIODS BY INDUCTION: 16 (EN) LAC

This B2 IV star is a single-lined spectroscopic binary (Lee 1910, Struve & Bobrovnikoff 1925) and an eclipsing variable with orbital period $P_{orb} = 12\overset{d}{.}09684$ (Jerzykiewicz *et al.* 1978, Jerzykiewicz 1980). The primary eclipse is a partial transit; it lasts only $0\overset{d}{.}37$ and has a depth of about $0\overset{m}{.}04$ in B. The secondary eclipse has not been detected. The primary component of the system is the well-known β Cep-type star EN Lac. As most

3

other β Cep variables, EN Lac is multiperiodic: periodogram analyses of its light and radial velocity observations have revealed three sinusoidal terms with periods $P_1 = 0\overset{d}{.}16917$, $P_2 = 0\overset{d}{.}17079$, and $P_3 = 0\overset{d}{.}18171$. These terms are always present, but their amplitudes vary on a time scale of years (Fitch 1969, Jerzykiewicz *et al.* 1984).

The three terms are believed to represent normal pulsation modes because (1) this accounts for the multiperiodicity, and (2) the periods are too short to be explained otherwise.

Six fainter terms were recently found in the light variation of 16 Lac by Jerzykiewicz (1993b, 1993c). The data consisted of *UBV* observations obtained at the Lowell Observatory in the summer and autumn of 1965. Observations falling between the first and the fourth contact were, of course, omitted. The method was Lomb's (1976) least squares frequency analysis. The results can be summarized as follows.

While the 1965 *V* amplitudes (half-ranges) of the P_1, P_2, and P_3 terms amount to 18.0 ± 0.14, 9.6 ± 0.14 and 10.5 ± 0.14 mmag, the amplitudes of the six fainter ones range from 2.1 ± 0.14 to 0.7 ± 0.14 mmag. The strongest has a period equal to half the orbital period, a value to be expected if an "ellipticity effect" were present. However, neither the amplitude nor the phase of the observed variation agree with such an explanation. Two obvious possibilities of accounting for this term are: rotational modulation by a pair of spots placed on the opposite sides of the stellar surface or pulsation in a g-mode.

The second of the six faint terms has a period equal to $0\overset{d}{.}139$. It may be an overtone of any of the three strongest ones. The next term is the lowest order harmonic of the P_1 term, and the last three are the first order combination terms with frequencies $1/P_2 + 1/P_3$, $1/P_1 + 1/P_3$ and $1/P_1 + 1/P_2$.

The standard deviation of the nine-term least squares fit to the *V* observations amounts to 3.0 mmag, a number close to the mean error of a photoelectric observation of good quality. Thus, the light variation of 16 Lac outside eclipse is satisfactorily accounted for by the nine sinusoidal terms.

2.2. ANOTHER EXAMPLE OF DERIVING PULSATION PERIODS BY INDUCTION: HD 74560 = HY VEL

This is one of the seven stars used by Waelkens & Rufener (1985) to define the class of "mid-B variables." These objects have spectral types from B3 to B8, MK luminosity classes from V to III, and low values of the projected rotation velocity. They vary in light with amplitudes of a few hundredths of a magnitude and periods between one and three days.

Waelkens & Rufener (1985) suggested that variability of mid-B stars is caused by pulsations in g-modes of high radial order. An observational proof of this suggestion was provided by Waelkens (1991), who found all seven above-mentioned stars to be multiperiodic. Hence, Waelkens (1991) proposed to call them "slowly pulsating B stars."

Waelkens' (1991) data consisted of observations in the seven filter Geneva system. In the case of HD 74560, there were 408 observations obtained between 1976 and 1989. However, three or less observations per year were taken before 1981 and none in 1982, 1984, 1987 and 1988. The average density of observations was about 0.2 per night in 1986, 0.5 per night in 1981, 1983 and 1985, and 1.3 per night in 1989. The analysis was carried out by means of the phase dispersion minimization (PDM) method of Stellingwerf (1978). Three sinusoidal terms were derived, with periods equal to $1^{d}.5511$, $1^{d}.6455$ and $2^{d}.3571$. Their amplitudes amount to 13.5, 6.5 and 4.0 mmag in V, and 22.5, 9.8 and 6.5 mmag in U. The standard deviations of the three-term least squares fits range from 7.2 mmag in V to 11.8 mmag in U. These numbers significantly exceed the mean error of a photoelectric observation of good quality. Thus, the three terms do not account for all of the light variation of HD 74560.

In the remaining six cases the situation is similar. In fact, there is a rough correlation between the number of periods and the standard deviation of the least squares fits: HD 143309 and HD 160124, for which Waelkens (1991) found the largest numbers of periods, also have the greatest residual standard deviations, while HD 177863, with only two periods, has the smallest.

The question of what the unaccounted component of the light variation of the mid-B variables is due to is difficult to answer. Unfortunately, the data are not available for an independent analysis. Perhaps it is erratic in character, with a time scale of the order of a day or shorter. For HD 74560 this possibility is suggested by the fact that in the yearly PDM periodograms (Waelkens' Fig. 17), the depths of the minima corresponding to the dominant period are inversely proportional to the average density of observations: the 1986 minimum is the deepest, while the 1989 one is the shallowest.

2.3. DERIVING PULSATION PERIODS BY DEDUCTION: THE 53 PER VARIABLES

The term "53 Per variables" has been introduced by Smith (1980b) to denote the sharp-lined late O to mid B stars that show low-order line profile variations with a time scale of the order of a day. Smith (1977, 1980a) attributed the variations to g-modes of low l. Assuming a specific model, usually an $l = 2$ sectorial mode, he was then able to affix phases in the pulsation cycle to the observed line profiles. According to Smith (1980a), a few profiles suffice to determine the period, provided that only one mode is visible at a time.

Applying this deductive procedure to half a dozen stars, Smith (1980b) derived periods that ranged from about 5 to 45 hours. He concluded that the periods are unstable, the dominant one switching approximately every month to another value, with period ratios 2:1 seen most frequently.

The problem is that neither the values of the periods nor the period switching could be confirmed photometrically. An attempt by Smith *et al.* (1984) to reconcile the line profile and photometric observations of 53 Per itself was not entirely successful. The two photometric periods found by these authors, equal to about $1\overset{d}{.}7$ and $2\overset{d}{.}1$, were different from the ones determined earlier by Smith (1980b). In addition, the photometric fits were far from satisfactory, leaving a component of the light variation unaccounted for. An investigation by Balona & Engelbrecht (1985) of three 53 Per stars accessible from the southern hemisphere resulted in detecting light variability, but with periods different from those found by Smith (1980b) and, in addition, with amplitudes much smaller than in 53 Per. A similar result was obtained for the northern 53 Per variable ι Her by Chapellier *et al.* (1987). One 53 Per variable, ζ Cas, was found constant in light to within ± 1 mmag by Jerzykiewicz (1993a), and to within ± 2 mmag by Sadsaoud *et al.* (these Proceedings).

These discrepancies show the line-profile periods to be spurious. Thus, the assumption of nonradial pulsations, on which Smith's deductive method is based, may be questioned. The issue is somewhat confused by the fact that 53 Per itself is similar in its photometric behaviour to the mid-B variables. However, in this respect the prototype clearly differs from all other members of the group it is supposed to represent.

3. Empirical Pulsation Constants and Pulsation Modes

For two B stars, the pulsation constants can be derived directly from the definition $Q = P\sqrt{<\rho>}$, where P is the period and $<\rho>$ is the (equilibrium) mean density. These stars are the primary component of Spica (α Vir A), the β Cep variable that has apparently stopped pulsating (Lomb 1978, Balona 1985, Sterken *et al.* 1986), and EN Lac. In the first case, the mean density can be computed from the mass and radius obtained by combining interferometric and spectroscopic orbits (Herbison-Evans *et al.* 1971, Code *et al.* 1976). In the second case, the photometric and spectroscopic orbital elements lead to a mass-radius relation such that $<\rho>$, and hence Q, are virtually independent of the assumed mass (Pigulski & Jerzykiewicz 1988). For the three strongest terms, mentioned in Subsection 2.1, the pulsation constants are $Q_1 = 0.0335 \pm 0.0023$, $Q_2 = 0.0339 \pm 0.0023$, and $Q_3 = 0.0360 \pm 0.0024$.

A comparison with the results of linear nonadiabatic calculations of Dziembowski & Pamyatnykh (1993) shows that these three numbers are all very close to the theoretical Q values for the lowest radial order acoustic modes of low spherical harmonic order l. Thus, one of the three strongest terms in the light variation of 16 Lac may represent pulsation in the radial fundamental mode. However, additional information is required to find out which of them, if any, it is. One possibility is to use multicolour observations.

Such data contain information about l because the wavelength dependence of the light amplitude and phase is determined by the surface amplitudes and phases of the eigenfunctions.

The first attempt to take advantage of this possibility was made by Stamford & Watson (1977). An improved and extended version of this work has been recently published by Watson (1988). Using an analytic formula, originally derived by Dziembowski (1977), Watson (1988) expressed the light variation, $\Delta m(l, t)$, as a sum of two cosine terms, $A_1 \cos 2\pi t/P$ and $A_2 \cos(2\pi t/P + \Psi_T)$, where P is the pulsation period, and Ψ_T is the phase shift between local effective temperature variation and local radius variation. A_1 and A_2 are functions of l, the ratio of local fractional effective temperature amplitude to local fractional radius amplitude, henceforth denoted $\mathcal{B}$, and parameters obtainable from model stellar atmospheres. Adopting stationary model atmospheres of Kurucz (1979), Watson (1988) was able to use the expression for $\Delta m(l, t)$ to obtain the colour amplitude to visual amplitude ratio, A_{U-V}/A_V, and the colour phase minus visual phase difference, $\Phi_{U-V} - \Phi_V$, as a function of l, Ψ_T, and $\mathcal{B}$. Since the nonadiabatic eigenfunctions for models of β Cep stars were not available at the time, all Watson (1988) could do was to allow Ψ_T and $\mathcal{B}$ to vary within suitable limits in order to trace outlines of "areas of interest" in the $\Phi_{U-V} - \Phi_V$, A_{U-V}/A_V plane, that is, areas corresponding to a given l that can be occupied by β Cep stars.

For 16 Lac, the observed A_{U-V}/A_V and $\Phi_{U-V} - \Phi_V$ values, obtained from the UBV data mentioned in Subsection 2.1, place the P_1 term in the $l = 0$ area, and the P_2 and P_3 terms in the $l = 2$ or 3 area. Thus, the P_1 pulsation is the radial fundamental mode. Moreover, the close P_1, P_2 doublet is not caused by rotational splitting of two m states belonging to the same l, but results from a coincidence of two modes of different l.

Given the nonadiabatic eigenfunctions (Dziembowski & Pamyatnykh 1993), the "areas of interest" in the $\Phi_{U-V} - \Phi_V$, A_{U-V}/A_V plane reduce to points. Apart from l, the coordinates of the points are determined by the equilibrium parameters. In this case, comparison with the observations may yield not only l, but also T_{eff}, $\log g$, etc. This approach, especially promising for multiperiodic variables, has been explored by Cugier et al. (these Proceedings).

4. T_{eff}, $\log g$ and Q from Photometric Indices

In the following discussion, I shall use the effective temperatures and surface gravities derived from the Strömgren photometric indices, c_0 and β. From the several temperature and gravity calibrations of these indices available in the literature, I chose the recent one of Napiwotzki et al. (1993; henceforth NSW). I obtained the NSW T_{eff} and $\log g$ values by means of a FORTRAN

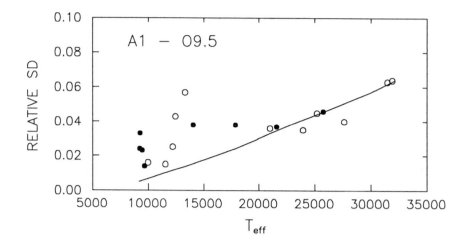

Fig. 1. Relative standard deviations of the empirical effective temperatures of Code *et al.* (1976) and propagation of photometric errors. Filled circles represent stars that were used as calibration standards by NSW. Solid line shows the relative standard deviation in the NSW effective temperature, produced by standard deviations of 20 mmag in c_0 and 5 mmag in β.

program kindly provided by Dr. Napiwotzki, using as data the c_0 and β indices from Balona (1984), Clausen & Giménez (1991), Davis & Shobbrook (1977), Giménez *et al.* (1990), and Shobbrook (1978, 1985). A comparison of the NSW effective temperatures with the empirical ones of Code *et al.* (1976) shows a good systematic agreement. In most cases the NSW effective temperatures agree with the empirical ones to within one standard deviation (SD) of the latter. However, the SD's of the empirical effective temperatures are quite large. This can be seen from Fig. 1, where the relative SD's are plotted as a function of T_{eff}. Taking into account the fact that effective temperatures derived from photometric indices (on the Strömgren or any other system) must be at least as uncertain as the empirical T_{eff} values, I conclude from this figure that the minimum relative SD of a photometric T_{eff} amounts to about 3 percent between 9000 and 15000 K, 4 percent between 15000 and 25000 K, and 5 to 6 percent between 25000 and 32000 K. It can also be seen from Fig. 1 that the SD's of c_0 and β would have to be as large as 20 and 5 mmag, respectively, to produce a 4 percent relative SD in the photometric T_{eff} at 25000 K. Since the SD's of c_0 and β (or similar indices in other systems) are normally smaller than the above-mentioned values, improving the effective temperature scale of B stars would require better empirical T_{eff} values than those available at present.

A comparison of NSW surface gravities with the empirical ones (that is, the ones derived directly) is presented in Fig. 2. The objects shown are: the components of the detached double-lined eclipsing binary CW Cep (Clausen

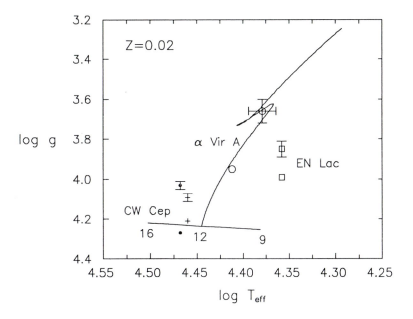

Fig. 2. Comparison of the photometric and empirical surface gravities in the log $T_{\rm eff}$ −log g plane. Symbols with error bars were plotted using the empirical values of log g. For α Vir A, two values of $T_{\rm eff}$ were used: the empirical (open circle with error bars) and the NSW one (open circle). For the other stars, that is, the A and B components of CW Cep (points and plus signs, respectively) and EN Lac (squares), the NSW temperatures were used. Theoretical ZAMS in the range from 9 to 16 M$_\odot$ and a 12 M$_\odot$ evolutionary track for X = 0.7 and Z = 0.02, computed by means of the Warsaw-New Jersey stellar evolution code (see Dziembowski & Pamyatnykh 1993), are shown with solid lines.

& Giménez 1991), α Vir A and EN Lac. I selected these systems because they illustrate the situations in which photometric indices of the components can be reliably derived from the combined light photometry: CW Cep has components of nearly equal brightness, Spica can be resolved, and 16 Lac has components that differ in brightness so much that the secondary's contribution is negligible. The empirical values of log g for the components of CW Cep and for α Vir A could be computed directly from the masses and radii. In the case of EN Lac, I used the mass–radius relation mentioned in Section 3. The value shown in Fig. 2 corresponds to the mass of 11 M$_\odot$, while the error bar covers the range from 8.5 to 14.5 M$_\odot$.

As can be seen from Fig. 2, empirical surface gravities are in all cases larger than the photometric ones. However, the sample is small. In addition, this figure shows that the popular method of deriving stellar masses and radii by comparing their positions in the log $T_{\rm eff}$ − log g plane with the theoretical evolutionary tracks may yield unreliable results. Masses obtained in this way for the two components of CW Cep and for α Vir A from the photometric log g values would be about 10 percent too large, while the radii would be too small by 10, 20 and 30 percent for CW Cep B, CW Cep A and α Vir A, respectively. As far as the masses are concerned, the situation would be even worse if the photometric $\log g$ values agreed with the empirical ones: the discrepancy would then increase to 20 percent.

Masses and radii obtained from the evolutionary tracks in the $\log T_{eff} - \log g$ plane and the photometric T_{eff} and $\log g$ values can be used to derive pulsation constants. Clearly, such photometric pulsation constants must be treated with caution.

A mass obtained from an evolutionary track in the $\log T_{eff} - \log g$ plane is sometimes combined with T_{eff} and $\log g$ in order to compute the luminosity, which is then used as the ordinate in the H-R diagram. Apart from its dubious accuracy, the procedure is based on a circular argument. A recent example of an application of this incorrect procedure can be found in a PhD thesis from a well known European university.

The vicious circle just mentioned can be broken by adding information, for example, the absolute magnitude. However, the two components of Spica are the only B stars for which the distance has been derived directly (Herbison-Evans et al. 1971), so that their absolute magnitudes can be obtained from the observed magnitudes and the text-book definition. For other early-type stars, methods of varying degrees of indirectness are used, ranging from main sequence fitting for galactic clusters and associations to MK classification for field stars. In practice, the absolute magnitudes for plotting early-type stars in the H-R diagram are often obtained from photometric indices calibrated by means of the main sequence fitting procedure. Of course, the resulting plot is not really an H-R diagram at all, but a $\log T_{eff} - \log g$ diagram in disguise.

Since the empirical T_{eff} value of the fainter component of Spica is not known, α Vir A is the only B star that can be plotted in the H-R diagram without recourse to indirect methods. Whether HIPPARCOS will help to change this situation remains to be seen.

5. Conclusions

(1) Since the pulsation mechanism for the β Cep and mid-B variables is now known (see Dziembowski, these Proceedings), nonadiabatic eigenfunctions can be computed so that these objects will become important targets for asteroseismology. Obviously, asteroseismology should be attempted in the first place for stars which have (some of) their equilibrium parameters known directly. For example:
- α Vir A (if it recovers) and EN Lac,
- HD 92024, which is an eclipsing binary similar to 16 Lac and, in addition, a member of NGC 3293 (Engelbrecht & Balona 1986, Jerzykiewicz & Sterken 1992, Balona 1994),

- other β Cep and mid-B variables in open clusters with well known distances,
- the three β Cep stars with the empirical effective temperatures determined by Code *et al.* (1976), namely, β CMa, β Cru A and ϵ Cen.

(2) More and better empirical T_{eff} values of OB stars are needed.

(3) In order to better calibrate the photometric $\log g$ scale, good line profile and photometric data are needed for OB components of spectroscopic eclipsing binaries with known orbits.

Acknowledgements

An IAU travel grant is gratefully acknowledged. This research has been supported by KBN grant Nr 2 P304 001 04.

References

Balona, L.A.: 1984, *Mon. Not. Roy. Astr. Soc.* **211**, 973.

Balona, L.A.: 1985, *Mon. Not. Roy. Astr. Soc.* **217**, 17P.

Balona, L.A.: 1994, *Mon. Not. Roy. Astr. Soc.* , in press.

Balona, L.A. and Engelbrecht, C.A.: 1985, *Mon. Not. Roy. Astr. Soc.* **214**, 559.

Chapellier, E., Le Contel, J.M., Valtier, J.C., Gonzalez-Bedolla, S., Ducatel, D., Morel, P.J., Sareyan, J.P., Geiger, I. and Antonelli, P.: 1987, *Astron. Astrophys.* **176**, 255.

Clausen, J.V. and Giménez, A.: 1991, *Astron. Astrophys.* **241**, 98.

Code, A.D., Davis, J., Bless, R.C. and Hanbury Brown, R.: 1976, *Astrophys. J.* **203**, 417.

Davis, J. and Shobbrook, R.R.: 1977, *Mon. Not. Roy. Astr. Soc.* **178**, 651.

Dziembowski, W.: 1977, *Acta astr.* **27**, 203.

Dziembowski, W.A. and Pamyatnykh, A.A.: 1993, *Mon. Not. Roy. Astr. Soc.* **262**, 204.

Engelbrecht, C.A. and Balona, L.A.: 1986, *Mon. Not. Roy. Astr. Soc.* **219**, 449.

Fitch, W.S.: 1969, *Astrophys. J.* **158**, 269.

Giménez, A., García, J.M., Rolland, A. and Clausen, J.V.: 1990, *Astron. Astrophys. Suppl.* **86**, 259.

Herbison-Evans, D., Hanbury Brown, R., Davis, J. and Allen, L.R.: 1971, *Mon. Not. Roy. Astr. Soc.* **151**, 161.

Jerzykiewicz, M.: 1980, in Hill, H.A. and Dziembowski, W.A., eds., *Nonradial and Nonlinear Stellar Pulsation*, Springer: Berlin, 125.

Jerzykiewicz, M.: 1993a, *Astron. Astrophys. Suppl.* **97**, 421.

Jerzykiewicz, M.: 1993b, *Acta astr.* **43**, 13.

Jerzykiewicz, M.: 1993c, *Acta astr.* **43**, 182.

Jerzykiewicz, M. and Sterken, C.: 1992, *Mon. Not. Roy. Astr. Soc.* **257**, 303.

Jerzykiewicz, M., Borkowski, K.J. and Musielok, B.: 1984, *Acta astr.* **34**, 21.

Jerzykiewicz, M., Jarzębowski, T., Le Contel, J.M. and Musielok, B.: 1978, *Inf. Bull. Var. Stars* , 1508.

Kurucz, R.L.: 1979, *Astrophys. J. Suppl.* **40**, 1.

Lee, O.J.: 1910, *Astrophys. J.* **32**, 307.

Lomb, N.R.: 1976, *Astrophys. Space Sci.* **39**, 447.

Lomb, N.R.: 1978, *Mon. Not. Roy. Astr. Soc.* **185**, 325.

Napiwotzki, R., Schönberner, D. and Wenske, V.: 1993, *Astron. Astrophys.* **268**, 653 (NSW).

Pigulski, A. and Jerzykiewicz, M.: 1988, *Acta astr.* **38**, 401.

Shobbrook, R.F.: 1978, *Mon. Not. Roy. Astr. Soc.* **184**, 43.

Shobbrook, R.F.: 1985, *Mon. Not. Roy. Astr. Soc.* **214**, 33.
Smith, M.A.: 1977, *Astrophys. J.* **215**, 574.
Smith, M.A.: 1980a, in Fischel, D., Lesh, J.R. and Sparks, W.M., eds., *Current Problems in Stellar Pulsation Instabilities*, NASA TM 80625, 391.
Smith, M.A.: 1980b, in Hill, H.A. and Dziembowski, W.A., eds., *Nonradial and Nonlinear Stellar Pulsation*, Springer: Berlin, 60.
Smith, M.A., Fitch, W.S., Africano, J.L., Goodrich, B.D., Halbedel, W., Palmer, L.H. and Henry, G.W.: 1984, *Astrophys. J.* **282**, 226.
Stamford, P.A. and Watson, R.D.: 1977, *Mon. Not. Roy. Astr. Soc.* **180**, 551.
Stellingwerf, R.F.: 1978, *Astrophys. J.* **224**, 953.
Sterken, C. and Jerzykiewicz, M.: 1993, *Space Sci. Rev.* **62**, 95.
Sterken, C., Jerzykiewicz, M. and Manfroid, J.: 1986, *Astron. Astrophys.* **169**, 166.
Struve, O. and Bobrovnikoff, N.T.: 1925, *Astrophys. J.* **62**, 139.
Waelkens, C.: 1991, *Astron. Astrophys.* **246**, 453.
Waelkens, C. and Rufener, F.: 1985, *Astron. Astrophys.* **152**, 6.
Watson, R.D.: 1988, *Astrophys. Space Sci.* **140**, 255.

Discussion

Smith: I have two comments:

(1) I think you have been a little hard on the "deductive" (line profile fitting) method of inferring NRP models. The principal problem you describe derived not so much from the methodology as from my own over-optimism in the early days in translating spectrophotometric errors into pulsation phase errors from too few observations, particularly when multiple modes are present. As for 53 Per itself, both line profile and photometric data in the mid-1980's and also recent photometric data are in very good agreement with the original two-period result determined by Buta and Smith in 1979. The two methodologies (photometric and line profiles) should be seen as complementary and indeed they seem to be in agreement whenever the amplitudes are large.

(2) The second point is semantical only: I would like to appeal to the photometric community to avoid using physical mechanisms in defining variable star classes. There are two reasons for this. The first is historical, namely, that 53 Per stars were discovered by northern observers via line profile variations, whereas the "SPB" stars were discovered by southern observers photometrically. However, more fundamentally, it is perhaps ill-considered to define a class of variable stars in terms of an expected physical mechanism (which will be surely wrong in some cases, e.g., for "ellipsoidal variable" binaries). This is why astronomers have fallen back on a prototypical star's name in their nomenclature. In this case, the term "53 Per" still fits as a prototype because the two periods found years ago are still the values found in modern analyses (e.g. Huang *et al.*, these Proceedings).

Jerzykiewicz: I agree that using physical mechanisms in defining variable star classes should be avoided. This is why I refer to the Waelkens & Rufener stars as the mid-B variables, the term they introduced in their discovery paper. However, using 53 Per as a prototype of this class of variable stars would be historically incorrect and, in addition, misleading. The point is that your 53 Per stars, except for 53 Per itself, are photometrically different from the mid-B variables.

Balona: I would like to point out that it is very dangerous to claim multiperiodicity unless you have a well-sampled data set. In the 3 best-studied 53 Per stars the photometry seems to suggest that there is only one period which is coherent over many seasons. The other periods change amplitude and period in the matter of weeks (i.e., they are not coherent). Under these circumstances it will be very difficult to obtain the correct eigenfrequencies for all except the single coherent pulsation.

Jerzykiewicz: I agree. Historically, the first β Cep star in which multiperiodicity (called "the beat phenomenon" at the time) has been discovered was β CMa. The discovery, made by Meyer in 1934, consisted in showing that two short periods, both close to 6 hours, account for the long period modulation of the radial velocity amplitude. In addition, only one of the periods was seen in the line width variation. Thus, there was no doubt that the two periods are physically distinct. Later, these periods were also found in the light variation (see, e.g., Sterken & Jerzykiewicz 1993 for the references and further details). As far as multiperiodicity of the mid-B variables is concerned, photometry is all we have so far.

Harmanec: In a search for multiperiodicity one must be cautious about the way of prewhitening the data for individual periods in cases when variations are significantly nonsinusoidal. Otherwise the multiperiodicity can only mean a Fourier decomposition of finite data set.

Jerzykiewicz: I agree. However, consistent prewhitening with sinusoids is also OK since harmonics, if they are present, will show up in the later runs anyway. Of course, the data should be adequate.

Henrichs: (1) Are the amplitude changes in the different periods that you mentioned for 16 Lac typical or exceptional among these stars?

(2) You showed in your diagram that when the amplitude of one period increased, the amplitudes of the other periods decreased at the same time. Can one say something quantitatively about these amplitude changes, for instance: did the sum of the squares of the amplitudes remain constant?

Jerzykiewicz: (1) They are believed to be an exception rather than a rule. However, apart from the extreme cases like Spica or 16 Lac, the amplitude changes may escape detection if adequate observations over a number of years are not available. Long-term stability of the periods and amplitudes of the well known multiperiodic β Cep star 12 (DD) Lac has been recently studied by Pigulski (these Proceedings).

(2) No, in 1977 all three amplitudes were by a factor of about 2 smaller than in 1965.

Bolton: (1) Be careful in equating 53 Per with SPB stars until we have equivalent data for both types of stars.

(2) I agree with Dr. Jerzykiewicz's comment that the so called 53 Per stars around the β Cep strip are different from the cooler 53 Per stars. The line profile survey carried out by Mike Fieldus and I shows that LPV has a sharp blue edge near $\log T_{\mathrm{eff}} = 4.3$ for $v \sin i \leq 100$ km s^{-1}.

Jerzykiewicz: Thank you.

Percy: One problem of identifying 53 Per stars with SPB stars is that most photometry is being done in the southern hemisphere and most spectroscopy in the northern. A good approach is to study one or two stars simultaneously with both techniques: 53 Per and ϵ Per are good candidates.

Jerzykiewicz: I agree.

NONADIABATIC OBSERVABLES
IN β CEPHEI STAR MODELS

H. CUGIER[1], W.A. DZIEMBOWSKI[2] and A.A. PAMYATNYKH[2,3]

[1] *Astronomical Institute, Wrocław University, Kopernika 11, 51-622 Wrocław, Poland*
[2] *N. Copernicus Astronomical Center, Bartycka 18, 00-716 Warsaw, Poland*
[3] *Institute of Astronomy, Pyatnitskaya 48, 109017 Moscow, Russia*

Results of the recent stability surveys (Dziembowski and Pamyatnykh 1993; Gautschy and Saio 1993) leave no doubts that the opacity mechanism is responsible for oscillations observed in β Cephei stars. The linear nonadiabatic analysis used to determine the instability domains in the HR diagram, yields also quantities that may be compared with observations. These nonadiabatic observables are evaluated from the complex eigenfunctions $y(r)$ and $f(r)$ describing variations of the radial displacement and the bolometric flux, respectively. Both y and f are very nearly constant within the stellar atmosphere. The eigenfunctions describing the horizontal displacement and variations of thermodynamical quantities may be expressed in terms of y and f. Since the linear eigenfunctions may be arbitrarily normalized, there are only two real independent observables. We may choose them to be $\tilde{f} = \mathrm{abs}(f/y)$ and $\psi = \arg(f/y)$. Using static atmosphere models, with the inertial term included in the effective gravity, we may evaluate amplitude ratios and phase differences for integrated changes in directly measured parameters (Dziembowski 1977, Stamford and Watson 1981, Watson 1988).

Balona and Stobie (1979) showed that the amplitude ratio *vs* phase difference diagrams for colors and luminosity are useful to identify the spherical harmonic degree, l, of an observed mode. We reexamined diagnostic value of such diagrams making use of $\tilde{f}$ and ψ for unstable low-l modes from the survey of Dziembowski and Pamyatnykh (1993). In Fig. 1a we show results of model calculations for the V and the 150 nm bands. Employing the satellite ultraviolet data turned out to be exceptionally revealing. An identification of the l value, with good photometric data should be unambiguous. In Fig. 1b we show that the radial velocity data combined with the UBV photometry may also be used for the same purpose. The model points corresponding to various spherical harmonic degrees occur in well separated domains. One may see that there is no ambiguity in assigning the l value to the observed modes.

Determination of the l-value is not the only use of the nonadiabatic observables. The plots in Fig. 1a clearly show that, especially in the case of radial pulsations, the data can be used to constraint mean stellar parameters. As well, that they may be used to distinguish between the fundamental and the first overtone pulsators. We believe that, in addition to precise fre-

L. A. Balona et al. (eds.), Pulsation, Rotation and Mass Loss in Early-Type Stars, 15–16.

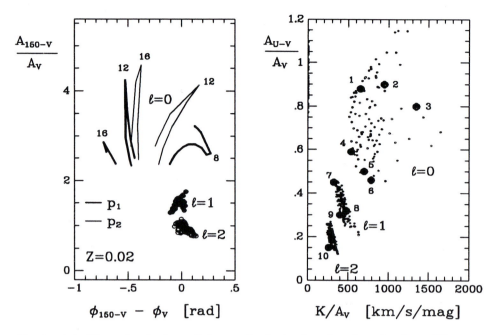

Fig. 1. (a) The ratio of color to light amplitudes is plotted against the phase difference
for unstable modes in β Cephei models. For the case of radial pulsations, the plotted lines
connect points, separately for p_1 (fundamental) and p_2 modes, corresponding to sequence
of models, from the instability onset to the moment of hydrogen exhaustion in the core.
The value of mass is given in $M_\odot$. Points plotted for the nonradial modes cover the same
range of stellar models.
(b) The ratio of color to light amplitudes is plotted against the ratio of radial velocity to
light amplitudes for the same models (small dots) and for β Cephei stars (big dots). The
numbers from 1 to 10 correspond to the dominat mode of pulsation of the following stars:
16 Lac, ξ^1 CMa, BW Vul, γ Peg, β Cep, λ Sco, KP Per, 12 Lac, 15 CMa, and β Cru,
respectively.

quency measurements, the nonadiabatic observables should be regarded as
important data for asteroseismology. Thus, future efforts in this field should
rely on *multicolor* photometry and/or employment of spectroscopic data.

This work was supported by the research grants No. 2 1185 91 01 and
2 1241 91 04 from the Polish Scientific Research Committee (KBN).

References

Balona, L.A. and Stobie, R.S.: 1979, *MNRAS* **189**, 649
Dziembowski, W.A.: 1977, *Acta Astr.* **27**, 203
Dziembowski, W.A. and Pamyatnykh, A.A.: 1993, *MNRAS* **262**, 204
Gautschy, A. and Saio, H.: 1993, *MNRAS* **262**, 213
Stamford, P.A. and Watson, R.D.: 1981, *Astrophys. Space Sci.* **77**, 131
Watson, R.D.: 1988, *Astrophys. Space Sci.* **140**, 255

MULTIWAVELENGTH STUDIES OF β CEPHEI STARS

H. CUGIER[1], A. PIGULSKI[1], G. POLUBEK[1] and R. MONIER[2]

[1] *Astronomical Institute, Wrocław University, Kopernika 11, 51-622 Wrocław, Poland*
[2] *IUE Observatoty VILSPA, P.O. Box-Apartado 50727, Madrid 28080, Spain*

As first pointed out by Moskalik and Dziembowski (1992) all β Cephei stars lie within the domain of H–R diagram where κ–mechanism effectively drives pulsations in the stellar layers with $T \approx 2\text{x}10^5$ K. For most of these objects a chemical composition described by X = 0.70 and Z = 0.02 is sufficient to account for the pulsations, cf. Dziembowski and Pamyatnykh (1993). Recently, Cugier, Dziembowski and Pamyatnykh (1993) have investigated how the present knowledge about nonadiabatic observables of β Cephei stars affects methods of identification of the spherical harmonic degree, l. They found that good photometric and radial velocity data should result in unambiguous identification of l. Cugier, Dziembowski and Pamyatnykh also concluded that nonadiabatic observables can be used to obtain mean stellar parameters of pulsating stars.

We report here, as examples, the studies of δ Ceti and BW Vulpeculae. The above mentioned analysis of the ground-based photometric data of δ Cet taken from Jerzykiewicz et al. (1988) indicates: $l = 0$, p_2, $\log T_{\text{eff}} = 4.346$ and $\log g = 3.73$. Figure 1 shows that indeed only a model with $l = 0$ is able to explain the observed flux behaviour of δ Cet in the satellite ultraviolet region. Futhermore, the observed phases of flux maximum as a function of wavelength offer the possibility to determine the effective temperature of β Cephei stars with high precision as Fig. 2 shows for δ Cet.

In Fig. 3 the observed light ranges for BW Vul are compared with the nonadiabatic model ($l = 0$, p_1, $\log T_{\text{eff}} = 4.29$ and $\log g = 3.71$). As one can see, a very good agreement exists even for this star, which is rather extreme case among β Cephei stars considering its large light and radial-velocity amplitudes.

Acknowledgements

This work was supported by the research grant No. 2 1241 91 04 from the Polish Scientific Research Committee (KBN).

References

Cugier, H. Dziembowski, W.A. and Pamyatnykh, A.A.: 1993, this issue
Dziembowski, W.A. and Pamyatnykh, A.A.: 1993, *MNRAS* **262**, 204
Jerzykiewicz, M., Sterken, C. and Kubiak, M.: 1988, *A&AS* **72**, 449
Moskalik, P. and Dziembowski, W.A.: 1991, *A&A* **256**, L5

17

L. A. Balona et al. (eds.), Pulsation, Rotation and Mass Loss in Early-Type Stars, 17–18.

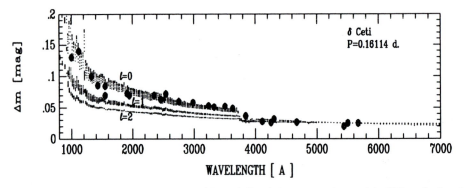

Fig. 1.   Calculated light ranges Δm (dotted lines) in comparison with UV and visual observations (filled circles) for δ Cet. All nonadiabatic models with $l=0$, 1 and 2 have the same period $(P = 0.16114\,\mathrm{d})$

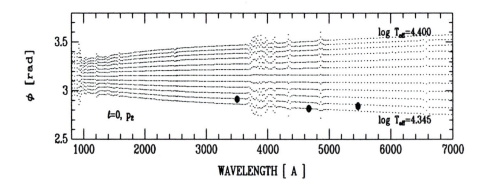

Fig. 2.   The observed (filled circles with error bars) phases of flux maximum for δ Cet are plotted together with nonadiabatic calculations (dots) corresponding to p_2 mode of $l = 0$. Stellar models (all with the period equal to 0.16114 d) are labelled by $\log T_{\mathrm{eff}}$ values. The step in $\log T_{\mathrm{eff}}$ is equal to 0.005 dex.

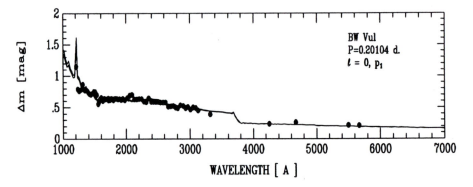

Fig. 3.   The best-fit nonadiabatic model compared with the IUE observations of BW Vul.

NONLINEAR PULSATIONAL BEHAVIOUR
OF BW VULPECULAE

P. MOSKALIK[1] and J.R. BUCHLER[2]

[1] N. Copernicus Astronomical Center, Bartycka 18, 00-716 Warsaw, Poland
[2] Department of Physics, University of Florida, Ganesville, FL32611, USA

BW Vulpeculae is a large amplitude β Cephei-type star, pulsating with a single period of 0.20104 day. Its nonlinear modeling was first attempted by Pesnell & Cox (1980). Because the true instability mechanism was not known then, the pulsation amplitude was imposed in that study in an arbitrary fashion.

The driving of β Cephei-type pulsations is now well understood (e.g. Dziembowski & Pamyatnykh 1993) and the hydrodynamical modeling can be done today in a fully self-consistent way. For this purpose we study 120-zone envelope models extending down to $T = 3 \times 10^7$K. We use the OPAL opacities (Iglesias, Rogers & Wilson 1992) for the metallicity of $Z = 0.02$. The models are constructed as to lie on the evolutionary tracks of Dziembowski & Pamyatnykh (1993) and are chosen to match BW Vul's pulsation period and $\log T_{\mathrm{eff}}$ ($\log T_{\mathrm{eff}} = 4.362 - 4.366$; Heynderickx 1991). The radial pulsations of the models are then studied with the relaxation hydrocode (Stellingwerf 1974), which yields strictly periodic full amplitude solutions (limit cycles).

In most of the models of this survey both the fundamental mode and the first radial overtone are linearly unstable. However, the overtone limit cycle has in all studied cases too small an amplitude to square with observations. In addition, this limit cycle is always unstable towards the excitation of the fundamental mode. Therefore, BW Vul almost certainly cannot be a first overtone pulsator.

For the fundamental mode, the hydrodynamical calculations reproduce the observed light and radial velocity curves reasonably well (Fig. 1). In particular, the large amplitudes of both curves are correctly modeled, although the lightcurve amplitude is somewhat too large. The jump and the following standstill in the velocity curve, which are the most outstanding observed features, are also reproduced. The standstill is caused by an emerging shock wave which originates at the bottom of the He$^+$ ionization zone (Fig. 2). The strong compression ocurring in the shock (14-fold increase of ρ in less then $0.02P$) results in an almost five-fold increase in the Rosseland-mean opacity. The sudden jump in the optical depth contributes to the formation of the apparent discotinuity in the observed radial velocities.

We find that the shock wave is stronger for more centrally condensed models in the secondary contraction phase. The resultant standstill is lo-

L. A. Balona et al. (eds.), Pulsation, Rotation and Mass Loss in Early-Type Stars, 19–20.

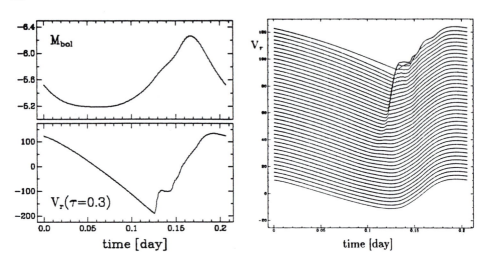

Fig. 1. Variations of bolometric magnitude (top) and radial velocities at constant $\tau = 0.3$ (bottom) for a BW Vul model of $M = 11 M_\odot$, $\log(L/L_\odot) = 4.1457$ and $\log T_{\text{eff}} = 4.3658$. The model is in the Main Sequence expansion phase and pulsates in the fundamental mode with $P = 0.2068$ day. The velocity amplitude is 234 km/s (after correcting with the factor of 17/24) and is close to the observed value of 210 km/s (Furenlid *et al.* 1987). The standstill in the velocity curve is due to an emerging shock wave. The theoretical lightcurve amplitude is 1.06 mag. This is close to the far-UV amplitude (1.16 mag), but larger than the estimated bolometric amplitude of 0.75 mag (Barry *et al.* 1984). No standstill appears in the lightcurve.

Fig. 2. Radial velocities *versus* time for the outer zones of the BW Vul model of Fig. 1. The shock wave propagating outwards originates at the bottom of the He$^+$ ionization zone.

cated in this case nearly exactly in the middle of the raising branch of the velocity curve. Such behaviour is in a better agreement with what is actually observed. This result suggests, that BW Vul is most likely evolved beyond the Main Sequence expansion phase.

The results presented here are preliminary. The full discussion of our modelling will be presented elswhere.

References

Barry, D.C., Holberg, J.B., Forrester, W.T., Polidan, R.S. and Furenlid, I.: 1984, *ApJ.* **281**, 766.

Dziembowski, W.A. and Pamyatnykh, A.A.: 1993, *MNRAS* **262**, 204

Furenlid, I., Young, A., Meylan, T., Haag, C. and Crinklaw, G.: 1987, *ApJ.* **319**, 264

Heynderickx, D.: 1991, Ph. D. Thesis, Katholieke Universiteit Leuven.

Iglesias, C.A., Rogers, F.J. and Wilson, B.G.: 1992, *ApJ.* **397**, 717

Pesnell, W.D. and Cox, A.N.: 1980, *Space Sci. Rev.* **27**, 337

Stellingwerf, R.F.: 1974, *ApJ.* **192**, 139

STABILITY OF THE PULSATION OF 12 (DD) LACERTAE

A. PIGULSKI

Astronomical Institute, Wrocław University, Kopernika 11, 51-622 Wrocław, Poland

Abstract. The results of the study of period changes of all pulsational components in the multiperiodic β Cephei-type star, 12 Lac, are summarized.

12 (DD) Lacertae is a well-known multiperiodic β Cephei-type star. Jerzykiewicz (1978) found 6 short-term periodic components in the variations of light and radial-velocity of the star. Five of them have independent frequencies, the frequency of the last is the sum of two others. Moreover, three of the components form an almost equidistant in frequency triplet, which was usually explained in terms of rotational splitting (Jerzykiewicz 1978, Smith 1980).

Only for the component with largest amplitude ($P_3 = 0.19309$ d, see Fig.1 for the designation of pulsational components, values of periods, P, and frequencies, f) the changes of the pulsation period were relatively well studied. All these studies indicated that this period decreased, although the last study (Ciurla 1987) found it to be increasing since about 1970.

In the course of our analysis of period changes in 12 Lac we have been able to obtain for the first time the O-C diagrams for *all* six periodic components (Fig.1). The main results of our study are following:

1. All periods change. The O-C diagram for the strongest component (C_3) have a wave-like shape. This confirms the result obtained by Ciurla (1987).
2. P_3 and P_4 behave similarly. The changes of P_6 follow the changes of P_3 and P_4, as is expected from the fact that $f_6 = f_3 + f_4$ is a combinatory frequency.
3. P_2 changes inversely to P_3. A similar effect was found by Shobbrook (1972) for two strongest pulsation components of β CMa.
4. Neither the evolutionary nor the light-time effect is able to explain the changes of all periods. Even a combination of these two effects does not solve the problem. We conclude therefore, that some unknown effect causes observed period changes. A degree of regularity seen in these changes (points **1–3** above) suggests that this effect probably distinguishes modes with different l and/or m.

The full results of this study will be published elsewhere.

Acknowledgements

This work was supported by the research grant No. 2 P304 001 04 from the Polish Scientific Research Committee (KBN), and by a travel grant from the International Astronomical Union.

L. A. Balona et al. (eds.), Pulsation, Rotation and Mass Loss in Early-Type Stars, 21–22.
© 1994 *IAU. Printed in the Netherlands.*

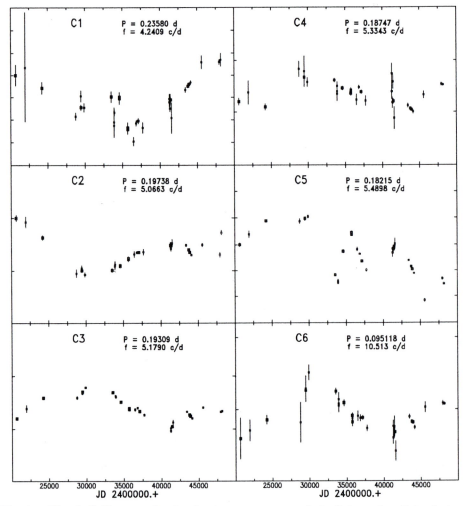

Fig. 1.   The O-C diagrams for 6 pulsating components of the light and radial-velocity variations of 12 Lac. For all components the corresponding periods and frequencies are given. The ordinate is O-C (in days); the markers are spaced by 0.05 d.

References

Ciurla, T.: 1987, *Acta Astronomica* **37**, 53
Jerzykiewicz, M.: 1978, *Acta Astronomica* **28**, 465
Shobbrook, R.R.: 1973, *M.N.R.A.S.* **161**, 257
Smith, M.A.: 1980, *Ap.J.* **240**, 149

PERIOD VARIATIONS IN THE BETA CEP STAR α LUP

E. CHAPELLIER and J.C. VALTIER
Observatoire de la Côte d'Azur, department Fresnel, B.P. 229, 06304 Nice cedex 4

α Lup (HR 5469) is a multiperiodic beta Cep star. According to Shob-brook (1979) and Lampens & Goossens (1982) the main period (P=0.25985 day) corresponding to the radial mode is constant. The amplitude of the secondary period, attributed to a non radial mode, is so small that it does not affect the determination of maximum dates.

In order to determine the phase lag between light and radial velocity, we compared the photometric observations from Van Hoof, analysed by Lampens & Goossens (1982) with the radial velocity measurements from Rodgers & Bell (1962). We obtained a phase lag of 0.0579 day (0.22 period).

The O−C diagram shows clearly a period increase of 0.28 sec. around the year 1974.

The observations can be fitted by two ephemeris:
1955 - 1974 : Ml = 2435000.0082 + 0.25984595 E
1974 - 1982 : Ml = 2442000.2593 + 0.25984916 E

In some beta Cep stars, period changes have been explained by a binarity effect. Although α Lup is not considered as a binary, some variations of the gamma axis exist. In 1955 the value of the γ-axis measured on 80 lines by Pagel (1956) was 4 km/s while the 9 lines measured by Mathias et al (1993) in 1989 ranged from 5.3 to 9.5 km/s with a mean value of 7.25 km/s. Such a 3.25 km/s difference would involve a 0.25 sec. increase of the pulsational period which is consistent with the observed variation.

α Lup is only 200 parsecs away, so it should be interesting to check by interferometry the existence of a companion. If the binary hypothesis is the good one, the orbital period should be greater than 50 years with large eccentricity.

References

Breger, M. : 1967, *Monthly Notices of the RAS* **136**, 51
Heynderickx, D. : 1992, *Astronomy and Astrophysics* **96**, 207
Hutchings J.B., Hill G. : 1980, *Astronomy and Astrophysics, Supplement Series* **42**, 135
Lampens, P., Goossens M.: 1982, *Astronomy and Astrophysics* **115**, 413
Lesh, J.R. : 1978, *Astrophysical Journal* **219**, 947
Mathias, P., Aerts, C., De Pauw, M., Gillet, D., Waelkens, C.: preprint
Pagel, B.E.J. : 1956, *Monthly Notices of the RAS* **116**, 10
Rodgers, A.W., Bell, R.A.: 1962, *Observatory* **82**, 26
Shobbrook, R.R. : 1979, *Monthly Notices of the RAS* **189**, 571

L. A. Balona et al. (eds.), Pulsation, Rotation and Mass Loss in Early-Type Stars, 23–24.
© 1994 IAU. Printed in the Netherlands.

DATE	Nights	O − C	References
35297.2722	14	0.0002	Pagel (1956)
37511.9398	1	0.0008	Rodgers & Bell (1962)
38931.2155	3	−0.0021	Breger (1967)
41713.1290	1	0.0007	Lesh (1978)
42882.4400	2	0.0049	Hutchings & Hill (1980)
43602.2294	13	0.0210	Shobbrook (1979)
47224.7865	12	0.0657	Heyndericks (1992)

O−C diagram of α Lup

PERIOD VARIATIONS IN THE BETA CEP STAR β CRU

J.C. VALTIER, E. CHAPELLIER and J.M. LE CONTEL
Observatoire de la Côte d'Azur - department Fresnel

β Cru (HR 4853) is a multiperiodic β Cep star. The main period has been found constant between 1958 and 1973 by Cuypers (1982). In order to extend the basis of the ephemeris, we computed two values of radial velocity maxima from individual measurements published by Campbell & Moore (1928) and Sahade & Albarracin (1952). In lack of simultaneous spectroscopic and photometric observations, we used an arbitrary value equal to 0.25 period for the phase lag.

The ephemeris (assuming a constant main period) :
Ml = 2422000.0177 + 0.19118464 E takes roughly into account all the spectroscopic and photometric observations from 1921 to 1973.

The figure shows the corresponding O−C diagram for the data published by Cuypers. Most of the dispersion is due to the multiperiodicity of β Cru, nevertheless a clear oscillation is present around the mean value. The corresponding period seems to be around 7 or around 14 years with an amplitude of 0.02 day.

At the other hand β Cru was discovered as an interferometric binary (Popper 1968), and Heintz (1957) found gamma axis variations with a period of 7 or 8 years and a range of 15 km/s.

Assuming that β Cru is a member of a binary system with an approximately circular orbit, and using the values given by Heintz, the movement of the variable star around the center of gravity of the system would lead to a sinusoidal oscillation in the O-C diagram with an amplitude of 0.02 day.

Although more observations are necessary to precise the value of the orbital period, we can conclude that the observed oscillation in the O−C diagram is due to the binarity and that the intrinsic pulsational period of the star remained constant over half a century.

References

Campbell, W.W., Moore: 1928, *Publ. Lick Obs.* **16**, 188
Cuypers, J.: 1987, *Astronomy and Astrophysics* **127**, 186
Heintz, W.D.: 1957, *Observatory* **77**, 200
Pagel, B.E.J.: 1956, *Monthly Notices of the RAS* **116**, 10
Popper, D.M.: 1968, *Astrophysical Journal* **151**, 51

L. A. Balona et al. (eds.), Pulsation, Rotation and Mass Loss in Early-Type Stars, 25–26.
© 1994 IAU. Printed in the Netherlands.

Sahade, J., Albarracin, J.: 1952, *Astrophysical Journal* **116**, 654
Shobbrook, R.R.: 1979, *Monthly Notices of the RAS* **189**, 571
Van Hoof, A. : 1962, *Zeitschrift fuer Astrophysik* **54**, 244
Watson, R.D. : 1971, *Astrophysical Journal* **170**, 345

DATE	Nights	O – C	References
22830.1400	-	−0.0014	Campbell & Moore (1928)
32200.1050	-	0.0044	Sahade & Albarracin (1952)
35225.5960	7	−0.0015	Pagel (1956)
36328.5328	3	−0.0089	Van Hoof (1962)
36651.4528	14	0.0002	Van Hoof (1962)
37028.2878	6	0.0103	Van Hoof (1962)
37414.2849	6	0.0056	Van Hoof (1962)
37738.5215	7	−0.0069	Van Hoof (1962)
40321.0480	3	−0.0026	Watson (1971)
41780.9379	5	0.0014	Shobbrook (1979)

O−C diagram of β Cru

HIGH-RESOLUTION SPECTROSCOPY OF BETA CEPHEI STARS

C. WAELKENS, H. VAN WINCKEL and K. DE MEY

Instituut voor Sterrenkunde

Celestijnenlaan 200B, B-3001 Leuven (Belgium)

Abstract.

We give a progress report on an observational program intended to determine detailed chemical abundances of β Cephei stars and constant stars with similar temperature and gravity. There is some evidence that non-variable stars have a lower metal content than variables, as the recently found pulsation mechanism would suggest.

1. Introduction

The long-lasting enigma of the pulsation mechanism of the β Cephei stars now seems to be solved, with the inclusion of the new opacities in the theoretical models (Dziembowski, this conference). An interesting corollary of the metal-opacity induced instability is that, since a three-fold increase of the metal opacity sets a limit from stable to unstable models, then the metal abundance is a critical parameter for the behavior of a star in the β Cephei instability strip. For instance, the β Cephei phenomenon should be absent in the Magellanic Clouds, as indeed seems to be corroborated by observation (Sterken and Jerzykiewicz 1988, Balona 1993).

Observational tests for the mechanism can also be devised in our Galaxy. Is the difference between constant and pulsating stars one of metal abundance? Is there a relation between metal abundance and pulsation amplitude? In order to address these questions, we have undertaken a high-resolution spectroscopic study of galactic β Cephei stars and of constant stars with similar photometric colors. We present here a first report on this project.

2. Observations

High-resolution spectra for northern β Cephei stars have been obtained with the Aurélie spectrograph at the 1.5m Telescope at Haute-Provence Observatory in France, during October 1992. High-resolution spectra for southern β Cephei stars have been obtained with the CES spectrograph at the 1.4m CAT Telescope at ESO, Chile, during April 1992 and January and February 1993.

Optical photometric parameters, such as the X-index of the Geneva Photometric System, are sensitive temperature estimators in the range defined by the β Cephei stars. A fortunate circumstance is also that the instability

L. A. Balona et al. (eds.), Pulsation, Rotation and Mass Loss in Early-Type Stars, 27–28.

strip occurs in the temperature range where lines from three different ion-izations stages of silicon can be observed (Si II-IV), so that also accurate spectroscopic estimates of the effective temperature are possible. On the other hand, optical photometric colors are only slightly sensitive to gravity in this spectral range.

It is somewhat unfortunate that iron lines are scarce in optical spectra of early-B-type stars. The best line seems to be the FeIII line at 4165 Å which is unblended and occurs in the linear part of the curve of growth.

3. First Results and Prospects

The observational temperature parameters (Geneva X index, uvby photome-try, Si-line ratios) correlate well with directly derived effective temperatures and with effective temperatures derived from model atmospheres. However, temperatures derived from published non-LTE analyses appear to be system-atically too high when compared with those obtained with direct methods.

When the equivalent width of the FeIII-4165 line is plotted as a function of effective temperature, it appears that on average the equivalent width is smaller for non-variable stars than for the β Cephei stars. The mean effect amounts to some 30 %, and individual exceptions occur. The large-amplitude variables are not characterized by unusually strong iron abundances.

The next step in the investigation will be a quantitative analysis of the abundances of the metals.

4. The References

Balona, L.A., 1993, Monthly Notices of the RAS256, 425
Sterken, C., Jerzykiewicz, M., 1988, Monthly Notices of the RAS235, 565

A SEARCH FOR BETA CEPHEI STARS IN LMC AND SMC *

H. KJELDSEN and D. BAADE

European Southern Observatory
Karl-Schwarzschild-Str. 2, D-85748 Garching bei München, Germany

1. Testing the κ mechanism

The β Cephei instability is believed to be caused by the κ-mechanism. Model calculations, based on revised metal opacities, show this mechanism to be strongly sensitive to metallicity (e.g., Dziembowski, W. A. & Pamyatnykh, A. A., 1993: *MNRAS*, 262, 204). It appears that [Fe/H] must be larger than the solar value in order to drive the β Cephei pulsations. For this reason, the models predict that no β Cephei stars should be found in the Magellanic Clouds. Balona (1992: *MNRAS*, 256, 425; 1992: *ASP Conf. Ser.*, Vol. 30, 155; 1993: *MNRAS*, 260, 795) has tested this prediction by observing three young clusters in the SMC and LMC (NGC 330, NGC 2004 and NGC 2100). He found no β Cephei stars down to a magnitude of $V \approx 16$; the threshold for detection of variability on time-scales of a few hours was 10 mmag. Most Galactic β Cephei stars have amplitudes greater than this, but we might expect β Cephei stars in the Magellanic Clouds (if they exist) to have smaller amplitudes, due to the different environment. To test this, we have observed two young clusters NGC 371 (SMC) and NGC 2122 (LMC) with high sensitivity.

2. Observing technique and data reduction

The data were collected on 2×6 nights in November-December, 1992 using the 0.91m Dutch telescope at La Silla. We did time-resolved high-precision CCD photometry using a 385×578 pixel GEC CCD and typical exposure times of 120 sec. In total we got 1660 usefull time-series frames. All time-series observations were done through a Bessel V-filter. We also collected UBV, Hβ and uvby-Strömgren photometry for both clusters, in order to place the observed stars in the colour-magnitude diagram. The data were reduced using the MOMF software (Kjeldsen, H. and Frandsen, S., 1992: *PASP*, 104, 413), which was developed specifically for high-precision differential photometry. We have calculated the photometric precision as a function of relative magnitude for NGC 371 and find that, for most stars, the precision is close to the theoretical limit set by photon, sky and detector noise.

* Based on observations obtained at the European Southern Observatory, La Silla, Chile

L. A. Balona et al. (eds.), Pulsation, Rotation and Mass Loss in Early-Type Stars, 29–30.
© *1994 IAU. Printed in the Netherlands.*

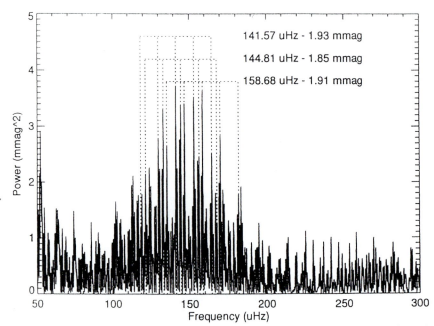

Fig. 1. Amplitude-scaled power spectrum for star # 565 in NGC 371 (SMC). The spectrum can be explained by three close frequencies convolved with the window function of the observations. The amplitudes are below 2 mmag and frequencies correspond to 12.23, 12.51 and 13.71 c/day (periods: 1.96, 1.92 and 1.75 hours).

3. Results

To search for variability, we examined the amplitude spectra for the ~ 100 brightest stars in each cluster (down to $V \approx 16$). We find no clear evidence for variability on timescales of 1–6 hours above a noise level of 1–4 mmag. The lack of pulsating stars in these metal-poor clusters gives strong support to the κ-mechanism being the cause of β Cephei pulsations.

Two stars in NGC 371 show some indication of pulsations at the level of a few mmag. The signals are close to our sensitivity limit and may simply be peaks in the noise distribution. However, both stars (#565 and #642) are located in the β Cephei instability strip in the CM-diagram ($-5 \lesssim M_V \lesssim -4$). If real, the pulsations have amplitudes of 2–3 mmag and periods close to two hours (see Figure 1). We note that Galactic β Cephei stars, in comparison, have longer periods (3–6 hours).

Acknowledgements: We thank Tim Bedding and Søren Frandsen for stimulating discussions and help during this project.

(The full results of this project will be submitted to Astronomy & Astrophysics when the analysis has been completed.)

PHOTOMETRIC VARIABILITY IN EARLY B STARS I.

53 ARIETIS

SALVADOR GONZÁLEZ BEDOLLA

Instituto de Astronomía, UNAM, México Apartado Postal 70-264, C.P. 04510, México, D.F. México

Known as well as β CMa stars, the classical β Cep stars are a group of variables of spectral type between B0 to B2 and luminosity classes II to IV with light and radial velocity periods between 3 and 7 hours and radial velocity ranges usually < 50 km/sec. Their photometric amplitudes are usually < 0.1 mag in the visible. The light curves are approximately sinusoidal (however, multiple modes are frequently detected with low amplitudes), which lag about 90° behind the radial velocity curve and the color changes in the blue and visual spectral regions are small. Color and luminosity show that maximum temperature coincides with minimum radius. Probably not all the stars in this spectral range are variables; most investigators searching in this region for β Cep variables have observed that a large number of stars in this zone are constant in light. Also, the β Cep stars are not different spectroscopically from nonvariable stars of the same MK type.

Recently, however, new β Cep stars have been detected and with the new types of variables discovered near this zone, like the faint OB stars (mostly found in associations, Hill 1967), the "53 Per" variables, the "ultrashort" period, the "slow" and Be variables, it is now clear that the classical box of instability for B stars does not have a well delimited border. Hence, with the small number of new groups of variables discovered, one is led to the conclusion that some of the stars in these new "groups" must be monitored in order to know if they are really different from the classical β Cep stars.

With the aim of detecting new variables of the types mentioned above, we present here a possible new variable star situated near the zone discussed above; this star, 53 Ari (B2V, V=6.1) is located near the red edge of the classical instability box where the β Cep stars are. The comparison stars were C_1 = HR948 (B8V, V=5.9) and C_2 = HR972 (A0IV-V, V=4.9); the observations were made on the nights of Oct. 7, 1980 and Oct. 19 and 20, 1982 at the National Astronomical Observatory in San Pedro Mártir, Baja California, México, with the 1.5m telescope, the Johnson B filter and a DC photoelectric photometer equipped with 1P21 photomultiplier tube cooled with dry ice; a voltage to frequency converter provided digital read out of the output. The observational sequence followed was, uninterruptedly: C_1, V, C_2, V, C_1, ... on all nights, each observation consisted at least of five 10-second integrations of the star followed by one of the sky.

The result of subtracting the mean of the magnitudes of the comparison

L. A. Balona et al. (eds.), Pulsation, Rotation and Mass Loss in Early-Type Stars, 31–32.
© 1994 IAU. Printed in the Netherlands.

stars from the magnitude of the variable star (ΔB) are plotted and shown in the figure along with ($C_1 - C_2$). The accuracy of each observation is 0.004 mag; time is reported in UT and its accuracy is of 1 minute.

The light curve obtained for 53 Ari shows that on the night of Oct. 7, 1980 the star was probably variable: the light curve shows a clear oscillation with a $\approx$0.02 mag amplitude and a period between 0.5 and 1 h with a modulation that indicates that other frequencies are probably present. The dispersion for the difference $C_1 - C_2$ on this night (0.008 mag) is only 40% of the total amplitude of the light curve. In contrast, on Oct. 19, 1982, the star was constant since the dispersion for ΔB and $C_1 - C_2$ is the same (0.008 mag) and apparently there is no clear oscillation for the light curve of the star observed. The star was also constant during most of Oct. 20, 1982, but, surprisingly, showed a sudden, short and bright little increase (like a flash) of 0.01 mag approximately in a elapsed time of around one hour.

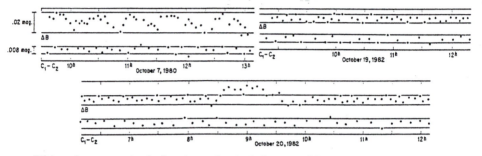

This photometric behavior of 53 Ari resembles in some sense 53 Psc (B2IV, V=5.9) which is located on the red edge of the "instability box" with a controversial photometric behavior where a peculiar amplitude modulation or transient variability may better account for the observations made for this star (Jerzykiewicz and Sterken, 1990) and the observed amplitude variation could be related to the location of 53 Psc at the low temperature border of the β Cep instability strip where the mechanism responsible for the instability is perhaps not efficient enough to maintain a stable pulsation (Le Contel et al. 1988). In this case, another possibility is that 53 Psc presents long-term changes in amplitudes similar to those pointed out with different β CMa stars (Chapellier, 1986). These suggestions could be applied to 53 Ari too. However, the little short flash observed on Oct. 20, 1982 is not explained with these ideas. In any case, a continuous and careful photometric monitoring for 53 Ari will be very useful in order to confirm the observations presented here. Recently, differential photoelectric Strömgren photometry was made for 53 Ari. The reductions are in progress.

Chapellier, E., 1986, *Astron. Astrophys. Suppl. Ser.*, **64**, 275.
Hill, G., 1967, *Ap. J. Suppl.* **130**, 263.
Le Contel et. al., 1988, IBVS No. 3131.
Jerzykiewicz, M. and Sterken C., 1990, *Astron. Astrophys.* **227**, 77.

MULTIPERIODICITY IN LIGHT VARIATIONS OF 53 PERSEI: RESULTS FROM OPTICAL PHOTOMETRY IN 1990 OCTOBER –1991 JANUARY

L. HUANG, Z. GUO and J. HAO
Beijing Astronomical Observatory, Beijing, China

J.R. PERCY
Erindale Campus, The University of Toronto, Mississauga, Canada

M.S. FIELDUS*
Department of Astronomy, The University of Toronto, Toronto, Canada

R. FRIED
Braeside Observatory, Flagstaff, U.S.A.

R. K. PALOVSKI, H. BOŽIĆ and Ž. RUŽIĆ
Hvar Observatory, Zagreb University, Zagreb, Croatia

M. PAPARÓ and B. VETÖ
Konkoly Observatory, Budapest, Hungary

and

R. DUKES
Physics Department, The College of Charleston, Charleston, U.S.A.

1. Introduction

The B type star 53 Persei was discovered in 1977 by Smith (1977) as the prototype of a separate group of B-type variables showing light and line profile variability. The physical cause of the variability was thought to be nonradial pulsation (NRP) (see, e.g. Smith et al. 1984). However, the NRP model for this star has been questioned by Balona (1986) who suggested the rotational modulation (RM) model to explain the variability. In order to resolve the long lasting debate about 53 Persei, a campaign was initiated to organize coordinated optical photometry and spectroscopy from the ground, and Far-UV photometry from Voyager in 1991 January. This paper presents the results of period analysis on the groundbased UBV data. In another paper, Smith & Huang (1994) report the new identification of pulsation modes using Voyager Far-UV photometry combined with the results from optical observations. Some preliminary results from APT $uvby$ observations taken at a single site are also cited for comparison.

* Deceased in 1992

L. A. Balona et al. (eds.), Pulsation, Rotation and Mass Loss in Early-Type Stars, 33–36.

2. Observation and data reduction

Four observatories (Beijing in China, Braeside in U.S.A., Hvar in Croatia, and Konkoly in Hungary) participated in the campaign. Two A type stars, 47 Persei and HR 1482, were adopted as comparison and check star, respectively. All observations were corrected for atmospheric extinction and transformed to the standard UBV system. The observational uncertainties estimated from the standard deviations of the check star measurements are 0.011, 0.012, and 0.020 mag for V, B and U, respectively. The $uvby$ measures were made using the 0.75 m Automatic Photoelectric Telescope (APT) of the Four College Consortium. The same comparison and check stars were used. The data were reported as the differential magnitudes between 53 Persei and the comparison. The standard deviations of the check minus comparison measures were 0.009, 0.008, 0.010, and 0.008 mag for u, v, b, and y, respectively.

3. Period analysis and results

We first calculated the clean spectra of data using a CLEAN algorithm (Roberts, Lehár, & Dreher 1987) and then, taking the frequencies at the highest peak in clean spectra as the trial ones, we computed the best least-squares fits of sinusoids using the main program of PERIOD (Breger 1990) which can examine up to seven trial frequencies simultaneously without prewhitening the data.

The clean spectra for V, B, and U data shows the same distinct feature: the power is highly concentrated in a narrow range near 0.5 c/d (cycles/day). These clean spectra in the 0 –2 c/d range are displayed in Figure 1. Computations by using PERIOD have shown that frequencies of the best 7 frequency fits to different data only coincide for the two strongest ones: 0.40 c/d in all V, B, and U and 0.60 c/d in V and B. Using the mean frequencies and phases derived from V and B as fixed parameters, we determined the final fits to individual datasets as listed in Table 1, where in parentheses the formal errors are given in units of the last figure of significance. In Figure 2 phase diagrams are plotted for data phased with the primary and then phased with the secondary frequency 0.462 c/d (top) and the same data but prewhitened with the primary and then phased with the secondary frequency 0.603 c/d (bottom). Obviously, the two frequencies are coherent over many cycles during the campaign. The clean spectra and the multi frequency fits to the $uvby$ data exhibit the same two strongest frequencies. Since Since the UBV and $uvby$ observations partly overlap in time, the minor difference between the two determinations may be caused by systematic errors and we take them as resulting from the same physical process.

The frequencies near 0.46 and 0.60 c/d derived in this study are in good

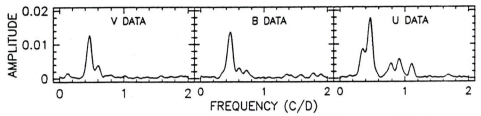

Fig. 1. Clean spectra for light variations of 53 Persei in 1991 January

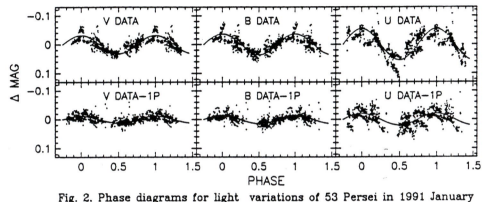

Fig. 2. Phase diagrams for light variations of 53 Persei in 1991 January

agreement with 0.464 and 0.595 c/d found by Smith and coworkers (Smith & McCall 1978; Buta & Smith 1979; Smith and Buta 1979; Smith et al. 1984) from observations obtained in 1977–1983. The difference between them is comparable to the errors. We believe that the pair of frequencies derived from observations more than ten years apart represent the same phenomenon. Therefore, we have confirmed, mainly on the more reliable basis of multi-longitude observations, the stable multiperiodicity in 53 Per and the stability duration is extended from 5.5 years (Smith et al. 1984) to about 13 years. This pivotal result strongly supports the viewpoint that NRP is present in this archetypical star.

TABLE I

Final 2-frequency fits for the UBV campaign datasets

Data	Amplitude (mag)	Zero-point (mag)	Residuals (mag)	Fixed Parameters (see text)
V	0.0317(6)	4.8447(16)	0.0101	Mean frequency (c/d):
	0.0122(21)			$f_1 = 0.4620(8)$, $f_2 = 0.6030(27)$
B	0.0363(22)	4.8074(7)	0.0108	Mean phases(cycles):
	0.0145(11)			$\phi_1 = 0.233(8)$, $\phi_2 = 0.665(28)$
U	0.0544(14)	4.2376(73)	0.0216	Mean Periods (days):
	0.0205(20)			$P_1 = 2.164(4)$, $P_2 = 1.658(7)$
$B - V$	0.0046(29)	−0.0374(23)	0.0068	Epochs of Light Maxima:
	0.0021(33)			$T_{1max} = $ HJD 2448261.724(19)
$U - B$	0.0179(36)	−0.5697(80)	0.0209	$T_{2max} = $ HJD 2448261.501(53)
	0.0058(31)			

Acknowledgements

LH, ZG and JH are grateful to the National Natural Science foundation of China for financial support. KP, HB and ZR were supported by a research grant from the Croatian Ministery of Science.

References

Balona, L.A.: 1986, 'Stellar Pulsation' in A.N. Cox et al., ed(s)., *Berlin: Springer*, 83,
Breger, M.: 1990, *Comm. Asteroseismology* **20**, 1
Buta, R.J., Smith, M.A.: 1979, *Astrophysical Journal* **232**, 213
Roberts, D.H., Lehár, J, Dreher, J.W.: 1987, *Astronomical Journal* **93**, 968
Smith, M.A.: 1977, *Astrophysical Journal* **215**, 574
Smith, M.A., Buta, R.J.: 1979, *Astrophysical Journal* **232**, L193
Smith, M.A. et al.: 1984, *Astrophysical Journal* **282**, 226
Smith, M.A., McCall, M.L.: 1978, *Astrophysical Journal* **223**, 221
Smith, M.A., Huang, L. 1994, this Proceedings

NRP MODE TYPING FOR 53 PERSEI: RESULTS FROM VOYAGER PHOTOMETRY

MYRON A. SMITH

IUE/CSC Observatory, 10000A Aerospace Rd., Lanham-Seabrook MD 20706, USA

and

LIN HUANG

Beijing Observatory, Beijing, China

1. Introduction

Huang *et al.* (1994) have conducted an extensive photometric campaign on the prototypical nonradially pulsating (NRP) star 53 Per and confirmed the pair of frequencies at 0.46 cy d^{-1} (dominant) and 0.60 cy d^{-1} originally reported by BS79. In an analysis of these two modes, SP79 underestimated the effects of the Balmer jump on this star's color variation, leading them to an erroneous conclusion that geometric effects dominate the color variations and also that the modes are described by indices l=3, $-m = 3$ and 2. Herein we describe results from *Voyager 2* observations obtained during Lin's optical campaign. The amplitudes derived for these Far-UV data using the optical ephemeris provide for the first time a large enough wavelength baseline to discriminate in favor of thermal effects over geometrical ones in producing NRP light variations. In addition, they allow the dominant mode to be constrained to $l = 2$ (or 1).

2. Observations

Eleven $\sim$1 hr observations made over three days in 1991 January were of sufficient quality to permit reliable extraction, binning, calibration, and descattering of UVS *Voyager* spectra by the techniques of Holberg and Watkins (1992). As expected for bright stars, the deviations of our data from a two-sinusoid fit suggest an error of $\pm$3%. We have extracted 100Å bandpasses centered at λ1180 and λ1585. The fit to our data gives amplitudes of 0.147, 0.035 mags. at λ1180 and 0.098, 0.022 mags. for λ1585, respectively, for the two modes. To determine a modal l identification for 53 Per's primary mode, we can simply ratio the Far-UV amplitudes with Huang *et al.*'s U, V observations.

3. Theoretical NRP Light Variations and Conclusions

In linear adiabatic theory the amplitude of an NRP wave on a star is described by the spherical harmonic function, Y_m^l. Light variations for all

L. A. Balona et al. (eds.), Pulsation, Rotation and Mass Loss in Early-Type Stars, 37–38.

Table 1: Light Variation Terms, Theory and Observation

	l Mode	$\alpha + \beta$	γ	$\gamma F'_T T'_P P'_R$	Sum	Pred. Rat.	Obs.
V:	1	0	.6771	9.17	9.17		
	2	-1.155	.2878	12.35	11.20		
	3	-.314	.0326	2.84	2.53		
							3600Å/V
U:	1	0	.6778	15.87	15.87	1.7	
	2	-1.163	.2896	21.47	20.31	1.8	1.7 ±.2
	3	-.337	.0348	5.24	4.90	1.9	
							1565Å/V
1565:	1	0	.7138	25.4	25.4	2.9	
	2	-1.446	.3592	42.15	40.70	3.6	3.0 ±1.
	3	-.910	.0942	22.43	21.52	8.5	
							1180Å/V
1180:	1	0	.7374	26.4	26.4	3.1	
	2	-1.660	.4094	50.83	49.17	4.4	4.6±1.
	3	-1.40	.1437	36.20	34.80	13.8	

m-orders of a given l are the same. They are comprised of three terms: two geometrical (surface area, surface normal) and one compressional/thermal. These terms can be written as disk-integrated combinations of the Legendre and limb darkening functions and their derivatives; BS79 (eqns. 37-39) define them as coefficients α_λ, β_λ, and γ_λ. These are computed and shown in cols. $3 - 4$ of Table 1. (Limb darkening coefficients used for their computation are taken from J. Lester, who kindly calculated them from an ATLAS9 model for 53 Per). Column 5 multiplies col. 4 by parameters representing (via the chain rule) the flux variation per radius change for the temperature term. Col. 6 is the sum (3) + (5) (opposing signs), *viz.* the predicted amplitude in mags. divided by the fractional radius change, ϵ, and 1.086.

Table 1 shows the clear dominance of the thermal (γ) term for light and color variations because of 53 Per's long period and because of the wavelength sensitivity of γ. A comparison of the observed and predicted amplitude ratios shows agreement only for $l = 2$ (or 1). Separate line profile arguments (namely the near absence of RV variations at the half-power points, the large line width variations) can be used to narrow the modal identification to $l = 2$. This value is also more in keeping with this star's observed large light variations than is $l = 3$.

References

Huang, L. *et al.* 1994, these proceedings.

Buta, R. J., and Smith, M. A. 1979, ApJ, 232, 213; "BS79."

Holberg, J., and Watkins, R. 1992, Voyager Data Analysis Hdbk., Vers. 1.2, Univ. Ariz., unpub.

Smith, M. A., and Buta, R. J. 1979, ApJ, 232, L193; "SB79."

PHOTOMETRIC AND SPECTROSCOPIC OBSERVATIONS
OF THE 53 PER STAR : ζ CAS

H. SADSAOUD[1]*, D. LE CONTEL[2], E. CHAPELLIER[2],
S. GONZÁLEZ-BEDOLLA[3] and J.M. LE CONTEL[2]

[1] *C.R.A.A.G, B.P. 63, Bouzareah, 16340 Alger, Algeria*

[2] *O.C.A, URA 1361, B.P. 229, 06034 NICE-Cedex 04, France*

and

[3] *Instituto de Astronomia, UNAM, Apdo Postal 70-264, 04510, Mexico D.F, Mexico*

ζ Cas (HR 153, m_v=3.66, $v\sin i$=18 kms^{-1}) was classified as a 53 Per variable by Smith (1980). He detected variations in the profile of the 4552 Å Si III line with a period of 0.9 day. The radial velocity and the luminosity were observed to be constant by Abt & Levy (1978) and by Jerzykiewicz (1993) respectively.

1. Observations and analysis

We performed :
Photometric observations at San Pedro Martir Observatory (Mexico) (October 22 to 29, 1990) using uvbyβ Strömgren filters and two comparisons C_1=HR 123 (m_v=4.73, B8Vn), C_2=HR 144 (m_v=5.08, B7III). The star was constant in all filters.

Spectroscopic observations with the Aurélie spectrograph (S/N= 300, resolution 50000) at the O.H.P (France) in October 1990 (one night) and in October 1991 (6 nights).

In the radial velocity (RV) data, we detect two types of variations: first night to night and year to year variations over 2 kms^{-1} and secondly fluctuations over 1 kms^{-1} during each night. After detrending for the night to night variation, we analysed the 1991 data for period determination. Only two peaks can be retained: P_1= 0.173, P_2= 0.215 day with K_1= 0.149, K_2=0.155 kms^{-1} respectively. We are not able to determine which of these two periods (corresponding to frequencies separated by 1 cycle/day) is the good one.

The full width at half maximum (FWHM) of the 4552 Å Si III line varies from one night to an other within 0.08 Å, but no correlation is observed with the night to night RV variations. Short time scale variations within 0.04 Å are also observed. After detrending for the night to night variation, a peak appears in the power spectrum at 0.27 day.

39

L. A. Balona et al. (eds.), Pulsation, Rotation and Mass Loss in Early-Type Stars, 39–40.
© 1994 IAU. *Printed in the Netherlands.*

2. Discussion

Two time scales of variability are present in our RV and FWHM data: night to night changes and short period variations. The lack of correlation between long term variability in the RV and in FWHM suggests that two different phenomena are responsible for these variations.

The long term RV variations could be due to a binary motion. As our observations cover only 6 nights, we reanalysed the Abt & Levy (1978) data. In the periodogram the largest peak is present at 11.53 days with K= 1.80 kms^{-1}. More observations are needed to confirm the orbital period, the amplitude and to determine the orbital elements.

Night to night changes in FWHM could be due to a high degree non radial gravity mode or to surface variations modulated by the stellar rotation. These variations are similar to those observed in 53 Per (Le Contel et al, 1989) and in ι Her (Le Contel et al, in preparation).

The 0.27 d period takes into account the FWHM observations. This period does not appear clearly in the Fourier analysis of the RV measurements, probably due to the fact that the RV amplitudes are of the order of the dispersion on the points. Smith's (1980) period equal to 0.9 d is not detected but such a value would be difficult to find in our observations.

Using Shobbrook calibration (1985), the pulsational constant corresponding to a 0.27 d period is Q=0.049 d. Due to its length this period corresponds to a non radial mode. That is confirmed by the importance of line profile deformations compared to the RV variations.

Considering the B2 IV classification of ζ Cas we should expect this star to be a β Cephei star. However the photometric indices (β=2.625 and C_o=0.152) put it slightly outside the cool border of the instability strip (Sterken & Jerzykiewicz, 1993). The length of the period and the nature of the observed variations also exclude this hypothesis.

As the only clearly variable parameter in our observations is the shape of the profiles, ζ Cas is a typical 53 Per star according to the definition of this group by Smith.

*Mr. H. Sadsaoud acknowledges financial support from the French Ambassy in Algeria.

References

Abt, H.A., Levy S.G: 1978, *Astrophysical Journal, Supplement Series* **36**, 241

Jerzykiewicz, M.: 1993, *Astronomy and Astrophysics, Supplement Series* **97**, 421

Le Contel, J.M., Chapellier, E., Le Contel D., Rau, G., Endignoux, A., Valtier, J.C., Mevollon, M.: 1989, *Acta Astron.* **39**, 227

Shobbrook, R.R.: 1985, *Monthly Notices of the RAS* **214**, 33

Smith, M.A.: 1980, *"Nonradial and Nonlinear Stellar Pulsation"* eds. H.A. Hill and W.A. Dziembowski, New York: Springer Verlag , 60

Sterken, C., Jerzykiewicz, M.: 1993, *preprint* ,

HD 37151: A NEW "SLOWLY PULSATING B STAR"*

P. NORTH and S. PALTANI

*Institut d'Astronomie de l'Université de Lausanne,
CH-1290 Chavannes-des-Bois, Switzerland*

Abstract. We report the discovery of non-radial pulsations with at least 4 periods in the B7V star HD 37151. This result is based on 465 photometric measurements spanning 12 years. In addition, 30 high-resolution spectra were taken in the Mg II λ4481 region: they show a small projected rotational velocity and slight variations of the line profiles. Although this star was once classified B8Vp(Si), all available data show it is a normal B7V star. It is the second coolest SPB star found to date.

1. Photometry

HD 37151, classified B8Vp(Si) by Abt & Levato (1977) and a member of the Orion OB1 association, was discovered to be multiperiodic in the course of a photometric survey of Ap stars in clusters (North 1984, 1987). 465 photometric maesurements made in the Geneva system during the last 12 years with the Swiss telescope at ESO La Silla, Chile, confirm the multiperiodicity and show at least 4 frequencies. These are listed in Table 1 together with the amplitudes for the 3 Geneva passbands [U], [B] and V. The long-term stability of the periods is attested by the presence of the alias frequencies $\nu_i \pm 1$ yr^{-1} in the periodograms. The [U] amplitude is twice as great as the V one, like in the other SPB stars (Waelkens 1991), and the length of the periods points to non-radial oscillations.

2. Spectroscopy

30 high-resolution spectra taken in October 30 – November 5, 1989 at ESO La Silla with the CAT telescope show a clear variability of the Mg II λ4481 and He I λ4471 lines. The frequency ν_1 is seen in the second moment (m_2, see e.g. Balona 1986) of these lines, while ν_2 is present in their first and third moments. The time series is too short to show all photometric frequencies.

One spectrum taken in the Si II λ4128 − 4130 region shows that HD 37151 is **not** an Si star, confirming the lack of photometric peculiarity in the Geneva and Δa systems (Joncas & Borra 1981) and in the UV (Shore & Brown 1987).

The spectra show that $v \sin i = 28$ kms^{-1}, which nicely confirms the suspicion that all SPB stars rotate slowly (Waelkens 1987, 1991).

* Based on observations made at European Southern Observatory, La Silla, Chile and supported in part by the Swiss National Foundation for Scientific Research

L. A. Balona et al. (eds.), Pulsation, Rotation and Mass Loss in Early-Type Stars, 41–42.
© 1994 IAU. Printed in the Netherlands.

TABLE I

Coefficients of the function least-squares fitted to the lightcurve:
$m(t) = A_o + A_1 cos(\omega_1 t + \phi_1) + A_2 cos(\omega_2 t + \phi_2) + A_3 cos(\omega_3 t + \phi_3) + A_4 cos(\omega_4 t + \phi_4)$
where t = HJD-HJD$_o$ and HJD$_o$ = 2446948.734. The r.m.s. residual
scatter around the fitted curve (absolute photometry) is given too.

Period [d] Frequency [d^{-1}]	bandpass	A_o	A_i	$\phi_i/2\pi$	σ_{res}
$P_1 = 0.804397$	[U]	7.2860	0.01826	0.103	
$\nu_1 = 1.24317$	[B]	6.3176	0.01040	0.107	
	V	7.3798	0.00899	0.115	
$P_2 = 0.847333$	[U]		0.02124	0.103	
$\nu_2 = 1.18017$	[B]		0.01210	0.099	
	V		0.01034	0.097	
$P_3 = 0.904769$	[U]		0.01523	0.758	
$\nu_3 = 1.10525$	[B]		0.00870	0.759	
	V		0.00740	0.744	
$P_4 = 0.959113$	[U]		0.01607	0.238	0.0100
$\nu_4 = 1.04263$	[B]		0.00951	0.234	0.0068
	V		0.00789	0.241	0.0061

3. HD 37151 among the other SPB stars

A comparison of the physical parameters of HD 37151 (obtained from the
X and Y Geneva parameters through the calibration of North & Nicolet
1990) with those of the other SPB stars show that it is the second coolest
one, with $T_{eff} = 12865 K$ and $M = 3.13 M_\odot$, just after HD 123515 which has
$T_{eff} = 11944 K$ and $M = 2.88 M_\odot$. It is therefore a useful object for defining
the empirical red edge of the instability strip of the SPB's.

References

Abt, H.A. and Levato, H.: 1977, *Publications of the ASP* **89**, 797
Balona, L.A.: 1986, *Monthly Notices of the RAS* **219**, 111
Joncas, G. and Borra, E.F.: 1981, *Astronomy and Astrophysics* **94**, 134
North, P.: 1984, *Astronomy and Astrophysics, Supplement Series* **55**, 259
North, P.: 1987, *Astronomy and Astrophysics, Supplement Series* **69**, 371
North, P. and Nicolet, B.: 1990, *Astronomy and Astrophysics* **228**, 78
Shore, S.N. and Brown, D.N.: 1987, *Astronomy and Astrophysics* **184**, 219
Waelkens, C.: 1987, in A.N. Cox, W.M. Sarks and S.G. Starrfield, ed(s)., *Stellar Pulsa-tions. A Memorial to J.P. Cox*, Lecture Notes in Physics, Springer-Verlag, 75
Waelkens, C.: 1991, *Astronomy and Astrophysics* **246**, 453

THE PECULIAR BINARY SYSTEM HR 8891 (ET AND)

R. KUSCHNIG[*†∘1], W. W. WEISS[*†∘1], N. PISKOUNOV[†∘2],
T. RYABCHIKOVA[†∘2], T.J. KREIDL[*3], and M. ALVAREZ[*4],
S.G. BEDOLLA[*4], S.J. BUS[*3], Z. GUO[*5], J. HAO[*5], L. HUANG[*5],
F. KUPKA[†1], D. Le CONTEL[∘8], J.M. Le CONTEL[∘8], D.J. OSIP[*3],
K. PANOV[*6], N. POLOSUKCHINA[*7], J.P. SAREYAN[*8],
H. SCHNEIDER[∘9], J.C. VALTIER[†8], M. ZBORIL[∘10],
J. ŽIŽŇOVSKÝ[*10], J. ZVERKO[*10]

Contributions to photometry (*), spectroscopy (†) and mapping (∘)

[1]Institute for Astronomy, Vienna, [2]Astronomical Institute of the Russian Academy of Sciences, Moscow, [3]Lowell Observatory, Flagstaff, [4]Instituto de Astronomia – UNAM, Mexico [5]Beijing Astronomical Observatory, [6]Institute of Astronomy, Sofia, [7]Crimean Astrophysical Observatory, Ukraine, [8]Observatoire de la Côte d'Azur, Nice, [9]Universitätssternwarte, Göttingen, [10]Astronomical Institute of the Slovak Academy of Sciences, Tatranska Lomnica

1. Introduction

ET And is a binary system with a B9 Si star as the main component ($P_{orb} = 48.308^d$, e=0.46). Controversial claims in the literature concerning pulsation with periods ranging from few minutes to few hours and with variable amplitudes indicated a challenging target and motivated us to organize several photometric and spectroscopic observing campaigns. The problem with pulsation of ET And is that T_{eff} and log g put this star in the cool domain of Slowly Pulsating B-type (SPB) stars, but the pulsation periods would be too short by a factor of about four, relatively to the shortest hitherto known periods for SPB stars.

2. Rotation and pulsation

Period and shape ($P_{rot} = 1.618^d$, $a_{rot} = 23$ mmag(B)) of the rotation light curve did not change during the last three decades. Second, there are strong evidence that the short time scale variations attributed in the literature to ET And are due to pulsation of the main comparison star HD 219891 (HR 8870, A5 V). If this suspicion can be corroborted, a serious conflict between observation and current theory of stellar opacities will be eliminated and HD 219891 can be identified as a new δ Sct type variabel. Two pulsation periods ($P_1 = 2.38^h$, $a_1 = 4.1$ mmag(B), $P_2 = 3.55^h$, $a_2 = 1.8$ mmag(B)) are sufficient to reproduce the observed amplitude spectrum to a noise level of 0.3 mmag.

A full account on our photometry is being prepared for publication (Weiss et al. 1993, A&A in prep.).

L. A. Balona et al. (eds.), Pulsation, Rotation and Mass Loss in Early-Type Stars, 43–46.

3. Atmospheric parameters and abundances

The spectrum of ET And ranging from 4000Å to 7500Å was observed in 17 overlapping segments with the AURELIE Coudé spectrograph (resolution = 0.15Å) of OHP. First, we tried to fit the observed Balmer line profiles (Hα to Hδ) with the help of Kurucz ATLAS 9 atmospheres and obtained an optimum fit for T_{eff}=11 500 K, log g=3.5, and v sini=80 kms^{-1}, in good agreement with photometrically derived estimates.

The abundance analysis indicates a 2 dex overabundance of Si. No significant enhancement of the iron-peak elements was found. The light elements, like He, C, Mg, are underabundant by a factor of about 10, and even more so in the case of He. The heavy elements, like Sr, Y, and Zr are overabundant and follow the even-odd abundance pattern.

A detailed spectroscopic analysis of ET is being prepared for publication (Kuschnig et al. 1994, A&A in prep.)

4. Silicon and Helium surface mapping

A series of 170 AURELIE spectra was available for mapping ET And. Instead of using the approximation formula for local line profiles we synthesized the spectral region for Si (4194Å to 4206Å), and for He (4022Å to 4027Å). The inversion from observed line profiles to the surface distribution was done with an interpolation in order to derive the local line profiles for any given point on the stellar surface. An angle of inclination i=80° gave the best fit to the observed line profiles.

The resulting image for silicon shows a strong equatorial spot-like concentration. The structure in longitude is particularly well reproduced because of the coverage of 20 phases. The largest local overabundance found is about 4 dex compared to solar.

The helium map is essentially defined by two spots close to the equator. They are found in a region were silicon is relatively depleted. Only in the two strongest He spots comes the abundance close to solar.

Also this part of our investigation is being prepared for publication (Piskounov et al. 1994, A&A in prep.).

5. Acknowledgments

This project was supported in Austria by the FWF (projects No. 6927 and 8776), the BMfWF (project *Modellieren von radialen und nicht-radialen Pulsationen*), the ÖAW (project *Asteroseismology*), in USA by NSF grant No. AST-8716971 and the U.Hawaii (Mauna Kea Observatory), in China by NNSF grant No. F-108. The authors greatfully acknowledge telescope time allocated to this project from their institutions.

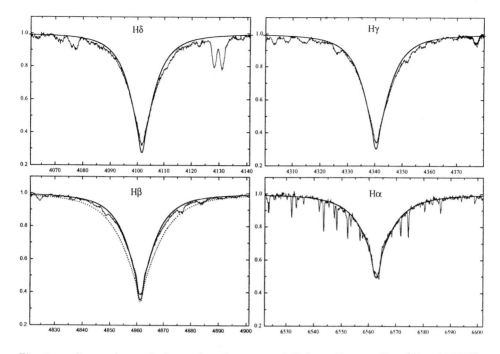

Fig. 1. Comparison of observed and computed Balmer line profiles (T_{eff}=11 500K, log g=3.5 (solid line), log g=4.0 (dotted)

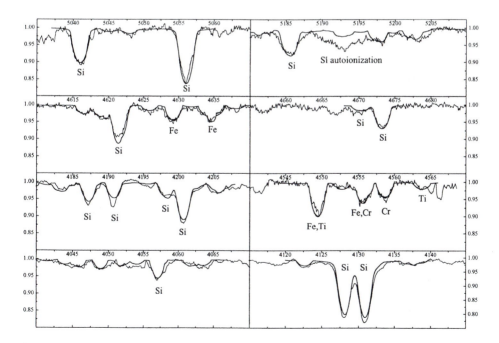

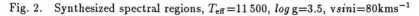

Fig. 2. Synthesized spectral regions, T_{eff}=11 500, log g=3.5, v$sini$=80kms^{-1}

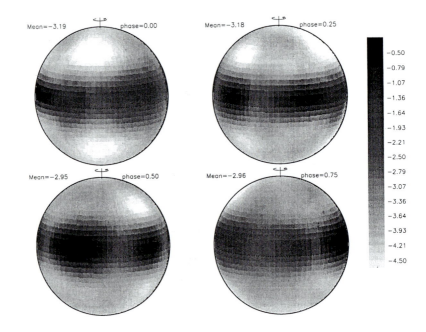

Fig. 3. Silicon surface distribution map for ET And

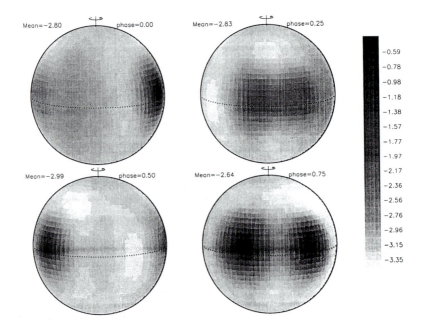

Fig. 4. Helium surface distribution map for ET And

SPECTROSCOPICAL ANALYSIS OF δ SCUTI STARS

E. SOLANO
VILSPA-INSA. P.O. Box 50727. 28080 Madrid (Spain)

and

J. FERNLEY
VILSPA-ESA. P.O. Box 50727. 28080 Madrid (Spain)

δ Scuti are pulsating stars with a characteristic period of several hours, lying in the A2-F5 spectral range and placed on or a little above the main sequence. Unlike the well-known "classical" RR Lyrae or Cepheids, which show a homogeneous pulsational behaviour (i.e. radial modes with large amplitudes), δ Scuti stars have peculiarities: their amplitudes vary from millimagnitudes to tenths of magnitude and their pulsation modes can be both radial and non-radial.

67 δ Scuti stars have been observed in different campaigns. The aim of this work is to calculate the physical parameters associated with each star ($T_{eff}, v \cdot \sin i, [M/H]$) using different methods to get values as accurate as possible and to establish possible correlations between these parameters and the pulsational behaviour of the stars.

1. Measurement of rotation

Two methods of measuring rotational velocities have been studied: The first, described in Sletteback (1975), is based on the existing linear relationship between the full width at half maximum of a spectral line and the rotational velocity of the star while the second (Gray,1975,1992) uses properties of the Fourier transform of the rotation profile to calculate rotational velocities. After different tests, we decided to used the second one because it provides more accurate values and it allows us to distinguish between rotation and macroturbulence.

The influence of rotation on pulsation has been discussed by some authors (e.g. Mc Namara,1985). The idea pointed out by this author that large pulsational amplitudes are only present in stars with rotational velocities $\leq 40~kms^{-1}$ is confirmed with our larger sample of spectra. A theoretical explanation can be found in Dziembowski (1988).

2. Measurement of effective temperature

The H_β line profile is well known to be a good indicator of the effective temperature in the range where the δ Scuti lie because it is not affected

L. A. Balona et al. (eds.), Pulsation, Rotation and Mass Loss in Early-Type Stars, 47–48.

by the gravity . A grid of H_β synthetic profiles were built using Kurucz's models (1979).

The values calculated in this way have been compared with those ones from photometric calibrations. As each one of them is based on different color indexes $(c_1 - (b - y), c_1 - \beta)$ and on different standards stars, comparisons with our set of values will give information about which calibrations show systematic errors (i.e. Philip & Relyea, 1979) and which agree assuming a random error of $\pm$ 100 K (i.e. Moon & Dworetsky, 1985).

3. Spectral restoration and metallicity calculation

Due to instrumental limitations or to phenomena which take place in the proper star (rotation, turbulences...), the observed spectrum can appear broadened with the subsequent loss in contrast. Essential in the metallicity calculation is to get values of equivalent width as accurate as possible; therefore some spectral restoration techniques have to be used. After different tests, it was seen that the most suitable technique for our purposes is the one developed by Jansson (1970) based on the non-linear constrained algorithms.

The metallicity calculations and comparisons with values from photometric calibrations are still in progress.

References

Dziembowsky,W: 1988, *Acta Astronomica* **38**, 61.

Gray,D.F.: 1975, *Astrophysical Journal* **202**, 148.

Gray,D.F.: 1992, *The observation and analysis of stellar photospheres*, University Press: Cambridge, 380.

Jansson P.: 1984, *Deconvolution. Applications in spectroscopy*, Academic Press: London, 183.

Mc Namara D.H.: 1985, *P.A.S.P.* **97**, 715.

Moon T., Dworetsky M.: 1985, *M.N.R.A.S.* **217**, 305.

Philip A., Relyea L.: 1979, *Astronomical Journal* **84**, 1743.

Sletteback A.: 1985, *Astrophysical Journal Supplement* **29.**, 137.

STACC

SØREN FRANDSEN

Institute for Physics and Astronomy, Aarhus University, Denmark.

1. Small Telescope Array with CCD Cameras

At the 1992 GONG Meeting in Boulder (ASP Conf. Ser. Vol. 42) an attempt was done to form a group of people interested in multiobject, multisite observations with CCD cameras on moderate size telescopes ($D \approx 1m$). The motivations is the need for uninterrupted, long time strings of photometry of certain types of variables. Observations of single objects by networks of photoelectric photometers have been practiced for some time. The extension to multiobject, differential CCD photometry not only multiplies the output, but also provide better signal/noise results. Many of the objects of interest (δ-Scuti's, β-Cephei's) are found close together in open clusters, which already serve as testbeds for stellar evolution theories. Nevertheless, *suitable target fields are difficult to find.*

An informal group has been formed, and a first small campaign was organized this year. We tried to get three sites to observe one southern open cluster, but got only two (see below).

A call for a 1994 campaign, where we tried to get a northern sky target, did not lead to any final project. The principal reason was that no obvious target of the kind we would like to study could be located. The large format CCD's coming into use will alleviate the problem considerably, because a major problem is the small field of view of CCD cameras nowadays.

2. The first STACC campaign

The participants were S. Frandsen, M. Viskum, Aarhus University, Denmark, observing from ESO, and L. Balona, C. Koen, SAAO, South Africa. The primary target for the campaign was the open cluster NGC 6134, where five δ-Scuti stars had been found earlier (Kjeldsen and Frandsen 1989). Only four of these could be observed simultaneously. At the beginning of the night, before NGC 6134 was observable, we monitored another similar open cluster NGC 2660. The outcome was data for 10 nights, some complete and of good quality, some obtained under difficult conditions. Even differential photometry can fail, when the seeing is 3 arcs and the full moon is centered in the slit of the dome and illuminates the telescope.

L. A. Balona et al. (eds.), Pulsation, Rotation and Mass Loss in Early-Type Stars, 49–50.
© 1994 *IAU. Printed in the Netherlands.*

50

3. NGC 6134

We can confirm the variables discovered earlier (one with an amplitude of 1-2 mmag). We show a spectrum from the combined data for this low amplitude variable. One gets an impression of the good window function obtained by the combination of data from two sites.

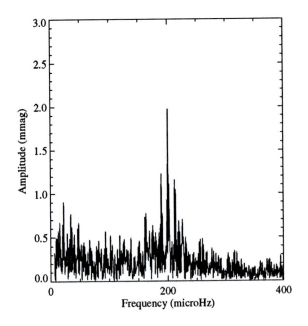

Fig. 1.   A spectrum of a low amplitude δ-Scuti star in NGC 6134 illustrating the low noise of the time series obtained

4. NGC 2660

A suspicion (Frandsen et al. 1989) that this cluster also would have several δ-Scuti members was verified. A first inspection reveals
—   4 δ-Scuti stars
—   2 Eclipsing binaries
—   3 Long period variables
The amount of data is much less than for NGC 6134, the cluster is quite distant (three times the distance to NGC 6134) and the stars quite faint.

References

Frandsen, S., Dreyer, P. & Kjeldsen, H.: 1989, A&A **215**, 287
Kjeldsen, H. & Frandsen S.: 1989, *The ESO Messenger* **57**, 48

TAMS COMPANIONS OF CLASSICAL CEPHEIDS

NANCY REMAGE EVANS
The Institute of Space and Terrestrial Science
4850 Keele St
North York, M3J 3K1, Ontario, Canada

IUE low resolution spectra are an excellent way to determine the temperatures of the hot companions of binary Cepheids. The spectral types of the companions are derived by comparing the spectra with the spectra of standard stars. Absolute magnitudes are calculated from the magnitude difference between the two stars and the absolute magnitude of the Cepheid. In this study, eight binaries containing a Cepheid and a hot companion evolved beyond the ZAMS are discussed.

The binary components are compared with isochrones using several values of main sequence core convective overshoot: Stothers and Chin (1991): no convective overshoot, Schaller, et al. (1992): moderate convective overshoot, and Bertelli, et al. (1986): full convective overshoot. Half the systems are well matched by evolutionary tracks or isochrones with little or no convective overshoot. Four of the systems, however, cannot be matched by any current isochrone (with any overshoot value), in the sense that the companions are too cool for the isochrone appropriate to the Cepheid. Figure 1 shows the theoretical HR diagram for the sample with the four "renegade" systems identified (KN Cen, RW Cam, SV Per, and AW Per), compared with the evolutionary tracks of the Geneva group (Schaller, et al, 1992). Even for the full overshoot models, the mass difference between the two stars is too large to be consistent with companions at the TAMS (Evans, 1993).

Furthermore, these systems occur throughout the whole Cepheid period or mass range, indicating that the cause is not a mass dependent parameter. Rotation is a possible explanation.

This research was supported by a Natural Sciences and Engineering Research Council (NSERC), Canada to NRE.

References

Bertelli, G., Bressan, A., Chiosi, C., and Angerer, K.: 1986, *Astronomy and Astrophysics, Supplement Series* **66**, 191
Evans, N. R. 1993 *Astrophysical Journal*, submitted
Schaller, G. Schaerer, D., Meynet, G., and Maeder, A.: 1992, *Astronomy and Astrophysics, Supplement Series* **96**, 269
Stothers, R. B. and Chin: 1991, *Astrophysical Journal* **381**, L67

L. A. Balona et al. (eds.), Pulsation, Rotation and Mass Loss in Early-Type Stars, 51–52.
© 1994 *IAU. Printed in the Netherlands.*

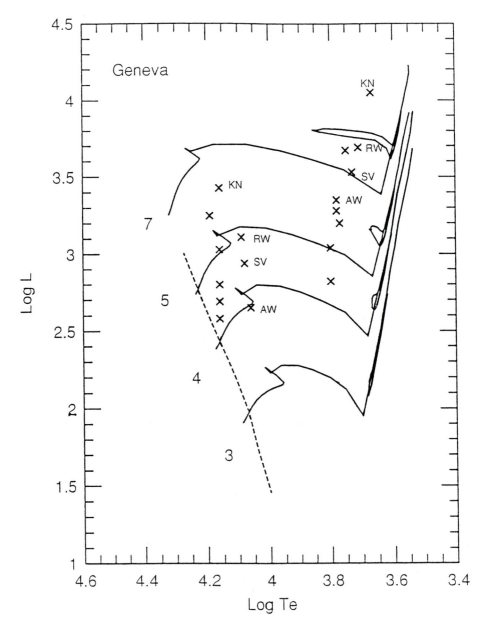

Figure 1. Binary systems containing a Cepheid and a hot companion evolved beyond the ZAMS compared with evolutionary tracks with moderate convective overshoot. The components of the four systems with surprisingly cool companions are identified. Note that in all four cases the mass difference between the two components is too large to be consistent with companions at the very end of their main sequence lifetimes.

PULSATION OF WOLF–RAYET STARS: WR40*

H. SCHNEIDER[1], M. KIRIAKIDIS[1], W.W. WEISS[2],
W. GLATZEL[1] and K.J. FRICKE[1]

[1] *Universitäts–Sternwarte, Geismarlandstr. 11, D-37083 Göttingen, FRG*
[2] *Institut für Astronomie, Türkenschanzstr. 17, A-1180 Wien, Austria*

1. Observations

In the past several Wolf–Rayet stars have been investigated photometrically by different authors to search for fast pulsations. So far all results were negative. However, in 1992 Blecha, Schaller & Maeder published a 627sec period with a semi–amplitude of about 2.5mmag for the Southern Wolf–Rayet star WR40 (HD96548).

In order to check the significance of this result, WR40 was observed during three nights in May/April 1993 at La Silla/Chile using the ESO 1m telescope equipped with a standard photometer. The observations were carried out using a Johnson B filter and an integration time of 20sec.

The typical observing sequence for detecting rapid oscillations is to continuously monitor the program star, interrupted only approximately every 20 minutes for checking the centering of the star in the photometer aperture. This technique relies on a stable atmosphere, but even under moderate conditions a clear detection of pulsations with amplitudes as small as about 1mmag is possible.

The data were corrected for extinction and in addition small changes in the sky transparency (> 30 minutes) were removed. The atmospheric quality of the nights was good, only the first night was moderate. Nevertheless, small erratic fluctuations in the sky transparency could not be removed completely. As a consequence of this (and the observation mode) we can not make statements about pulsations with $f < 4h^{-1}$.

From the amplitude spectra of the individual nights it is evident that no pulsations with $f > 4h^{-1}$ above the noise level (≈ 0.1mmag, ≈ 0.25mmag in the first night, respectively) are detectable. In the amplitude spectrum of the merged data some peaks exceed the level of 0.1mmag (noise level ≈ 0.08mmag), but they are unequivocally not significant. In order to reduce the influence of time gaps we applied the CLEAN algorithm to the amplitude spectra. Again, no significant peaks could be found.

A detailed analysis of these observations is in preparation (Schneider & Weiss, 1993) and will be published elsewhere.

* based on observations made at ESO, La Silla, Chile

L. A. Balona et al. (eds.), Pulsation, Rotation and Mass Loss in Early-Type Stars, 53–54.
© 1994 IAU. Printed in the Netherlands.

2. Theory

As models for Wolf–Rayet stars we have used generalized inhomogeneous helium burning main sequence stars where the chemical composition $(X, Y, Z) = (0.2, 0.78, 0.02)$ has been adopted for the stellar envelope to account for the hydrogen observed in WR40. The hydrogen profile was chosen such that shell burning did not occur. The stability analysis is based on complete stellar models which are constructed using the Göttingen stellar evolution code. Opacities are taken from the latest version of the OPAL library.

In a first step, stability is tested with respect to infinitesimal radial perturbations, where convection is treated in the frozen–in approximation and the linear nonadiabatic stability analysis is performed using the Riccati method. All ordinary acoustic modes are damped and the ε–instability of the ordinary fundamental mode is not present. In addition to the ordinary modes we find a set of strange modes, whose frequencies decrease with mass. These strange modes are associated with extremely strong instabilities above $4.6 M_\odot$ having growth rates in the dynamical range.

As a result of the linear stability analysis we predict fast, in most cases multiperiodic, variability of Wolf–Rayet stars with periods below 30 minutes. The strong dependence on mass of the periods of the unstable strange modes could provide a powerful tool to determine masses of WR stars from observed pulsation periods. Their extremely high growth rates indicate that strange modes play an important role in driving WR winds.

In selected cases we have followed the evolution of the strange mode instabilities discovered into the nonlinear regime using the dynamical version of the Göttingen stellar evolution code. For moderate growth rates of the instability its final result is a nonlinear finite amplitude pulsation.

3. Conclusions

From our 11.4 hours of observations no pulsations of WR40 can be established unambiguously above the noise level of about 0.1 mmag. On the other hand, stellar models appropriate to match the parameters of WR40 show a large variety of simultaneously unstable modes. At present we are not able to provide a satisfactory explanation for this discrepancy. However, we speculate that a large number of simultaneously excited modes could result in a distribution of power over a frequency interval, which simulates an increased noise level in the amplitude spectrum.

References

Blecha, A., Schaller, G., Maeder, A.: 1992, *Nature* **360**,320
Glatzel, W., Kiriakidis, M., Fricke, K.J.: 1993, *MNRAS* **262**,L7
Schneider, H., Weiss, W.W.: 1993, in prep.

THE NEW OPACITIES AND B-STAR PULSATIONS

W.A. DZIEMBOWSKI

Nicolaus Copernicus Astronomical Center, Bartycka 18, 00-716 Warsaw, Poland

Abstract. Over thirty years ago Baker & Kippenhahn (1962) demonstrated that an instability driven by the opacity mechanism is the cause of Cepheid pulsations. Recently it has been shown that the same mechanism is responsible for oscillations observed in β Cephei, SPB and perhaps in other variable B-type stars. The search for the driving mechanism in hot stars began in the late sixties with no success until the opacities calculated with the OPAL code (Iglesias, Rogers and Wilson 1990, 1992) became available. The crucial new feature in the opacity is the local maximum at $T \approx 2 \times 10^5$ K caused by iron lines which was ignored in earlier calculations. Recently, stellar opacity data from an independent project (OP) became available (Seaton *et al.*, 1993). The agreement between the two opacity data is satisfactory.

In B stars the opacity mechanism drives two distinct categories of normal modes. The one relevant to β Cep stars encompasses low order p- and g-modes with periods 0.1–0.3 d. The other includes high-order g-modes with periods ranging up above 4 d. Excitation of such modes may explain most of the slow variability observed in B stars. The theoretical instability domain in the H-R diagram is very sensitive to metal abundance. For the standard value of $Z = 0.02$, the total instability domain in the main sequence band extends from spectral type O9.5 to B9. In types later than B2 only high-order g-modes are unstable.

1. The Opacity Mechanism

For many years, explaining the cause of β Cephei variability has been a major challenge to stellar pulsation theory. The problem is now solved, but we owe the solution to progress in opacity calculations and not to new astrophysical ideas. In fact, if the OPAL opacities were available the problem would have been solved many years ago (Baker & Dziembowski 1969, unpublished).

What drives pulsation in β Cep stars is the opacity mechanism—the same as in classical pulsating variables—except that it relies on a different opacity bump. The spontaneous excitation of an oscillation mode is caused by a particular behaviour of the opacity. In the high-pressure phase of oscillation, the Rosseland mean opacity, κ, increases in the outer layers. This causes a capturing of part of the radiative flux and converting its energy into pulsation energy. The net energy gain during one cycle, W, usually called *the work integral*, is given by

$$W = - \int d^3\mathbf{x} \nabla_{\mathrm{ad}} \oint dt \frac{\delta p}{p} \delta \boldsymbol{\nabla} \cdot \mathbf{F},$$

where δ denotes the Lagrangian perturbation, p is the gas+radiation pressure, $\mathbf{F}$ is the radiative flux and $\nabla_{\mathrm{ad}} = \left(\frac{d\ln p}{d\ln T}\right)_{\mathrm{ad}}$. The expression is valid if the amplitude growth rate is much longer than the period. However, locally the departure from adiabaticity may be arbitrarily large. Furthermore, it applies

L. A. Balona et al. (eds.), Pulsation, Rotation and Mass Loss in Early-Type Stars, 55–66.

both to radial and nonradial perturbations. The subsequent formulae are *strictly* valid only for radial perturbation. They remain *approximately* valid for all nonradial modes considered in this review, but we have to remember that there is an implicit spherical harmonic factor in the perturbed quantities. We have

$$\delta \nabla \cdot \mathbf{F} = \frac{1}{4\pi r^2} \frac{d\delta L_r}{dr},$$

where L_r is the local luminosity, and

$$\frac{\delta L_r}{L} = \frac{dr}{d\ln T} \frac{d}{dr} \frac{\delta T}{T} - \frac{\delta\kappa}{\kappa} + 4\left(\frac{\delta T}{T} + \frac{\delta r}{r}\right).$$

The opacity mechanism may work only if the second term is large. Depending on the form of the $\kappa(T, \rho)$-dependence, this term may contribute either to a mode excitation or to its damping. In ordinary terrestrial conditions the first term dominates and its net effect is always a wave damping.

The plots shown in Fig. 1 will help to remind us how the opacity mechanism works in Cepheids. We see in the bottom panel that the dominant positive contribution to driving arises in a narrow zone around $\log T \approx 4.65$. In this zone δL_r decreases outwards, while both $\delta\kappa/\kappa$ and κ_T increase. The direction of change is opposite in the zone around $\log T \approx 4.85$ where most of the negative contribution to W arises. Let us note, parenthetically, that simplified analytical models such as Baker's one-zone model assuming constant κ_T, miss the most important aspect of the opacity mechanism.

The local maximum of κ_T at $\log T \approx 4.75$ is caused by He II ionization. A bigger maximum at $\log T \approx 4.1$ occurring in the H ionization zone is much less significant for the mode stability despite the fact that the eigenfunctions $\delta T/T$ and $\delta\kappa/\kappa$ are big and rapidly varying. This is because the thermal time scale of the H ionization zone is much shorter than the pulsation period and therefore the tendency towards thermal equilibrium prevails, which enforces $\nabla \cdot \mathbf{F} \approx 0$. The third maximum located at $\log T \approx 5.3$ is the new opacity feature causing instability in β Cep models. The contribution to W from this deep layer is very small because the $\delta T/T$ amplitude is very small there.

The occurrence of a local maximum of κ_T at the proper location within the stellar model is the condition for mode excitation through the opacity mechanism. There are corresponding maxima in the opacity itself occurring at somewhat higher temperatures. We will refer to the three opacity maxima seen in Fig. 1 as the H, the He II and the metal bump. In different models, the three bumps occur at approximately the same temperature as in the model used in Fig. 1, but at different geometrical depths. The He II bump is the primary cause of pulsations in classical Cepheids, W Vir, RV Tau, RR Lyr and δ Sct stars. This has been convincingly demonstrated in a large number of papers following the pioneering work of Baker & Kippenhahn (1962). In cooler stars, the H bump is located in deeper layers and consequently it

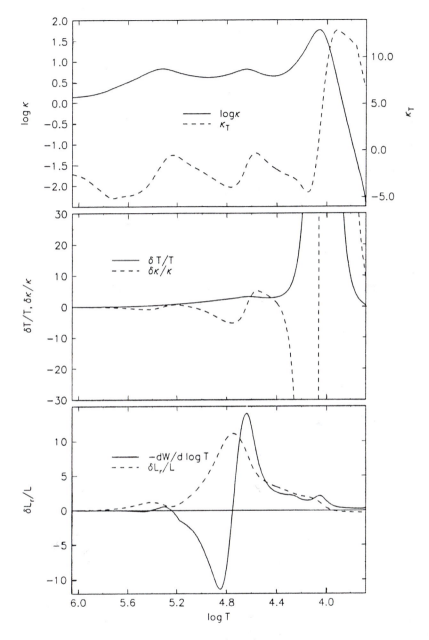

Fig. 1. Opacity κ, opacity derivative $\kappa_T = \left(\frac{\partial \ln \kappa}{\partial \ln T}\right)_\rho$, eigenfunctions describing relative changes of selected parameters (temperature $\frac{\delta T}{T}$, opacity $\frac{\delta \kappa}{\kappa}$, local luminosity $\frac{\delta L_r}{L}$) and the differential work integral $dW/d \log T$ (arbitrary units, positive in driving zones) in the envelope of a δ Cep star model ($M = 6M_\odot$, $\log(L/L_\odot) = 3.32$, $\log T_{\text{eff}} = 3.75$). Pulsation characteristics refer to the fundamental radial mode (period 5.54 day) and to the phase of maximum pressure. The abscissa value $\log T = 6$ corresponds to the fractional radius value $r/R = 0.26$.

may play a dominant role in driving pulsations in cooler stars. In several studies, the critical role of the H bump in Mira models has been confirmed. In oscillating DA and DB White Dwarfs the excitation takes place in the H and He II ionization zones, respectively. However, because of the critical role of convective transport in these cases, the driving effect cannot be regarded as the classical opacity mechanism.

2. The OPAL and OP Opacities

In the Los Alamos opacities, which were commonly used for stellar structure calculations until 1991, there is no bump for $\log T > 4.8$. At $\log T \approx 5.3$ and the corresponding density, the κ value is nearly a factor three less than in the new opacities used in Fig. 1. How was it possible that such a big effect was overlooked? Why was it that even after Simon (1982) had pointed out that an opacity bump located at $\log T \approx 5.3$ would help to resolve the two outstanding problems in stellar pulsations, the Los Alamos team (Magee, Merts & Hubner, 1984) decidedly dismissed such a possibility? It was perhaps because intuition suggests to us that strong lines are important and weak lines are an unimportant. However, since the κ coefficient is the harmonic mean of monochromatic opacities, the truth may be the exact opposite.

The opacity calculations are complicated, and the accuracy requirements are high. It is therefore very fortunate that the task of re-examination of stellar opacities was undertaken by two independent groups. The Lawrence Livermore National Laboratory group developed a sophisticated opacity code named OPAL. The other was a large international group involved in the "Opacity Project" (OP). I use the names OPAL and OP to denote values of the Rosseland mean opacity published by the two respective groups.

The existence of a large bump in the Rosseland mean opacities for temperatures near 2×10^5 K, caused mostly by iron lines, was already demonstrated in an early version of the OPAL opacities (Iglesias, Rogers & Wilson 1990, Rogers & Iglesias 1992). This qualitatively new feature was found thanks to an improved treatment of atomic transitions which specifically included term structures in electron configurations. The most recent version of the opacities (Iglesias, Rogers, & Wilson 1992) takes into account spin-orbit interactions for the heavier elements and results in a further enhancement of the bump. An early version of the OPAL opacities has already established the instability of certain modes in B-star models (Cox et al. 1992; Kiriakidis, El Eid & Glatzel 1992; Moskalik & Dziembowski 1992). However, only calculations employing the improved version (Dziembowski & Pamyatnykh 1993; Gautschy & Saio 1993) explain the occurrence of the β Cep domain in the H-R diagram without invoking abnormally high metal abundances.

The OP opacities became available only very recently (Seaton et al. 1993).

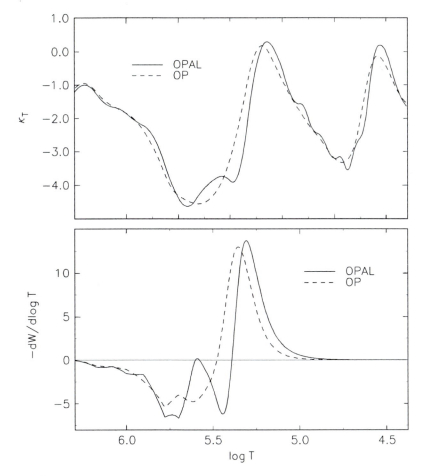

Fig. 2. The opacity derivative κ_T, and differential work integral $-dW/d\log T$ for the fundamental radial mode in the envelope of a β Cep star model ($M = 12 M_\odot$, $\log(L/L_\odot) = 4.40$, $\log T_{\text{eff}} = 4.24$) calculated with two different opacity data. The maximum abscissa value of $\log T = 6.29$ corresponds to the fractional radius value $r/R = 0.71$. The p_0, $l = 0$ mode is unstable only in the model calculated with OPAL opacities.

The approach adopted in this project was different. In particular, plasma effects were ignored in calculating atomic radiative properties. This leads to a considerable simplification in computation and allows a more accurate treatment of electron transitions. In the case of B-star envelopes calculated with the same chemical composition, the agreement between the results obtained with the two opacities is very encouraging. The comparison was done with a test version of OP calculated for the composition adopted in the OPAL opacities. In the standard version, the three iron-group elements (Cr, Mn and Ni) are added and this leads to a significant improvement relative to the OPAL opacities.

The important difference between the two opacities consists of a slight displacement of the metal bump towards higher temperatures. This is shown in Fig. 2, where the critical opacity derivative in a β Cep star model is plotted. In the same figure we plot the differential work integral for the fundamental radial mode. In this case the bump displacement causes mode destabilization. A more general consequence of switching from the OPAL to OP opacity data is a displacement of the instability domain in the H-R diagram toward higher T_{eff} and L , which is a straightforward consequence of the bump displacement. More details on comparison of results obtained with the two opacities are given in the contribution by Pamyatnykh et al. (these Proceedings).

3. Unstable Oscillation Modes

In models of β Cep type stars with the standard value of the metal abundance parameter $Z = 0.02$, the instability of radial modes is restricted to the fundamental mode and the first overtone. The latter is unstable only for stellar masses $M \geq 10 M_{\odot}$. Simultaneously, there is always a large number of unstable nonradial modes. They come in two distinct classes. The first class, characterized by periods similar to those of unstable radial modes, encompasses low-order p- and g-modes. The second class, present only at sufficiently high l- values, encompasses a range of high-order g-modes with significantly longer periods.

As an illustration, we consider the 12 $M_{\odot}$ model calculated with the OP opacities used in the previous section. In this model we have two unstable modes (p_1 and p_2) for $l = 0$. The g_1 and p_1 modes are unstable for $l = 1$. All these modes have periods in the range 0.146–0.192 d, which is typical for β Cep stars. At $l = 8$ we have one unstable mode in this period range. The mode may be classified as p_0, because most of its kinetic energy comes from the acoustic cavity where its radial displacement eigenfunctions have no node, or as g_2 because there are two nodes in the gravity-wave cavity close to the boundary of the convective core. Modes from g_3–g_{14} are stable and those from g_{14}–g_{25} are again unstable. The second instability range corresponds to the 0.42–0.76 d period range. Plots of the pressure eigenfunction and the differential work integral for three selected modes as shown in Fig. 3 will help us understand the origin of the two instability ranges. The crucial aspect is the shape of the pressure eigenfunction. For both p_0 and g_{21} and unlike g_{10}, the amplitude is large only in the outer layers where driving occurs. The amplitude behaviour in the interior($\log T > 6$) for the two former modes is vastly different, but the contribution to W from these layers is negligible in both cases.

At lower l-values the modes from g_{14} to g_{25} have an eigenfunction behaviour similar to those at $l = 8$, and thus favorable for excitation. However, their

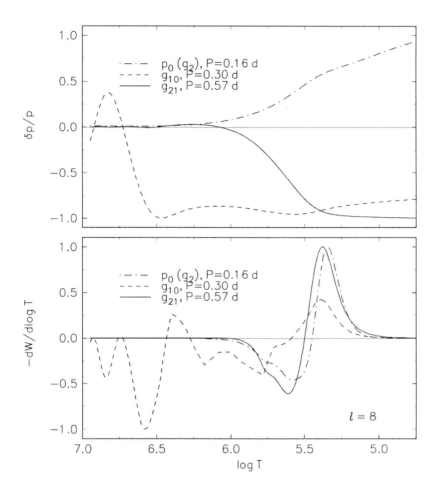

Fig. 3. The radial dependence for the $\delta p/p$ eigenfunction and the differential work integral for selected $l = 8$ modes but in a different $\log T$-range, corresponding to the r/R-range 0.35–0.993. The model is the same as used in Fig. 2.

periods, which at fixed radial order are proportional to $\sqrt{l(l+1)}$, are too long for efficient driving in the metal bump region. The low-l, high-order g-modes are unstable in less luminous star models, because in such stars the metal bump is located in deeper layers and consequently its thermal time scale is longer. Excitation of such modes has been proposed (Dziembowski & Pamyatnykh 1993; Gautschy & Saio 1993; Dziembowski, Moskalik & Pamyatnykh 1993) as an explanation of the slow variability observed in mid and late B-type stars.

4. The B-star Instability Strip

The instability domains in the H-R diagram are shown in Fig. 4. The plots are based on a stability survey (Pamyatnykh *et al.*, in preparation), for low-*l* modes. Stellar models were calculated with the OP opacities for the abundance parameter values $X = 0.7$ and $Z = 0.02$.

Only high-order *g*-modes are unstable in the lower domain. The periods of the unstable modes range from 0.5 to above 4 d. This range is consistent with periodicities observed in variable objects lying in this part of the H-R diagram. The name SPB stars was used after Waelkens (1991) who discovered most of these objects and who demonstrated that they are *g*-mode pulsators. The instability of high-order *g*-modes is present in more luminous and hotter stars but only at higher degrees, *l*. It is possible that the spectral line changes observed in some B (mainly Be) stars are caused by the excitation of such modes. The role of *g*-modes in Be stars should be clarified not only in the variability context, but also as a possible cause of their activity.

The β Cep stars occur only in a small part of the B-star instability strip. Their observational domain in the H-R diagram is limited to the advanced main-sequence phase of 9–18 $M_\odot$ stars. The lack of objects in the post-MS phase is naturally explained by the speed of evolution. There is a paucity of pulsating stars near the ZAMS predicted by theory in the mass range 8–10 $M_\odot$. I do not know whether this reflects an inadequacy in the linear stability calculations or the difference in the nonlinear behaviour between low- and high-mass stars.

The B-star instability strip widens into the supergiant region and may even merge with the extension of the δ Cep instability strip (Zalewski, in preparation). Thus it seems possible that the widespread variability of extreme supergiants may also be explained by the opacity mechanism.

5. Prospects and Problems

Identification of the instability mechanism of β Cep stars does not eliminate these stars from the list of interesting astronomical objects. The solution of the puzzle of their pulsations, which may not seem very interesting, certainly has a bright side. We can now extract much more interesting information about these stars and the systems they live in from the observational data. The occurrence of β Cep stars in a stellar system may now be used to infer information about age and chemical composition.

Having credible linear nonadiabatic models for oscillations in these stars, we may use the information contained in the amplitude ratios light/colour and light/(radial velocity) and the corresponding phase differences to infer the stellar parameters (Cugier *et al.*, these Proceedings). These data in combination with the frequencies should be regarded as valuable asteroseismo-

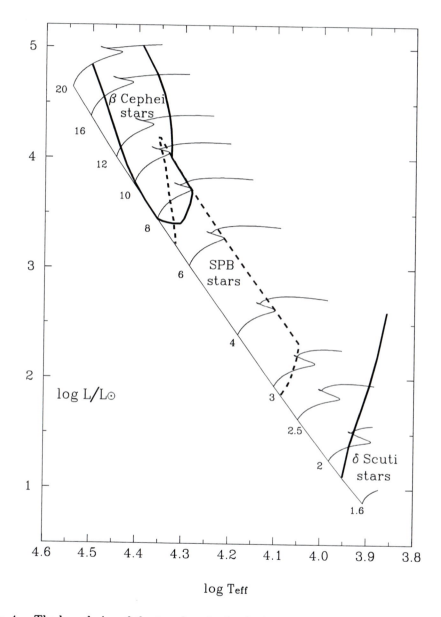

Fig. 4. The boundaries of the two domains in the H-R diagram where low-l mode are unstable due to the driving effect of the metal opacity bump. Within the boundaries there are models having at least one unstable mode with $l \leq 2$. In the upper (β Cep) domain the unstable modes are low-order p- and/or g-modes and in the lower (SPB) domain the unstable modes are high-order g-modes. The blue edge of the low luminosity extension of the Cepheid instability strip is also shown (δ Sct stars).

logical information. Numerous multimodal objects are particularly attractive targets. The same applies to SPB stars, though their long periods make these objects more difficult for observation. It is important that in the case of B-type pulsators, the predictions of stellar pulsation theory are not affected by the lack of a satisfactory description of convective transport. A thin convective layer does appear in the driving zone of more luminous B stars, but the estimated convective flux is far less significant than in the case of Cepheid-type pulsators.

Our understanding of B star oscillations is still not satisfactory. We know almost nothing about what happens at finite amplitudes. Nonlinear studies of β Cep models started only very recently (Moskalik & Buchler, these Proceedings). The linear theory predicts instability of many modes. We do not know which of them will be present in the nonlinear development and what their amplitudes will be or whether the amplitudes will be constant or varying in time. If the latter is true, we do not know whether the variability will be periodic or chaotic. We do not even know whether for a specified stellar model there is a unique answer to these questions. The problems are not specific to B- type pulsators, but because convection is least significant in these stars, these are perhaps the best objects for exploring complex nonlinearities in stellar pulsation theory.

One of the biggest puzzles is the origin of activity in Be stars. The possibility that the excitation of g-modes plays a role remains to be explored. A necessary first step which is still within the framework of the linear theory, is to include effects of rotation in the stability calculations.

Acknowledgements

I am grateful to Alosha Pamyatnykh for Fig. 4 and many very helpful conversations. The assistance of the LOC and IAU in financing my participation in the Symposium is gratefully acknowledged.

References

Baker, N. and Kippenhahn, R.: 1962, *Zeit. Astrophys.* **54**, 114.
Cox, A.N., Morgan, S.M., Rogers, F.J. and Iglesias, C.A.: 1992, *Astrophys. J.* **393**, 272.
Dziembowski, W.A., Moskalik, P. and Pamyatnykh, A.A.: 1993, *Mon. Not. Roy. Astr. Soc.* in press.
Dziembowski, W.A. and Pamyatnykh, A.A.: 1993, *Mon. Not. Roy. Astr. Soc.* **262**, 204.
Gautschy, A. and Saio, H.: 1993, *Mon. Not. Roy. Astr. Soc.* **262**, 213.
Iglesias, C.A., Rogers, F.J. and Wilson, B.G.: 1990, *Astrophys. J.* **360**, 221.
Iglesias, C.A., Rogers, F.J. and Wilson, B.G.: 1992, *Astrophys. J.* **397**, 717.
Kiriakidis, M., El Eid, M.F. and Glatzel, W.: 1992, *Mon. Not. Roy. Astr. Soc.* **255**, 1p.
Magee, N.H., Merts, A.L. and Huebner, W.F.: 1984, *Astrophys. J.* **283**, 264.
Moskalik, P. and Dziembowski, W.A.: 1992, *Astron. Astrophys.* **256**, L5.
Rogers, F.J. and Iglesias, C.A.: 1992, *Astrophys. J.* **79**, 507.

Seaton, M.J., Yu Yan, Mihalas, D. and Pradhan, A.K.: 1993, *Mon. Not. Roy. Astr. Soc.* in press.
Simon, N.R.: 1982, *Astrophys. J. Let.* **260**, 87.
Waelkens, C.: 1991, *Astron. Astrophys.* **246**, 453.

Discussion

Percy: Could you comment on the pulsation of B supergiants? Which modes (periods) are excited?

Dziembowski: So far, Zalewski (in preparation) has considered only radial modes. In addition to the fundamental mode and first overtone, he found unstable strange modes i.e. modes which have no counterpart in the adiabatic approximation. Their periods are always shorter than the fundamental mode period. These modes are responsible for merging the high-luminosity extension of the Cepheid and B-star instability strips.

Balona: The new opacities clearly show that there exists a region of instability for non-rotating mid-B stars. Recent observations of mine of two open clusters, NGC 3293 and NGC 4755 have failed to find any pulsating stars in the predicted strip. One may speculate that the relatively high rotational velocity in the cluster stars may be damping the pulsations. Can you say anything about the role of rotation in these stars?

Dziembowski: Our stability surveys were done ignoring effects of rotation. The results are not directly applicable to the case when the g-mode frequency is comparable with the rotation frequency. When the equatorial velocity exceeds some 50 km s^{-1} the properties of low-l high-order g-modes are significantly changed by rotation. I cannot say whether such modes will be stabilized.

Waelkens: It is a significant observational fact that SPBs are slow rotators. All stars for which I found multiperiodic photometric variations had been selected on photometric grounds only, but turned out to be narrow-lined, with $v \sin i$ values of the order of 30 km s^{-1} or lower. Apparently, pulsations are suppressed in rapid rotators, or they have such large l-values that they are not seen in the photometry.

Henrichs: What do you mean by your conclusion that the origin of activity in Be stars might be understood ?

Dziembowski: We have established that in all B stars, there is a mechanism converting the radiative flux energy directly into mechanical energy. It is a crucial step and, admittedly, the easiest, on the road to explaining the activity. The next step lies in the domain of the nonlinear theory. One may imagine, for instance, dynamo action of the g-modes. An asymmetry between prograde and retrograde modes, which is expected if rotation effects are significant, would lead to a net field amplification.

Henrichs: Do you expect an instability strip in the Magellanic Clouds to occur with the new opacities?

Dziembowski: We (Dziembowski, Moskalik & Pamyatnykh 1993) checked that with the metal abundance parameter $Z = 0.01$, which is above the current upper limit for the SMC and perhaps also for the LMC. The β Cep domain disappears, while the SPB domain is still present. We are currently working on precise criteria for both types of instability.

Kjeldsen: Concerning short periodic pulsators (β Cep stars) in the LMC and SMC, I want to point out that we (Kjeldsen & Baade, these Proceedings) have possibly detected two β Cep stars in the open cluster NGC 371 in the SMC. The amplitudes are 2–3 mmag ($P \approx 2$ h) and if these pulsations are real, it shows that even in environments with low [Fe/H] the κ-mechanism is working.

Dziembowski: This is a very interesting result, but if confirmed I would be more inclined to suspect that the environment is not as low [Fe/H] as we think.

Jerzykiewicz: You get a large number of g-modes excited in models of B-type stars. But your theory is linear, while we in fact observe real, non-linear objects. What would be the number of g-modes in the nonlinear case?

Dziembowski: At this stage I can say only that it will be greater than zero. However, this does not mean that an object lying within the instability strip must show detectable variability. One may imagine that the driving mechanism is saturated with one or more high-l modes, which even at large intrinsic amplitudes will not be seen.

CONVECTIVE ZONES IN THE ENVELOPE OF MASSIVE STARS

FRANK M. ALBERTS

Astronomical Institute 'Anton Pannekoek' and Center for High Energy Astrophysics
University of Amsterdam, Kruislaan 403, 1098 SJ Amsterdam, Netherlands

In the calculation of stellar models with the Cox–Stewart opacities no convective zones in the outer layers of massive stars appear. The new OPAL opacities (Rogers & Iglesias, 1992) show a significant bump in the opacity near temperatures of log $T = 5.2$. This opacity effect results in a small convective zone in the envelope of stars with mass ranging from 15 $M_\odot$ to 150 $M_\odot$, apart from possible convective zones caused by ionization. This was also briefly mentioned by Glatzel & Kiriakidis (1993). For stars on the main sequence this zone is small, about 1% of its radius on the zero age main sequence up to 7% at the onset of the core helium burning and contains a negligible amount of mass. For helium burning stars, however, this convective zone moves inward, keeping the same size but containing more and more mass.

The effect of the opacity on the quantity $\nabla_{\rm rad} - \nabla_{\rm ad}$ is shown in the figures. A contour is plotted where this quantity equals zero. The outlined regions are therefore convective zones by the Schwarzschild criterion. Figure 1 shows the convective regions for a 20 $M_\odot$ star (X=0.70, Z=0.02) as a function of time and radius. Figure 2 shows the convective regions for ZAMS stars as a function of mass in the range between 1 $M_\odot$ and 145 $M_\odot$. Only the outer parts of the star (0.9 $R_\star$ to 1.0 $R_\star$) are shown.

The models have been calculated with the stellar evolution code developed by Eggleton (1971, private communication) in which semi-convection is treated as a slow diffusion process. Mass loss and convective overshooting have not been taken into account.

Consequences of the effects of the presence of this convective zone for oscillatory motion in these stars are currently being investigated.

References

Eggleton P.P., 1971, MNRAS, **151**, 351
Glatzel W. and Kiriakidis M., 1993, MNRAS, **262**, 85
Rogers F.J. and Iglesias C.A., 1992, ApJS, **79**, 507

L. A. Balona et al. (eds.), Pulsation, Rotation and Mass Loss in Early-Type Stars, 67–68.

68

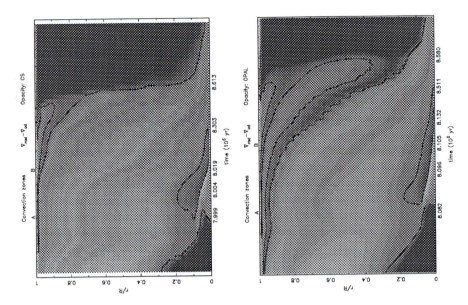

Fig. 1. Convective regions for a 20 M$_\odot$ star with the Cox-Stewart opacities (left) and the OPAL opacities (right). The x-axis displays the evolution in time. The letters at the top of the plot indicate the evolution phase: A – End of hydrogen core burning, B – Start of helium core burning. The dark shading corresponds to convective regions. Note that the horizontal scale is not linear

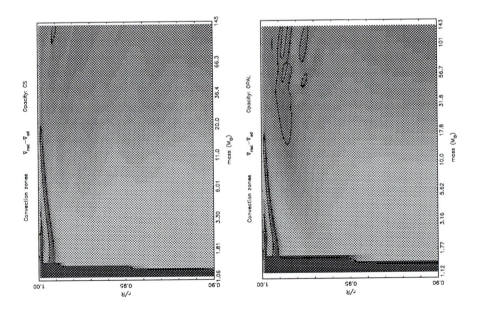

Fig. 2.   Convective regions for ZAMS stars, calculated with the Cox-Stewart opacities (left) and the OPAL opacities (right). Note that the horizontal scale is not linear.

g-MODE INSTABILITY

IN THE MAIN SEQUENCE B-TYPE STARS

W.A. DZIEMBOWSKI[1], P. MOSKALIK[1] and A.A. PAMYATNYKH[1,2]

[1] N. Copernicus Astronomical Center, Bartycka 18, 00-716 Warsaw, Poland
[2] Institute of Astronomy, 48 Pyatnitskaya St., 109017 Moscow, Russia

We show that the OPAL opacities, in addition to explaining the origin of β Cep stars pulsations, also predict existence of a large region in the Main Sequence band at lower luminosities, where high-order g-modes of low harmonic degrees, l, are unstable. The excitation mechanism remains the same and is due to the usual κ-effect acting in the metal opacity bump ($T \approx 2 \times 10^5 K$). The new instability domain nearly bridges the gap in spectral types between δ Sct and β Cep stars. Periods of unstable modes are in the range $0.4 - 3.5$ days for $l = 1$ and $l = 2$. We propose that this excitation mechanism causes photometric variability in the Slowly Pulsating B-type stars (SPB stars, Waelkens 1991) and perhaps in other B stars whose variability in the same period range has been reported.

Typically, there is a large number of modes simultaneously unstable in one model. Most of them have $l > 2$. Such modes are not likely to be detected photometrically but may be visible in line profile changes. The excitation of many high-l modes in a star may also cause a spurious contribution to the rotational $v \sin i$ values.

Sequences of unstable modes at each l exhibit periodically varying departure from equal spacing in period. This feature, first noted in White Dwarf g-mode spectra (calculated and measured), in present case is a probe of the region left behind the shrinking core (μ-gradient zone). We discuss prospects and difficulties of SPB star asteroseismology.

This is the abstract of the paper which will appear in *MNRAS*.

References

Waelkens, C.: 1991, *Astron. Astrophys.* **246**, 453

L. A. Balona et al. (eds.), Pulsation, Rotation and Mass Loss in Early-Type Stars, 69.
© 1994 IAU. Printed in the Netherlands.

OP *VERSUS* OPAL OPACITIES:
CONSEQUENCES FOR B STAR OSCILLATIONS

A.A. PAMYATNYKH[1,2], W.A. DZIEMBOWSKI[1],
P. MOSKALIK[1] and M.J. SEATON[3]

[1] *N. Copernicus Astronomical Center, Bartycka 18, 00-716 Warsaw, Poland*
[2] *Institute of Astronomy, 48 Pyatnitskaya St., 109017 Moscow, Russia*
[3] *Department of Physics and Astronomy, University College London,*
Gower St., London WC1E 6BT, UK

Radiative Rosseland-mean opacities for stellar envelopes from The Opacity Project, hereafter OP, became available (Seaton *et al.* 1993). Here we present a preliminary survey of pulsation properties of B-star models, obtained with this new opacity data and compare these properties with those of models built with the OPAL opacities (Iglesias, Rogers and Wilson 1992). The two opacity projects employ very different approximations in treatment of the plasma and atomic physics. In addition the OP tables are calculated including absorbtion due to three iron group elements ignored in OPAL tables, namely Cr, Mn and Ni.

Using standard codes for modelling stellar evolution and linear nonadiabatic oscillations we have surveyed properties of stars in the Main Sequence and early Post-Main Sequence phases, covering mass range from 3 to 16 $M_\odot$. This range corresponds to β Cephei stars (Dziembowski and Pamyatnykh 1993) and Slowly Pulsating B-type (SPB) stars (Waelkens 1991; Dziembowski, Moskalik and Pamyatnykh 1993). We have studied pulsation properties of p- and g-modes of low spherical harmonic degrees l.

Occurence of unstable modes in a model sequence of $12 M_\odot$, which is typical for the β Cephei stars, is shown in Fig. 1. The difference between OPAL and OP results for the same mixture (G91, see Iglesias, Rogers and Wilson 1992) is noticable, but generally small. In particular OPAL opacities favour first overtone pulsations, while OP prefer the fundamental. The inclusion of the additional iron group elements (S92 mixture) shifts the Blue Edge of the instability strip to significantly higher effective temperature and consequently leads to broadening of the strip. The effect is caused by a slight displacement of the critical opacity bump towards higher temperatures, *i.e.* to deeper layers.

The β Cephei instability domains in the H-R diagram are shown in Fig. 2. With the OP opacities and $Z = 0.02$ we obtain a pleasing agreement with observational data. Thus, there is no longer a need to invoke high-Z values to explain pulsations in the hotter objects. The general displacement of the instability strip is not only to higher $T_{\rm eff}$ but also towards higher L.

L. A. Balona et al. (eds.), Pulsation, Rotation and Mass Loss in Early-Type Stars, 70–72.

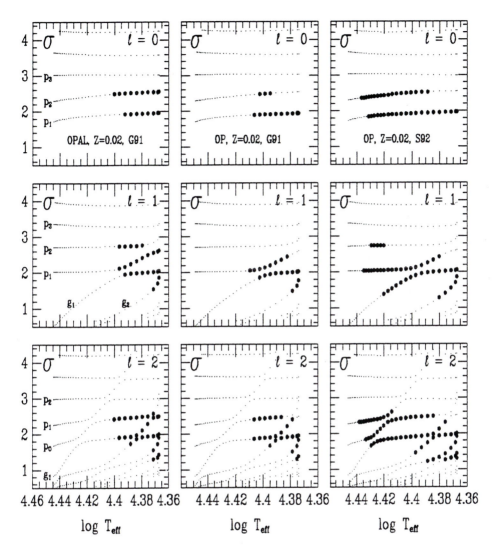

Fig. 1. Oscillation frequencies in units of $\sqrt{4\pi G\langle\rho\rangle}$ for low order p- and g-modes of spherical harmonic degrees $l = 0-2$, plotted against T_{eff}. The models correspond to $12M_\odot$ star in the Main Sequence expansion phase. Big dots represent pulsationlly unstable modes. The three consecutive columns display results for OPAL (G91 mixture), OP (G91 mixture) and OP (S92 mixture: Cr, Mn, Ni added) opacities, respectively.

In the case of the SPB stars the effect of switching to the OP opacities is also significant. The high-order g-mode instability domain in the H-R diagram is now extended towards higher luminosities. Furthermore, the longest periods for the unstable modes are even longer than before. Excitation of modes with $P > 4$ days is now possible. This improves the agreement with observations accommodating longer period SPB stars like HD 27563.

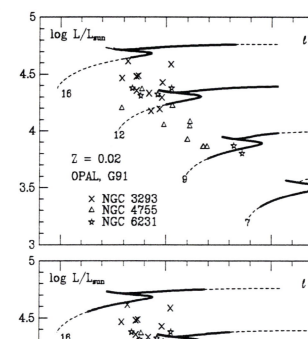

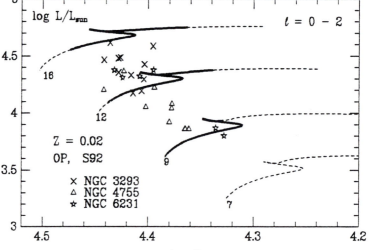

Fig. 2. Evolutionary tracks for stellar masses of $7M_\odot$, $9M_\odot$, $12M_\odot$ and $16M_\odot$, calculated with the OPAL (top) and OP/S92 (bottom) opacities. Thick lines correspond to models, which are vibrationally unstable to low degree modes ($l = 0-2$). The practical Red Edge of the instability strip can be identified with the lowest $T_{\rm eff}$ points of the Main Sequence track, because the Post-Main Sequence evolution is very fast. Cluster β Cephei stars are shown for comparison (Balona 1993, Balona and Koen 1993, Heynderickx 1991).

References

Balona, L.A.: 1993, *MNRAS* , in press

Balona, L.A. and Koen, C.: 1993, *MNRAS* , in press

Dziembowski, W.A., Moskalik, P. and Pamyatnykh, A.A.: 1993, *MNRAS* , in press

Dziembowski, W.A. and Pamyatnykh, A.A.: 1993, *MNRAS* **262**, 204

Heynderickx, D.: 1991, Ph. D. Thesis, Katholieke Universiteit Leuven

Iglesias, C.A., Rogers, F.J. and Wilson, B.G.: 1992, *Astrophys. J.* **397**, 717

Seaton, M.J., Yu Yan, Mihalas, D. and Pradhan, A.K.: 1993, *MNRAS* , in press

Waelkens, C.: 1991, *Astron. Astrophys.* **246**, 453

THE STABILITY OF MASSIVE STARS

W. GLATZEL, M. KIRIAKIDIS and K.J. FRICKE

Universitäts-Sternwarte, Geismarlandstr. 11, D-37083 Göttingen, FRG

1. Introduction

An investigation of the stability properties of stellar models describing massive stars is motivated observationally by the necessity to explain the observed Humphreys - Davidson (HD) limit and the variability of the most massive stars known, i.e. the existence of luminous blue variables (LBVs). Theoretically, a determination of the upper mass limit for stable stellar objects together with its physical explanation and interpretation is of fundamental interest.

2. Stability analysis

The stability analysis is based on complete stellar models obtained from standard stellar evolution calculations with mass loss taken into account according to an empirical mass loss rate. Opacities were taken from the latest version of the OPAL library. Stability of stellar models is tested with respect to infinitesimal radial perturbations. For further details we refer the reader to Glatzel & Kiriakidis (1993) and Kiriakidis et al. (1993).

Massive stars are found to suffer from strange mode instabilities with growth time scales in the dynamical range. Compared with these, instabilities driven by the the classical ε and κ mechanisms are negligible both with respect to range and strength. For solar chemical composition the domains of unstable stellar models in the HRD are shown in Fig.1.

Recently, Stothers & Chin (1993, hereafter SC) have argued that the mean adiabatic exponent becoming smaller than $4/3$ is the origin of LBV instability. For their most unstable model (initial mass: $120 M_\odot$) we have redone their analysis using identical parameters. Concerning stellar evolution our object becomes significantly less massive and its effective temperature stays above $\log T_{eff} = 3.83$ (lowest temperature given by SC: $\log T_{eff} \approx 3.65$). To cover the range of low temperatures reached in the calculations of SC we have also tested the stability of envelope models with appropriate masses and luminosities in this regime.

Contrary to SC both for the complete stellar models and the envelope models the mean adiabatic exponent always stays well above $4/3$. As $\langle \Gamma_1 \rangle < 4/3$ is only a sufficient condition for instability, we have performed an adiabatic stability analysis of both complete stellar models and the envelope

73

L. A. Balona et al. (eds.), Pulsation, Rotation and Mass Loss in Early-Type Stars, 73–74.
© 1994 *IAU. Printed in the Netherlands.*

74

models. We find dynamical instability only below much lower effective temperatures ($\log T_{eff} < 3.56$) than those given by SC. Moreover, as evolutionary tracks do not reach sufficiently low temperatures these low critical temperatures can only be derived on the basis of envelope models. We emphasize that even the higher critical temperatures determined by SC would be too low to serve as an explanation for the LBV phenomenon or the existence of the HD limit.

HRD, Z=0.02

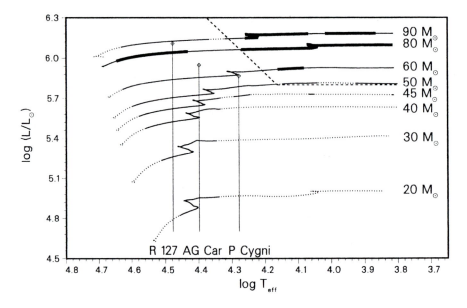

Fig. 1. HRD containing the evolutionary tracks of eight stars (dotted lines) with solar initial chemical composition and the initial masses indicated except for the $90 M_\odot$ track which corresponds to $(X, Y, Z) = (0.73, 0.26, 0.01)$. Unstable phases are denoted by solid lines, thick lines correspond to dynamical growth rates. Together with the observed positions of some LBVs the location of the HD limit is shown as a dashed line. The domain of instability covers the range of observed LBVs and extends to the main sequence for masses above $80 M_\odot$. For low masses the instability domain is continuously connected to the β Cepheid instability strip. Note the agreement of the horizontal branch of the HD limit and the corresponding boundary of the instability domain. For high temperatures the instability domain is strongly metallicity dependent. It shrinks with decreasing metallicity and its blue edge ultimately becomes identical with the blue branch of the HD limit.

References

Glatzel, W., Kiriakidis, M.: 1993, *MNRAS* **263**, 375
Kiriakidis, M., Fricke, K.J., Glatzel, W.: 1993, *MNRAS* **264**, 50
Stothers, R.B., Chin, C.: 1993, *ApJ* **408**, L85

MODE IDENTIFICATION IN PULSATING STARS

C. AERTS
Instituut voor Sterrenkunde, Celestijnenlaan 200 B, B–3001 Leuven, Belgium.

Abstract. During the past twenty years, different methods have been developed to identify the modes in non-radially pulsating stars. Before the introduction of high-resolution spectrographs with sensitive detectors, identifications were obtained from photometric observations. More recently, mode identification is obtained by means of spectroscopic methods. In this paper, we present an overview of the different mode-identification techniques currently used and we describe their accuracy to identify the modes present in different kinds of pulsating stars. By means of some applications of the moment method, we show that this method deserves far more attention than it has received until now.

1. Introduction

The displacement field generated in the photosphere of a star by an non-radial pulsation (hereafter called NRP; see e.g. Smeyers 1984, Unno *et al.* 1989) leads to periodic variations of observable physical quantities. By comparing the observed variations with those predicted by theory, one may hope to determine the most important parameters that appear in the expression for the displacement field. Specifically, mode-identification techniques try to assign values to the spherical wavenumbers (ℓ, m), the degree and the azimuthal number of the spherical harmonic Y_ℓ^m that describes the NRP mode. We will discuss various techniques in Section 2.

A caveat for many analyses is that the theoretical framework which is used only applies to the slow-rotation approximation, i.e. the case where the effects of the Coriolis force and the centrifugal forces may be neglected in deriving an expression for the components of the Lagrangian displacement. It is, however, not allowed to describe an oscillation mode for a rotating star in terms of a single spherical harmonic, and so to ascribe a single set of wavenumbers (ℓ, m) to a mode (see e.g. Lee & Saio 1990, Lee *et al.* 1992; Aerts & Waelkens 1993). The Coriolis force introduces a transverse velocity field that is of the same order of magnitude as the pulsation velocity for a non-rotating star if the ratio Ω/ω of the rotation frequency to the pulsation frequency approaches unity. It is clear that this last condition may be met in several stars discussed in the literature.

In this paper, we will summarise the different methods commonly used to identify modes in Section 2, where we also list their advantages and disadvantages. Some applications of the moment method will be discussed in Section 3. Finally, we present some concluding remarks in Section 4.

L. A. Balona et al. (eds.), Pulsation, Rotation and Mass Loss in Early-Type Stars, 75–86.
© 1994 IAU. Printed in the Netherlands.

2. Mode-identification Techniques

2.1. Q-VALUES

By comparing the observed values of the period P and the pulsation constant $Q \equiv P\sqrt{\bar{\rho}/\bar{\rho}_\odot}$, where $\bar{\rho}$ is the average density, with the ones calculated from theoretical models, it should be possible to estimate the degree and the order of the pulsation mode. While this technique works well for radial pulsations, it is of low power for NRPs. Indeed, the errors on the physical parameters introduce uncertainties on Q that are often larger than the difference in theoretical Q-values associated with different pulsation parameters.

2.2. PHOTOMETRIC AMPLITUDES

This method is based on the photometric variations of a pulsating star and was introduced by Dziembowski (1977), refined by Stamford & Watson (1981), Watson (1988) and Heynderickx (1991), and also applied by Cugier & Boratyn (1992) and Cugier et al. (these Proceedings). Heynderickx (1991, 1992) has made a thorough study of a large group of β Cephei stars and slowly pulsating B stars (SPBs, see Waelkens 1991) for which he determined the degree of the pulsation modes. To achieve this, a theoretical expression for the photometric amplitude of a non-rotating, pulsating star is derived as a function of the wavelength λ and of the degree ℓ of the pulsation mode. By calculating this expression for different values of ℓ and for different values of λ, and by comparing the results with the observed photometric amplitudes at the same wavelengths, one finds the value of ℓ that best fits the observations. The azimuthal number m cannot be determined with this method because a non-rotating star is considered. The method appears to work well for β Cephei stars and SPBs (see Heynderickx 1991).

It would be interesting to extend the method of photometric amplitudes to a rotating pulsating star and apply it to photometric observations of the rapidly rotating Be stars. Such an analysis might help to decide whether or not the observed photometric and spectroscopic variations in these stars are due to NRP or to rotating spots.

The main difficulty with this method seems to be linked with the uncertainties on the stellar parameters needed to calculate the theoretical amplitudes. In some cases, inaccurate stellar parameters might lead to a misidentification of the mode.

2.3. LINE-PROFILE FITTING

The velocity field caused by the NRP(s) leads, through Doppler displacement, to periodic variations in the profiles of spectral lines. After having been first detected in the β Cephei stars, line-profile variations (hereafter called LPVs) have been observed in other kinds of variable OB-type stars as well (Petrie & Pearce 1962, Smith 1977, Baade 1982, Vogt & Penrod 1983,

Baade & Ferlet 1984, Baade 1990, among others).

Since Osaki (1971) computed theoretical line profiles for various NRPs, the identification of modes from spectroscopic observations has become a popular technique. The identification of NRP modes from LPVs is achieved by line-profile-fitting on a trial-and-error basis. The idea is to compare the observed LPVs with those predicted by theoretical calculations. Until recently, this technique was used by most authors; examples are given by Baade (1984), Smith (1977,1983,1985,1989), Vogt & Penrod (1983), among others.

The main disadvantage of the method is the large number of free parameters that appear in the velocity expression due to the NRP. Indeed, the NRP model may be successful in reproducing the line profiles for different sets of input parameters, implying that the fitting technique does not necessarily lead to a unique solution. We also mention that the apparent quality of some fits is suspect in the sense that in modelling, one usually neglects temperature variations, which obviously must affect the profiles.

Some problems appear in applying the line-profile-fitting technique. It often appears necessary to assume that some modes temporarily disappear and reappear afterwards in order to obtain good fits over a large time scale. Also, the values found for the intrinsic profile sometimes have to be varied from one night to another in order to obtain reliable fits. Such assumptions are often introduced in an *ad hoc* fashion and cast doubt on the reliability of the model. In case of rapidly rotating stars, one usually assumes equator-on geometries and high-degree sectorial modes because these are the only ones that can produce the observed "moving bump phenomenon". On the other hand, high-degree modes are almost never seen in slowly rotating stars. Finally, it is always mentioned, but most often not taken into account, that one uses an expression for the pulsation velocity that is related to one spherical harmonic. This is, however, only valid in the case of a non-rotating star.

To overcome some of these problems, Kambe & Osaki (1988) have made a study of theoretically generated LPVs in rapidly rotating stars due to an NRP for moderate to high ℓ-values ($\ell = 2, \ldots, 8$). They also used, however, a pulsation velocity expression derived for a non-rotating star. On the other hand, they also considered purely toroidal modes in a non-rotating star and studied their associated LPVs. The authors find that moving bumps can be produced for modes with $\ell > 4$ in the following situations :

1. spheroidal sectorial modes with a large ratio of the horizontal to the vertical velocity amplitude at small inclination angles,
2. toroidal sectorial modes with intermediate inclination angles,
3. spheroidal tesseral modes with $\ell - 1 = |m|$ and an almost equator-on geometry.

It should be noted, however, that pure toroidal modes only become time-dependent in a rotating star. It is then not appropriate to use an expression for the velocity field associated with toroidal modes that relies on the solutions of order zero in the rotation frequency. Expressions for the toroidal displacement fields that are correct up to second order in Ω/ω have been computed by Smeyers *et al.* (1981).

2.4. DOPPLER IMAGING

In more recent studies, emphasis is especially laid on the spectacular LPVs of rapidly rotating OB stars, such as Be stars. Indeed, it has been recognised that the line profiles of rapid rotators allow a Doppler Imaging (DI) of the stellar surface (Vogt *et al.* 1987), so that a mapping of the pulsation velocity over the surface of variable stars should become possible (Baade 1987). Gies & Kullavanijaya (1988) presented an objective criterion based on DI to determine the periods and pulsation parameters of the modes in rapidly rotating stars. Fourier analysis of the LPVs at each wavelength point yields the periods of the variations by frequency peaks in the resulting periodogram. The true periods are distinguished from alias patterns with Robert's CLEAN algorithm (Roberts *et al.* 1987) or with the Akaike Information Criterion (Kambe *et al.* 1990). The azimuthal number m is obtained by considering the number of phase changes at each signal frequency versus the line position (see also Gies, these Proceedings).

It is usually assumed that equator-on-viewed sectorial modes appear. Moreover, although the method is developed for rapidly rotating stars, one uses an expression for the pulsation velocity based on the slow-rotation approximation. Although these assumptions seem to be rather restrictive, the DI technique is by now often applied, mostly to identify the modes claimed to be present in Be stars; examples are given by Baade (1988, μ Cen), Kambe *et al.* (1990, 1993, ζ Oph), Reid *et al.* (1993, ζ Oph), and Floquet *et al.* (1992, EW Lac).

Some attention should be paid to the fact that the DI technique is based on the usual slow-rotation approximation. Indeed, Lee & Saio (1990) and Aerts & Waelkens (1993) have shown that rotation can significantly affect the eigenfunction in a pulsating star. Aerts & Waelkens find that the effects of the toroidal correction terms due to the rotation become important when the ratio of the pulsation period to the rotation period becomes larger than 20 per cent. Therefore, one should be careful when applying DI to rapidly rotating stars for which this ratio is larger than 20 per cent.

An experiment in cooperation with A. Reid has been conducted to test the implications of the slow-rotation approximation when applying DI. Hereto, two sets of line profiles were generated for a non-radially pulsating star with the velocity expression taking into account the effects of the Coriolis force (see Aerts & Waelkens 1993). Application of DI in the usual slow-rotation approximation shows that a misidentification of both the period and the wavenumbers (ℓ, m) may appear in case of a low-degree non-sectorial input mode; for the test case of an $\ell = 2, m = -1$ input mode, the DI technique wrongly identified a high-degree sectorial mode ($\ell = -m = 5$ or 6). For a given $\ell = -m = 8$ mode, DI recovered the correct parameters (see Reid & Aerts 1993).

The findings of Reid & Aerts are based upon only two examples. It is clear that a more extensive range of NRP and stellar parameters has to be considered in order to obtain a final conclusion about the accuracy of the DI technique in combination with the CLEAN algorithm. In any case, the findings of Aerts & Waelkens and of Reid & Aerts could have some implications for mode identifications obtained thus far with the DI technique. Therefore, a re-evaluation of the mode identification based on DI seems necessary in case of rapidly rotating stars for which the effects of the rotation cannot be neglected.

2.5. THE MOMENT METHOD

To overcome the problems of line-profile fitting, Balona (1986a,b;1987;1990) proposed an alternative method of pulsation-mode identification from LPVs. The method is based on the time variations of the first few moments of a line profile. The periodograms of the moments can immediately be interpreted in terms of the periods that are present and in terms of the NRP parameters. The basic idea is to compare the observed variations of the moments with theoretically calculated expressions for these variations in case of various pulsation modes, and so to determine the mode that best fits the observations. This is achieved through the construction of a discriminant, which is based on the amplitudes of the moments.

At first sight, it may seem that by considering only a few numbers to characterise the profiles, one looses much of the richness of the observation. However, it turns out that much of the information necessary to determine the pulsation modes is contained in the first three moments. Moreover, the mode identification is based on the variation of the moments, and so is only marginally affected by uncertainties concerning temperature variations and the intrinsic profile.

The method as proposed by Balona (1986b, 1987) is developed in the case that the projected rotation velocity is much larger than the pulsation velocity. However, we suggest to combine Balona's approach with the velocity field as presented by e.g. Aerts & Waelkens (1993) since they explicitly take into account Coriolis correction terms, which might be important in identifying the correct modes. The combination of Balona's approach and the velocity field of Aerts & Waelkens may then provide a powerful tool to identify the modes in rapidly rotating stars.

The moment method in case of slowly rotating stars has been introduced by Balona (1990). His discriminant is a one-dimensional function based on the first two moments. It turns out that this discriminant has some limitations because no error on the velocity amplitude v_p is allowed. The moment method for slow rotators has been refined by Aerts *et al.* (1992), who have proposed a more accurate discriminant to obtain the mode identification. Indeed, their two-dimensional discriminant is based on the leading amplitudes of the first three moments and is less sensitive to errors on the estimates of both the velocity amplitude and the inclination angle. Together with the mode identification, this discriminant also gives estimates of the pulsation amplitude v_p and the inclination angle i of the star.

Aerts *et al.* (1992) also present a method to determine the projected rotation velocity v_Ω and the width of the Gaussian intrinsic profile σ from the other terms of the moments, once a mode identification has been obtained by the discriminant. Recently, the moment method for slow rotators has been generalised to a multiperiodic pulsation by Mathias *et al.* (1993), who use essentially the same discriminant as Aerts *et al.* for each of the present modes separately.

The accuracy of the moment method is fully described by De Pauw *et al.* (1993), who studied the shape of the moments, the ability of the discriminant to obtain a correct mode identification, and the accuracy of the estimates of the velocity parameters for a large set of theoretically generated profiles with all kinds of NRP parameters. They conclude that a quantitative determination of the pulsation mode is possible for $\ell < 4$; for higher-degree modes the discriminant is not very convenient, but these modes are distinguished from those with a lower degree. In their paper, De Pauw *et al.* also studied the influence of the noise of observed spectra and of the number of available line profiles on the mode identification; their paper is a useful guide for users of the moment method in the formulation given by Aerts *et al.* and Mathias *et al.*

3. Applications of the Moment Method

The moment method in the slow-rotation approximation has recently been applied to β Cephei stars (see Aerts 1993). We have considered these stars to be ideal test cases for the moment method since they are regular pulsators that usually have a long record of photometric observations. We can then rely on photometric studies for period determinations and sometimes also for estimates of the degree ℓ of the mode (see e.g. Heynderickx 1991, Cugier *et al.* these Proceedings).

A first star that has been analysed with the moment method is δ Ceti, a well-known radially pulsating star. The analysis of this star (Aerts *et al.* 1992) has shown that the moment method works well for a star exhibiting such a simple pulsation. By means of illustration, we show in Fig. 1 some discriminants for δ Ceti, representing the four best solutions in the parameter space (ℓ, m, v_p, i). The best solution for (ℓ, m) is the one for which the

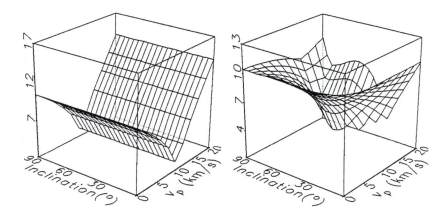

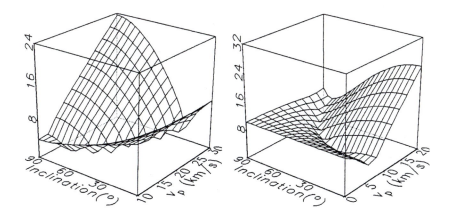

Fig. 1.   Four discriminants for the β Cephei star δ Ceti representing the best solutions in the parameterspace (ℓ, m, v_p, i)

discriminant $\Gamma_\ell^m(v_p, i)$ reaches the lowest minimum. This minimum then also gives the best values for v_p and i for this given set (ℓ, m). In case of δ Ceti, the discriminants point towards a radial mode with $v_p = 10.6\,\text{km s}^{-1}$; the second to fourth best solutions have minima that are at least a factor two higher than the minimum of the radial solution (see Fig. 1 and Aerts *et al.* 1992). It should be stressed that the theoretical profiles constructed for δ Ceti with the set of parameters found by the discriminant do not lead to perfect fits, because the temperature effects become important in studying the shape of the line profiles. We recall that the moment method is less sensitive to temperature variations, because it is based on the time *variations* of the moments of the line profiles.

Two more β Cephei stars, that were thought to be monoperiodic, were analysed with the moment method: β Cephei and KK Velorum. The analyses have shown, however, that both stars pulsate in more than one mode (Aerts 1993). In β Cephei, the two newly discovered modes seem to be a sectorial $\ell = 2$ one and an axisymmetric $\ell = 2$ one, both having an amplitude that is much smaller than the one of the well-known radial mode. The two additional modes lead to beat-periods of 7.4, 2.2, and 3.0 days (Aerts *et al.*, in preparation). The mode identification of these two newly discovered modes is uncertain because of the small amplitudes. KK Velorum exhibits a second radial mode, besides the already reported $\ell = 4$ g-mode (Heynderickx 1991, Waelkens & Aerts 1991). We note that the theoretically calculated instability region for β Cephei stars obtained with the new opacities is not only confined to the p-mode region, but extends to the g-modes as well (Dziembowski & Pamyatnykh 1993; Dziembowski, these Proceedings). KK Velorum may thus be an interesting object for further testing the pulsation mechanism for β Cephei stars.

Several multiperiodic stars were also analysed with the moment method (Aerts 1993). The analysis of α Lupi (Mathias *et al.* 1993) has shown that the moments of different spectral lines can be combined in applying the moment method. A problem for the analysis of LPVs of multiperiodic stars are the long beat periods that appear due to the interaction of the different modes, making it very difficult to obtain a full cover of the total interaction period. For such cases, it is very useful to combine photometric and spectroscopic analyses.

In principle, the method should also apply to less regular and/or longperiod variables, such as e.g. the SPBs. These multiperiodic stars are known to pulsate in g-modes (see Waelkens 1991) but are very slow rotators so that the formalism as presented by Mathias *et al.* (1993), which neglects the effects of the Coriolis force, still applies for them.

4. Conclusions

It is clear that the introduction of accurate detectors into spectroscopy has been a very important issue for the mode identification in pulsating stars. Indeed, the photometric methods are less informative on the identification of the modes, since information on the total energy of the star is used, i.e. we have to work with information that is an integration of a physical quantity over the stellar hemisphere. However, photometric observations are very useful to find accurate periods and also allow an estimation of temperature effects. Moreover, in case of long beat periods, it is very useful to combine photometric and spectroscopic results since a full cover of the beat period may be very difficult to obtain spectroscopically.

Contrary to photometric measurements, spectroscopic observations offer a detailed picture of the visible stellar surface and allow determination of all the parameters that appear in the velocity expression. Due to the richness of the NRP model, it has become clear that quantitative identification techniques are preferred to the direct line-profile-fitting technique.

For slow rotators, the moment method is by far the best one available thus far. Rapid rotators can be analysed with the two different quantitative techniques presented in the literature, although we urge caution in using them both in their actual formulation. The moment method should be generalised by combining Balona's (1987) approach with a velocity field that takes into account the effects of the rotation on the velocity expression due to an NRP; the DI technique should be tested more thoroughly on synthetically generated LPVs for all kinds of NRP parameters as is being done at the moment by e.g. Reid (PhD Thesis, in preparation).

Finally, after having obtained the most likely set of all the velocity parameters, theoretical LPVs should be generated in order to check if their evolution during a pulsation cycle is consistent with the observed ones. We stress that perfect fits should not be expected since the used pulsation models neglect many physical processes, such as temperature effects. While these effects may have a minor influence in studying the relative variation of the line profiles, they cannot be neglected in studying the absolute shape of the observed line profiles.

Acknowledgements

I am indebted to Christoffel Waelkens for his helpful comments on the manuscript and I also thank Andy Reid for explaining many aspects of the Doppler Imaging technique.

References

Aerts, C.: 1993, *Ph.D. Thesis*, Katholieke Univ. Leuven, Belgium.

Aerts, C., De Pauw, M. and Waelkens, C.: 1992, *Astron. Astrophys.* **266**, 294.

Aerts, C. and Waelkens, C.: 1993, *Astron. Astrophys.* **273**, 135.

Baade, D.: 1982, *Astron. Astrophys.* **105**, 65.

Baade, D.: 1984, *Astron. Astrophys.* **135**, 101.

Baade, D.: 1987, in Slettebak, A. and Snow, T.P., eds., *Physics of Be stars*, Cambridge Univ. Press: Cambridge, 361.

Baade, D.: 1988, in Cayrel de Strobel, G. and Spite, M., eds., *The Impact of Very High S/N Spectroscopy on Stellar Physics*, Kluwer: Dordrecht, 193.

Baade, D.: 1990, in Baade, D., ed., *ESO Workshop on Rapid Variability of OB-Stars: Nature and Diagnostic Value*, ESO: Garching, 217.

Baade, D. and Ferlet, R.: 1984, *Astron. Astrophys.* **140**, 72.

Balona, L.A.: 1986a, *Mon. Not. Roy. Astr. Soc.* **219**, 111.

Balona, L.A.: 1986b, *Mon. Not. Roy. Astr. Soc.* **220**, 647.

Balona, L.A.: 1987, *Mon. Not. Roy. Astr. Soc.* **224**, 41.

Balona, L.A.: 1990, in Osaki, Y. and Shibahashi, H., eds., *Progress of Seismology of the Sun and Stars, Lecture Notes in Physics 367*, Springer: Berlin, 443.

Cugier, H. and Boratyn, D.A.: 1992, *Acta Astr.* **42**, 191.

De Pauw, M., Aerts, C. and Waelkens, C.: 1993, *Astron. Astrophys.* in press.

Dziembowski, W.A.: 1977, *Acta Astr.* **27**, 203.

Dziembowski, W.A. and Pamyatnykh, A.A.: 1993, *Mon. Not. Roy. Astr. Soc.* **262**, 204.

Floquet, M., Hubert, A.M., Janot-Pacheco, E., Mekkas, A., Hubert, H. and Leister, N.V.: 1992, *Astron. Astrophys.* **264**, 177.

Gies, D.R. and Kullavanijaya, A.: 1988, *Astrophys. J.* **326**, 813.

Heynderickx, D.: 1991, *Ph.D. Thesis*, Katholieke Univ. Leuven, Belgium.

Heynderickx, D.: 1992, *Astron. Astrophys. Suppl.* **96**, 207.

Kambe, E., Ando, H. and Hirata, R.: 1990, *Publ. Astr. Soc. Japan* **42**, 687.

Kambe, E., Ando, H. and Hirata, R.: 1993, *Astron. Astrophys.* **273**, 435.

Kambe, E. and Osaki, Y.: 1988, *Publ. Astr. Soc. Japan* **40**, 313.

Lee, U. and Saio, H.: 1990, *Astrophys. J.* **349**, 570.

Lee, U., Jeffery, C.S. and Saio, H.: 1992, *Mon. Not. Roy. Astr. Soc.* **254**, 185.

Mathias, P., Aerts, C., De Pauw, M., Gillet, D. and Waelkens, C.: 1993, *Astron. Astrophys.* in press.

Osaki, Y.: 1971, *Publ. Astr. Soc. Japan* **23**, 485.

Petrie, R.M. and Pearce, J.A.: 1962, *Pub. Dom. Astrophys. Obs.* **12**, 1.

Reid, A.H.N., Bolton, C.T., Crowe, R.A., Fieldus, M.S., Fullerton, A.W., Gies, D.R., Howarth, I.D., McDavid, D., Prinja, R.K. and Smith, K.C.: 1993, *Astrophys. J.* in press.

Reid, A.H.N. and Aerts, C.: 1993, *Astron. Astrophys. Letters* in press.

Roberts, D.H., Lehár, J. and Dreher, J.W.: 1987, *Astrophys. J.* **238**, 946.

Smeyers, P.: 1984, *Theoretical Problems in Stellar Oscillations: Proc. 25th Liège International Astrophysical Colloquium*, Univ. Liege: Liege, 68.

Smeyers, P., Craeynest, D. and Martens, L.: 1981, *Astrophys. Space Sci.* **78**, 483.

Smith, M.A.: 1977, *Astrophys. J.* **215**, 574.

Smith, M.A.: 1983, *Astrophys. J.* **265**, 338.

Smith, M.A.: 1985, *Astrophys. J.* **297**, 224.

Smith, M.A.: 1989, *Astrophys. J. Suppl.* **71**, 357.

Stamford, P.A. and Watson, R.D.: 1981, *Astrophys. Space Sci.* **77**, 131.

Unno, W., Osaki, Y., Ando, H., Saio, H. and Shibahashi, H.: 1989, *Nonradial Oscillations of Stars*, 2nd edition, Univ. Tokyo: Tokyo.

Vogt, S.S. and Penrod, G.D.: 1983, *Astrophys. J.* **275**, 661.

Vogt, S.S., Penrod, G.D. and Hatzes, A.P.: 1987, *Astrophys. J.* **496**, 127.

Waelkens, C.: 1991, *Astron. Astrophys.* **246**, 453.

Waelkens, C. and Aerts, C.: 1991, in Heber, U. and Jeffery, C.S., eds., *The Atmospheres of Early-Type Stars: Lecture Notes in Physics 401*, Springer: Berlin, 159.

Watson, R.D.: 1988, *Astrophys. Space Sci.* **140**, 255.

Discussion

Smith: I agree with some of the shortcomings of the line-profile-fitting technique, but perhaps not all of them! Certainly, when confronted with multiple modes in line profiles, one should try to obtain the most reliable ephemeris by an "objective" period-finding technique. I prefer a Doppler Imaging technique (e.g. with CLEAN) to the moment method because it is easier to understand in terms of the derived NRP parameters and it does not assume a Gaussian intrinsic profile (Gaussians tend to mask the appearance of bumps for $m \leq 4$ modes).

Once one has derived periods from whatever technique, one wants to put some physics back into the LPVs because they contain so much information. The treatment of profiles by a short series of mathematical functions will be unable to explain e. g. the asymmetric wings to the continuum, the ratio of narrow/broad line phase widths, the profile footprints, the appearance/disappearance of profile bumps and even signatures of NRP nonlinearities. So perhaps what is needed is to combine the strengths of both objective and profile-fitting methods, in addition to light and color curves.

Aerts: I disagree with your first comment since the variations of the moments are immediately interpreted in terms of the periods and the NRP parameters, contrary to the CLEANed periodograms that are integrated quantities without a physical meaning necessary to obtain the most likely periods when using the CLEAN algorithm. The variations of the moments are hardly affected by the assumption of a Gaussian intrinsic profile, since we consider the time variations of the normalised moments, the normalisation factor being the equivalent width of the line profiles.

As I concluded, we should indeed generate theoretical LPVs with the most likely set of parameters found by an objective method. This should be done however *after* the mode identification and not as a method to obtain the most likely parameters.

Baade: Have you tested how easily the moment method gets confused by the presence of shock effects ?

There would be some interest in applying the moment method to Be stars because a comparison of their Hα emission line profiles with the models of Poeckert & Marlborough often permits one to make a coarse estimate of the inclination angle. Therefore, one may pre-select targets seen at a low inclination which should show much stronger evidence of tesseral modes, if any. From the same arguments it follows that the Doppler Imaging method does not always require the assumption of equator-on views.

Aerts: So far, we only applied the moment method to LPVs of β Cephei stars. In these stars, only very weak shock waves are present, if any (ν Eri, α Lupi). Our analyses have shown that such small shocks do not affect the accuracy of the method. We plan to apply the moment method to 12 Lac and BW Vul in the near future. This should give an idea about the effect of stronger shocks on the mode identification with the moment method.

It would indeed be interesting to apply the moment method to the rapidly rotating stars, such as the Be stars. However, the method should then first be generalised by taking into account the toroidal correction terms as well as temperature variations. The accuracy of the estimate of inclination angle using the Doppler Imaging technique needs to be tested by numerical simulations.

Henrichs: May I strongly suggest that you change or annotate your terminology of "slow" and "rapid" rotators. Traditionally, rapid rotators are stars with high $v \sin i$, and those stars might be slow or rapid rotators in your description. I suggest that you discriminate between slow/rapid rotators in the gravitational sense and slow/rapid rotators in the pulsational sense.

Aerts: OK. To avoid confusion: in this talk I have always used the term rapid rotator in the pulsational sense, i.e. for stars that have $\Omega/\omega > 20$ per cent, no matter how their rotation velocity relates to their break-up velocity.

Jerzykiewicz: You mentioned that the pulsation constant for KK Velorum is 0.050 d. This is quite unusual for a β Cephei star. How did you get your value ?

Aerts: We got the Q-value from photometry as well as from spectroscopy. We mention that the Q-value for the main mode is indeed exceptionally large and points towards a g-mode, but the Q-value of the second mode is 0.025 d which is a medium value in case of a p-mode.

Amplitude Variations in Rotationally Split Multiplets

J. Robert Buchler[1], Marie-Jo Goupil[2] & Thierry Serre[1]

[1] *University of Florida,* [2] *Observatoire de Paris*

We have extended a formalism that was developed for radial pulsators (Buchler & Goupil 1984) to the nonradial ones (Goupil & Buchler 1993). For an $\ell = 1$ triplet, for example, the modal amplitudes satisfy the nonlinear *amplitude equations*

$$\frac{da_-}{dt} = (\kappa_- - i\Delta\omega_-)a_- + \gamma_- a_+^* a_0^2 - 2\beta_{--}a_-|a_-|^2 - 2\beta_{-0}a_-|a_0|^2 - 4\beta_{-+}a_-|a_+|^2$$

$$\frac{da_0}{dt} = \kappa_0 a_0 + 2\gamma_0 a_0^* a_+ a_- - 2\beta_{0-}a_0|a_-|^2 - 3\beta_{00}a_0|a_0|^2 - 2\beta_{0+}a_0|a_+|^2$$

$$\frac{da_+}{dt} = (\kappa_+ + i\Delta\omega_+)a_+ + \gamma_+ a_-^* a_0^2 - 4\beta_{+-}a_+|a_-|^2 - 2\beta_{+0}a_+|a_0|^2 - 2\beta_{++}a_+|a_+|^2$$

The complex amplitudes a_m, $m = -1$, 0, $+1$, represent the slowly varying amplitudes of the multiplet in which a phase-factor $\exp(i\omega t)$ has been removed. The asterisks denote complex conjugation. $\Delta\omega_\pm$ is the rotationally induced linear frequency split of the triplet. The κ_m are the linear growth-rates, $\beta_{mm'}$ and γ_m are the nonlinear coupling coefficients, calculable from the stellar hydrodynamic equations, $\beta_{mm'}$ are the usual cubic saturation coefficients, and γ_m are the coupling coefficients specific to the resonance considered here, *viz.* $\omega_- \approx \omega_0 \approx \omega_+$. We note that angular momentum constraints limit the types of couplings that could otherwise occur for this type of resonance.

The assumptions that underlie the derivation of these particular ampltitude equations are, first, that the stellar rotation is slow, second, that we can disregard magnetic fields, and that we can ignore coupling between multiplets.

These amplitude equations allow us to examine the nonlinear behavior of the amplitude multiplets in the Fourier spectrum in a slowly rotating, pulsating star. Large amplitude asymmetries are found to be possible in a given multiplet (which are *not* due to projection effects). Both steady and time-dependent amplitudes occur, depending on the coupling coefficients between the components of the multiplet.

Steady as well as modulated amplitudes can occur depending on the stellar model:
1. <u>Steady Amplitudes</u>, with $A_m = |a_m|$ can be of 3 types
   - single nonzero amplitude **(singlet**, either $m = -1, 0,$ *or* $+1$))
   - two nonzero amplitudes **(doublet**, $m = -1$ *and* $m = +1$, $A_0 = 0$)
   - three nonzero amplitudes **(triplet**, with $m = -1, 0$, and $m = +1$.

   Here, phase-lock causes the three peaks to be equidistant in frequency.
2. <u>Time-Dependent Amplitudes</u>
   - periodic amplitude modulations **(unsteady triplet)**
   - irregular amplitude modulations **(intermittent triplet)**

L. A. Balona et al. (eds.), Pulsation, Rotation and Mass Loss in Early-Type Stars, 87–88.
© 1994 IAU. Printed in the Netherlands.

In general, one can show that there is no 'energy conservation', *i.e.* the sum of the pulsation energies in the components of the multiplet is not constant. However, for some parameter values it may be approximately constant.

We stress that the nature and aspect of the solutions do *not* depend the scale of the growth-rates; the latter merely affect the time-scale of the amplitude modulation. However, the nature and aspect of the solutions depend sensitively on the values of the nonlinear coupling coefficients and on the *relative* values of the growth-rates.

Figure 1 is a typical example of the amplitudes a steady three-mode pulsation with the concomitant Fourier spectrum. The amplitudes are constant in time and phase-lock causes the Fourier peaks to be equally spaced (even when they are not in the linear case).

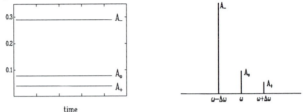

Fig. 1: Triple-mode Fixed-point,   and   Typical Fourier Spectrum for ℓ=1 Multiplet

Figures 2 show examples of unsteady amplitudes. The time-dependent amplitudes (over short observing runs lead to a Fourier spectrum similar to Figure 2, but with differing results from run to run).

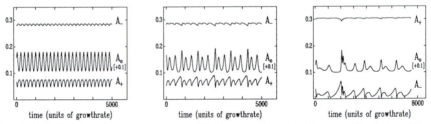

Fig. 2: Amplitudes (A_0 shifted up by 0.1) *left to right*: periodic, irregular, intermittent

A parameter study has revealed that the amplitude equations can display a variety of different types of behavior. For example, the amplitude ratios for the different peaks are sensitive to the parameters values. Amplitude asymmetries (which are intrinsic, *i.e.* not due to projection effects) occur quite naturally. Ab initio computations of the nonlinear coefficients for a given stellar model, white dwarf, main sequence or other, are possible, in principle, but have not yet been carried out. Work is in progress to determine what useful constraints a comparison with observations puts on the coefficients.

References: Buchler J. R. & Goupil M.-J. 1984, ApJ 279, 394
              Goupil M.-J. & Buchler, J. R., 1993, AA, submitted.

Work supported in part by NSF.

LINE PROFILE VARIATIONS IN Be STARS

DOUGLAS R. GIES

Department of Physics and Astronomy
Georgia State University, Atlanta, GA 30303, U.S.A.

Abstract. The periodic variations in the photospheric lines of Be stars could result from rotation or nonradial pulsations (NRP). Here I review the spectroscopic results on ten Be stars which have been targeted in recent multiwavelength campaigns. Most of the stars display variability consistent with an $l = -m = 2$ NRP mode with periods on the order of a day, and the associated light curves are consistent with the pulsation periods, amplitudes, and phases derived from spectroscopy. The variability is strongest in the line wings which only agrees with the NRP prediction that horizontal motions should exceed vertical motions for g-mode pulsations. High frequency, non-periodic variations are also observed, but I argue that these are of circumstellar origin and do not represent high-order NRP. The NRP periods are also found in variations of the Hβ emission line but these probably result from photospheric and not disk variations. However, the same periods occur in the C IV $\lambda1550$ P Cygni line, which strongly suggests that NRP modulates the local stellar wind.

1. Introduction

Ten years have past since Vogt & Penrod (1983) showed how nonradial pulsations (NRP) could explain the blue-to-red moving absorption bumps observed in the spectral lines of the O9.5 V*enn* star ζ Oph. Penrod (1986) found similar line profile variations (*lpv*) in a high S/N spectral survey of some 20 Be stars, and he speculated that NRP could be the missing factor leading to formation of Be envelopes. There has been substantial progress over the past decade in studies of *lpv* in selected targets (see the reviews of Baade 1987, 1992; Fullerton 1991; Smith 1986; Walker 1991), and several recent studies are particularly noteworthy (cf. Yang *et al.* 1990 [ζ Tau]; Floquet *et al.* 1992 [EW Lac]; Bossi *et al.* 1993 [28 Cyg]; Reid *et al.* 1993 [ζ Oph]; Kambe *et al.* 1993 [ζ Oph]). This substantial effort has demonstrated the widespread occurrence of *lpv* among Be and other early-type stars, but the specific periodic content, amplitudes, duration, and origin remain controversial. Here I describe new results from a series of multiwavelength campaigns on Be stars designed to study rapid variability and mass loss.

2. Be Star Observing Campaigns

The close agreement between time scales of variation observed in *lpv*, photometry, and UV wind lines provided the motivation for a series of coordinated observing campaigns to search for a connection between *lpv* and mass loss in Be stars. Here I shall highlight the preliminary results of the campaigns with special emphasis on the spectroscopy. Other contributions in

89

L. A. Balona et al. (eds.), Pulsation, Rotation and Mass Loss in Early-Type Stars, 89–99.

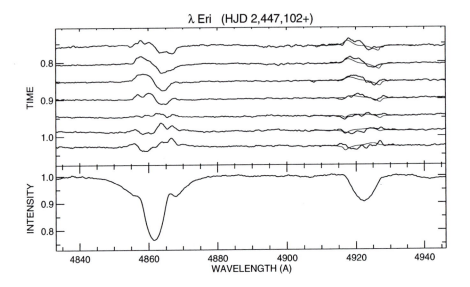

Fig. 1.   The time evolution of the spectrum of λ Eri during the night of 1987 November 3.
The lower panel shows the average spectrum from the run, and the upper panel illustrates
individual difference spectra aligned with the time of observation (intensity scaled to the
same units as the time axis). The thin line superimposed on the He I λ4921 difference
profiles shows the predictions of the NRP model for a single $l = -m = 2$ mode.

these Proceedings describe in more depth IUE high and low dispersion spectroscopy (Peters *et al.*), optical photometry (Percy), polarimetry (McDavid
et al.), and high S/N spectroscopy (Hahula & Gies). The line profile data
were obtained in six campaigns on a total of ten targets, and some 1200 individual spectra have been recorded. Whenever possible we observed a large
portion of the optical spectrum, but in all cases the spectral range included
the Hβ and He I λ4921 lines.

An example of the observations is illustrated in Figure 1 which shows the
time evolution of the spectrum of λ Eri over one night. This example shows
two kinds of variability that we found in virtually every Be star. First, there
is an overall change in the broad shape of the difference profile which is seen
in both Hβ and He I λ4921. This type of low-order variation generally shows
up as a periodic signal in time series analyses (see below). Second, there are
narrower sub-features that often display blue-to-red motion but which generally are irregular in appearance and time evolution. These narrow features
are not noise artifacts, since they have an amplitude much greater than the
measurement errors and display similar patterns in both Hβ and He I λ4921.
In the following sections, I will describe the characteristics and possible ori-
gins of both kinds of variability.

TABLE I

Periodic Signals in He I λ4921 and UV Fluxes

Object	P (d)	$-m$	T_0(NRP)	R	P_{UV} (d)	ϕ_{min}	Δm_{UV}
o And	1.48±0.33	0	7011.6018	0.056	1.57	0.36	0.10
λ Eri	0.71±0.12	2	7104.5320	0.173	0.70	0.01	0.20
ω Ori	...	...	...	...	1.40	...	0.05
28 Cyg	0.64±0.06	2	7790.3188	0.088	0.69	0.73	0.20
η Cen	0.61±0.04	2	8347.2906	0.082	0.61	0.75	0.16
48 Lib	...	...	...	...	0.40	...	0.07
ζ Tau	0.80±0.06	2	8538.5296	0.066	0.81	0.36	0.17
ψ Per	...	...	...	...	1.04	...	0.05
2 Vul	1.27±0.15	2	8879.8170	0.086	0.64	0.23	0.07
KY And	0.79±0.60	2	8879.8975	0.070	...	...	...

3. Low-Order Variations

The profile data were analyzed using the time series methods described in Gies & Kullavanijaya (1988). This involves the Fourier transformation of the line intensity data for time series defined at each wavelength position across the profile. The resulting periodograms can be used to find the periodic signals, power distribution, and and number of cycles of variation across the profile (Fig. 2). The last quantity leads to the modal order m in the context of the NRP model (based on the assumptions that $l = |m|$, since these modes are least prone to cancellation effects from competing surface elements, and that there is a correspondence between radial velocity and stellar longitude, i.e., the Doppler imaging approximation which is valid for inclinations near $i \approx 90°$). We formed a ratio R of the integrated semi-amplitude to the equivalent width to provide a simple estimate of how much of the profile is modulated by the periodic signal.

The time series results are summarized in Table 1 (details in Hahula & Gies, these Proceedings). We find evidence of periodic variability in 7 of the 10 stars, and the periods we obtain correspond in most cases to those determined from UV and optical photometry. None of these seven stars show any evidence of multiple periodicities. This is a surprising result since other workers have found additional signals (usually with shorter periods) and since high-order variations are clearly visible in the profiles (Fig. 1). Evidently, such high-order variability is not periodic in the stars of our sample. The seven periodic variables all have $m = -2$ except for o And which has $m = 0$. Finally, the strength of the periodic lpv (as measured by the ratio R) appears to be correlated with photometric amplitude, Δm_{UV}.

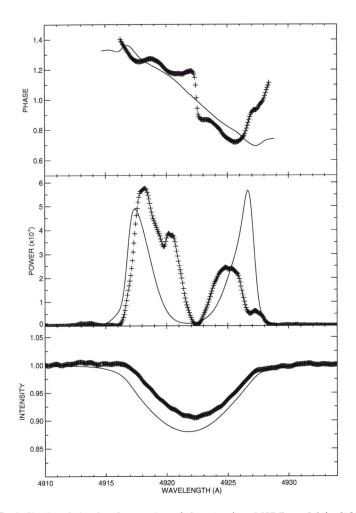

Fig. 2. Periodic signals in the observations (*plus signs*) and NRP model (*solid lines*) for the He I λ4921 profile in λ Eri. *Lower panel:* average profile. *Middle panel:* power distribution across the profile at the signal frequency of 1.4 d^{-1}. *Upper panel:* complex phase of the periodogram across the profile at the signal frequency. Phase becomes indeterminate where the power vanishes (in the line core and wings).

4. NRP Model

Two basic "clocks" could lead to these periodic signals, pulsation or rotation. The key difference between the models is that there are velocity fields associated with NRP that are absent in the rotation model. Thus an analysis of *lpv*, which is particularly sensitive to velocity fields through the Doppler effect, can play a critical role in distinguishing between acceptable models. Here I will explore what predictions about *lpv* and photometric variability

follow from the NRP model using a light curve and profile synthesis program I have written based on the NRP formulation of Buta & Smith (1979).
This code is a surface integration scheme that represents the pulsations as
Legendre polynomials. This is strictly only applicable to nonrotating stars,
and additional terms are needed in rotating stars (Lee & Saio 1990; Aerts &
Waelkens 1993). Nevertheless, the models should give a first-order approximation of the expected lpv. The code includes the geometric Roche distortion
of a rapidly rotating star and the temperature variations that result from
gravity darkening and NRP (Lee et $al.$ 1992).

   I have applied the code to the case of λ Eri to present a representative
example of NRP variability (I plan to make detailed models for each target).
The stellar parameters for λ Eri are taken from Smith et $al.$ (1991): $R_e =$
$7\ R_\odot$, $M = 12\ M_\odot$, $i = 90°$, and $v \sin i = 310$ km s^{-1}. For spheroidal modes,
the ratio of horizontal to vertical velocity amplitude k follows from the
assumed mass, radius, and corotation frequency, and from the adopted NRP
mode $l = -m = 2$ and period $P = 16.84$ hours, I obtain $k = 24.6$ (essentially
all horizontal motion). If the oscillations are adiabatic, then the temperature
variation follows from k (Buta & Smith 1979): $dT/T = 58\ dR/R$. The only
free parameter in the model is the vertical velocity semi-amplitude A_r which
I set at $A_r = 0.65$ km s^{-1} based on a rough fit of the observed lpv. This
leads to a horizontal velocity semi-amplitude $A_h = 16.0$ km s^{-1}, fractional
radius semi-amplitude $dR/R = 0.13$ per cent, and a fractional temperature
semi-amplitude $dT/T = 7.4$ per cent. The model calculations of surface
fluxes, velocity fields, and profiles are illustrated for several NRP phases in
Figure 3.

   The predictions of the NRP model can now be compared directly with
the observed lpv in λ Eri. Synthetic difference profiles are superimposed on
the observed He I λ4921 profiles in Figure 1. The models do a reasonable
job of reproducing the low-order variations, but as noted above there is still
an additional high-order (non-periodic) component of variability. I made a
time series analysis of the model profiles in the same way as the observations
(results in Fig. 2). This exercise verifies that the assumed period and mode
used in the model are successfully identified in the periodogram. The chosen
velocity amplitude for the NRP mode yields lpv with an amplitude similar
to that observed: $R(\text{model}) = 0.11$ vs. $R(\text{obs}) = 0.17$.

   A key test of the NRP model lies in the power distribution of the periodogram across the profile. The model predicts that there will be more
power in the line wings because of the dominance of horizontal motions for
the large k value associated with the $m = -2$ mode, and this trend is indeed
found in the power distributions (middle panel of Fig. 2). The power distributions observed in the periodograms of most of the targets appear to be
double-peaked or flat. In the rotational modulation picture, "spots" act to
change the flux distribution in a rotationally broadened profile. Clearly, such

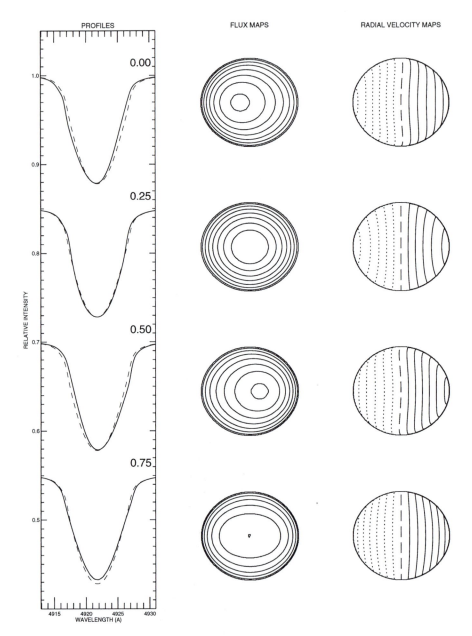

Fig. 3.   NRP profiles, flux distributions, and velocity fields for the λ Eri model. The
left column illustrates the calculated He I λ4921 profiles (*solid lines*) relative to a model
with no pulsation (*dashed lines*) for four NRP phases. The central column gives the UV
flux distributions across the visible hemisphere of the star for the same NRP phases
(as isophotal contours in equal steps equivalent to 1/8 maximum intensity). The star is
brightest at phase 0.25. The right column shows radial velocity maps for the same NRP
phases (in contour steps of 50 km s^{-1}; dotted lines for $V_r < 0$, dashed lines for $V_r = 0$,
and solid lines for $V_r > 0$).

a "spot" would have its maximum influence on the profile when it crosses the central meridian since the star is brightest near the center of the disk due to limb darkening. Thus the rotation model predicts that the signal power should attain a maximum at line center. This is a definitive test, and the observations clearly support the NRP model and appear to rule out a "spot" origin in most cases.

The NRP model also succeeds in reproducing the photometric variations. The full amplitude of the light curve at 1450 Å is $\Delta m_{1450}(\text{model}) = 0.21$ (close to the observed value, $\Delta m_{1450}(\text{obs}) = 0.20$), and this declines in the B-band to $\Delta m_B(\text{model}) = 0.09$ (slightly larger than the value $\Delta m_b(\text{obs}) = 0.05$ found by Balona $et\ al.$ 1992 during the time of the campaign). Minimum light occurs at NRP phase 0.75 when a cool sector crosses the meridian. The observed phase of minimum light $\phi_{\min}$ (based on the spectroscopic ephemeris) is 0.01 from the UV light curve (which covers only 2 cycles), but the more extensive b-band photometry from Balona $et\ al.$ (1992) yields $\phi_{\min} = 0.77$ as expected in the NRP model. The other targets with well defined UV light curves also have $\phi_{min} \approx 0.75$. The observed sense of the phase relation between the profile and photometric variations indicates that the adiabatic dT/T relation of Buta & Smith (1979) is probably valid for low-order NRP in Be stars (Lee $et\ al.$ 1992).

5. Hβ Variability with Similar Periods

If NRP is the source of the photospheric lpv, then it is important to search for the occurrence of these periods in other phenomena associated with mass loss. I have made a preliminary time series analysis of the Hβ emission lines in the campaign targets, and in most cases, the derived periods are the same as found in the photospheric lines. However, the properties of these periodograms suggest that the source of the variability lies in the photospheric variations and not in the emission components. The Hβ periodograms of the non-shell stars (λ Eri, 28 Cyg, 2 Vul, and KY And) are very similar to the corresponding He I λ4921 periodograms. This indicates that the two lines vary in unison, and, consequently, the Hβ lpv are probably entirely due to NRP effects on the underlying photospheric line. In the shell stars (η Cen, 48 Lib, and ζ Tau), the power distribution is concentrated at the central shell component, and the complex phase undergoes an abrupt 0.5 cycle change through line center. The shell lines are presumably formed at some distance from the star where the gas motions are mainly orbital (nearly tangential to the line of sight). The blue (red) portion of the shell feature corresponds to gas projected on the approaching (receding) semi-circle of the rotating star. Thus as a hot, bright sector crosses the disk of the star, the blue half of the shell feature would first absorb more flux until the wave crosses the meridian when the red half would absorb more. Consequently, the apparent the stellar surface, no matter what is the origin of this material. I agree that the data exclude simple spheroidal models. You should also check for radial velocity changes; if you forget about them, your Doppler imaging is fooled. Note that Štefl *et al.* (these Proceedings) found radial velocity variations of the whole line of full amplitude 70 km s^{-1} (for η Cen).

Gies: I agree that the narrow feature *lpv* may have an origin in circumstellar material, but that is only part of the story. The broad, low-order *lpv* can be explained successfully with the NRP model using amplitudes that produce photometric variations that are consistent with those observed. Thus I think it would be a mistake to try to explain all the observed *lpv* on variations in the circumstellar gas. You are right to promote the measurement of radial velocities, since some Be stars are members of binary systems and it is important to remove Keplerian motion before attempting any time series analysis. None of our targets are known short-period binaries. However, it is also important to realize that low-order NRP will produce lines that have large apparent radial velocity variations (see my reply above to Balona).

Waelkens: Coming back to Luis Balona's statement, I want to point out that the ratio k of horizontal to vertical displacement is fixed by theory and is not a free parameter. For a 0.7 day pulsation period, k should be order unity, and then the vertical displacement would be large, causing larger photometric variations. A larger k would correspond to a much longer pulsation period in the corotating frame, and then the ratio of pulsation to rotation period is much too large to use the usual expressions for nonradial modes.

Gies: Yes, k is fixed by theory and is given by

$$ k = \frac{GM/R^3}{\sigma_{\mathrm{cr}}^2} $$

where σ_{cr} is the corotating frequency. For the λ Eri parameters I have used, the corotation period is 74 hours and thus $k = 24.6$. The corotation pulsation period is larger than the rotation period (27.4 hours), so clearly expressing the NRP waves as simple spheroidal functions is only an approximation. I plan to revise my NRP code in the near future to better account for the influence of rotation on the eigenfunctions.

Smith: I have three comments for you.

(1) Erratic bumps in line profiles have indeed been noticed almost as soon as *lpv* were discovered in various Bn and Be stars; for example, in ϵ Per and in λ Eri. I think there are a variety of possible explanations for them, such as nonlinearities in the NRP (e.g. in ϵ Per) and non-LTE effects associated with possible localized heating on the star's surface as the wave sweeps by. But who knows?

(2) I am very surprised that you get such good agreement with adiabatic NRP theory as prescribed by Buta & Smith (1979). Actually, that paper was written in the context of 53 Per itself, a slow rotator but with a corotating period rather similar to that of λ Eri. Ironically, even new optical/UV photometry do not remove the now-famous discrepancy between the light and *lpv* amplitudes, and probably Dziembowski's non-adiabatic correction is needed!

(3) For clarity I would suggest that you find a way of normalizing the power functions in wavelength that you compute to remove the effects of "phase clustering" in your observing window.

Gies: (1) I agree that there may be many sources for the erratic bumps that appear in profiles. My suggestion that the narrow features originate in clumps in the inner disk was motivated by the sharp appearance of these features in Hβ, but other possibilities need to be explored.

(2) The Buta & Smith prescription for NRP is successful in matching model and observed *lpv* and photometric amplitudes and their phase relationship in the best studied targets.

(3) Unfortunately, with a limited duration of observations, there is usually a non-uniform coverage of the NRP cycle, and the power distribution may partially reflect the NRP phase distribution of the observations. The best solution is to obtain long observing runs covering many cycles of variability.

ACTIVITY OF THE Be STAR 28 CYGNI: 1985 – 1991

Ž. RUŽIĆ, V. VUJNOVIĆ, K. PAVLOVSKI, H. BOŽIĆ
Hvar Observatory, Zagreb University, Croatia

J.R.PERCY, A. ATTARD
Department of Astronomy, University of Toronto, Canada

and

P. HARMANEC, J. HORN, P. KOUBSKÝ, K. JUZA
Astronomical Institute, Ondřejov, Czech Republic

1. Introducing the star 28 Cyg

28 Cyg (V1624 Cyg, HD 191610, HR 7708; B2e, $v \sin i = 310$ km s^{-1}) has been the target of several observational projects, and in 1988 of a large international campaign. This attention was inspired by several photometric studies and especially by the 1985 nearly simultaneous optical and UV spectroscopic monitoring by Peters & Penrod (1988). They found that the line-profile variations were controlled by two frequencies, 1.45 c/d, and 7.43 c/d, which they identified with sectorial pulsations of modes $l = 2$, $m = +2$ and $l = 10$, $m = +10$. Rapid changes (0.5 to 1 hr) of the C IV wind profile were found; its equivalent width appeared to correlate with the phase of the $l = 2$ mode. Pavlovski & Ružić (1990) - who independently analysed Hvar 1985 UBV photometry of 28 Cyg - found periodic light variations with a double-wave light curve and a frequency of 1.54 c/d. However – because of the residual scatter around the mean light–curve – the authors tentatively suggested possible multiperiodicity (1.54, 1.33, and 0.95 c/d).

Bossi, Guerrero & Zanin (1993) obtained series of reticon He I λ6678 line profiles and challenged the multimodal pulsations claimed by previous studies. They found periodic changes both in the equivalent width (f = 1.36 c/d) and shape of the profile (f = 2.72 c/d). Also the equivalent width and symmetry of the Hα emission line varied with f = 1.32 and 2.11 c/d, respectively. The authors located the source of this activity into a region close to the star. Hahula & Gies (1991) published preliminary results of a large observational study of line-profile variations of 28 Cyg. They confirmed the frequency of 1.54 c/d, and attributed it to $l = 2$ non-radial pulsation mode.

L. A. Balona et al. (eds.), Pulsation, Rotation and Mass Loss in Early-Type Stars, 102–103.
© 1994 IAU. Printed in the Netherlands.

2. New period analysis: data 1985 - 1991

Since 1985, the Be star 28 Cyg was regularly observed at Hvar Observatory and the University of Toronto Campus. Also, some additional data were supplied from the Automatic Photometric Telescope at MtHopkins in 1990 and 1991.

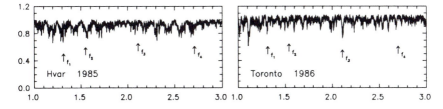

Fig. 1 ϑ–statistics for observations of 28 Cyg in 1985 and 1986, respectively. Frequencies identified by previous authors are denoted by arrows.

The ϑ–statistics and power spectra were calculated for each season and each observing station, respectively. For the 1985 season, complemented by additional Hvar and Toronto observations, the CLEANed spectrum (algorithm by Roberts et al. 1987) shows two prominent peaks at 1.32 and 1.55 c/d, in accordance with the original result. However, an examination of frequency spectra from different seasons reveals that the 1.55 c/d frequency, also found by Hahula & Gies from spectroscopy, does not remain the most prominent one. It is present, however, in almost all frequency spectra, along with 1.36, 2.11 and 2.72 c/d found by other investigators (Fig. 1).

One possible interpretation is that the variations of 28 Cyg are controlled by two principal frequencies: f_a=1.45 c/d and f_b=0.09 c/d. Note that the following relations hold within the accuracy of the frequency determinations: f_a-f_b=1.36 c/d, f_a+f_b=1.54 c/d, $2(f_a$-$f_b)$=2.72 c/d, $2+f_b$=2.11, and $6+f_a$=7.43 c/d. This could indicate that f_a is the principal frequency whose amplitude is being modulated in about a 10-day cycle and, perhaps, also secularly. Another possibility is that some kind of stochastic variations disturbs single- or multiple-periodic changes of the star at certain epochs. A large, well-organized multilongitude photometric campaign would help to solve the problem.

References

Bossi, M., Guerrero, A., Zanin, A.: 1993, *A&A* **269**, 343
Hahula, H., Gies, A.: 1991, *BAAS* **23**, 1379
Pavlovski, K., Ružić, Ž.: 1990, *A&A* **236**, 393
Peters, G.J., Penrod, G.D.: 1988, in *Decade of UV Astronomy with IUE* ESA SP-281, Vol. 2, 117
Roberts, D.H., Lehar, J., Dreher, J.W.: 1987, *AJ* **93**, 968

SEISMOLOGY OF SOUTHERN Be STARS

EDUARDO JANOT-PACHECO and NELSON VANI LEISTER
Depto. Astronomia, IAG-USP, Brazil

1. The survey: search for rapid variability in Be stars

We have started in 1990 a search for moving bumps in the HeI λ 667.8 nm of mainly southern, bright Be stars. The objects of our sample have been selected on the basis of photometric variability (Cuypers et al., 1989). High resolution (R$\geq$ 30,000), high signal-to-noise ratio (S/R$\geq$ 300) spectroscopic observations have been performed at the brazilian Laboratório Nacional de Astrofísica with a CCD camera attached to the coudé spectrograph of the 1.60 m telescope (e.g. Table I). Several hundred spectra have been taken during the last three years. Photometric observations simultaneous with spectroscopy were made on the same site in July 1992 with a two-channel photometer (Stromgren b filter) and a CCD camera (Johnson B filter) installed at two 0.60 m telescopes. The idea is try to disentangle the controversy between NRP and RM models with the help of simultaneous spectroscopy and photometry.

TABLE I

Some Be stars observed at L.N.A.

Star	Epoch	Type of Observation	Remarks
α Eri	1990-93	Spectroscopy	$l = \|m\| \sim 10$
	1990-93	Photometry	
	1993	CCD	
λ Eri	1990	Spectroscopy	bumps?
	1993	Photometry	
	1993	CCD	
η Cen	1990-93	Spectroscopy	$l = \|m\| \sim 14$
ζ Oph	1990-93	Spectroscopy	$l = \|m\| \sim 7$

2. Some results

Moving subfeatures are seen in $\sim$ 50% of the stars in our sample. α *Eri* is discussed elsewhere in this meeting (Leister et al., 1993a this volume).

L. A. Balona et al. (eds.), Pulsation, Rotation and Mass Loss in Early-Type Stars, 104–105.
© 1994 IAU. Printed in the Netherlands.

- η **Cen (B2 IVe)-** Subfeatures ($< 1\%$ of the continuum were detected moving from blue to red through the line profiles. A detailed analysis of the data is presented elsewhere (Leister et al. 1993b). The line profile variations have been examined in terms of nonradial oscillations: from the acceleration of the bumps at the line center and the average time delay between sub-features crossings. We found $l = |m| = 14 \pm 4$, the error being essentially due to the 10% formal error in $V\sin i$ (Slettebak 1982). Data are also compatible with spheroidal *tesseral* NRP modes with $l \cong 7$, $m \cong 6$ (see Kambe and Osaki 1988). The travelling period of bumps is greater than the stellar rotation period, showing that NRP are *retrograde* in the corotating frame. Bumps can be followed over a range of at least $\pm$ 400 kms^{-1} which is larger than $V\sin i$ of the star. The azimuthal NRP degree remained unchanged during the time span of our observations. This standing character of the overall stellar pattern from 1990 to 1992 shows that the (presumed) NRP phenomenon in η Cen can be stable during at least two years. This result must be taken into account in future theoretical work.

- ζ **Oph (O9.4 Vn)-** The analysis of observational data on this object followed the same script as for η *Cen* (see above). Bumps are quite preeminent in this object. Preliminary results were presented by Janot-Pacheco et al. (1991). A quick look on the velocity residual for 1990 suggests $l = |m| \cong 7 \pm 2$.

This campaign on analysis of rapid variability among Be stars is jointly conducted by Departamento de Astronomia, IAG/USP (Brazil), and by DASGAL Observatoire de Paris-Meudon (France)

Acknowledgements

This research is supported in part by FAPESP and CNPq (Brazil), and by CNRS (France).

References

Cuypers J., Balona L.A. and Marang F.: 1989, *A&A* **81**, 151
Janot-Pacheco E., Leister N.V., Quast G.R., Torres C.A.P.C.O.,: 1991, *Rapid Variability of OB Stars: Nature and Diagnostic Value, ESO workshop*, ed. D. Baade , 45
Kambe E.,Osaki Y.: 1988, *PASP* **40**, 313
Leister N.V., Janot-Pacheco E., Buck M.T.C., Dias M.P., Hubert A.M., Hubert H., Floquet M. and Briot D.: 1993a, *this volume* ,
Leister N.V., Janot-Pacheco E., Hubert A.M., Hubert H., Floquet M. and Briot D.: 1993b, *A&A submited* ,
Slettebak A.: 1982, *ApJS* **50**, 55

RAPID SPECTROSCOPIC AND PHOTOMETRIC VARIABILITY IN THE Be STAR ALPHA ERIDANI

N.V.LEISTER,E.JANOT-PACHECO,M.T.C.BUCK,M.P.DIAS
Depto. Astronomia, IAG-USP, Brazil

and

A.M.HUBERT,H.HUBERT,M.FLOQUET and D.BRIOT
Dasgal, Obs. de Paris-Meudon, France

1. Introduction: α *Eri* a variable Be star

We present here a preliminary report on the analysis of simultaneous photometric and spectroscopic observations of the Be star α *Eri*, taken during 1992 at the Laboratório Nacional de Astrofísica. α *Eri* (Archenar, HR472, HD10144, B3-4 III-IV) is the brightest Be star in the sky. It shows activity cycles in a time scale of a few years. Balona et al. (1987) presented a comprehensive study of the star based on photometric and spectroscopic observations. They found a periodicity of 1.26 days in both radial velocity and light variations and argued that a spotted star rotating at this period is the simplest working model that explains the observations.

2. Observations, Reductions and Discussion

We obtained about 110 high resolution, high signal-to-noise spectra of α *Eri* in 1992 July 14,15 and 16 at the L.N.A. with a EMI CCD camera attached to the coudé spectrograph of the 1.60m telescope. Simultaneous photometric observations through the Strömgrem b filter were made with the 0.60m Zeiss reflector on the same site. Spectra were reduced at Instituto Astronômico e Geofísico, Universidade de São Paulo using the "eVe" package. Period search in photometric data were performed through standard FFT and DFT algorithms.

Moving bumps are clearly seen in the residuals (individual spectra minus mean profile).We get $l= |m| = 10 \pm 2$ for α *Eri* in July 1992. The travelling period of bumps is $P_{bumps} = 2\pi V\sin i/a$, which gives in the present case $P_{bumps} = 1.5 \pm 0.3$ days. If the rotational period of the star is 1.26 days (see above), then the NRP are retrograde. The stellar surface pattern due to NRP is reproduced around a star with a periodicity $P_p = P_{bumps}/m$, which in our case is 0.15 ± 0.06 d. Small brightness variations due to this changing surface pattern are expected to occur (e.g. Balona 1987). They should show up in the simultaneous photometric data that will be modulated

L. A. Balona et al. (eds.), Pulsation, Rotation and Mass Loss in Early-Type Stars, 106–107.

with a period P_p (see below). Moving bumps can be followed in our July 1992 spectra over a range $\sim \pm 350$ kms^{-1} which is considerably larger than $V\sin i$ (225 kms^{-1}, Slettebak 1982) for α Eri. This is clearly seen in the "mean absolute deviation" of the spectra (Walker 1991): the region where line profile variability occur is quite evident and extends from about -300 kms^{-1} up to 350 kms^{-1}. This kind of phenomenon has also been reported in other Be stars (e.g. Floquet et al. 1992) and deserves further investigation, as it can be an important clue to the *Be phenomenon*.

Photometric data obtained in 1992 July 14,15 and 16 have been analysed through Fourier transform techniques to search for periodicities. $P_p = 0.15 \pm 0.06$ d deduced from the spectral analysis corresponds to a frequency $\nu_p = 6.7 \pm 2.4$ c/d. Unfortunately, the time span of the observations for each night is less than about four hours. It is thus not possible to find the above frequency. However, the periodogram shows peaks at multiples of this (supposedly fundamental) frequency around 13, 20, 23 and 33 c/d, with 85% success probability for the F distribution for the ensemble of them.

3. Conclusions

Spectral analysis of α Eri suggests the presence of nonradial pulsations with $l = |m| = 10 \pm 2$ in July 1992. This could be the result of the interference of various modes: they cannot be easily distinguished with the kind of analysis presented here. Simultaneous photometric data shows brightness oscillations at frequencies compatible with the first four harmonics of the above determined NRP. However, the power spectrum resolution fails to establish precisely the frequency values and the number of pulsations present in the star. A more extensive analysis of α Eri based on observations spanning about two years will be presented elsewhere.

Acknowledgements

This research is supported in part by FAPESP and CNPq (Brazil), and by CNRS (France).

References

Balona L.A.: 1987, *M.N* **224**, 41

Floquet M., Hubert A.M., Janot-Pacheco E., Mekkas A., Hubert H. and Leister N.V.: 1991, *A&A* **264**, 177

Slettebak A.: 1982, *ApJS* **50**, 55

Walker G.A.H.: 1991, *Rapid Variability of OB Stars: Nature and Diagnostic Value*, ESO workshop, ed. D. Baade , 27

A SURVEY OF LINE PROFILE VARIATIONS
IN NON-EMISSION LINE B0-B5 III-V STARS

M. S. FIELDUS* and C. T. BOLTON

David Dunlap Observatory, University of Toronto
Richmond Hill, Ontario, L4C 4Y6, CANADA

1. Introduction

Some investigators have attributed the photometric and spectral line profile variations (*lpv*) that are common among the B stars to nonradial pulsation while others have attempted to explain them by rotation of photospheric or circumstellar structures with respect to our line of sight. One of the problems in resolving this debate has been our lack of knowledge of how these variations depend on fundamental stellar characteristics. Surveys of *lpv* have covered the Be and a few Bn stars (D. Penrod, unpublished), the O stars (Fullerton 1990), and B8-B9.5 main sequence stars (Baade 1989), but noone has carried out a systematic search for *lpv* among the near main sequence, non-emission line early- and mid-B stars. This paper describes preliminary results from such a survey.

2. Observational Program

Our survey sample consists of forty-eight non-emission line B0-B5 III-V stars. We excluded Be stars from the sample to eliminate *lpv* caused by circumstellar material and avoid duplication of earlier work. We selected the stars to cover as wide a range of $v \sin i$ as possible at each spectral type and luminosity class. However, the sample is biased towards lower $v \sin i$ because of the exclusion of Be stars. We excluded known β Cephei stars, double-line spectroscopic binaries, and large-amplitude single-line spectroscopic binaries. We included some small-amplitude single-line spectroscopic binaries to see if they are line profile variables whose variations have been misinterpreted and a couple of 53 Per stars that have not been well studied.

The spectra of the sharp-lined stars were recorded with either reticon or CCD detectors on the coudé spectrographs on the 1.2-m telescope of the Dominion Astrophysical Observatory ($\lambda/\Delta\lambda \approx 40,000$)and the 2.7-m telescope at McDonald Observatory ($\lambda/\Delta\lambda \approx 26,000$) and the échelle spectrograph on the 1.88-m telescope at the David Dunlap Observatory

---

* Deceased 1992 July 4. Guest observer at McDonald Observatory and Dominion Astrophysical Observatory

L. A. Balona et al. (eds.), Pulsation, Rotation and Mass Loss in Early-Type Stars, 108–109.
© 1994 *IAU. Printed in the Netherlands.*

$(\lambda/\Delta\lambda \approx 60,000)$, The spectra of the broad-lined stars were recorded with a CCD detector on the cassegrain spectrograph of the 1.88-m telescope at the David Dunlap Observatory $(\lambda/\Delta\lambda \approx 8500)$.

The final data set consists of an average of 15 spectra for each of the forty-eight stars in the survey. The He I $\lambda447.1$ nm and Mg II $\lambda448.1$ nm lines were the primary lines observed in the survey, but many other lines were observed in the échellograms. The typical star in our survey has spectra with $\langle S/N \rangle = 600$ in the continuum near the Mg II line, but the $\langle S/N \rangle$ ranges from 350 to 1600 for the stars in our sample. The observing program was designed to obtain a modest number of high S/N spectra to detect *lpv*, measure its amplitude, and characterize the "mode" of the variability. In most cases the spectra are too widely and irregularly spaced to allow us to measure the time scale or period(s) of the variations.

3. Preliminary Results

The work on this project has been delayed by the tragic death of the first author. To date, we have analyzed temporal variance spectra for thirty-nine of the forty-eight program stars. This sample is more biased toward low $v \sin i$ than the complete sample because the hotter broad-lined stars in our sample have not yet been analyzed.

We have not detected any *lpv* in stars with $\log T_{eff} > 4.3$ and $v \sin i < 100$ km s^{-1}. Nearly all of the cooler slowly rotating stars are line profile variables. We are tempted to claim that $\log T_{eff} \approx 4.3$ is the blue edge of the 53 Per instability strip, but we note that 53 Per-like *lpv* have been observed in the O9 V star 10 Lac (Smith 1977, 1978; Fullerton 1990). We haven't found any other correlation between the occurrence or amplitude of the *lpv* in the Mg II line and T_{eff}, $\log g$ (or its proxy, M_{bol}), or $v \sin i$.

Three of the stars in our sample have particularly interesting *lpv*. η UMa appears to be constant in one set of observations with $\langle S/N \rangle = 1000$, but weak *lpv* is detectable in a second set of spectra with $\langle S/N \rangle = 1600$. The moderately slow rotator ($v \sin i = 75$ km s^{-1}) HD 49567 has large amplitude *lpv*. Variations characteristic of both low- and high-order NRP modes are visible in its lines. Finally, the lines of the rapid rotator ($v \sin i = 332$ km s^{-1}) HD219688 appear to vary in a low-order mode, but rather than swaying from side-to-side, they show a single bump that moves across the line profile from blue to red.

References

Baade, D.: 1989, *A. & A.* **222**, 200
Fullerton, A.: 1991, *Ph.D. thesis* University of Toronto,
Smith, M. A.: 1977, *Ap.J.* **215**, 574
Smith, M. A.: 1978, *Ap.J.* **224**, 927

MUSICOS 92 OBSERVATIONS OF θ^2 TAURI

E.J. KENNELLY, G.A.H. WALKER and THE MUSICOS TEAM
Department of Geophysics and Astronomy, University of British Columbia,
Vancouver, B.C., V6T 1Z4, Canada

MUSICOS (MUlti-SIte COntinous Spectroscopy) is an international collaboration interested in areas of astronomical research requiring continuous, high-resolution spectroscopy. The MUSICOS stategy has been 1) to organize international campaigns (Catala et al 1993), 2) to build a prototype fibre-fed, high-resolution, echelle spectrograph (Baudrand & Böhm 1992), and 3) to install duplicates of the MUSICOS spectrograph on 2m class telescopes around the world. Stages 1 and 2 have been completed. Preliminary results on θ^2 Tau from the second MUSICOS campaign are presented here.

The δ Scuti star θ^2 Tau is a spectroscopic binary and a member of the Hyades cluster. Stellar evolution models of the binary system have been derived from well-constrained observational parameters (Królikowska 1992). θ^2 Tau is an excellent candidate for stellar seismology as extensive photometric observations have indicated that its oscillations are multiperiodic (Breger et al 1989). However, the theoretical frequency spectrum is rich and the identification of oscillation modes from photometry is difficult. To circumvent this problem, multi-site spectroscopic observations were planned. With nearly complete coverage of the Doppler-broadened line-profile variations, the frequencies and modes of oscillation can be determined.

Time series observations of θ^2 Tau were obtained from sites in China, OHP, the Canary Islands, and Kitt Peak satisfying requirements of high resolution ($R > 30,000$, minimum 20,000), high signal-to-noise ($S/N > 500$, minimum 100) and good time sampling ($t_{exp} < 600$ sec, maximum 1200 sec). Nearly complete coverage was obtained during four days of observation.

Low- and high-degree modes of oscillation appear as variations in radial velocity and line-profile shape. Frequencies established by previous photometric campaigns at 13.2, 13.7 and 14.3 cycles per day were rederived from the radial velocity variations. The line-profile variations were analyzed using a two-dimensional Fourier technique (Kennelly et al 1992) to reveal modes of oscillation possibly as high as $|m| = 8$ at frequencies between 12 and 17 cycles per day (Figure 1).

Much of the credit for the success of this campaign belongs to the MUSICOS observers: Claude Catala, observer at the 1.5m OHP telescope, Bernard Foing, Zhao Fuyuan, and Jiang Shiyang, observers at the 2.16 Xinglong telescope, Eric Houdebine, observer at the 4.2m WHT telescope, and Jim Neff, observer at the 1.5m McMath telescope. Other participants in the cam-

L. A. Balona et al. (eds.), Pulsation, Rotation and Mass Loss in Early-Type Stars, 110–111.
© 1994 *IAU. Printed in the Netherlands.*

paign were J. Baudrand, T. Böhm, J. Butler, B. Carter, A. Collier Cameron, G. Cutispo, J. Czarny, J.-F. Donati, K.K. Ghosh, L. Huang, D. Rees, M. Rodono, M. Semel, T. Simon, A. Welty, and D. Zhai.

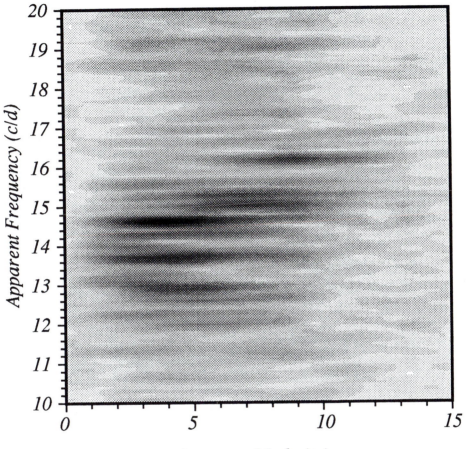

Figure 1. Fourier Representation of Line-Profile Variations. Residual variations within line profiles were mapped onto a system of angular coordinates and transformed in both time and space.

References

Baudrand, J. and Böhm, T.: 1992, *Astronomy and Astrophysics* **259**, 711

Breger, M., Garrido, R., Huang, L., Jiang, S., Guo, Z., Freuh, M., and Paparo, M.: 1989, *Astronomy and Astrophysics* **214**, 209

Catala C. et al: 1993, *Astronomy and Astrophysics* **275**, 245

Kennelly, E.J., Walker, G.A.H., and Merryfield, W.J.: 1992, *Astrophysical Journal, Letters to the Editor* **400**, 71

Królikowska, M.: 1992, *Astronomy and Astrophysics* **260**, 183

TIME-SERIES ANALYSIS
OF THE O4 SUPERGIANT ζ PUPPIS

ANDY REID

Department of Physics and Astronomy, University College London, Gower Street,
London WC1E 6BT, United Kingdom.

Abstract. Preliminary results of a time-series analysis on a sequence of high dispersion, high-signal-to-noise, optical echelle spectra of the O4 supergiant, ζ Puppis (HD 66811), are presented. Lines of He I λ5876Å, He II λ5411Å, N IV λ6381Å and C IV λλ5801, 5812Å are analysed; they show variations in line shape and equivalent width, with 'bumps' and 'dips' seen moving from blue to red within the line profile. The qualitative pattern of variability seen in all these lines is similar.

A Fourier technique, incorporating the iterative CLEAN algorithm, was used on the individual wavelength samples. A periodicity of 8.5-hrs was detected in He II, and strongly suspected in the remaining absorption lines. This periodicity seems to rule out wind variability as the principal origin for the optical line profile variations seen in ζ Puppis.

1. Summary of the Time-Series Analysis

The technique of Baade (1988) and Gies & Kullavanijaya (1988), which includes the CLEAN algorithm of Roberts et al. (1987), is applied to time-series spectra constructed from the individual wavelength samples. The resultant CLEANed power spectra are assembled into data cubes of line velocity (x-axis) *vs.* frequency (y-axis) *vs.* power (z-axis). Although the periodograms show considerable aliasing structure, a period of 8.5-hrs is found across the line profile of He II λ5411Å. In addition, the alias frequencies either side of this 8.5-hr period are detected across the line profiles of the remaining lines of study. This indicates that the 8.5-hr modulation may be present within all the lines presented here, but that the sampling pathology has caused the CLEAN algorithm to 'lock' onto the alias frequency, giving erroneous periods.

Plotting the sinusoid half-amplitude of the 8.5-hr period as a function of line velocity, shows that the strength of the variation is concentrated in the line wings with the exception of N IV λ6381Å, where power is spread about the line profile more evenly. This pattern of variability suggests a strong azimuthal component within the origin of the 8.5-hr modulation.

2. Origin of the Variations

The origin of the line profile variations (lpv) seen within the optical absorption lines of ζ Puppis is controversial. Baade (1986, 1991) has attributed these variations to photospheric non-radial pulsation (NRP). However,

L. A. Balona et al. (eds.), Pulsation, Rotation and Mass Loss in Early-Type Stars, 112–113.
© 1994 IAU. Printed in the Netherlands.

Fullerton (1990) attributes much of the lpv of the O supergiants to variability within the stellar wind.

The coherent periodicity across the line profile would rule out stochastic variations in the wind as the cause of the lpv. Moreover, the period indicated in this study, which Baade (1986) also found, is much shorter than the estimated rotation period of ζ Puppis (5.075-days, Moffat & Michaud 1981; 5.21-days, Balona 1992), and indeed, is shorter still than the critical rotation period of 3.4-days (estimated in the Roche approximation using stellar parameters given by Prinja et al. 1992, and accounting for radiation pressure). Evidently, the apparent 8.5-hr modulation does not represent the corotation of a single wind inhomogeneity about the star. This short period does not rule out the possibility of a cylindrical pattern of features within the wind or photosphere which would then produce the regular variation within the line profiles as the star rotates. This would mean inhomogeneities spaced at regular intervals, which whilst highly unlikely, cannot be entirely ruled out. The NRP hypothesis is a much more attractive explanation; cylindrical symmetry of both velocity and temperature variations upon the photosphere is a natural consequence of NRP. In addition, the appearance of strong amplitude in the line wings, can be explained as a pulsation mode possessing a significant horizontal pulsation velocity.

Further analysis, which is in progress, is required to identify if the frequencies found in the line profiles are due to a combination of a 8.5-hr modulation and temporal sampling. The analysis presented here is preliminary and caution is therefore advised in the interpretation of these findings.

References

Baade D.: 1986, 'Observational properties of non-radial oscillations in early-type stars and their possible effect on mass loss: the example of ζ Puppis (O4 If)' in D.O. Gough, ed(s)., *Seismology of the Sun and the Distant Stars*, Reidel: Dordrecht, p. 465

Baade, D.: 1988, 'Doppler imaging of variable early-type stars' in G. Cayrel de Strobel & M. Spite, ed(s)., *IAU Symposium 132: The Impact of Very High S/N Spectroscopy on Stellar Physics*, Kluwer: Dordrecht, p. 193

Baade, D.: 1991, 'Regular variability of optical lines in ζ Pup' in D. Baade, ed(s)., *ESO Workshop Proceedings No. 36, Rapid Variability of OB-Stars: Nature & Diagnostic Value*, ESO: Munich, p. 21

Balona, L. A.: 1992, *MNRAS* **254**, 404

Fullerton, A. W.: 1990, *Ph.D. Thesis*, University of Toronto: Toronto

Gies, D. R., & Kullavanijaya, A.: 1988, *ApJ* **326**, 813

Moffat, A. J., & Michaud, G.: 1981, *ApJ* **251**, 133

Prinja, R. K., et al.: 1992, *ApJ* **390**, 266

Roberts, D. H., Lehár, J., & Dreher, J. W.: 1987, *AJ* **93**, 968

PHOTOSPHERIC AND CIRCUMSTELLAR VARIABILITY
OF THE RAPIDLY ROTATING O STAR HD 93521

A. W. FULLERTON

Bartol Research Institute, University of Delaware, Newark, DE 19716–4793 U.S.A.

D. R. GIES

Dept. of Physics & Astronomy, Georgia State University, Atlanta, GA 30303 U.S.A.

and

C. T. BOLTON

David Dunlap Observatory, P.O. Box 360, Richmond Hill, Ont. CANADA L4C 4Y6

1. HD 93521: Pulsation, Rotation, and Mass Loss

HD 93521 is an O9.5 V star with several well-documented peculiarities, including: an extremely large projected rotational velocity (400 km/s according to Conti & Ebbets 1977); absorption line profile variations (Fullerton 1990); weak emission at Hα (Irvine 1989); evidence for a rotationally-modified stellar wind in its UV resonance lines (Massa 1992; Howarth & Reid 1993); and stellar wind variability in the form of discrete absorption components (Prinja & Howarth 1986). Thus, HD 93521 provides an ideal laboratory for studying the interaction between rapid rotation and variability in both photospheric layers and the circumstellar environment.

2. Photospheric Variability

We obtained a time series of 36 high S/N (350), moderate resolution (7000) spectra of the He II λ4686 and He I λ4713 lines of HD 93521 with the coudé spectrograph of the 2.1m telescope at the McDonald Observatory on 2 nights in 1987 March. These observations show recurrent, regularly spaced "bumps" with semi-amplitude ~1% of the continuum that traverse the He I line profile from blue to red every 4.8 hours. Analysis of these variations by means of the periodogram technique developed by Gies & Kullavanijaya (1988) indicates that a single sinusoidal frequency of 13.65 ± 0.04 d^{-1} (i.e., period 1.76 ± 0.01 hours) extends across the He I λ4713 profile. In contrast, similar variations are barely visible in the He II line, and we could not detect the 1.76-hour period in our observations of it. Simultaneous photometry did not indicate statistically significant light variations with amplitude greater than 0.002 magnitudes in the Strömgren *y* filter.

The simplest interpretation of the variability in He I λ4713 is in terms of a single mode of nonradial pulsation (NRP), with period in the observer's

L. A. Balona et al. (eds.), Pulsation, Rotation and Mass Loss in Early-Type Stars, 114–115.

frame of 1.76 hours. The variation of pulsation phase with position in the line profile indicates that a high-degree, prograde mode is excited ($m = -9 \pm 1$), while the large amplitude suggests that the mode is sectorial ($\ell = |m|$). Photometric variations are not expected from such a mode because of the nearly perfect cancellation of temperature variations and surface-area distortions between adjacent sectors of the visible disk of the star. Since the NRP amplitude is greatly reduced in He II λ4686, we infer that this line must be formed either farther from the photosphere than He I λ4713, or closer to the hot, polar caps of the rotationally distorted stellar surface.

3. Circumstellar Variability

At various epochs we have also obtained less intensive time series observations of the He I λ5876 and Hα lines of HD 93521, both of which are diagnostics of the circumstellar environment. The cores of these lines are blue shifted by $\sim$50 km/s, and both exhibit small blue- and red-emission peaks that are displaced by $\sim$500 km/s from line center. Although the temporal sampling of our data is fragmentary, changes in the shape, strength, and emission V/R ratio are evident from night to night. We cannot determine if the circumstellar variations are cyclical from the present data.

The morphology of these profiles confirms that a weak disk persists around HD 93521, as had been deduced previously from the peculiar UV resonance line profiles of this star (Massa 1992; Howarth & Reid 1993). The blue shift of the line centers and the most common V/R asymmetries suggest that material is flowing slowly outward through disk, perhaps in the manner described by the "Wind-Compressed Disk" model of Bjorkman & Cassinelli (1993). We are planning further optical and UV observations of HD 93521 to investigate whether the circumstellar variability is related to the underlying NRP.

Acknowledgements

AWF thanks the LOC for financial support to attend this Symposium.

References

Bjorkman, J. E., and Cassinelli, J. P.: 1993, *Astrophysical Journal* 409, 429
Conti, P. S., and Ebbets, D.: 1977, *Astrophysical Journal* 213, 438
Fullerton, A. W.: 1990, Ph.D. thesis, University of Toronto
Gies, D. R., and Kullavanijaya, A.: 1988, *Astrophysical Journal* 326, 813
Howarth, I. D., and Reid, A. H. N.: 1993, *Astronomy and Astrophysics*, in press
Irvine, N. J.: 1989, *Astrophysical Journal* 337, L33
Massa, D. L.: 1992, in L. Drissen, C. Leitherer, A. Nota, ed(s)., *Nonisotropic and Variable Outflows from Stars*, ASP: San Francisco, 84
Prinja, R. K., and Howarth, I. D.: 1986, *Astrophysical Journal, Supplement Series* 61, 357

PULSATION IN RAPIDLY ROTATING STARS

MAURICE J. CLEMENT

Department of Astronomy, University of Toronto, Toronto, ON, Canada M5S1A7

Abstract. The line-profile variables observed on the upper main sequence have been interpreted by some astronomers to be the manifestation of nonaxisymmetric oscillations. More specifically, most of these variables can be modelled by prograde or corotating equatorial waves. In the absence of rotation, these waves have surface velocity distributions which are given simply by spherical harmonics. Unfortunately, the corresponding velocity fields in the presence of rotation are much more difficult to calculate. In this paper, I will summarize what is known about the effect of rapid rotation on the normal mode eigenfunctions of main sequence stars. The principal conclusions are as follows: Low-order, axisymmetric modes couple very strongly to rotation and their velocity distributions are very much different from those of their zero-rotation counterparts. On the other hand, higher-order (shorter wavelength), nonaxisymmetric modes couple only weakly to rotation and, therefore, retain many of the spherical harmonic properties that they possess in the absence of rotation.

1. Stellar Normal Modes: A Review

1.1. THE DISCRETE SPECTRUM OF NONRADIAL MODES

The discrete modes are a measure of the *global* stability of the system. For spherical symmetry, the eigenfunctions are separable in the coordinates and their angular dependence is given by spherical harmonics:

$$f_k(r)\, P_\ell^m(\cos\theta)\, e^{im\varphi} \quad \begin{cases} k & : \text{ radial order} \\ \ell & : \text{ polar order} \\ m & : \text{ azimuthal order} \end{cases}$$

In this case, the eigenfrequencies are degenerate in m. Rotation, however, lifts the degeneracy and each mode involves a mixture of spherical harmonics. In general, the discrete modes fit into the following classification scheme:

1. *Fundamental (f) modes*:
   These are analogs of the divergence-free Kelvin modes of an incompressible, uniform density sphere. They occur only in the lowest radial order which corresponds normally to an absence of radial nodes in the eigenfunction although there are exceptions in the limit of very high mass concentrations.

2. *Pressure (p) modes*:
   Pressure is the main restoring force in this case. The modes are longitudinal (acoustic) waves whose motion is primarily radial and large only near the surface. The pressure variations are relatively large.

L. A. Balona et al. (eds.), Pulsation, Rotation and Mass Loss in Early-Type Stars, 117–127.
© 1994 IAU. Printed in the Netherlands.

3. *Gravity (g) modes*:

Gravity is the main restoring force; the motion is more horizontal than for the p-modes and can be large even in the deep interior. The pressure variations are small.

4. *Toroidal (t) modes*:

The inertial forces of rotation are the restoring agents for these waves. The motion is mainly horizontal with very small variations in pressure and density.

The p- and g-modes form two separate spectra with k running from 0 to ∞ for a given ℓ and m. In the first case, the eigenfrequency *increases* with k whereas, in the second case, it *decreases*.

1.2. THE CONTINUOUS SPECTRUM OF NONRADIAL MODES

Stars can have branches of *continuous* modes; i.e., in a particular range of frequency, *every* value is an eigenvalue of the system. Thus, the modes are not discrete. (Computationally, they become strictly continuous only in the limit of zero grid spacing.)

Continuous modes appear in other areas of physics (e.g., geophysics). In astronomy, they were first noticed by Aizenman & Perdang (1973) in some secular stability studies of main sequence stars. The associated eigenfunctions have a *nonanalytic* behavior along 2D-surfaces in the interiors of the models.

Perdang (1976, 1977) has shown that the stellar stability equations generally allow a class of continuous modes if one drops the "smooth and continuous" restriction. Sometimes, one can transform variables in such a way as to expose singularities in the equations and these give rise to nonanalytic solutions, not all of which are necessarily physical. That is, continuous modes may be introduced artificially into a problem by a poor choice of variables. For example, Clement (1981) encountered nonphysical continuous modes as a result of reducing the number of independent variables to make the problem more numerically tractable.

Continuous modes which are real are believed to be a measure of the *local* stability of the equilibrium state. Indeed, Perdang has shown that there is a branch of continuous modes associated with convection zones in stars.

2. Methods for Computing the Normal Modes

At one time or another, I have tested or employed all of the following techniques but in this paper only the results of methods 2 and 3 are summarized.

1. *Collocation and Least Squares Fitting*:

Trial functions are fitted to the partial differential equations. The idea is conceptually simple but the solutions are crude and they don't always converge with an increasing number of coefficients. Also, spurious non-physical solutions commonly appear.

2. *Variational/Tensor Virial Methods*:
   For analytic trial functions, all solutions are real physical solutions of the equations of motion and all discrete modes appear with no continuous branches. But the solutions are not exact – they are only the best ones for the chosen basis functions. Furthermore, the "ideal" set of basis functions that should be used for rotating configurations is unknown at the present time.

3. *Exact 2D/3D Solutions of the Equations of Motion*:
   These are exact solutions but, unfortunately, the method permits both continuous and discrete modes and the former, if present, can hide the latter.

4. *Hydrodynamical Simulations of Oscillating, Rotating Stars*:
   In principle, one can include convection, turbulence, and nonadiabatic effects in a natural way, but the method is computationally intensive and certainly overkill if only the eigenfunctions are required.

3. Axisymmetric Modes

3.1. DIRECT INTEGRATION OF THE EQUATIONS OF MOTION

For a first look at the axisymmetric modes, I solved the difference equations directly on a two-dimensional grid. Refer to Clement (1981) for the details. But, in summary, solutions of the form

$$\xi(\mathbf{r}, t) = \xi(r, \theta)\, e^{i\omega t}$$

were found numerically for 15 $M_\odot$ main sequence models as well as for some polytropes to test the effects of different density gradients. The independent variables at each mesh point were (i) ξ_n (the normal component of ξ), (ii) $\nabla \cdot \xi$, and (iii) $\delta\phi$ (the potential perturbation). The elimination of two of the displacement components reduces the matrix inversion required at each mesh point to order 3 and greatly speeds up the calculation. The eigenfrequency ω follows from the requirement that the determinant of the 3×3 matrix vanish at the equator.

Unfortunately, this choice of variables introduces singularities into the equations and results in the appearance of continuous modes. If this calculation were to be repeated today with faster computers, it would probably be advisable to avoid the continuous modes by choosing a more straightforward set of independent variables.

3.2. RESULTS FOR FOUR LOW-ORDER MODES

Rotation has a strong effect on the spatial distribution of the eigenfunction amplitude. Fig. 1 illustrates two modes which are purely radial in the absence

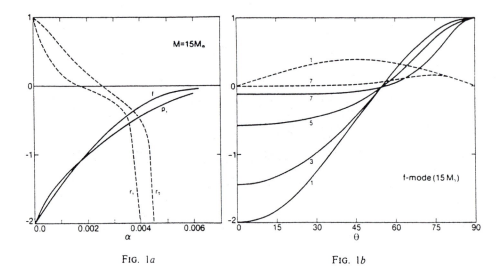

Fig. 1. (a) The effect of rigid rotation on the normalized eigenfunction amplitudes at either the pole or the equator of an $M = 15\,M_\odot$ main sequence model. *Dashed curves,* the normal or vertical component of the displacement, ξ_n, at the equator relative to ξ_n at the pole for the r_0 and r_1 normal modes; *solid curves,* ξ_n at the pole relative to ξ_n at the equator for the f and p_1 modes. The rotation parameter α is equal to $\Omega^2/8\pi G$. (b) The surface distributions of the horizontal and vertical displacements of the $M = 15\,M_\odot$ f-mode. *Dashed curves,* ξ_h as a function of polar angle; *solid curves,* ξ_n as a function of polar angle. The model numbers attached to each curve vary from 1 ($\alpha = 0$) to 7 ($\alpha = 0.005$ or $v_{eq} = 378$ km s^{-1}). All the displacements are normalized with respect to the equatorial ξ_n.

of rotation: the lowest-order radial mode r_0 and its first overtone r_1. They both end up in the limit of rapid rotation with most of their amplitude confined to the equatorial zones. This same transformation with rotation is experienced by two initially $\ell = 2$ modes (f and p_1). These results are confirmed by completely independent variational calculations applied to the case $m = 0$. The variational method is described briefly in the next section.

4. Nonaxisymmetric Modes

4.1. THE VARIATIONAL METHOD

As applied to rapidly rotating stellar models, this approach and some results for upper main sequence stellar models can be found in Clement (1981, 1984, 1986, 1989). The method involves three basic steps:

1. Linearize the equations of motion by setting $\delta \mathbf{r} \equiv \boldsymbol{\xi}(\mathbf{r},t) = \boldsymbol{\xi}(\mathbf{r})e^{i\omega t}$:

$$-\omega^2 \mathbf{A}(\boldsymbol{\xi}) + \omega \mathbf{B}(\boldsymbol{\xi}) + \mathbf{C}(\boldsymbol{\xi}) = 0 \ .$$

2. Multiply by the complex conjugate ξ^* and integrate over the equilibrium volume to obtain

$$-\omega^2 a + \omega b + c = 0 .$$

The (Hermitian) symmetry properties of the operators $\mathbf{A}, \mathbf{B}$, and $\mathbf{C}$ make this a *variational* equation. That is, ω is *stationary* for arbitrary variations of an eigensolution ξ (and conversely).

3. Choose a trial function involving linear combinations of "basis" functions or vectors $\mathbf{f}_i$:

$$\xi = \sum_{i=1}^{N} c_i \mathbf{f}_i(\mathbf{r}) .$$

The coefficients c_i can be determined variationally by requiring

$$\partial \omega / \partial c_i = 0 \quad (i = 1, N) .$$

The resulting N linear homogeneous equations will have a solution only if the determinant of the coefficient matrix vanishes. This condition yields N normal modes for each equilibrium model.

4.2. A SET OF BASIS VECTORS FOR VARIATIONAL CALCULATIONS WITH RAPID ROTATION

Ideally, the basis vectors $\mathbf{f}_i$ should form a complete set (preferably orthogonal), and they should span the vector space of all possible eigenfunctions. Obviously, for practical reasons, one wants a set which is capable of describing an eigenfunction with a linear combination of a *finite* (i.e., small) number of $\mathbf{f}_i$.

Inspired by Sabouti (1977a, 1977b, 1981), I tested a number of possibilities and finally chose a combination of three basis vector families: *p*-type, *g*-type, and *t*-type. They are approximately orthonormal and describe respectively *p*-modes, *g*-modes, and *t*-modes in the limit of small rotation (Clement 1989). With large rotation, however, the modes lose some of their unique character and become less distinguishable from one another. That is, the normal modes under the influence of rapid rotation can be represented accurately only by a mixture of *p*-type, *g*-type, and *t*-type basis vectors.

Each basis type is derivable from a scalar potential as follows:

1. *p*-type : $\rho \xi_p = \rho \nabla \phi_1$
2. *g*-type : $\rho \xi_g = \nabla \times (\nabla \times \phi_2 \mathbf{g}) + A \mathbf{g} \times (\nabla \times \phi_2 \mathbf{g})$
3. *t*-type : $\rho \xi_t = \nabla \times \phi_3 \mathbf{g}$

where $A \equiv (d\rho/dP)_{\text{ad}} - (d\rho/dP)$ measures the departure from adiabaticity and $\mathbf{g}$ is the local gravity. The *p*-type vectors generate acoustic, shear-free waves, the *g*-type are associated with isobaric waves, and the *t*-type yield constant density, horizontal waves. The three scalar potentials are given by

$$\phi_1(\mathbf{r}) = e^{im\varphi} \sum_{k=m}^{K} \sum_{\ell=m}^{k} a_{k\ell} \, r^k P_\ell^m(\cos\theta) ,$$

$$\phi_2(\mathbf{r}) = e^{im\varphi} P(\mathbf{r}) \sum_{k=m}^{K} \sum_{\ell=m}^{k} b_{k\ell} r^k P_\ell^m(\cos\theta),$$

$$\phi_3(\mathbf{r}) = i e^{im\varphi} \rho(\mathbf{r}) \sum_{k=m}^{K} \sum_{\ell=m}^{k} c_{k\ell} r^{k-1} P_{\ell+1}^m(\cos\theta).$$

Notice the presence of either the pressure or the density in two of the potentials. This ensures the correct behavior at a model's outer boundary. Also, note that the toroidal potential is *imaginary* and involves *odd* Legendre polynomials for *even* modes. In practice, the parameter K was chosen large enough to yield 8 or 9 radial orders and about 150 normal modes for each equilibrium model.

4.3. MIXING OF THE BASIS VECTOR TYPES

For slow rotation, a basis vector family describes more or less accurately its corresponding normal mode type. For example, the p-modes involve primarily p-type basis vectors as shown in Table I by the large value of p_{mix} (the fraction of the total amplitude due to p-type basis vectors). However, with rapid rotation, the basis types mix and, in particular, the motions of all the modes becomes more toroidal in their nature as indicated by the value of $t_{mix} = 1 - p_{mix} - g_{mix}$. On the other hand, in the absence of rotation, t_{mix} is zero because the toroidal vectors uncouple from the other basis types in that limit.

TABLE I

The p_{mix} and g_{mix} Factors for Some $(\ell = m = 4)$-Modes

Mode	factor	\multicolumn{5}{c}{$\Omega/(2\pi G\rho_c)^{1/2}$}
		0.00	0.02	0.04	0.06	0.08
p(1,4,4)	p_{mix}	0.92	0.93	0.90	0.72	0.58
	g_{mix}	0.08	0.05	0.01	0.08	0.07
f(0,4,4)	p_{mix}	0.74	0.73	0.73	0.46	0.43
	g_{mix}	0.26	0.21	0.15	0.15	0.12
g(1,4,4)	p_{mix}	0.36	0.28	0.20	0.12	0.11
	g_{mix}	0.64	0.59	0.54	0.48	0.40

4.4. Effect of Rapid Rotation on Nonaxisymmetric Eigenfunctions

Unlike the axisymmetric case, the rotational coupling of modes to adjacent spherical harmonics is not particularly strong. Aside from the motion becoming more toroidal as mentioned in the preceding section, the modes maintain their basic zero-rotation spatial distribution. There is some focusing of the amplitude into the equatorial regions as illustrated in Fig. 2, but the change is not so dramatic as in the $m = 0$ case because the polar amplitude is always zero for the nonaxisymmetric modes regardless of the rotation rate. Refer to Clement (1989) for more details. I should emphasize that this conclusion applies only to the lowest half dozen radial orders that I have been able to compute. The effect of rotation on very high-order modes remains unknown.

g(1,6,−6)

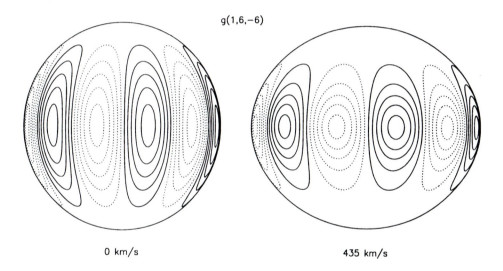

0 km/s   435 km/s

Fig. 2. The azimuthal velocity component of the $\ell = 6$ prograde sectorial g-mode for zero rotation on the left and rapid rotation ($v_{eq} = 435$ km s^{-1}) on the right. The solid contours indicate positive velocity and the dashed curves, negative velocity. There is a significant focusing of amplitude toward the equatorial zones by rotation.

4.5. Line Profile Variations

The prograde sectorial g-modes appear to be the most likely source of the line profile variations observed in various B-stars on the main sequence. Some of the periods are quite long, indicating perhaps the presence of high radial orders. However, one must be careful here because rotation destabilizes the retrograde g-modes (Clement 1989), especially the $\ell = m$ sectorial modes. Thus, the eigenfrequency ω can pass through zero, making the modes prograde with very long periods if the rotation rate is just above the neutral

stability point. Here, the modes become secularly unstable and their amplitudes may grow if there is a significant dissipation of energy (Friedman &
Schutz 1978).

As an example, Fig. 3 illustrates the effects of rotation on the line profile
variations due to the lowest radial order $\ell = 6$ sectorial g-modes. These
results are representative of orders up to $k = 6$, say. Again, I can say nothing
about the very high radial orders that appear to be present in some variables,
but note my remarks in the preceding paragraph. It should also be pointed
out that these line profiles are highly schematic or ideal in the sense that
they completely ignore atmospheric effects such as shocks which may give
profiles a sharper, less coherent appearance.

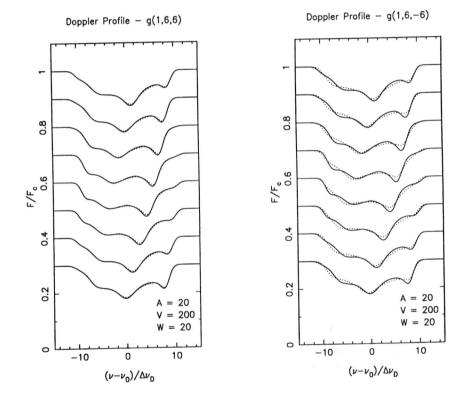

Fig. 3. Line profile variations due to the $\ell = 6$ sectorial g-modes in a 15 $M_\odot$ model ($m = 6$
on the left and $m = -6$ on the right). All the profiles are normalized in the same way;
that is, they are broadened with a rotational velocity of 200 km s^{-1} and incorporate an
intrinsic broadening in the Voigt function of 20 km s^{-1}. The solid curves were computed
with the zero rotation eigenfunctions (which are pure spherical harmonics) whereas the
dotted curves were found using the corresponding eigenfunction for a rapidly rotating
model ($v_{eq} = 435$ km s^{-1}).

That said, observe that even relatively fast rotation (namely, $v_{eq} = 435$
km s^{-1}), changes the profiles in only a modest way although the $m = -6$

mode does give rise to larger effects than the $m = 6$ one. Note that *both* the rotating modes are prograde with Ω/ω equal to 0.095 ($m = -6$) and 0.66 ($m = 6$). Thus, the mode which is retrograde at small rotation appears as a high-order, prograde mode at a somewhat faster rotation. One can push the 0.66 ratio up to any higher value just by choosing an appropriate slower rotation rate. [In the figure, the $m = 6$ solid-line profiles indicate bumps moving in the prograde direction, opposite to the correct sense; this was done only to make the comparison with the rotating eigenfunction easier.]

5. An Alternative Approach: Hydrodynamical Simulations

As part of a long term study of the interior dynamics of rotating stars, I am now tackling this problem with a 2D/3D hydrodynamics code. Conserving energy and modeling the subgrid-scale viscosity were major problems which have now been solved (Clement 1993). This approach can simulate convection, turbulence, pulsation, and rotation simultaneously. Fourier analyses of the velocity time series can be used to identify pressure, gravity, and Kelvin modes and to follow the evolution of the oscillation spectrum. For example, starting from arbitrary initial conditions, one can observe long period gravity modes quickly die out in the stellar core. Thus, it should be possible with this technique to extract the normal modes and their dependence on rotation.

6. Conclusions

Low-order, axisymmetric modes are strongly affected by rotation. In the limit of high angular velocity, the oscillation amplitude is confined to the equatorial regions. On the other hand, nonaxisymmetric modes and, in particular, the sectorial ($\ell = |m|$) modes do not "see" the rotational distortion of the star and retain many of their zero-rotation characteristics. On closer examination, the oscillation amplitude is focused more into the equatorial zones and the motion becomes more toroidal in its nature as shown by the larger mixing ratio for the toroidal basis vectors. The weak coupling of the nonaxisymmetric modes to adjacent harmonics in the presence of rigid rotation will probably carry over to differential rotation as well except perhaps in the extreme case where the angular velocity scale length becomes comparable to the pulsation wavelength.

7. Acknowledgements

This work has been supported in part by the Natural Sciences and Engineering Research Council of Canada and by the University of Toronto.

References

Aerts, C. & Waelkens, C.: 1993, *Astron. Astrophys.* **273**, 135.
Aizenman, M. & Perdang, J.: 1973, *Astron. Astrophys.* **23**, 209.
Clement, M. J.: 1981, *Astrophys. J.* **249**, 746.
Clement, M. J.: 1984, *Astrophys. J.* **276**, 724.
Clement, M. J.: 1986, *Astrophys. J.* **301**, 185.
Clement, M. J.: 1989, *Astrophys. J.* **339**, 1022.
Clement, M. J.: 1993, *Astrophys. J.* **406**, 651.
Friedman, J. L. & Schutz, B. F.: 1978, *Astrophys. J.* **222**, 281.
Perdang, J.: 1976, *Astrophys. Space Sci.* **44**, 177.
Perdang, J.: 1977, *Astrophys. Space Sci.* **52**, 313.
Sabouti, Y.: 1977a, *Astron. Astrophys.* **55**, 327.
Sabouti, Y.: 1977b, *Astron. Astrophys. Suppl.* **28**, 463.
Sabouti, Y.: 1981, *Astron. Astrophys.* **100**, 319.

Discussion

Balona: You show the rotational velocity in km s^{-1}, but fail to specify the pulsation period of your modes. The case which is of greatest interest is the one for which the ratio of rotation to pulsation frequency is no longer small; i.e., when it is of order unity, for example. Do your calculations include this domain?

Clement: My variational calculations cover radial orders up to 8. Of these, the sectorial g-modes with $m = -\ell$ (i.e., the ones which are always prograde) fall in the range $\Omega/\omega < 0.20$. However, the $m = \ell$ modes cover the whole Ω/ω domain because, for them, ω passes through zero at some rotation rate whereupon they become *prograde* with arbitrarily long periods. Refer to Clement (1989) for plots of eigenfrequency versus rotation rate.

Owocki: I may have missed a simple point, but could you give the critical angular velocity for the equilibrium models that you use. That is, what fraction of critical rotation are your assumed rates? Also, do you include oblateness and gravity darkening effects in your line profile calculations?

Clement: For the 15 $M_\odot$ models which I was using, the critical equatorial velocity is around 600 km s^{-1}. The fastest rotating model which I showed in my talk has a velocity of 435 km s^{-1} corresponding to an equatorial gravity which is 0.43 of the polar value. I did include oblateness and limb darkening effects in my calculations but not those of gravity darkening. I might include the latter effects in the future although limb darkening leads to only small changes in the profile variations.

Dziembowski: What is the cause of the continuous normal mode spectrum in the case of rigid rotation?

Clement: This is a good question. Rigid rotation should not in principle introduce continuous modes although differential rotation can. I believe that the branches of continuous modes reported in my 1981 paper are nonphysical and the result of a poor choice of independent variables. I eliminated two of the velocity components in order to make the problem more numerically tractable and this introduced singularities into the equations.

Aerts: The effect of rotation depends completely on Ω/ω. For $\Omega/\omega < 1$, we find that axisymmetric modes are hardly coupled (Aerts & Waelkens 1993), while nonaxisymmetric modes are strongly coupled. That is, the effects of rotation on LPV's are a lot larger for nonaxisymmetric modes than for axisymmetric ones if Ω/ω is small. So you should have given explicitly Ω/ω for the modes which you discussed.

Clement: I have to disagree somewhat with what you say here. Your paper with Waelkens includes only rotational perturbations up to first order in Ω. This excludes all second-order effects which I would argue are necessary for the axisymmetric modes since there are almost no first-order terms in that case. In particular, you do not include the rotational distortion of the equilibrium model. My numerical calculations are exact (i.e., they are not perturbation analyses) and were done with two completely independent methods: the variational approach and a direct solution of the equations of motion. Both techniques give the same result and indicate a strong rotational coupling for the axisymmetric modes as pointed out in my talk. For the nonaxisymmetric modes, on the other hand, the first-order terms are relevant and you are on safer ground. However, I believe that you have pushed your analysis too far. It is not correct to argue that if only Ω/ω is less than unity then your results are valid. You must also check that the centrifugal force at the equator is much less than the gravitational force. Also, the change in the eigenvalue should be much less than the zero-rotation value. Neither of these additional restrictions is satisfied by many of your models. In particular, I believe that your profile calculations for $\Omega/\omega = 0.5$ (and perhaps even for 0.2 for some modes) correspond to rotation rates close to or above the critical value. In such cases, you cannot ignore the oblateness of the model and other second-order effects. I think that if you restrict yourself to reasonable rates, consistent with a first-order perturbation analysis, you will find that the effect of rotation on the line profiles is not so large after all. Also, I recommend that you give up the practice of normalizing your profiles so that they have the same depth at a particular phase. This artificially enhances the effect of rotation as illustrated by the profiles in your paper which clearly have smaller and smaller equivalent widths with increasing rotation.

OVERSTABILITY OF NONRADIAL PULSATIONS
IN ROTATING EARLY TYPE STARS

HIDEYUKI SAIO

Astronomical Institute, Faculty of Science, Tohoku University, Sendai 980, Japan

and

UMIN LEE

*Department of Physics and Astronomy, University of Rochester, Rochester, NY 14627,
USA*

We present the result of nonadiabatic analysis for nonradial pulsations in uniformly rotating main sequence models.

The angular dependence of the amplitude of a nonradial pulsation mode with an azimuthal order m in a rotating star is represented by a sum of terms proportional to spherical harmonics $Y_l^m(\theta, \phi)$ with $l = |m|, |m|+2, \ldots$ (even mode) or $l = |m| + 1, |m| + 3, \ldots$ (odd mode; see e.g. Saio and Lee 1991 for detail). (In this paper we consider only even modes.) This property makes the analysis complex compared with the case without rotation, in which a single Y_l^m expresses the angular dependence of a given mode. In our numerical analysis the summation is truncated, in which only first two terms are taken into account. Lee and Saio (1987) give the differential equations for nonadiabatic nonradial pulsations in a uniformly rotating star. Treating the angular frequency of rotation as a free parameter, we applied the nonadiabatic analysis to a main-sequence evolutionary model, for which the effect of rotation is neglected.

The adopted main-sequence model has the following properties; $15 M_\odot$, $X = 0.70$, $Z = 0.03$, $\log L/L_\odot = 4.53$, $\log R/R_\odot = 1.00$, $\log T_{\rm eff} = 4.395$, and $X_{\rm center} = 0.159$. The location of the model on the HR diagram is close to ϵ Per, for which Gies and Kullavanijaya (1988) have found nonradial pulsations with $-3 > m > -6$.

Low order p- and g-modes are excited by the κ-mechanism which works at the opacity peak at $T \sim 2 \times 10^5$K (Dziembowski and Pamyatnykh 1993, Gautschy and Saio 1993). Since rotation modifies the amplitude distribution of a nonradial pulsation mode, the rotation influences the stability as well as the frequency spectrum especially for large $|m|$ (m is the azimuthal order of mode). Some of our results for $|m| = 6$ are presented in Fig. 1, which shows how the real parts of frequency, ω_r, (in the co-rotating frame) change with the angular frequency of rotation Ω. In this figure ω_r and Ω are normalized by $\sqrt{GM/R^3} = 7.79 \times 10^{-5}$s^{-1}.

A perturbation analysis in terms of Ω gives $\omega_r = -mC\Omega + O(\Omega^2)$, where C is a small number obtained from the displacement vector of a given nonradial

L. A. Balona et al. (eds.), Pulsation, Rotation and Mass Loss in Early-Type Stars, 128–129.

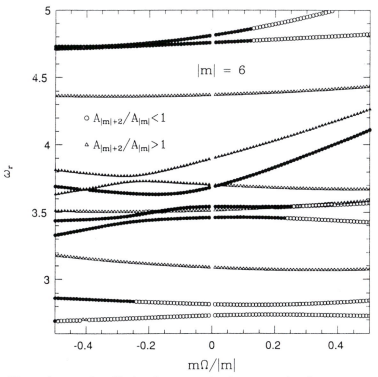

Fig. 1. The real part of oscillation frequency in the co-rotating frame versus $m\Omega/|m|$ for $|m| = 6$, where Ω is the angular frequency of rotation. Circules (triangles) indicate modes for which the amplitude of $l = |m|$ component ($A_{|m|}$) is larger (smaller) than that of $l = |m| + 2$ component ($A_{|m|+2}$). Filled symbols are for overstable modes. The modes with $m > 0$ are retrograde in the co-rotating frame.

pulsation mode for $\Omega = 0$. As seen in Fig. 1 the value of C differs from mode to mode, and for some mode C is negative. Since different modes have different $d\omega_r/d\Omega$, many avoided crossings exist in the plane of pulsation frequency versus rotation frequency.

Fig. 1 indicates that modes tends to be stabilized at large $m\Omega$. Some retrograde (in the corotating frame) modes ($m > 0$) which are overstable when Ω is small are stabilized when Ω becomes large enough.

In summary, prograde ($m < 0$) modes tend to become overstable when Ω is large. This property is more apparent for modes with larger $|m|$.

References

Dziembowski, W.A. and Pamyatnykh, A.A.: 1993, *M.N.R.A.S.* **262**, 204
Gautschy, A. and Saio, H.: 1993, *M.N.R.A.S.* **262**, 213
Gies, D.R. and Kullavanijaya, A.: 1988, *Ap.J.* **326**, 813
Lee, U. and Saio, H.: 1987, *M.N.R.A.S.* **225**, 643
Saio, H. and Lee, U.: 1991, '' in D. Baade, ed(s)., *Rapid Variability of OB-stars: Nature and Diagnostic Value*, ESO: Garching, 293

THE EVOLUTION OF ROTATING 15 M$_\odot$ STARS

S. SOFIA, J.M. HOWARD AND P. DEMARQUE

Astronomy Department, Yale University, New Haven, CT, USA

Abstract. Theoretical evolutionary sequences have been generated with the YREC code for a 15 M$_\odot$ star from the ZAMS to core helium exhaustion with a variety of physical assumptions, covering both rotating and non-rotating cases. The non-rotating models agree qualitatively with other models found in the literature. The addition of only rotational distortion has little effect on the models, while the full treatment of rotation results in additional mixing and theoretical tracks that are similar to models with small amounts of convective overshoot. Models which include only rotation have fair agreement with the observed main sequence surface rotation velocities, but rotate too rapidly during the post-main sequence phases. The addition of mass loss at the given rates helps this problem somewhat but does not appear to completely resolve it. Neither the non-rotating models nor the rotating models provide full agreement with the terminal-age main sequence band used by Maeder & Meynet (1987); this may be indicative that additional mixing processes are necessary or that a more recent TAMS, such as that of Stothers (1991), should be used.

1. Introduction

An understanding of the evolution of massive stars is important for many areas of astronomy. Massive stars are the most luminous stars; thus, they are the obvious first probes for observational tests of stellar evolution in other galaxies. The progenitors of type II supernovae are highly evolved massive stars, and accurate models of high mass stars are therefore needed to fully understand supernovae. While high mass stars are greatly outnumbered by low mass stars, they give off much more radiation, particularly in the blue and ultraviolet regions of the spectrum. Thus, massive stars strongly affect the UBV colors of galaxies. They are also the sites for heavy element nucleosynthesis, and so influence the chemical evolution of galaxies.

Unfortunately, the evolution of massive stars is at best a poorly understood subject. There are many discrepancies between theory and observation during the main sequence phase of evolution. It is difficult to pick out a "terminal age main sequence" (TAMS) on observed HR diagrams (Blaha & Humphreys 1989, Fitzpatrick & Garmany 1990, and Garmany & Stencel 1992). Theoretically and observationally determined mass-to-luminosity ratios also disagree; observationally determined masses for O and early B stars are in general about 50 per cent smaller than those calculated from models (Herrero *et al.* 1992).

The situation is no better for post-main sequence evolution. The observed blue-to-red supergiant ratio is significantly larger than has been predicted (Chiosi *et al.* 1978; Bressan *et al.* 1981; Brunish & Truran 1982a,b). This is probably related to another problem, that of post-red-supergiant (post-

L. A. Balona et al. (eds.), Pulsation, Rotation and Mass Loss in Early-Type Stars, 131–144.
© 1994 *IAU. Printed in the Netherlands.*

more appropriate, as these models are more consistent in producing blue loops. SN 1987A also provides evidence in favor of this.

2.2. CONVECTIVE OVERSHOOT

The classically defined edge of a convective region is the point at which $\nabla_{rad} = \nabla_{ad}$. Within the boundaries of this region, complete mixing occurs due to the turbulent motion of the material. At the edge of the region, the material is said to be stable against convection. However, it is not clear to what extent this boundary provides an end to mixing, and this point has important consequences on the evolution of the star.

2.3. MASS LOSS

Mass loss is a common feature of all stars of high luminosity. Theories for the physical mechanisms behind mass loss can be divided into two regions in the H–R diagram: those for early-type stars, and those for late-type.

There has been a wealth of evolutionary calculations performed for massive stars with mass loss (see reviews by de Loore 1980, 1981, 1982; Chiosi 1981a,b; Chiosi 1982; Maeder 1984a,b; and Chiosi & Maeder 1986, and references therein.) The mass loss rates are generally expressed as empirical formulations, and the mass of the model is decreased with time according to the assumed rate. While this approach is not entirely realistic and may neglect important aspects of the mass loss process, it avoids the problem of implementing a poorly understood physical model.

2.4. ROTATION

Rotation directly affects the structure and evolution of a star in two ways. It causes departures from spherical symmetry through the addition of centrifugal forces; this is sometimes referred to as the structural effects of rotation. Rotation can also give rise to both secular and dynamical instabilities, which act to redistribute angular momentum throughout the star. The redistribution of angular momentum will affect the structure via rotational distortion, and the associated mass motions will cause chemical mixing which can have profound effects on the evolution. Reviews on this subject include Tassoul (1978, 1984, 1990), Kippenhahn & Thomas (1981), Zahn (1983), and Schatzman (1984).

Several techniques for computing rotating stellar models have been developed over the years, with varying degrees of success. Kippenhahn & Thomas (1970) derived a method which was then adapted by Endal & Sofia (1976). This technique uses the mass contained within an equipotential surface, M_ψ, as the independent variable. The angular momentum distribution is used to solve for the shape of the equipotential surface, and distortion terms are calculated which are added to the standard stellar structure equations. The rotation-modified equations are then solved in the normal way. A modi-

fied version of this was devised (Endal & Sofia 1978, 1979) which attempted to solve the problem of angular momentum redistribution by various
instabilities (cf. Endal & Sofia 1978 and references therein). Using Prather's
(1976) code (modified by Seidel *et al.* 1987) as a starting point, Pinsonneault
(1988) created the Yale Rotating Evolution Code (YREC), which incorporates Law's (1980, 1981) treatment of structural distortions and Endal
& Sofia's (1976, 1978) concepts of rotationally-induced mixing. This technique has been applied to rotating models of the Sun (Pinsonneault *et al.*
1989) and, using solar calibrations, to the study of young open clusters in
the Galactic disk (Pinsonneault *et al.* 1990) and low metallicity halo stars
(Deliyannis *et al.* 1989; Deliyannis and Pinsonneault 1990; Pinsonneault *et
al.* 1991, 1992). The studies using this code have been very successful in
matching both global rotational and composition constraints. While this
does not guarantee that all of the assumptions used in this method are correct, the technique has considerable predictive power and flexibility. It has
been used with great success for stars with a variety of masses, metallicities,
and evolutionary status. Of course, there are always unanswered questions,
and so YREC is a continually evolving code.

3. The Evolution Code

The most recent version of YREC includes new opacities, up-to-date nuclear
reaction rates, and improved formulations for rotationally-induced instabilities (Pinsonneault 1993). The code used for this research also includes mass
loss and atmosphere tables generated from the latest Kurucz atmospheres
(Kurucz 1992).

3.1. EQUATION OF STATE

YREC uses a simplified equation of state with three regimes. In regions
where $\log T < 5.5$, the Saha equation is used to solve for a single ionization
state of hydrogen and heavier elements, and the single and double ionization states of helium. When $\log T > 6.0$, full ionization is assumed, and an
iterative equation of state calculation is performed using tabulated values
for the electrons, which are partially degenerate and partially relativistic.
A smoothed interpolation scheme is used in intermediate regions. For more
details, see Prather (1976).

3.2. OPACITIES

A number of different opacity tables are available for use with YREC. This
work used Kurucz molecular opacities (Kurucz 1991) for low temperatures
and the Lawrence Livermore (OPAL) opacities (Iglesias & Rogers 1991,
Rogers & Iglesias 1992) at all other temperatures. These two sets of opacities
are considered the best available today. The two sets are averaged in the

temperature range $4.0275 \leq \log T \leq 4.0325$, using a weighted ramp function to ensure a smooth transition between the tables.

3.3. NUCLEAR REACTIONS

The energy generation calculations include the individual rates for the PP chain, the CNO cycle, the triple-alpha process, and neutrino losses. The reaction rates in YREC have been checked and revised by John Bahcall so that they agree with the modern numbers in Bahcall (1989) and Bahcall & Pinsonneault (1992). The energy released per reaction has been updated to agree with the values of Bahcall & Ulrich (1988). In addition to these changes, Marc Pinsonneault has also introduced an implicit nuclear burning scheme which uses weighted mean reaction rates in convective cores.

3.4. CONVECTION AND THE MIXING LENGTH THEORY

The Schwarzschild criterion is used to determine stability against convection. Convective regions are assumed to be instantaneously, homogeneously mixed. If $\log T < 6.9$ in a convective region, mixing length theory is used to calculate the temperature gradient (see Prather 1976 for details); otherwise, the temperature gradient is assumed to be adiabatic.

3.5. SURFACE BOUNDARY CONDITIONS

One method of generating surface boundary conditions is to compute an Eddington grey model atmosphere. This works in most regions of the HR diagram, but can prove costly in computational time; also, there are some regions of the HR diagram where this technique as applied in YREC is problematic for massive stars. Tables of surface boundary conditions were constructed to resolve this difficulty.

3.6. SEMICONVECTION

YREC contains a formulation for helium semiconvection developed by Castellani *et al.* (1971) for the cores of horizontal branch stars. It is not clear if this is applicable to the helium-burning convective cores of massive stars. Preliminary studies showed that the extent of additional mixing is extremely small for blue and red supergiants, and even small amounts of extended mixing due to convective overshoot or rotationally-induced instablities eliminate it entirely. Thus, this effect is ignored in these models.

    YREC contains no formulation for hydrogen semiconvection. Since extended mixing due to convective overshoot or rotationally-induced instabilities probably eliminates most, if not all, semiconvection, it was not added to the modified version of YREC used for this project.

3.7. CONVECTIVE OVERSHOOT

Theoretical models of convective overshoot contain a great deal of uncertainty. The best results have been obtained by assuming arbitrary amounts of overshoot and attempting to match models to observations. The amount of convective overshoot is thus specified in the form of some fraction of the pressure scale height at the edge of the convective region, and material is mixed out to this distance. Three different overshoot parameters can be specified for extended mixing above a convective core, below and above an intermediate convective zone, and below a convective envelope. The temperature gradient is assumed to be non-adiabatic in the overshoot region; thus, this is truly convective overshoot as per the nomenclature of Zahn (1991).

3.8. MASS LOSS

Mass loss has been incorporated into YREC in a fairly simple manner: mass is removed from the stellar model during each time step at a rate supplied by the user. The mass loss rate is given in the parameterized form

$$\log \dot{M} = C_1 \log M + C_2 \log L + C_3 \log R + C_4 \log g + C_5 \log \eta + C_6 \qquad (1)$$

where C_1, C_2, C_3, C_4, C_5, and C_6 are constants, η is as per Reimers' formulation (Reimers 1975), and all other quantities are as usual, in solar units.

3.9. ROTATION

The treatment of rotation incorporated into YREC is conceptually the same as that used by Endal & Sofia (1976, 1978) but the technical details are different, as YREC was re-written using Prather's (1976) code as a starting point. YREC differs from all non-rotating codes excepting that of Endal & Sofia in that it accounts for the effects of both non-spherical geometry and the transport of angular momentum due to rotationally-induced instabilities. Details of this method are given in Endal & Sofia (1976) and Law (1980, 1981); the specific treatment as applied to the Yale code is given in Pinsonneault (1988).

We construct pre-main sequence (PMS) starting models as polytropes. For massive stars, an $n = 3$ polytrope works best. Starting models for rotation are based on observations for main sequence surface rotation rates (Fukuda 1982). Since the surface rotation should not change too much over the course of the CHB phase, it is reasonable to apply these rates to ZAMS models. Rigid rotation in the starting models is assumed.

3.10. MODEL PARAMETERS

The parameters used for these models were obtained from the recently published Yale standard solar model (Guenther *et. al.* 1992). The Anders-Grevesse mixture was used (Anders & Grevesse 1989), as is required for consistency if the OPAL opacities and Kurucz atmospheres are used. Rotation parameters were obtained in the same way, using observationally calibrated models of the rotating Sun. These parameters were supplied by Pinsonneault (1991). Mass loss is performed in the code using an empirically determined mass loss rate. Initially, the formulation of Waldron (1984) was chosen. Unfortunately, this gives mass loss rates that are much too high during the RSG phase, so it became necessary to match this parameterization with that of Reimers (1975).

4. Results of Evolutionary Calculations

4.1. THE EFFECTS OF ROTATION

Our results show that if rotational effects are included without the angular momentum redistribution, and the resulting mixing, there is very little effect on the evolution. The CHB and CHeB lifetimes are extended by approximately 1 per cent. The structure is virtually unchanged.

However, as can be seen in Fig. 1, the addition of rotational instabilities greatly affects the evolution of a 15 $M_\odot$ model. Initially, there is an increase in luminosity, and the main sequence extends further toward the red. The CHB lifetime is extended, as is the case for models including overshoot. The main sequence lifetime is now 8 per cent larger than for the nonrotating case. During the supergiant phase, however, the differences become larger. The rotating model undergoes several loops and for the most part is always significantly more luminous than its non-rotating counterpart. Unfortunately, none of these loops is long-lived, and the model ends its CHeB phase as a red supergiant. The inclusion of rotational instabilities extends the CHeB lifetime by 36 per cent. Rotational mixing also affects where the models spends its CHeB stages. Approximately a third of this phase is spent as a blue supergiant, with a log T_{eff} of approximately 4.1. After this, the star rapidly moves to the red side of the H–R diagram and spends the remaining time as a red supergiant with a log T_{eff} of about 3.5. This should decrease the blue-to-red supergiant ratio.

Obviously, substantial rotationally-induced mixing must be occurring to affect the evolution in such a manner. The additional mixing virtually eliminates all intermediate convection zones during the main sequence evolution. The hydrogen profile is smoothed out by rotational diffusion, while the nonrotating model has a step-like profile.

The internal transfer of angular momentum has important effects on the surface rotation velocity. Angular momentum is transferred from the core to the envelope, and the net result is a higher surface rotation velocity than for the case of no rotational instabilities. During the CHB phase of evolution, the additional angular momentum is sufficient to keep the outer portions of the model rotating at rates within the observed ones for the entire

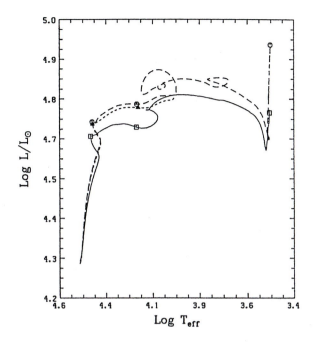

Fig. 1.   Evolutionary track for a model without overshoot, mass loss or rotation (solid), rotation with instabilities, but no overshoot or mass loss (medium dashes), and with mass loss and rotation with instabilities, but no overshoot (short dashes).

main sequence stage. Unfortunately, the results are not as satisfactory for the supergiant phase of evolution. Here, the additional angular momentum causes the model to rotate at over twice the observed rates for approximately half of the CHeB lifetime. After this, the surface rotation drops well below the observable limits.

4.2. ROTATION AND MASS LOSS

The evolutionary track for a model with both rotation (including instabilities) and mass loss shows that mass loss has little effect on the evolution during the main sequence phase. The CHB lifetime is extended by 9 per cent, or 1 per cent more than the extension due to rotation alone. Mass loss has progressively greater effects during the supergiant phase. Overall, the luminosity of the entire sequence is decreased.

The structural evolution during the CHB stage shows that the addition of mass loss causes only slight changes in the structure during this phase of evolution. The rotation rate is largely unaffected during main sequence evolution . As the total mass and angular momentum lost increase, however, the effects become more significant. The addition of mass loss will at least give closer agreement with the observed surface rotation rates during CHeB evolution.

The addition of mass loss does not significantly improve the agreement of the rotating models with the observed TAMS. The main sequence phase still does not extend far enough into the red, suggesting that an additional mixing process may be necessary. One possibility would be a combination of rotation, mass loss and convective overshoot. It is also probable that the TAMS used here is too red (Stothers 1993). As in the previous case, however, the present models are merely suggestive. More massive sequences are needed before any definite conclusions may be made.

5. Conclusions and Suggestions for Future Work

Given the work that remains to be done, this is merely a preliminary study of the effects of rotation on the evolution of massive stars. However,there are also a number of valid conclusions which can be drawn.

1. We have confirmed that use of the Schwarzschild criterion will not produce extended blue loops during core-helium-burning evolution.

2. The current models need additional mixing beyond that produced by 0.1 H_p of convective overshoot, to agree with the shape and extent of the upper main sequence. Greater amounts of mass loss might cause sufficient main sequence extension, but would be unrealistic given the observed mass loss rates. Another possible explanation is that the TAMS used here is too red. As has been pointed out by Stothers (1993), Mermilliod & Maeder (1986) used a conversion between spectral type and temperature (Böhm-Vitense 1981) which is now known to be too cool.

3. The effects of rotational distortion are fairly small for a ZAMS surface rotation of about 200 km s^{-1}, starting with a rigidly rotating pre-main sequence model.

4. The effects of additional mixing due to rotational instabilities in some sense mirrors that of a small amount of convective overshoot. It does not induce an extended blue loop during the supergiant phase of evolution.

5. A combination of mass loss and rotation produces main sequence and supergiant surface rotation velocities which are in good agreement with observation.

6. Rotation and mass loss alone are insufficient to provide good agreement with the observed shape and extent of the upper main sequence.

This study opens the door for a great deal of future work. The evolutionary tracks must be extended further, more complex combinations of processes must be considered simultaneously, and different metallicities must be explored. This paper only reports the beginning of what promises to be a very exciting avenue of research.

References

Anders, E. and Grevesse, N.: 1989, *Geochim. Cosmochim. Acta* **53**, 197.
Bahcall, J.N.: 1989, *Neutrino Astrophysics*, Cambridge Univ. Press: Cambridge.
Bahcall, J.N. and Pinsonneault, M.H.: 1992, *Rev. Mod. Phys.* **64**, 885.
Bahcall, J.N. and Ulrich, R.: 1988, *Rev. Mod. Phys.* **60**, 297.
Blaha, C. and Humphreys, R.M.: 1989, *Astron. J.* **98**, 1598.
Böhm-Vitense, E.: 1981, *Ann. Rev. Astron. Astrophys.* **19**, 295.
Bressan, A.G., Bertelli, G. and Chiosi, C.: 1981, *Astron. Astrophys.* **102**, 25.
Brunish, W.M. and Truran, J.W.: 1982a, *Astrophys. J.* **256**, 247.
Brunish, W.M. and Truran, J.W.: 1982b, *Astrophys. J. Suppl.* **49**, 447.
Castellani, V., Giannone, P. and Renzini, A.: 1971, *Astrophys. Space Sci.* **10**, 340.
Chiosi, C.: 1981a, in D'Odorico, S., Baade, D. and Kjar, K., eds., *The Most Massive Stars*, ESO: Garching, 27.
Chiosi, C.: 1981b, in Chiosi, C. and Stalio, R., eds., *Effects of Mass Loss on Stellar Evolution*, Reidel: Dordrecht, 229.
Chiosi, C.: 1982, in de Loore, C. and Willis, A.J., eds., *Wolf Rayet Stars: Observations, Physics, Evolution*, Reidel: Dordrecht, 323.
Chiosi, C. and Maeder, A.: 1986, *Ann. Rev. Astron. Astrophys.* **24**, 329.
Chiosi, C., Nasi, E. and Sreenivasan, S.R.: 1978, *Astron. Astrophys.* **63**, 103.
Cloutman, L.D.: 1978, *Bull. Amer. Astron. Soc.* **10**, 400.
Conti, P.S.: 1978, *Ann. Rev. Astron. Astrophys.* **16**, 371.
Deliyannis, C.P., Demarque, P. and Pinsonneault, M.H.: 1989, *Astrophys. J. Let.* **347**, 73.
Deliyannis, C.P. and Pinsonneault, M.H.: 1990, *Astrophys. J. Let.* **365**, 67.
Endal, A.S. and Sofia, S.: 1976, *Astrophys. J.* **210**, 184.
Endal, A.S. and Sofia, S.: 1978, *Astrophys. J.* **220**, 279.
Endal, A.S. and Sofia, S.: 1979, *Astrophys. J.* **232**, 531.
Fitzpatrick, E.L. and Garmany, C.D.: 1990, *Astrophys. J.* **363**, 119.
Fukuda, I.: 1982, *Publ. Astr. Soc. Pacific* **94**, 271.
Garmany, C.D. and Stencel, R.E.: 1992, *Astron. Astrophys. Suppl.* **94**, 211.
Guenther, D.B., Demarque, P., Kim, Y.-C. and Pinsonneault, M.H.: 1992, *Astrophys. J.* **387**, 372.
Herrero, A., Kudritzki, R.P., Vilches, J.M., Kunze, D., Butler, K., and Haser, S.: 1992, *Astron. Astrophys.* **261**, 209.
Iglesias, C.A. and Rogers, J.F.: 1991, *Astrophys. J.* **371**, 408.
Kippenhahn, R. and Thomas, H.-C.: 1970, in Slettebak, A., ed., *Stellar Rotation*, Reidel: Dordrecht, 20.
Kippenhahn, R. and Thomas, H.-C.: 1981, in Sugimoto, D., Lamb, D.Q. and Schramm, D.N., eds., *IAU Symp. 93: Fundamental Problems in the Theory of Stellar Evolution*, Reidel: Dordrecht, 237.
Kurucz, R.L.: 1991, in Crivellari, L., Hubeny, I. and Hummer, D.G., eds., *Stellar Atmospheres: Beyond Classical Models*, Kluwer: Dordrecht, 440.
Kurucz, R.L.: 1992, in Barbuy, B. and Renzini, A., eds., *IAU Symp. 149: The Stellar Populations of Galaxies*, Kluwer: Dordrecht, 225.
Law, W.-Y.: 1980, *Ph.D. Thesis*, Yale University.
Law, W.-Y.: 1981, *Astron. Astrophys.* **102**, 178.
de Loore, C.: 1980, *Space Sci. Rev.* **26**, 113.
de Loore, C.: 1981, in Chiosi, C. and Stalio, R., eds., *Effects of Mass Loss on Stellar Evolution*, Reidel: Dordrecht, 405.
de Loore, C.: 1982, in de Loore, C. and Willis, A.J., eds., *Wolf Rayet Stars: Observations, Physics, Evolution*, Reidel: Dordrecht, 343.
Maeder, A.: 1984a, in Chiosi, C. and Renzini, A., eds., *Stellar Nucleosynthesis*, Reidel: Dordrecht, 115.
Maeder, A.: 1984b, in Maeder, A. and Renzini, A., eds., *IAU Symp. 105: Observational Tests of the Stellar Evolution Theory*, Reidel: Dordrecht, 299.
Maeder, A. and Meynet, G.: 1987, *Astron. Astrophys.* **182**, 243.

Mermilliod, J.-C. and Maeder, A.: 1986, *Astron. Astrophys.* **158**, 45.

Pinsonneault, M.H.: 1988, *Ph.D. Thesis*, Yale University.

Pinsonneault, M.H.: 1991, private communication.

Pinsonneault, M.H.: 1993, in preparation.

Pinsonneault, M.H., Deliyannis, C.P. and Demarque, P.: 1991, *Astrophys. J.* **367**, 239.

Pinsonneault, M.H., Deliyannis, C.P. and Demarque, P.: 1992, *Astrophys. J. Suppl.* **78**, 179.

Pinsonneault, M.H., Kawaler, S.D. and Demarque, P.: 1990, *Astrophys. J. Suppl.* **74**, 501.

Pinsonneault, M.H., Kawaler, S.D., Sofia, S. and Demarque, P.: 1989, *Astrophys. J.* **338**, 424.

Prather, M.: 1976, *Ph.D. Thesis*, Yale University.

Reimers, D.: 1975, *Mem. Soc. R. Sci. Liège, 6ᵉ Ser.* **8**, 369.

Rogers, J.F. and Iglesias, C.A.: 1992, *Astrophys. J. Suppl.* **79**, 507.

Roxburgh, I.W.: 1978, *Astron. Astrophys.* **65**, 281.

Saslaw, W.C. and Scwarzschild, M.: 1965, *Astrophys. J.* **142**, 1468.

Schatzman, E.: 1984, in Maeder, A. and Renzini, A., eds., *IAU Symp. 105: Observational Tests of Stellar Evolution Theory*, Reidel: Dordrecht, 491.

Schwarzschild, M. and Härm, R.: 1958, *Astrophys. J.* **128**, 348.

Seidel, E., Demarque, P. and Weinberg, D.: 1987, *Astrophys. J. Suppl.* **63**, 917.

Sreenivasan, S.R. and Wilson, W.J.F.: 1978, *Astrophys. Space Sci.* **53**, 193.

Stothers, R.: 1991, *Astrophys. J.* **383**, 820.

Stothers, R.: 1993, private communication.

Tassoul, J.L.: 1978, *Theory of Rotating Stars*, Princeton Univ. Press: Princeton.

Tassoul, J.L.: 1984, in Maeder, A. and Renzini, A., eds., *IAU Symp. 105: Observational Tests of the Stellar Evolution Theory*, Reidel: Dordrecht, 475.

Tassoul, J.L.: 1990, in Willson, L.A. and Stalio, R., eds., *Angular Momentum and Mass Loss for Hot Stars*, Kluwer: Dordrecht, 7.

Walborn, N.: 1988, in Nomoto, K., ed., *IAU Colloq. 108: Atmospheric Diagnostics of Stellar Evolution*, Springer: Berlin, 70.

Waldron, W.L.: 1984, in Underhill, A.B. and Michalitsianos, A.G., eds., *The Origin of Nonradiative Heating/Momentum in Hot Stars*, NASA CP-2358, 95.

Zahn, J.P.: 1983, in Hauck, B. and Maeder, A., eds., *Astrophysical Processes in Upper Main Sequence Stars: 13th Saas-Fee Course*, Geneva Observatory: Sauverny, 253.

Zahn, J.P.: 1991, *Astron. Astrophys.* **252**, 179.

Discussion

Smith: Would the introduction of differential rotation due to pre-ZAMS contraction accelerate or decelerate the differential rotation you currently find in your post-ZAMS calculations?

Sofia: Differential rotation due to pre-ZAMS contraction would slightly enhance the amount of differential rotation in the MS and post-MS stages of evolution, and consequently, slightly enhance the effects of rotation.

Moss: Uniform rotation may not be the appropriate solution for a convective region if rotation causes anisotropy of turbulence and related angular momentum transport (cf. Rudiger 1989). Inter alia, this might influence the critical angular momentum distribution *throughout* the interior of low-mass stars after any Hayashi turbulence has disappeared.

Sofia: The existence of differential rotation in convective regions is not only possible, but actually is known (from helioseismology) to exist on the solar convective envelope. To begin with, the simplification was made as merely a conservative assumption. Subsequently, observations have shown that the departure from the assumption is very small, at least in the solar case.

Harmanec: Your models predict a decrease of equatorial rotational velocity during evolution away from the ZAMS to the TAMS. Can you envisage a reasonable change of your input physics which would lead to increase of rotation during the MS evolution?

The problem of excessive number of blue supergiants may have a simple solution: that many of them are MS objects with extended shells, i.e., Be stars. I have suggested this in various contexts several times previously.

Sofia: Since the moment of inertia increases moderately during the MS evolution, the only way you can get an increase of the surface velocity is if the star reaches the ZAMS with considerable differential rotation, which causes an outward transport of angular momentum during the MS evolution. Such a scenario cannot be dismissed. However, at the present time, we do not have any independent evidence in support of its existence.

Dziembowski: (1) You said that in your stellar evolution code you apply the Schwarzschild criterion for onset of convection in chemically inhomogeneous zones. Does your code mix everything (entropy, elements, angular momentum) if such zones are found unstable?

(2) I think that the best physical justification exists for mixing entropy and not elements because the instability is not of a dynamical type but rather vibrational.

Sofia: (1) We only mix the chemical elements. The entropy gradient is computed in the conventional (MLT) fashion, and because of the solid body rotation assumption, the specific angular momentum increases as r^2.

(2) The possibility of angular momentum transport without mixing due to oscillations is interesting and very real. Unfortunately we have not yet explored how to incorporate it in our evolution codes.

Henrichs: Which opacities did you use for your calculations?

Sofia: The OPAL (Lawrence Livermore) opacities.

Marlborough: Do your calculations with mass loss include the entropy loss to the system required to transport the matter to infinity?

Sofia: Yes.

Heap: I wonder if all the observed problems you describe are real, in particular, the shape of the upper main sequence and the lack of a post-MS gap. The problems may be on the observational side: in deriving the surface parameters (T_eff, $\log g$) from the spectrum via NLTE models with winds. Until we can do that with reliability, we can't really say where the ZAMS is or where the post-MS is on the HR diagram.

Sofia: It is always tempting for the theorist to accept the observer's results without a great deal of questioning. Indeed, it might be inappropriate for us to suggest that our failure to solve a given perceived problem is due to faulty observations rather than shortcomings of our models. Of course, our continued inability to explain a given observation may lead us to begin to doubt its existence. However, only other (better) observations can disprove it.

ROTATIONALLY INDUCED TURBULENT DIFFUSION IN EARLY B-TYPE STARS: THEORY AND OBSERVATIONS

P. DENISSENKOV*
Max-Planck-Institut für Astrophysik,
Karl-Schwarzschild-Strasse 1, 85740 Garching, Germany

Available observational data indicate that some kind of additional mixing is present in radiative envelopes of O and early B-type stars. Because of these stars are known to be the fastest rotators amongst all normal stars, the most probable candidate for the role of additional mixing in their interiors is rotationally induced turbulent diffusion discovered by Zahn (1983, 1992). Recently Denissenkov (1993a,b) calculated evolution of massive main sequence (MS) stars with turbulent diffusive mixing in order to explain atmospheric abundance peculiarities in OB-stars. This note summarizes the main results which have been obtained.

Early B-type stars with masses around $10\,M_\odot$

1) The observational correlation of the N overabundance with the relative stellar age found earlier by Lyubimkov (1984) and confirmed recently by Gies & Lambert (1992) has been theoretically reproduced.

2) A possible explanation of the atmospheric microturbulence in B-stars has been proposed for the first time (Fig. 1).

3) It has been shown that turbulent diffusion in MS B-stars can transport Na synthesized in the convective core into the radiative envelope, which could be responsible for the anomalous Na excesses observed in F-K supergiants.

Luminous OB-stars with spectroscopic masses 10 to $50\,M_\odot$

One of the most unexpected contradictions between theory of evolution and observations of massive MS stars revealed during the last few years seems to be the so-called "mass discrepancy". It has been found that masses of luminous OB-stars determined from spectroscopic analysis M_{sp} are systimatically higher than those obtained by comparing the stars' location in the HR diagram with theoretical evolutionary tracks M_{ev}. Besides, in some of these stars large atmospheric helium abundances $\varepsilon_{He} \geq 0.16$ have been detected, what in addition gave rise to the "helium discrepancy" problem (see Herrero et al. 1992 and references therein).

---

* On leave from the Astronomical Institute of the St. Petersburg University (Russia), as an Alexander von Humboldt Fellow

L. A. Balona et al. (eds.), Pulsation, Rotation and Mass Loss in Early-Type Stars, 145–146.
© *1994 IAU. Printed in the Netherlands.*

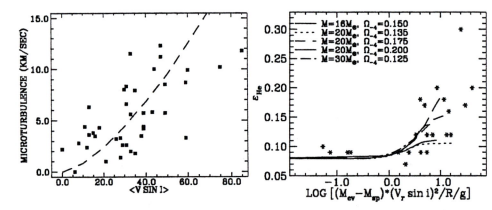

Fig. 1.   Correlation between microturbulent and projected rotational velocities in the atmospheres of B-stars (squares) and theoretical dependence of the turbulent diffusion velocity at the surface of a model star on its rotation rate (dashed line). The atmospheric microturbulence in B-stars seems to be identical with Zahn's rotationally induced turbulence.

Fig. 2.   Comparison of the observational (points) and theoretical (curves) dependences of the atmospheric helium abundance in OB-stars on the parameter characterising the efficiency of turbulent diffusive mixing.

Denissenkov (1993b) has proposed a model for massive MS stars that quantitatively accounts for these discrepancies. The radiative envelope of the model consists of two zones being mixed by rotationally induced turbulent diffusion on the MS. The rate of mixing in the outer zone is assumed to be substantially lower than that in the inner zone. Both, the mass and helium discrepancy, have been shown to be due to helium enrichment in the envelope produced by turbulent diffusion. In particular, the theoretical dependence of $\varepsilon_{\mathrm{He}}$ on the parameter $\log[(M_{\mathrm{ev}} - M_{\mathrm{sp}})(V \sin i)^2 / R/g]$, which estimates the efficiency of turbulent diffusive mixing, has turned out to approximate very well the corresponding observational correlation in OB-stars (Fig. 2).

References

Denissenkov, P.A.: 1993a, *Preprint MPA*, **733**
Denissenkov, P.A.: 1993b, *Preprint MPA*, in press
Gies, D.R., Lambert, D.L.: 1992, *Astrophys. J.*, **387**, 673
Herrero, A., Kudritzki, R.P., Vilchez, J.M., Kunze, D., Butler, K., Haser, S.: 1992, *Astronomy and Astrophysics*, **261**, 209
Lyubimkov, L.S.: 1984, *Astrofizika*, **20**, 475
Zahn, J.P.: 1983, *Astrophysical Processes in Upper Main Sequence Stars*, 13th Saas-Fee Course, eds. B. Hauck and A. Maeder, 253
Zahn, J.P.: 1992, *Astronomy and Astrophysics*, **265**, 115

EFFECTS OF ROTATION ON THE MAIN SEQUENCE
EVOLUTION OF A $5M_\odot$ STAR

J. FLIEGNER and N. LANGER
MPI für Astrophysik, D-85740 Garching, Germany

The way rotation influences the main sequence evolution of early type stars depends strongly on their internal angular momentum distribution. Their convective core mass is not always decreased as a consequence of a reduced "effective mass" due to rotation, since rotation laws close to uniform specific angular momentum may increase ∇_{rad} and thereby the convective core mass (Clement). In addition, rotationally induced mixing processes may redistribute angular momentum and chemical elements inside the stars (e.g. Endal & Sofia 1978).

To investigate the effects of rotation for a 5 $M_\odot$ star, we calculated the main sequence (MS) evolution for 3 different values of the total angular momentum, i.e. 0, 1.2×10^{51} and $6.1 \times 10^{51} g cm^2 s^{-1}$ (corresponding to the sequences R0, R1 and R1M, and R2M, where rotational mixing is suppressed for sequence R1; cf. Fig. 1). All sequences are started with a fully convective, rigidly rotating model at the Hayashi-line. The effects of rotation are treated as in Pinsonneault et al. (1989).

On the ZAMS, sequence R1 develops a convective core of $0.90M_\odot$ compared to $0.95M_\odot$ in the non-rotating model. Since its luminosity is slightly smaller, the evolutionary time-scale is nearly unchanged (cf. McGregor & Gilliland 1986). The models calculated with rotational mixing of angular momentum and chemical elements (R1M & R2M) develop larger convective core masses (cf. Fig. 2): During the pre-MS evolution rotationally induced mixing establishes a homogeneous distribution of specific angular momentum on a short time-scale. An even larger effect on the main sequence life time is caused by rotational mixing of hydrogen into the convective core (cf. Fig. 2 & Fig. 3). In total, the duration of central hydrogen burning is extended from $8.3 \times 10^7 yr$ for the non-rotating model (R0) to $1.4 \times 10^8 yr$ for model R1M and $2.5 \times 10^8 yr$ for model R2M.

In summary, rotationally induced mixing may support or even supersede convective overshooting as cause of an observationally required main sequence widening (cf. Fig. 1). Note that in our rotating models no alteration of the surface abundances occurs on the MS, in agreement with recent observations of Gies & Lambert (1992). However, from Fig. 3 large effects for both, surface abundances and internal evolution in the post MS phases can be anticipated.

L. A. Balona et al. (eds.), Pulsation, Rotation and Mass Loss in Early-Type Stars, 147–148.
© *1994 IAU. Printed in the Netherlands.*

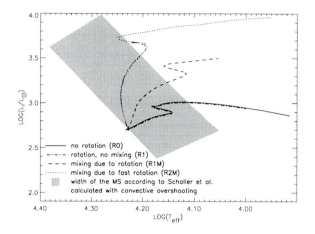

Fig. 1. The evolution of our models in the HR-diagram. The pre-MS phase is omitted.

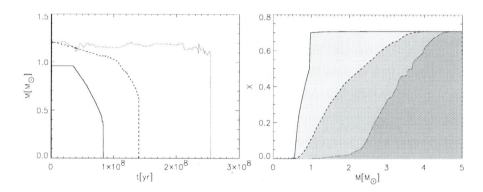

Fig. 2. The evolution of the convective core mass during central hydrogen burning. Lines are coded as in Fig. 1.

Fig. 3. The hydrogen profile at hydrogen exhaustion. Lines are coded as in Fig. 1.

References

Clement, Maurice J.: 'Differential Rotation and the Convective Core Mass of Upper Main Sequence Stars', *Preprint*

Endal, A.S. and Sofia, S.: 1978, *ApJ* **220**, 279

Gies, Douglas R. and Lambert, David J.: 1992, *ApJ* **387**, 673

MacGregor, K.B. and Gilliland, R.L.: 1986, *ApJ* **310**, 273

Pinnsonneault, M.H., Kawaler, S.D., Sofia, S. and Demarque, D.: 1989, *ApJ* **338**, 424

Schaller, G., Schaerer, D., Meynet, G., Maeder, A.: 1992, *A&A* **96**, 269

THE ABUNDANCE ANOMALIES IN A ROTATIONALLY EVOLVED 20 M⊙ OBN STAR MODEL

D. ERYURT, H. KIRBIYIK, N. KIZILOUĞLU, R. CİVELEK
Middle East Technical University, Physics Department
06531 Ankara, Turkey

and

A. WEISS
Max-Planck-Institut für Physik und Astrophysik, Garching, Germany

Walborn (1970, 1971) observationally found that OB stars showed nitrogen and carbon abundance anomalies in their atmospheres. Schönberner et al. (1988) found that OBN stars had helium enriched atmospheres. Several different suggestions have been made so far to explain the abundance anomalies in these stars (Paczynski 1973; Walborn 1976; Bolton and Rogers 1978). Maeder (1987) and Weiss et al. (1988) call attention to some possible effects of rotation in mixing chemical elements in Wolf-Rayet stars and B-supergiant stars respectively. They suggested that rotational mixing could explain the high He/H ratios and the high N/C and N/O ratios in the progenitor of 1987A Supernova. Saio et al. (1988) in an attempt to explain the nitrogen and helium enhancement in the progenitor of SN 1987A invoked extensive mixing in their models. They pointed out that such an effective mixing may be produced by rapid rotation. Such observations and ideas were the motivation for the present study.

20 M⊙ LMC star has been evolved for both normal and fully mixed cases up to $t_{ev} = 6.56 \times 10^6$ years. It has been shown that rotation has induced mixing and found that He^3 was enriched at the surface.

We also found that helium 4 and nitrogen were enriched and among them the enrichment of nitrogen was found to be the most conspicuous. The enrichment in N^{14} was about two hundred times that of normal rotating model. Besides $He^4/H^1, N^{14}/C^{12}$ and N^{14}/O^{16} ratios were also found to be rather large at surfaces of fully mixed star models compared to the normal star models. As a result of rotational mixing it has become possible to increase H^4 at the surface by about 61% and thus the surface abundance value of Helium-4 was obtained as 0.423 at $t_{ev} = 6.56 \times 10^6$ years, while the corresponding value of the abundance was 0.260 for the normal rotating star. The fully mixed models showed enrichments of helium and nitrogen and depletion of carbon and oxygen.

L. A. Balona et al. (eds.), Pulsation, Rotation and Mass Loss in Early-Type Stars, 149–150.
© *1994 IAU. Printed in the Netherlands.*

Schönberner et al. (1988) have observed and analyzed four galactic OBN stars and found significant helium enrichment amounting to as mass fraction of about 0.4 which agrees well with our finding $He^4 = 0.423$ quoted above. High values that we have found in relation to N/C and N/O do agree well with Walborn's results (1988). Although evolutionary positions of our last model and the 1987A Supernova are not the same, the trend in our results seemed to contribute to the explanation of the chemical abundance analysis in the above mentioned supernova. Early UV observations of the circumstellar material around SN 1987A (Panagia et al. 1987; Fransson et al. 1989) already indicated that nitrogen was overabundant with respect to carbon and oxygen which is also in good agreement with our finding in this work. It has also been claimed (Allen et al. 1989) that IR observations gave the fractional abundance of helium as 0.4 in the circumstellar envelope of SN 1987A, which also matches with our findings.

Conclusion has been that we could transport CNO processed material to the surface of the star, by a mixing mechanism resulted from rotation, and thus contributing to the explanation of chemical abundance anomalies in the atmospheres of OBN stars which are almost all on the main sequence; and our results might have some implications on the observed chemical abundances in SN 1987A.

References

Allen D.A., Meikle W.P.S., Spyromillio J., 1989, *Nature*, **342**, 403

Bolton C.T., Rogers G.L., 1978, *ApJ*, **222**, 234

Fransson C., 1989, *ApJ*, **336**, 429

Maeder A., 1987, *A& A*, **178**, 159

Paczynski B., 1973, *Acta Astr.*, **23**, 191

Panagia N., et al., 1987, *IAU Circ.*, **4514**

Saio H., Nomoto K., Kato M., 1988, *Nature*, **334**, 508

Schönberner D., Herrero A., Becker S., et al., 1988, *A& A*, **197**, 209

Walborn N.R., 1970, *ApJ*, **161**, L149

Walborn N.R., 1971, *ApJ*, **164**, L67

Walborn N.R., 1976, *ApJ*, **205**, L419

Walborn N.R., 1988, Atmospheric diagnostics of stellar evolution., In: K-Nomoto (ed.), *Proc. IAU Collowq.*, **108**. Springer, Berlin, pp. 70-78

Weiss A., Hillebrandt W., Truran J.W., 1988, *A& A*, **197**, L11 Hauck B., Maeder A., (eds.) p. 253

MAIN-SEQUENCE BROADENING, Be STARS,
AND STELLAR ROTATION IN h AND χ PERSEI[*]

J. DENOYELLE

Royal Observatory, Ringlaan 3, B-1180 Brussels, Belgium.

and

C. AERTS and C. WAELKENS

Instituut voor Sterrenkunde, Celestijnenlaan 200 B, B–3001 Leuven, Belgium.

1. Introduction

The double cluster h and χ Persei is one of the richest clusters containing early-B stars, and therefore is important for observational and theoretical studies on the fundamental parameters of massive stars. The colour-magnitude diagram of the double cluster shows an important scatter (see Figure 1). It has long been known that h and χ Persei are extremely rich in Be stars (Slettebak 1968). Our previous contention (Waelkens et al. 1990) that the large-amplitude variable stars we discovered are also Be stars, could be confirmed for a few objects. Rotation velocities for stars in h and χ Persei are usually high, which is not surprising in view of the large fraction of Be stars.

2. Rotation and main-sequence broadening

Rotation can affect colour-magnitude diagrams of open clusters in (at least) two ways :
1. Colours and magnitudes are affected by gravity darkening and geometric effects. These effects tend to brighten and redden the stars with respect to non-rotating stars (Maeder & Peytremann 1970, 1972; Collins 1987).
2. Very rapid rotation may thoroughly mix the stellar interior and so lengthen the main sequence life time (Maeder 1987). This effect would imply blueward evolution.

Rotation velocities for stars in h and χ Persei have been obtained by Slettebak (1968, 1985). We have also determined $v \sin i$ values for several cluster members, using the high-resolution spectrograph Aurélie of the Observatoire de Haute-Provence. We have adopted Slettebak's $v \sin i$ values to ours, since a systematic difference between his and our $v \sin i$-values was found for the stars common to both data sets and since our velocity determination is based on profiles with a higher resolution.

[*] Based on observations obtained at the Observatoire de Haute-Provence, France

L. A. Balona et al. (eds.), Pulsation, Rotation and Mass Loss in Early-Type Stars, 151–152.

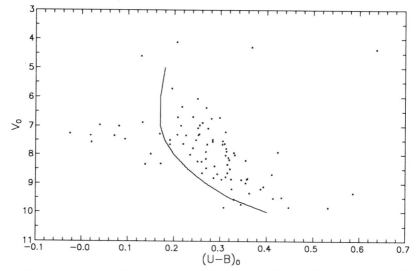

Fig. 1.   Colour-magnitude diagram for stars in h and χ Persei. Most stars on the left of the ZAMS are Be stars.

3. Discussion

The histogram of the observed $v \sin i$ values cannot be explained by effects of different inclination angles alone. Star-to-star variations of the equatorial rotation velocities are definitely present.

The average $v \sin i$ value for stars on the left of the ZAMS is 193 km/s, while it is 136 km/s for stars on the right of the ZAMS. This result may support the suggestion by Maeder (1987) that the blue-straggler phenomenon can be due to mixing induced by rapid rotation. However, the distance to the ZAMS is not correlated with $v \sin i$.

Also for stars on the right of the ZAMS, there is no clear relation between $v \sin i$ and colour excess. It is then not possible to accurately determine the effects of rotation on colours and magnitudes for these stars.

We then have to conclude that the effects of rotation are not sufficient to explain the observed main-sequence broadening in h and χ Persei.

References

Collins, G.W., 1987, In *Physics of Be stars*, Proceedings of the 92nd Colloquium of the IAU, eds. A. Slettebak & T.P. Snow, Cambridge University Press, p.3

Maeder, A., 1987, A&A 178, 159

Maeder, A., Peytremann, E., 1970, A&A 7, 120

Maeder, A., Peytremann, E., 1972, A&A 21, 279

Slettebak, A., 1968, ApJ 154, 933

Slettebak, A., 1985, ApJ 59, 769

Waelkens, C., Lampens, P., Heynderickx, D., Cuypers, J., Degryse, K., Poedts, S., Polfliet, R., Denoyelle, J., Van Den Abeele, K., Rufener, F., Smeyers, P., 1990, A&AS 83, 11

2. MAGNETIC FIELDS

OBSERVATIONS OF MAGNETIC FIELDS IN B STARS

DAVID A. BOHLENDER
Canada-France-Hawaii Telescope Corporation
PO Box 1597
Kamuela, Hawaii
USA 96743

Abstract. Globally ordered magnetic fields are known to exist in non-degenerate stars with spectral types between approximately F0 and B2. Among the B stars, and in order of increasing effective temperature, these include the Bp Si stars, helium-weak stars, and the helium-strong stars. These rather remarkable objects present us with an excellent opportunity to quantitatively examine the possible effects of magnetic fields on the photospheres, winds, and circumstellar environments of hot stars. In this paper we review some of the observations of the magnetic fields and field geometries of magnetic B stars, and also briefly discuss the success of attempts to measure magnetic fields in hotter OB and Be stars. We point out some of the interesting observational similarities of the helium-weak and helium-strong stars to Be and other hot stars, including their spectroscopic and photometric variability, variable winds as demonstrated by the UV resonance lines of C IV and Si IV, and their non-thermal radio emission. Continuing work also suggests that a considerable fraction of the rapidly rotating magnetic helium-peculiar stars are in fact variable Be and Be shell stars.

1. Introduction

Magnetic fields in stars other than the sun have been known to exist since Babcock's (1947) discovery of a strong field in the peculiar A star 78 Vir. Today several peculiarity classes are recognized among the upper main sequence magnetic stars and in order of increasing temperature these include the SrEuCr Ap stars, the Si Ap stars and the helium-peculiar stars.

The magnetic fields of these peculiar magnetic stars have a much different nature than fields in the sun and solar-type stars (Saar 1990). Rather than being locally complex, the magnetic fields appear to be globally ordered and have much simpler magnetic geometries. The observed magnetic fields are also generally variable in early-type stars and such variations are interpreted in terms of the Oblique Rotator Model (OBM) in which a usually predominantly dipolar magnetic field has its magnetic axis inclined to the star's rotation axis by some angle, β. As the star rotates, the orientation of the magnetic field to the observer changes and results in a variation of the observed field strength.

Since this review is concerned with magnetic fields in B stars we will limit our discussion for the most part to the helium-peculiar stars, although the hottest Si Ap stars are actually late B-type objects. The former group includes the B3–late B helium-weak stars with abnormally weak helium lines (Borra *et al.* 1983) and the B2–B3 helium-strong stars which have neutral helium lines too strong for their colours (Bohlender *et al.* 1987).

L. A. Balona et al. (eds.), Pulsation, Rotation and Mass Loss in Early-Type Stars, 155–166.

2. Measuring Magnetic Fields in B Stars

Several excellent reviews in the literature discuss in some detail the various techniques for measuring magnetic fields of early type peculiar stars (Landstreet 1980, 1992; Mathys 1989). Unfortunately, not all techniques used for cooler Ap stars are applicable to magnetic B stars. For example, magnetic B stars tend to rotate more rapidly than cooler Ap stars so that it is generally not possible to estimate surface magnetic field strengths by searching for Zeeman broadening of line profiles, a technique frequently employed for late type stars (e.g. Gray 1988). Even if a B star has a low $v \sin i$, the small number of useful spectral lines in these hot stars makes such an approach difficult. A few techniques that have been used successfully to measure magnetic fields in B stars are discussed below.

2.1. THE EFFECTIVE OR LONGITUDINAL MAGNETIC FIELD

Babcock's pioneering magnetic field measurements consisted of a high resolution spectrograph coupled with a Zeeman analyser that produced two simultaneous spectra of a star in right and left circularly polarized light (Babcock 1951). The shift between the two profiles is proportional to a weighted line-of-sight component of the magnetic field (the *effective* or *longitudinal* magnetic field) as well as the Zeeman sensitivity of the spectral line. This classical "photographic" technique consists of measuring these small shifts for many lines, but the technique is now employed with CCDs instead of photographic plates. Mathys (1988) has published some beautiful polarized line profiles for the helium-weak star HD 175362 and has used these to obtain a magnetic curve for this strongly magnetic star, as well as other helium-weak and helium-strong stars (Mathys 1991). Clearly, this technique requires very high signal-to-noise ratio data as well as moderately high spectral resolution to be successful. It is also only useful for slowly rotating stars, of which there are a limited number among B stars. In addition, since peculiar stars often have metal abundances that vary by as much as a factor of 100 over the surface, the magnetic field may not be sampled uniformly over the observed disk of the star which can make interpretation of the magnetic field structure problematical.

A less restrictive instrument for measuring longitudinal magnetic fields in hot stars is the Balmer line polarimeter, first developed for stellar applications by Angel & Landstreet (1970). A typical 2-channel photoelectric polarimeter uses narrow band (≈ 5 Å) filters to measure the polarization in the red and blue wings of various Balmer lines (usually Hβ) or more recently He I λ5876 (Bohlender *et al.* 1987). Mathys (1989) has demonstrated that the longitudinal magnetic field strength is proportional to the differential polarization measured between the two wings. Since pressure broadening dominates these line profiles, magnetic field measurements with a Balmer line polarimeter have the advantage that they can be performed even for rapidly rotating stars and since hydrogen and helium are generally much more evenly distributed over the surfaces of magnetic stars than are heavier elements the magnetic field is sampled much more uniformly over the surface of the star.

2.2. THE SURFACE MAGNETIC FIELD

Surface magnetic fields have been measured in only a few B stars. Babcock's star (HD 215441), a Bp Si star, presents an ideal case: this object has a very large magnetic field and rotates very slowly as far as B stars are concerned ($P_{rot} = 9^d4871$). With a sophisticated line synthesis program that includes the effects of a magnetic field and non-uniform surface abundances on the line profiles of a star Landstreet et al. (1989) modelled the resolved Zeeman splitting observed in spectra of the Si III multiplet 2 lines to measure dipolar, quadrupolar and octupolar polar field strengths of 67, 55, and 30 kG respectively. (This multiplet is perhaps the best suited as a surface magnetic field diagnostic in the visible spectral region of B stars since the three lines have a range of Zeeman sensitivities and Zeeman splitting patterns.) The λ4574 profile also demonstrates the remarkable uniformity of the surface field of the star: the intensity between the resolved Zeeman components of the line, a simple triplet, approaches the continuum very closely which can only occur if the range in surface magnetic field strength is very small.

The majority of magnetic B stars rotate too rapidly to have resolved Zeeman components or even enhanced broadening of magnetically sensitive lines. However, a related technique that has had some success in measuring surface fields in moderately rotating magnetic B stars is the observation of differential magnetic intensification of spectral lines in a multiplet. The principle behind this technique has been discussed by Babcock (1949) and is illustrated in Fig. 1 for the Si III multiplet 2 lines. Each line in the multiplet is located on the flat portion of the curve of growth. As the magnetic field strength increases, each line splits into its multiple Zeeman components (whose relative strengths and splittings are illustrated) and as the field continues to increase the separations between individual σ and π components eventually exceeds the thermal broadening of the line. The result is a desaturation of the line and an increase in the equivalent width—an effect very similar to that caused by microturbulence. The important difference is that the degree of intensification depends on the Zeeman structure of the line profile as well as the magnetic field strength so that the line strengthening is different for each line in the multiplet. If the quality of the data is high enough so that the continuum can be located with high precision and blended lines eliminated from the analysis then this technique, in principle, can be carried out for any star regardless of $v \sin i$ if the magnetic field is large enough. Surface magnetic field strengths have been derived in this manner

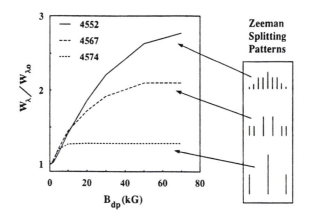

Fig. 1. Differential magnetic intensification of the Si III multiplet 2 lines. Calculated equivalent widths of each line in the multiplet relative to the equivalent width with no magnetic field present are plotted as a function of the polar field strength of a dipolar magnetic field. The equivalent width of the λ4574 line reaches a maximum for a relatively small field while the other two lines continue to intensify.

for only a few magnetic B stars (Bohlender 1989) but additional data is currently being analysed (Bohlender & Landstreet 1994).

3. Magnetic Field Observations

Typical effective magnetic field curves for two of the hottest known magnetic helium-strong stars are shown in Fig. 2. Each star has a field that varies sinusoidally with time but, as these examples show, in some cases the field strength variations are approximately symmetric about a mean value of zero, while in other cases the sign of the field does not change. In the ORM scenario the sinusoidal curves suggest field geometries dominated by a dipolar component and a non-reversing magnetic field is indicative of a small inclination of the field axis to the rotation axis.

Not all magnetic B stars have sinusoidal magnetic field curves. A few have apparently constant fields (Bohlender *et al.* 1987) almost certainly caused by an i or β near zero, but of more interest are stars with variable, non-sinusoidal field curves. The most notable of this small group of objects is the helium-strong star HD 37776 and its magnetic field curve based on data obtained by Thompson & Landstreet (1985) is plotted on their derived ephemeris for the positive extremum of the field in Fig. 3. As part of a detailed study of this star, Bohlender & Landstreet (1994) have attempted several model fits to the observed field curve and the fit for one of these is illustrated as the solid line in the figure and described by the accompanying magnetic parameters. HD 37776 is the first known case of a star whose magnetic field geometry is dominated by a quadrupolar component.

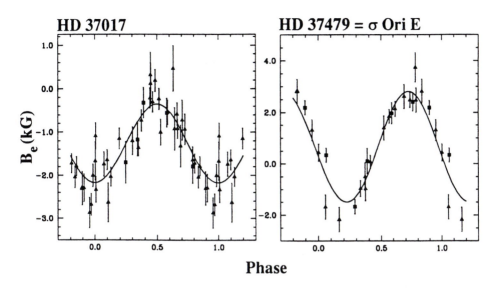

Fig. 2. Typical effective magnetic field curves for two helium-strong stars. The sinusoidal field variations imply magnetic geometries dominated by a dipolar field component.

4. Magnetic Fields and Stellar Winds

Besides having large (> 1 kG) effective magnetic fields, the helium-weak and helium-strong stars are spectroscopic and photometric variables, have magnetically controlled stellar winds as demonstrated by variable UV resonance lines of C IV and Si IV (Shore & Brown 1990) and variable Hα emission (Walborn 1982; Bolton *et al.* 1986) and are also non-thermal radio sources (Linsky *et al.* 1992). The helium-peculiar stars, therefore, display many of the same phenomena seen in OB stars and present us with an excellent opportunity to quantitatively examine the effects of magnetic fields on the atmospheres and winds of hot stars.

Gravitational and radiative diffusion processes interact with the magnetic field and the stellar wind and give rise to peculiar, non-uniform abundances, usually approximately axisymmetric with the magnetic axis. Helium abundances can vary by more than a factor of 5 over the surface of a star while metal abundance anomalies can be even more pronounced. Fig. 4 shows an example of helium and silicon line profile variations in HD 37776. These large abundance variations create structural changes in the atmosphere of magnetic stars which in turn lead to photometric variations of typically several hundredths of magnitudes. (e.g. Bohlender 1988, Bohlender & Landstreet 1990a, 1994; Landstreet *et al.* 1989).

Shore & Brown (1990) have produced a phenomenological model for the winds and magnetospheres of the helium-peculiar stars based on observa-

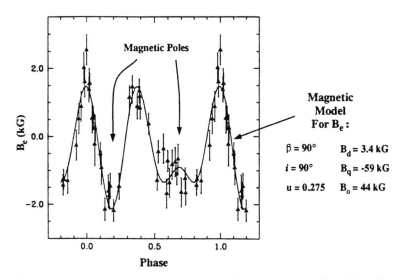

Fig. 3. The peculiar magnetic field curve of the helium-strong star HD 37776. The model fit given by the solid curve through the data points is given by the parameters to the right of the figure. The magnetic axis for the model crosses the line of sight at the two phases indicated.

tions of the UV line profile variations. They suggest that these stars have mass outflows restricted to the magnetic pole regions and hot circumstellar plasma trapped in the equatorial regions of the magnetic field. The variety of UV resonance profiles observed and the nature of the variability of the profiles are then a consequence of various oblique rotator geometries, and inclinations of the rotation axis to the line of sight. Linsky *et al.* (1992) have extended this picture somewhat to explain the radio emission from the magnetic peculiar stars in general.

Bolton and his collaborators have obtained extensive Hα observations of several helium-peculiar stars to investigate cooler regions of their magnetospheres. Early results for the prototypical helium-strong star σ Ori E were discussed by Bolton *et al.* (1986) and more detailed modelling has been presented by Short & Bolton (these Proceedings). In Fig. 5 we show an extensive collection of Hα spectra and their residuals for the helium-strong star δ Ori C. This is obviously another case of a magnetic Be star and we believe the emission is produced by wind material trapped near the magnetic equator. If we assume that the star's magnetic field forces the emitting material into corotation with the photosphere then the peak emission occurs at about 3.4 R_* above the photosphere and extends to about 5.2 R_*.

A similar, but less extensive data sample is shown for the helium-weak star 36 Lyn in Fig. 6. Also illustrated is our complete set of magnetic measurements for this star which permit a very precise period determination

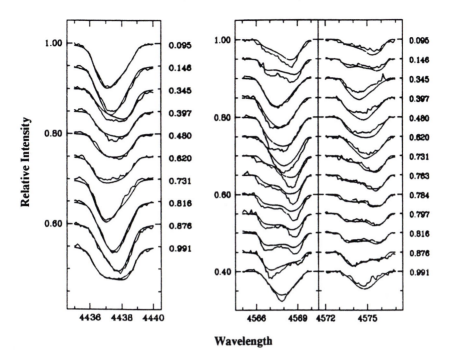

Fig. 4.   Observed and modelled helium $\lambda 4437$ and silicon $\lambda\lambda 4567$ and 4574 line profile variations in the helium-strong star HD 37776. For the model fits, generated without regard to the effects of the strong surface field predicted by the model in Fig. 3, the helium and silicon abundances vary by more than a factor of 5 and 70 respectively. Evidence of differential magnetic intensification in the silicon lines is apparent despite the high $v \sin i$ of 95 km s^{-1}.

of 3.83483 d. Additional Hα observations are being obtained for this inter-esting star, but it is already clear that it undergoes two brief shell phases precisely when the magnetic equator crosses the line of sight to the observer and when the C IV and Si IV line profiles increase markedly in strength (Shore *et al.* 1990). Together, these observations again suggest that we are seeing cool and hot material trapped in the magnetosphere of this object. As of this writing, 36 Lyn is the coolest star for which we have evidence of magnetically confined circumstellar material.

5. Other Magnetic Field Observations

Could magnetic fields in other OB stars contribute to the extensive vari-ability we observe in these objects? Several unsuccessful attempts have been made to measure magnetic fields in Be and Oe stars (Barker 1986). In no case, however, were more than two or three observations obtained for a sin-gle program star. We decided to pursue the question of magnetic fields in

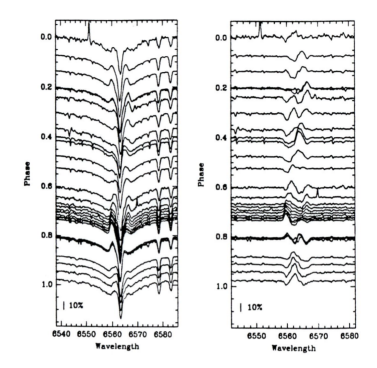

Fig. 5. Hα emission variability of the helium-strong star δ Ori C. The right panel displays the residual profiles after the mean of the profiles on the left has been subtracted from each observation. A rotation period of 1.4778 d has been derived from these data.

Be stars with more rigour for several objects that Harmanec (1984) has suggested might be related to the helium-strong stars because of their well defined light curves and stable periods. One of these objects was o And and the result of several seasons of Hβ polarization measurements is shown in Fig 7. No field was positively detected for this object or any of the other program targets. Fig. 7 also shows the magnetic curve for the bright Ap star ε UMa. This object has the weakest magnetic field detected so far among the magnetic Ap stars and has an amplitude of only 96 G (Bohlender & Landstreet 1990b). It is possible that a field of this magnitude could be lost in the noise of the measurements of o And but a considerable investment of 4-m telescope time will be needed to establish this.

We are also carrying out a long-term survey of magnetic field measurements in O stars. To date observations of more than 50 OB stars with spectral types earlier than B2 have not yielded a single positive detection of a magnetic field at the 300 G level. This survey does, however, suffer from a selection effect: we have so far avoided strong emission-line objects since emission effectively dilutes the photospheric polarization signal and makes

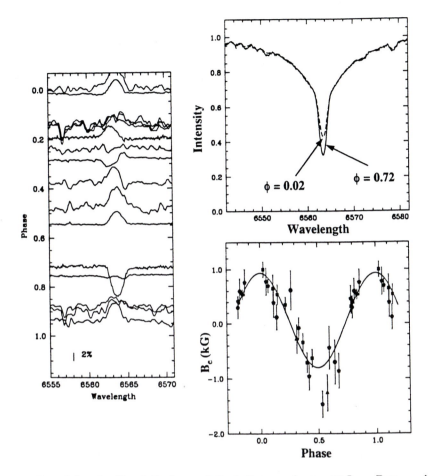

Fig. 6.   Top right: the Hα shell phase of the helium-weak star 36 Lyn. Bottom right: magnetic field observations phased on a period of 3.83483 d. Left: residual Hα profile variations after the mean of all spectra has been subtracted from each observation. Two shell episodes occur precisely when the magnetic equator crosses the observers line-of-sight.

magnetic field measurements even more difficult. However, we have already seen that the hottest known magnetic stars have Hα emission! Magnetic O stars may have even stronger emission because of their higher mass-loss rates, although the question arises as to how much material a star can constrain in its magnetosphere. In any case, given the tentative detection of a magnetic field in β Cep (Henrichs *et al.* 1993) and the suggestion that fields on the order of 100 G in some O stars (Kaper *et al.* 1994) are needed to explain the phase relationship between discrete absorption components and variable Hα emission components, the current survey should likely be extended to include emission line stars.

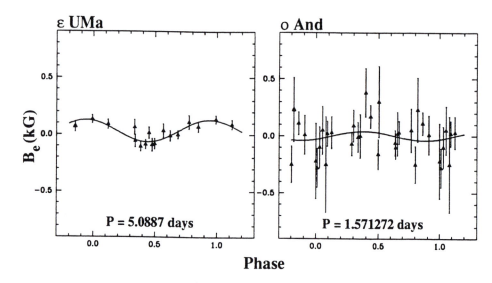

Fig. 7.   Magnetic field measurements of the Be star *o* And phased on the photometric period of the star. The least-squares sinusoidal fit to the observations indicated by the solid curve is *not* statistically significant. On the left, the very small magnetic field variation of the bright Ap star ε UMa is shown to illustrate the smallest magnetic field that has been positively detected in a magnetic star.

References

Angel, J.R.P. and Landstreet, J.D.: 1970, *Astrophys. J. Let.* **160**, 147.

Babcock, H.W.: 1947, *Astrophys. J.* **105**, 105.

Babcock, H.W.: 1949, *Astrophys. J.* **110**, 126.

Babcock, H.W.: 1951, *Astrophys. J.* **114**, 1.

Barker, P.K.: 1986, in Slettebak, A. Snow, T.P., eds., *IAU Colloquium 92, The Physics of Be Stars*, Cambridge Univ. Press: Cambridge, 38.

Bohlender, D.A.: 1988, *PhD Thesis*, Univ. Western Ontario: Canada.

Bohlender, D.A.: 1989, *Astrophys. J.* **346**, 459.

Bohlender, D.A., Brown, D.N., Landstreet, J.D., and Thompson, I.B.: 1987, *Astrophys. J.* **323**, 325.

Bohlender, D.A. and Landstreet, J.D.: 1990a, *Astrophys. J.* **358**, 274.

Bohlender, D.A. and Landstreet, J.D.: 1990b, *Astrophys. J. Let.* **358**, 25.

Bohlender, D.A. and Landstreet, J.D.: 1994, in preparation.

Bolton, C.T., Fullerton, A.W., Bohlender, D.A., Landstreet, J.D. and Gies, D.R.: 1986, in Slettebak, A. and Snow, T.P., eds., *IAU Colloquium 92, The Physics of Be Stars*, Cambridge Univ. Press: Cambridge, 82.

Borra, E.F., Landstreet, J.D. and Thompson, I.B.: 1983, *Astrophys. J. Suppl.* **53**, 151.

Gray, D.F.: 1988, *Lectures on Spectral-Line Analysis: F, G, and K Stars*, The Publisher: Arva, Ontario., 6-1.

Harmanec, P.: 1984, *Bull. Astron. Inst. Czechosl.*, **35**, 193.

Henrichs, H.F., Bauer, F., Hill, G.M., Kaper, L., Nichols-Bohlin, J.S. and Veen, P.M.: 1993 in *IAU Colloquium 139, New Perspectives on Stellar Pulsation and Pulsating Variable Stars*, in press.

Kaper, L., Henrichs, H.F., Ando, H., Bjorkman, K., Fullerton, A.W., Gies, D.R., Hirata, R., Kambe, E., McDavid, D. and Nichols, J.S.: 1994, *Astron. Astrophys.* submitted.

Landstreet, J.D.: 1980, *Astron. J.* **85**, 611.
Landstreet, J.D.: 1992, *Astron. Astrophys. Rev.* **4**, 35.
Landstreet, J.D., Barker, P.K., Bohlender, D.A. and Jewison, M.S.: 1989, *Astrophys. J.* **344**, 876.
Linsky, J.L., Drake, S.A. and Bastian, T.S.: 1992, *Astrophys. J.* **393**, 341.
Mathys, G.: 1988, *Astron. Astrophys.* **189**, 179.
Mathys, G.: 1989, *Fund. of Cosmic Physics*, **13**, 143.
Mathys, G.: 1991, *Astron. Astrophys. Suppl.* **89**, 121.
Saar, S.H.: 1990, in Stenflo, J.O., ed., *Solar Photosphere: Structure, Convection, and Magnetic Fields*, Kluwer: Dordrecht, 427.
Shore, S.N. and Brown, D.N.: 1990, *Astrophys. J.* **365**, 665.
Shore, S.N., Brown, D.N., Sonneborn, G., Landstreet, J.D. and Bohlender, D.A.: 1990, *Astrophys. J.* **348**, 242.
Thompson, I.B. and Landstreet, J.D.: 1985, *Astrophys. J. Let.* **289**, 9.
Walborn, N.R.: 1982, *Publ. Astr. Soc. Pacific* **94**, 322.

Discussion

Peters: I'm curious about the nature of the phase dependence in the photometric variations. Typically, do they correlate with extrema in field strengths?

Bohlender: No. The photometric variations are generally more closely correlated with the surface abundance geometries. These, of course, are in turn influenced by the magnetic field geometry, but also depend quite sensitively on the effective temperature as well as the age of the star.

Peters: Is it true that the equivalent width of the C IV wind line tends to anticorrelate with field strength?

Bohlender: This is certainly true in a few helium-weak stars and 36 Lyn is one example. It is more difficult to interpret the observations of the hotter helium-strong stars since some of the line variations observed in these objects may have a photospheric origin.

Bolton: Hunger has argued, I think persuasively, that the intensification of the C IV at the magnetic equator is due to photospheric abundance inhomogeneities.

Peters: In the Be stars that show short-term photometric variability the strength of the C IV line tends to correlate with the continuum flux.

Balona: Is there any conflict between the observed periods in the magnetic B stars and magnetic braking?

Bohlender: No. Walborn demonstrated that the $v \sin i$ distributions of helium-strong stars and non-magnetic B stars are indistinguishable, but Wolff also pointed out that magnetic braking in B stars is likely to be very inefficient because of the rapid evolution of such massive stars.

Henrichs: Is there any observational evidence that field strength depends on the line strength? I would expect this as a consequence of the steep radial dependence of the field strength over the atmosphere.

Bohlender: This is an interesting idea which has been investigated in late-type stars but, as far as I'm aware, not in upper main sequence magnetic stars. It might be possible to search for such an effect with a Balmer line polarimeter by making measurements of strongly magnetic stars with the interference filters sampling different portions of the line wings, or by looking at a widely separated pair of Balmer lines so that you sample a different level in the photosphere. My feeling, however, is that such an effect would be small when you consider the limited extent of the photosphere relative to the envelope of the star as a whole.

Baade: If magnetic fields are deduced from observations of a few line profiles, can one be confident that no confusion with other causes of line profile variability occurs? I am thinking in particular of the Be stars o And and LQ And mentioned by you.

Mathys: The best way to distinguish between various mechanisms possibly responsible for line profile variations and magnetic field effects is to observe in polarized light. The effect of a magnetic field will be different in different polarizations, contrary to the effect of pulsation, for instance.

Henrichs: In spite of your null detections among your OB-star survey, I would like to encourage you to continue this program, especially in view of the results by Grant Hill on β Cep, where he detected a significant field only after seven attempts. Because one expects the field to be at maximum once or twice per rotation period it might be worth while to concentrate first on the rapid rotators.

Bohlender: I'm in complete agreement, but it will require a significant amount of 4-m class telescope time since each data point in the various magnetic curves I've shown required on the order of 2 hours of observation.

Anandarao: There have been reports of rapid (on timescales of minutes) variability of emission lines ($H\alpha$ etc.) in several Be stars. Is there a possibility that magnetic fields present in these stars could be responsible for these variabilities?

Bohlender: We certainly see variations in $H\alpha$ emission on hourly timescales for the helium-peculiar stars and these are definitely a result of the strong magnetic fields of these stars. I suspect that rapid variability could be a result of small scale magnetic fields in Be stars but such complex fields will be virtually impossible to detect directly.

MASS LOSS, MAGNETIC FIELD
AND CHEMICAL INHOMOGENEITIES
IN THE HE-WEAK STAR HD 21699

S. HUBRIG
University of Potsdam,
Neues Palais, D–14415 Potsdam, Germany

and

G. MATHYS
ESO, Casilla 19001, Santiago 19, Chile

1. Introduction

HD 21699 is a He-weak star whose longitudinal magnetic field ranges from −1.3 to +1.1 kG (Brown et al. 1985). Its effective temperature, $T_{eff} = 15\,900$ K, and its gravity, $\log g = 4.0$, were determined by Wolff (1990) from photometry of the Balmer discontinuity and from measurements of the Hγ line. HD 21699 was reported to exhibit remarkable variations in the C IV ultraviolet resonance lines at 1548 and 1551 Å (Shore et al. 1987). These variations occur with the same period, $2\overset{d}{.}49$ (Percy 1985), as the magnetic variations and the photometric variations in the optical region. Similar UV line variations could be found only in two other He-weak stars, HD 5737 and HD 79158. They were interpreted as an evidence of the occurrence of outflowing winds controlled by the magnetic field. Furthermore, Shore et al. emphasized that no lines in IUE spectra other than C IV, and possibly Si IV, show any statistically significant variation. Recently, irregular brightening of a few hundredths of magnitude, superimposed over the periodic variations, were observed in the U photometric band (Vetö 1993); their origin is currently unknown.

Unfortunately, there are few optical spectroscopic studies of HD 21699 in the literature, and no information about the variability of the lines of helium and of iron-group elements. To decide whether the behaviour of HD 21699 is distinct from that of other typical He-weak stars, we looked in optical spectra for line profile changes which might be caused by inhomogeneous distribution of chemical elements on the stellar surface.

2. Observational results

HD 21699 was observed during 5 nights in March 1992 with the coudé spectrograph of the 2.2 m telescope of the Calar Alto Observatory. The detector

L. A. Balona et al. (eds.), Pulsation, Rotation and Mass Loss in Early-Type Stars, 167–168.
© *1994 IAU. Printed in the Netherlands.*

used was a GEC CCD with 1152×770 pixels of $22.5 \times 22.5 \ \mu m^2$. Spectra were taken in several wavelength ranges between 4610 and 5330 Å, at a reciprocal linear dispersion of 2.2 Å mm^{-1}, corresponding to resolving powers of $4 \ 10^4$ to $5 \ 10^4$. These spectra were reduced with MIDAS. To determine the radial velocities, the wavelengths of the line centres of gravity were measured by direct integration.

The observations are well distributed over the variation period. HD 21699 turns out to be a striking helium and silicon variable. The changes in the line profiles with rotational phase imply that these chemical elements are inhomogeneously distributed on the stellar surface. Our measurements of radial velocities as well as of equivalent widths suggest that helium is less abundant near the positive magnetic pole whereas silicon minimum coincides approximately with the negative longitudinal field extremum. Though they do not exhibit clear equivalent width variations, the lines of Cr II and of Fe II change markedly with rotation phase too. Their measured radial velocities vary between -8 and $+9$ km s^{-1}. At phase 0.57, splitting of these lines is definitely seen in our spectra. Such splitting is observable in some magnetic stars at rotational phases when the regions of enhanced element concentration appear at opposite stellar limbs.

Admittedly, with only observations at five different phases, we are unable to draw definite conclusions concerning the surface distribution of Fe, Cr, Si and He on HD 21699. More effort should be made to obtain a more detailed picture of the stellar surface. Such a study should rely on time series of high signal-to-noise, high resolution CCD spectra. From these data, magnetic and abundance maps for the inhomogeneously distributed elements should be derived simultaneously using a synthetic spectrum approach, taking into account the differential Zeeman effect affecting lines with different Landé factors. HD 21699 should lend itself well to this purpose, as its $v \sin i$, which we estimate of the order of 35 km s^{-1}, is sufficiently large so that Doppler effect on the line profiles can be exploited to extract geometrical information.

Comparison of abundance and magnetic maps with the inferred wind geometry is the most efficient way to investigate the effect of anisotropic mass outflows on the photospheric abundances of different chemical elements.

References

Brown, D.N., Shore, S.N. and Sonneborn, G.: 1985, *Astron. J.* **90**, 1354

Percy, J.: 1985, *Pub. Astron. Soc. Pacific* **97**, 856

Shore, S.N., Brown, D.N. and Sonneborn, G.: 1987, *Astron. J.* **94**, 737

Vetö, B.: 1993, 'Do Bp Stars Have "Flares"?' in M.M. Dworetsky, F. Castelli, R. Faragiana, eds., *Peculiar versus Normal Phenomena in A-Type and Related Stars*, ASP Conference Series, Vol. 44, 340

Wolff, S.C.: 1990, *Astron. J.* **100**, 1994

THE HE-STRONG STAR HD 96446: OBLIQUE ROTATOR
OR PULSATING MAGNETIC VARIABLE?

G. MATHYS

ESO, Casilla 19001, Santiago 19, Chile

HD 96446 is a B2p He-strong star, in which a large-scale organized magnetic field is present. From spectra recorded in both circular polarizations, various moments of this magnetic field have been repeatedly determined, among which the mean longitudinal magnetic field and the crossover. The mean longitudinal magnetic field $\langle H_z \rangle$ is the line-intensity weighted average over the visible stellar hemisphere of the line-of-sight component of the magnetic vector. The crossover, $v_e \sin i \langle x\, H_z \rangle$, is the product of the projected equatorial velocity, $v_e \sin i$, and of the mean asymmetry of the longitudinal magnetic field, $\langle x\, H_z \rangle$. The latter is the line-intensity weighted first-order moment about the plane defined by the line of sight and the stellar rotation axis of the component of the magnetic field along the line of sight.

Matthews & Bohlender (1991) have shown that the longitudinal magnetic field and the brightness of HD 96446 vary with the same period $P = 0\overset{d}{.}85137$. In agreement with the *oblique rotator model* (ORM), which accounts well for the observed magnetic and photometric variations of most Ap and Bp stars, they interpret this period as the period of rotation of the star.

The variations of $\langle H_z \rangle$ with rotation phase ϕ in HD 96446 are well approximated by a sinusoid, $\langle H_z \rangle = H_0 + H_1 \cos[2\pi(\phi + \phi_1)]$, with $H_0 = -(986 \pm 53)$ G, $H_1 = (242 \pm 73)$ G, and $\phi_1 = 0.617 \pm 0.049$. Such closely sinusoidal variations are typical of many Ap and Bp stars. Interpreted within the frame of the ORM, they suggest that the magnetic fields of most Ap and Bp stars are dominated by a single dipolar component, located close to the stellar centre.

For a stellar magnetic field consisting of a single dipole, whose centre coincides with the centre of the star, and whose axis makes an angle β with respect to the stellar rotation axis, approximate analytic expressions of the variations of $\langle H_z \rangle$ and of $v_e \sin i \langle x\, H_z \rangle$ with rotation phase can be derived. Both quantities vary sinusoidally, in phase quadrature with respect to each other. For non-saturated spectral lines (an excellent approximation in HD 96446) and a linear limb darkening law of the form $1 - u + u \cos\theta$ (with usual notations; $0 \le u \le 1$), one gets the following relation between the maximum $\langle H_z \rangle^+$ and the minimum $\langle H_z \rangle^-$ of the mean longitudinal field, and the extrema $v_e \sin i \langle x\, H_z \rangle^+$ and $v_e \sin i \langle x\, H_z \rangle^-$ of the crossover:

$$v_e \sin i \left(\langle x\, H_z \rangle^+ - \langle x\, H_z \rangle^- \right) / \left(\langle H_z \rangle^+ - \langle H_z \rangle^- \right) = 1.5\, v_e (8 - 3u)/(15 + u).$$

L. A. Balona et al. (eds.), Pulsation, Rotation and Mass Loss in Early-Type Stars, 169–170.

In HD 96446, no definite detection of the crossover was achieved. From this, we derive: $v_e \sin i \left[\langle x\, H_z \rangle^+ - \langle x\, H_z \rangle^- \right] \leq 16.3$ kG km s^{-1}. This is a very conservative $2\,\sigma$ upper limit, computed from the largest σ of all the determinations of the crossover performed for this star. On the other hand, $\langle H_z \rangle^+ - \langle H_z \rangle^- = 2\, H_1$. Thus, one can derive an upper limit of the equatorial velocity v_e, hence, as the rotation period is known, of the stellar radius. This upper limit is largest for $u = 1$. In this particular case, one finds $R \leq 1.21^{+0.52}_{-0.28}\, R_\odot$. This value, implausibly small for a B2 star, leads to question the validity of the model from which it is derived.

One possible shortcoming of this model might be that the actual field structure departs very significantly from a single dipole. However, the upper limit derived for the radius does not appear to depend critically on the field structure. For instance, the essentially sinusoidal variations of $\langle H_z \rangle$ observed in HD 96446 might be due to a quadrupolar field, in either of the following geometrical configurations: $i \approx 0$ (plausible as $v_e \sin i \leq 16$ km s^{-1}), or $\beta \approx 90°$ (then the actual rotation period of the star should be twice longer, i.e., $1^{\rm d}70$). In both cases, the same upper limit of the radius is derived: $R \leq 0.93^{+0.40}_{-0.22}\, R_\odot$.

Thus it appears that the variations of the magnetic field of HD 96446 *cannot be satisfactorily explained by the ORM*. Another mechanism should be sought. Pulsation is an obvious candidate, though it is at present unclear through which physical process it would induce variations of the longitudinal magnetic field. In relation to this, it is noteworthy that Matthews & Bohlender (1991) suspected the existence of a secondary periodicity ($P_2 \approx 6^{\rm h}2$) in the photometric variations of HD 96446, which they tentatively attributed to β Cephei-type pulsation. This second periodicity is not obvious in the $\langle H_z \rangle$ data, but the latter are neither accurate enough nor sufficiently numerous to be quite conclusive in that respect.

The evidence presented in this poster does not question the general validity of the ORM for the interpretation of the variations of the majority of Ap and Bp stars (which are mostly cooler than HD 96446). But it calls for a closer investigation of the variations of HD 96446 as well as of other He-strong Bp stars. In particular, it may well turn out that these stars are not so closely related to the cooler Ap and Bp stars as is generally believed. More observations (both of the magnetic field and of photometric variations) are needed to settle this issue.

References

Matthews, J.M. and Bohlender, D.A.: 1991, *Astron. Astrophys.* **243**, 148

THE CIRCUMSTELLAR ENVIRONMENT OF σ ORIONIS E

C. IAN SHORT AND C. T. BOLTON

David Dunlap Observatory, University of Toronto
Richmond Hill, Ontario, L4C 4Y6, CANADA

1. Introduction

The light and spectrum of the magnetic, chemically peculiar Be star σ Orionis E (=HD 37479) vary with the rotational period of 1.19081 d because of the effects of inhomogeneities in the photosphere and circumstellar matter (Walborn and Hesser 1976, Landstreet and Borra 1978, Hunger 1974). In this paper, we describe a model for the distribution of the Hα emitting material in the magnetosphere based on analysis of 24 high quality spectra covering the entire rotation period and published data from other sources.

2. Geometrical and Astrophysical Constraints on the Gas Distribution

There are two "eclipses" in the light curve that are caused by transits of dense clouds of circumstellar material across the stellar disk. These clouds are centered on the intersections of the magnetic and rotational equators at phase $\phi = 0.0$ and $\phi = 0.44$. The eclipse durations indicate that the clouds subtend 33° and 47° respectively at the star's center.

We have isolated the Hα line profiles of the circumstellar material by subtracting a photospheric profile derived by fitting the Hδ line profile with line blanketed model atmospheres. We obtain $T_{\rm eff} = 23250$ K, $\log g = 4.0$, $N_{\rm He}/N_{\rm H} = 1.0$, and $v \sin i = 150$ km s^{-1}. If the clouds are rotating rigidly, the Doppler widths of the circumstellar profiles observed when the dense clouds are in quadrature indicate that the radial dimensions of the clouds are $R(\phi = 0.0) = 7.0R_{\star}$ and $R(\phi = 0.44) = 6.4R_{\star}$. The Doppler width of the emission is nearly constant for most of the rotational period, so there must be gas present at the maximum radial extent over a broad range of rotational longitude.

Circumstellar emission is observed at $\Delta\lambda = \pm\lambda_{\rm H\alpha}v \sin i/c$ when the clouds are at quadrature, so they are "attached" to the photosphere. The absorption cores near $\Delta\lambda = 0$ observed during the eclipses indicate that the one of the clouds is occulted and one is transiting the stellar disk during each eclipse. Thus the clouds must be nearly co-planar and the inclination of the rotation axis must be $i \geq 82°$. The emission seen near $\Delta\lambda = 0$ for

L. A. Balona et al. (eds.), Pulsation, Rotation and Mass Loss in Early-Type Stars, 171–172.

$0.55 \leq \phi \leq 0.88$ indicates that magnetic equator has a modest inclination, $\beta \simeq 20°$, with respect to the rotational equator.

The shallow eclipses are either partial or else $\tau_{cont} < 1$ in the clouds. If the eclipses are total and the continuous opacity is due to HI bound-free transitions and Thomson scattering, the wavelength dependence of the eclipse depths in the Paschen continuum, yield $N_e(\phi = 0.0) \geq 5.47 \pm 0.06 \times 10^{10}$ cm^{-3} and $N_e(\phi = 0.44) \geq 3.52 \pm 0.03 \times 10^{10}$ cm^{-3}. Then the depth of the u-band eclipse yields $\tau_{H\alpha}(\phi = 0.0) \approx 3400$ in the radial direction. The blue (red) wings of the emission profiles obtained near $\phi = 0.3$ differ from the red (blue) wings of profiles $0.5P_{rot}$ later, so the clouds are also optically thick in the longitudinal coordinate. The clouds must be very thin perpendicular to the rotational equator because N_e derived from the last visible Balmer shell line during the eclipses (Groote and Hunger 1976) is 30 times greater than the lower limit derived from the light curve, and the Hα shell lines absorb $< 10\%$ of the photospheric flux. This suggests that $i = 90°$.

3. A Preliminary Model

These results show that the circumstellar gas is distributed in a thin disk with large blade-shaped clouds at the intersections of the magnetic and rotational equators. If $i = 90°$, the asymmetry in the line-of-sight magnetic field variation and the eclipse spacing can be explained by a symmetric magnetic dipole whose axis has a moderate inclination with respect to the rotational axis and center is displaced toward $\phi = 0.8$ and downward toward the hemisphere containing the negative magnetic pole. The radiatively driven wind from a broad region around the magnetic equator is channeled into the magnetic equator by the magnetic field where it compresses the gas into a thin disk. Recombined hydrogen ions drift outward and backward in the disk relative to the co-rotating frame due to the effects of centrifugal acceleration, radiation pressure, and wind compression. After many ionizations and recombinations they enter the leading edge of one of the clouds at the potential minima (Nakajima 1981) centered on the intersections of the equators. We believe that this process can explain many of the details in the light curves as well as the dissimilarity of the leading edges and similarity of the trailing edges of the clouds.

References

Groote, D. and Hunger, K.: 1976, *Astr. Ap.* **52**, 303
Hunger, K.: 1974, *Astr. Ap.* **32**, 449
Landstreet, J.D. and Borra, E.F.: 1978, *Ap. J. Lett.* **224**, L5
Nakajima, R.: 1981, *Tohoku Univ. Sci. Rept.* **2**, 129
Walborn, N.R. and Hesser, J.E.: 1976, *Ap. J. Lett.* **205**, L87

THE MAGNETIC FIELDS OF THE A AND B STARS

D. MOSS
Mathematics Department, The University, Manchester M13 9PL, United Kingdom

Abstract. The strong magnetic fields found in the Ap and Bp stars are reviewed, with critical attention to the competing dynamo and fossil theories for their origin. A number of difficulties for the dynamo theory are identified. Whilst the fossil theory is not free of problems, they appear less severe. From the modulation of certain spectral lines, rather weaker fields are deduced to be present above the surface of Be stars. The question of whether observed spectral transients might arise from magnetic fluctuations is discussed.

1. Introduction

Comprehensive reviews of the magnetic fields observed in the early type chemically peculiar Ap and Bp stars can be found in Moss (1986) and Landstreet (1992), so only a brief summary will be given here. Large scale magnetic fields, of strength from about 10^2 to 10^4G, are observed at the surfaces of these CP stars. These figures refer to an integrated measure of the field over the surface: local fields strengths can be somewhat higher, typically by a factor of 3 or 4. The lower limit appears to be instrumental. Statistically these stars are slow rotators compared to non-magnetic stars of similar spectral type. Nevertheless, rotation periods vary from less than one day to perhaps tens of years, so rapid rotation certainly does not preclude a strong surface magnetic field. Early suggestions of a marked correlation between field strength and period have proved unfounded, and there is at best a weak statistical relation. Typical field strengths do seem to increase somewhat with stellar mass, but there is no apparent relation between field strength and spectral type, nor any other stellar property. Strong fields appear early in the main sequence lifetime of these stars: the age of the OB1 association in Orion is estimated as about 5×10^6 years, and it possesses stars with kilogauss fields. The CP star magnetic fields vary strictly periodically, with magnetic and rotation periods equal.

Observations have nothing to say about the interior stellar magnetic fields and cannot, for example, rule out the possibility that strong internal fields are present in the interiors of even observably non-magnetic stars. There are only null field measurements for the Be stars, giving an upper limit to the poloidal field strength of about 100G. However, in these stars regular modulation at the rotation period of spectral lines formed in the wind regions is interpreted as indirect evidence for the presence of large scale fields, perhaps no more than of order 10G, corotating with the star (e.g. Barker 1982). Significant toroidal fields could also be present, but evade detection.

The CP star magnetic fields are ordered on a global scale, without substantial small scale components. In almost all cases the field geometry can be

L. A. Balona et al. (eds.), Pulsation, Rotation and Mass Loss in Early-Type Stars, 173–183.
© 1994 IAU. Printed in the Netherlands.

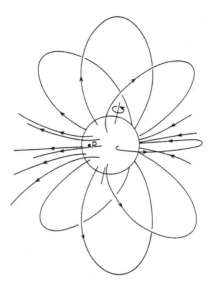

Fig. 1.　The geometry of the oblique rotator/displaced dipole model. P marks the nearer magnetic pole.

well approximated by a dipole, often with its centre displaced from the centre of the star along the axis of field symmetry. Field variations are well represented phenomenologically by the 'oblique rotator' (OR) model, in which this 'displaced dipole' (DD) field is inclined at angle $\chi > 0$ to the rotation axis, and rotates rigidly with the star. Thus the observed field variations are caused by the advection of a field that is invariant in the corotating frame, see Fig.1.

In this review, evidence for the competing fossil and contemporary dynamo theories of field origin is critically examined. We also discuss some ideas related to the inferred rapid changes in the fields believed to be present in the Be stars.

2. Field Origin

The fossil and contemporary dynamo theories are the two serious contenders to explain the origin of the CP star magnetic fields.

2.1. THE DYNAMO THEORY

Stars of a few solar masses on or near the main sequence have a convective core extending over 20% or less of the stellar radius, surrounded by a static envelope in radiative equilibrium. The dynamo theory of field origin proposes that a magnetic field is generated by a turbulent dynamo operating in the core. This field rises through the envelope to penetrate the surface, where it is observed. Certain statements about the nature of such a dynamo can be made immediately, independently of any particular dynamo theory.

1. The field must be steady in the rotating frame, otherwise skin effects would confine the field to the core (Moss 1980).
2. Stable nonaxisymmetric modes must be excited, as $\chi > 0$ is always observed.
3. The envelope differential rotation must be small, otherwise the nonaxisymmetric part of the field would be wound up and reconnection would occur, leaving only an axisymmetric field (Rädler 1986; Moss 1992).

Further, the differential rotation in the dynamo active core cannot be too large. This conclusion may depend slightly on the dynamo model, but follows basically from point 3 above. Adopting a mean field dynamo viewpoint, the dynamo then must be approximately an α^2 dynamo, and the relevant parameter is

$$C_\alpha = \frac{\alpha R_{\text{core}}}{\eta_{\text{turb}}} \sim \frac{\Omega_0 R}{u_t}. \tag{1}$$

Here it is assumed conventionally that $\alpha \sim \Omega_0 l$, $\eta_{\text{turb}} \sim \frac{1}{3} u_t l$, where Ω_0 is the angular velocity, $l \sim R_{\text{core}}$, R is the stellar radius and u_t a typical speed of the convective elements. With representative stellar parameters, $C_\alpha \gg 1$, and so it can be expected that a dynamo will operate.

If the turbulence is anisotropic and/or large-scale circulations are important in the core, then it is plausible that stable nonaxisymmetric dynamo modes are excited (Rüdiger & Elstner 1993; Barker & Moss 1993). The relevant questions are then

1. Can the field reach the stellar surface in the time available ($\lesssim 5 \times 10^6$ years for the youngest objects)?
2. If it can, will it be strong enough and of an appropriate geometry?

In a radiative envelope the classical magnetic diffusion time is $O(10^{10})$ years—clearly much too long. Magnetic buoyancy will cause flux tubes to rise on a relatively slow timescale, governed by the rate of diffusion of heat into the rising tubes. From Parker (1979) the rise time can be estimated to be

$$\tau_{\text{rise}} \sim \frac{Rp}{F_{\text{RAD}}} \frac{8\pi p}{B^2} \left(\frac{R_T}{\lambda} \right)^2 \text{ secs}, \tag{2}$$

where F_{RAD} is the radiative flux, R_T is the flux tube radius and λ is the temperature scale height. With values appropriate to the lower part of the radiative envelope, $\tau_{\text{rise}} \lesssim 5 \times 10^6$ years $B \gtrsim 10^7$ G if $R_T \sim 10^{-2}R$, that is flux tubes must be thin and strong.

Now we will try and estimate the possible strength of a dynamo field in the core. In an α^2 dynamo, the poloidal and toroidal field components are of comparable strength and nonlinear limitation will occur for $B^2/8\pi \lesssim \frac{1}{2}\rho u_t^2$, where $\mathbf{B}$ is the large scale part of the field. This may possibly be a considerable overestimate of the equipartition value of B^2. With typical values, $|\mathbf{B}| \lesssim 10^5$G. (An estimate for an $\alpha\omega$ dynamo is not very different.) Thus in order to reach the surface in the time available, flux tubes must have $R_T/R \ll 10^{-2}$: the field must rise as thin 'spaghetti', but reform into a 'clean', coherent, large scale quasi - dipolar field at the surface.

There is always the possibility that unknown instabilities might cause a rapid transport of field from the core to the surface: against that the development of a molecular weight gradient at the base of the radiative envelope as the star evolves will tend to stabilize the envelope against vertical motions.

A further restriction on surface field strengths can be obtained as follows. $R_{\text{core}}/R \lesssim 0.2$. If magnetic field lines connect the core and the surface, then flux conservation gives $R_{\text{core}}^2 B_{\text{core}} \geq R^2 B_{\text{surf}}$ (with equality only if no field lines close within the star). Given that $B_{\text{core}} \sim 10^5$G, then $B_{\text{surf}} \lesssim 4 \times 10^3$G, and the observed effective or longitudinal field is no more than about 10^3G.

Finally, dynamo theory appears to have some difficulty in explaining the lack of correlation between field strength and period: without an extra parameter some relation of the form $B = B(\Omega)$, and even perhaps $\chi = \chi(B)$, might be expected, but stars with periods differing by an order of magnitude can have similar surface fields. One suggestion is that the differential rotation profile in the envelope with which the star arrives on the main sequence might provide this extra degree of freedom: a stronger differential rotation would then give smaller nonaxisymmetric components at the stellar surface (Krause 1983). This would imply that the non magnetic A and B stars have strong internal differential rotation.

2.2. FOSSIL THEORY

Now the CP star flux is posited to be the remnant of that present in the interstellar medium from which the star contracted. Certainly there is enough flux present in the ISM and, indeed, a substantial proportion must be lost for star formation to occur (eg Mestel 1965). One possible hurdle to be overcome is the Hayashi turbulence during contraction to the main sequence, when the field might be tangled and destroyed or expelled from the star. It now seems that stars of several solar masses may avoid a significant Hayashi phase (eg Stahler et al. 1986; Shu et al. 1987), but in any case it is plausible that some field might survive in thin ropes, diffusing to a more uniform configuration as the turbulence dies away. (A hybrid version of the fossil theory would have a turbulent dynamo operating at this time, with the field being frozen into the stellar material as the convection ceased.) On the main sequence the global decay time is $O(10^{10})$ years – much greater

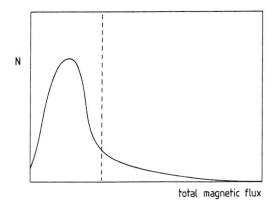

Fig. 2. A schematic distribution function for the magnetic flux on the zero age main sequence. Only stars in the tail of the distribution, to the right of the broken line, will be the observably magnetic stars.

than the main sequence lifetime. The initial magnetic flux, its orientation with respect to the rotation axis, and the local state of the ISM now provide the extra parameters needed to avoid relations of the form $B = B(\Omega)$. The relatively low incidence of observable magnetism among the A and B stars can be explained if the magnetic flux on the zero age main sequence has a distribution peaked at small values with a high flux tail, and if it only is the stars from this tail that become observably magnetic (see Fig. 2).

Detailed model calculations have elucidated some of the processes that influence the observable magnetic fields. The rotationally driven Eddington-Sweet circulation (symmetric about the rotation axis) influences the field geometry in at least two ways, provided that its mean magnetic Reynolds number is greater than order unity and that some measure of the mean magnetic field strength in the star is not large enough that the field can choke the flow. With conventional values for the resistivity and plausible field strengths, advection by the circulation is important during a main sequence lifetime for rotational periods of a few days or less. It is clear intuitively that, if there is a significant freezing in of the field into the stellar material, then the circulation will cause an increase in χ, see Fig 3a. A further effect depends on the value of χ. Suppose $\chi = 0$; in this case a sufficiently rapid circulation will drag field lines beneath the surface, whilst leaving the interior field little changed (Fig 3b and c). When $\chi = \pi/2$, consideration of Fig 3d shows that the circulation will tend to compress the field at the surface near to the equator, but *not* to bury it. The two regimes are separated by an angle $\chi_c \approx 55^o$. Clearly, the evolution of the field is quite a complex process, but these gross effects are confirmed by detailed model computations (eg Moss 1990).

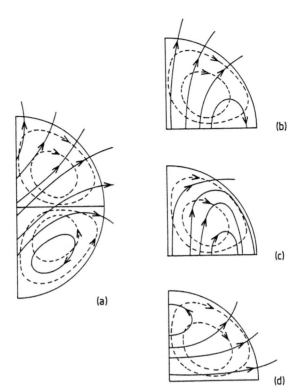

Fig. 3.   In each figure magnetic field lines are solid, circulation streamlines are broken and the rotation axis is vertical. Two quadrants of a cross-section of a sphere are shown in a), one in b), c) and d). a) Schematic relationship between modified Eddington-Sweet circulation and oblique magnetic field, $0 \leq \chi \leq \pi/2$. b) Schematic Eddington-Sweet circulation and magnetic field lines when $\chi = 0$, initial configuration. c) As b), but at a later time, showing reduction of surface flux. d) As b), but $\chi = \pi/2$.

Thus we can in principle find relations of the form $\chi = \chi(\text{conductivity}, \chi_0, \Omega_0, \mathbf{B}_0, t)$, $\mathbf{B}_{\text{surf}} = \mathbf{B}_{\text{surf}}(\text{conductivity}, \chi_0, \Omega_0, \mathbf{B}_0, t)$, where $\chi_0 = \chi(t = 0)$ and $\mathbf{B}_0 = \mathbf{B}(t = 0)$. With conventional values of the resistivity, significant changes in χ can occur in a main sequence lifetime. (If non-classical sources of resisitivity markedly increase the effective value, changes in χ may be much reduced.) The situation is complicated by the existence of other mechanisms affecting $\chi(t)$, such as internal motions driven dynamically by the non-symmetry of the magnetic distortion about the rotation axis (Mestel et al. 1981), and stellar wind torques (Mestel & Selley 1970), but in principle such relations could provide a test of the theory if the distribution of angles χ_0 were known or, alternatively, could give information about χ_0. Unfortunately the number of well determined values of χ is too small for this to be useful. The analysis of Mestel *et al.* (1981) and the models of Moss (1990 and references therein) are, however, consistent with interior field strengths not being greatly in excess of those seen at the stellar surfaces.

The stability of such large scale fields has attracted considerable attention, but no completely definitive answers. It is clear that a necessary condition for dynamical stability is that the topology be that of linked poloidal and toroidal fields. Even so, fields that are symmetric about the rotation axis may still be subject to instabilities (e.g. Tayler 1982). Comparatively few results are available for rotating nonaxisymmetric configurations—it may be significant that only $\chi > 0$ is observed.

In summary, according to the fossil theory, the fields that we observe in the CP stars are the primeval fields, after some expulsion, decay and advection. The picture is generally consistent with the OR/DD model. From a perhaps rather personal viewpoint, and with a *caveat* about stability, the evidence does at the moment appear to favour the fossil theory.

3. Be Stars

If the regular modulation seen in certain spectral lines is caused by a stellar wind in the presence of a corotating magnetic field, it is natural to try to explain irregular fluctuations in these lines as being associated with magnetic fluctuations. There seem to be two basic ideas.

By analogy to the OR model for the magnetic CP stars, the large scale field, $\mathbf{B}_b$ say, might be expected to be stable and unchanging. But what if Be stars have even a very thin sub-surface region that is unimportant for energy transport, but is the site of rapid turbulent motions. Could such motions influence the field, producing fluctuations $\mathbf{B}_1$? Given the existence of such a turbulent layer, there appear to be two possibilities. (This discussion does not depend on the nature of the origin of the large scale field $\mathbf{B}_b$.)

1. A dynamo operates in the turbulent region.

a. Consider a conventional mean field dynamo. Put $u_t = f_1 c_s$, $\Omega_0^2 R = f_2 g$, $\eta_{\text{turb}} = f_3 u_t d$, $\alpha = f_4 \Omega_0 d$, where c_s is the sound velocity, g the acceleration of gravity, d the depth of the turbulent layer and $f_i \lesssim 1$. Then

$$C_\alpha = \frac{\alpha d}{\eta_{\text{turb}}} = \frac{f_2^{1/2} f_4}{f_1 f_3} \left(\frac{d}{R} \frac{d}{H_p} \right)^{1/2} \ll 1, \tag{3}$$

where H_p is the pressure scale height and d/H_p is not large. Similarly $C_\omega = \Omega_0 d^2 / \eta_{\text{turb}} \ll 1$. Thus there is no conventional α^2 or $\alpha\omega$ mean field dynamo action.

b. Does a small scale 'fluctuation' dynamo operate? Simulations (e.g. Meneguzzi & Pouquet 1989) suggest that such a dynamo will be excited if $R_m = u_t l / \eta_{\text{true}} \gg 1$, where η_{true} is the normal, non-turbulent resistivity and l is a turbulent length scale. This condition is adequately satisfied in this case. However these numerical experiments do not have the extreme aspect ratios of very thin sub-surface turbulent zones of the type considered here, nor do they take account of the presence of the large scale background field which may alter the nature of the turbulence, so the situation is not altogether clear. Scaling from numerical simulations and the solar convection zone suggests that timescales could be of order one day.

2. There is no dynamo action, but high Rm turbulent motions in a relatively thin layer distort $\mathbf{B}_b$. To order of magnitude $(\mathbf{B}_b + \mathbf{B}_1)^2/4\pi\rho u_t^2$ will be less than unity. There are enough uncertainties that it is difficult to make good numerical estimates but, with $|\mathbf{B}_b|$ of order 100G, significant fluctuations with $|\mathbf{B}_1| \sim |\mathbf{B}_b|$ seem *a priori* plausible. However we can note that observations of the integrated longitudinal field in even the weaker field classical CP stars do not show corresponding variations, which may imply that this estimate for $|\mathbf{B}_1|$ is rather generous, and that really $|\mathbf{B}_1| < |\mathbf{B}_b|$. Even so, fluctuations could still be large enough to influence the wind.

Thus a combination of sub-surface turbulence and a stable background field might, in principle at least, provide a mechanism to generate fluctuations in the wind and hence in the spectral lines formed in the wind regions of Be stars. However, the presence of such a turbulent layer is not unambiguously established. B stars may be sufficiently massive that they do not have the sub-surface convection zone found in slightly less massive main sequence stars. With the recently available OPAL opacities, a thin sub-surface convectively unstable layer reappears in stars on the ZAMS with masses greater than about 15 solar masses (Alberts, these Proceedings). It is unclear how firmly this limiting mass is established. Very speculatively, it is possible that a shear instability near the stellar surface could result in such a turbulent layer, although the situation is not theoretically clearcut (see, e.g. Vauclair 1976 and the discussion in section 2.4 of Moss & Smith 1981). Note that rapid rotation favours such instabilities, but that a large scale magnetic field, if strong enough, might well inhibit them.

Another, quite different, suggestion is that Be stars are unstable to a number of g modes of nonradial oscillation. These modes are perturbed by the rapid rotation and may possess sufficient net helicity to provide an α-effect that could drive a dynamo (e.g. Dziembowski, these Proceedings). This idea is as yet unquantified: presumably any such dynamo would have to operate in a chaotic regime in order to explain the irregular spectral fluctuations.

Further, Smith (1989) has proposed that spectral transients arise from flare-like events, associated with regions of tangled magnetic field above the stellar surface.

At the moment none of these ideas are well developed. We should keep well in mind the indirect nature of the evidence for the magnetic fields in these objects: direct, rather than inferential, field detection would be very welcome!

References

Barker, D.M., Moss, D.: 1993, *Astron. Astrophys.* in press.
Barker, P.K.: 1982, in Jaschek, M. and Groth, H-G., eds., *Be Stars: Proc. IAU Symp. 95*, Reidel: Dordrecht, 485.
Krause, F.: 1983, in Soward, A.M., ed., *Stellar and Planetary Magnetism*, Gordon & Breach: New York, 205.
Landstreet, J.N.: 1992, *Astron. Astrophys. Rev.* **4**, 35.
Meneguzzi, M. and Pouquet, A.: 1989, *J. Fl. Mech.* **205**, 297.
Mestel, L.: 1965, *Quart. J. Roy. Astr. Soc.* **6**, 265.
Mestel, L., Nittmann, J., Wood, W.P. and Wright, G.A.E.: 1981, *Mon. Not. Roy. Astr. Soc.* **195**, 979.
Mestel, L. and Selley, C.S.: 1970, *Mon. Not. Roy. Astr. Soc.* **149**, 197.
Moss, D.: 1980, *Astron. Astrophys.* **91**, 319.
Moss, D.: 1986, *Physics Reports* **140**, 1.
Moss, D.: 1990, *Mon. Not. Roy. Astr. Soc.* **244**, 272.
Moss, D.: 1992, *Mon. Not. Roy. Astr. Soc.* **257**, 593.
Moss, D. and Smith, R.C.: 1981, *Rep. Prog. Th. Phys.* **44**, 53.
Parker, E.N.: 1979, *Cosmic Magnetism*, Clarendon Press: Oxford, 149.
Rädler, K-H.: 1986, in Guyenne, T.D., ed., *Plasma Astrophysics*, ESA SP-251, 569.
Rüdiger, G. and Elstner, D.: 1993, preprint.
Stahler, S.W., Palla, F. and Salpeter, E.E.: 1986, *Astrophys. J.* **308**, 697.
Smith, M.A.: 1989, *Astrophys. J. Suppl.* **71**, 357.
Shu, F.H., Adams, F.C. and Lizano, S.: 1987, *Ann. Rev. Astron. Astrophys.* **25**, 23.
Tayler, R.J.: 1982, *Mon. Not. Roy. Astr. Soc.* **198**, 811.
Vauclair, G.: 1976, *Astron. Astrophys.* **50**, 435.

Discussion

Henrichs: Stellar structure calculations of massive stars with the new OPAL opacities show that there is a convection zone at a fractional radius of about 0.98, the depth of which somewhat increases with mass. This convection zone did not appear in earlier calculations. See Alberts (these Proceedings). I wonder how this affects your considerations, in particular in the light of the observationally suggested correlation between mass and field strength.

Moss: This convection zone appears for main sequence masses somewhat in excess of those of the B stars, and any fields associated with such a convection zone seem unlikely to be related to the observed large scale fields in CP stars.

Mathys: Although detailed modelling is still lacking, there is growing evidence that the magnetic fields of most classical Ap and Bp stars do not have cylindrical symmetry about an axis passing through the centre of the star. Could you comment on the implications for the origin of the field?

Moss: The magnetic fields in my published A and B star models (Moss 1990 and references therein) naturally depart from cylindrical symmetry after time zero, even if they are initially cylindrically symmetric, provided that the initial inclination angle χ (β) is non zero. This results from the interaction of the meridional circulation with even an initially cylindrically symmetric field. Additionally, with the fossil field theory, there is no reason for fields on the age zero main sequence to be strictly cylindrically symmetric.

Megessier: Following the question from G. Mathys: is it possible to distinguish between the two magnetic field geometries—decentred dipole or a mixture of dipole and quadrupole? In the case of a decentred dipole, it is necessary to have the distance between the magnetic field and stellar centres as large as one third of the stellar radius. Is that conceivable?

Moss: I regard the displaced dipole and dipole plus quadrupole (multipole) descriptions as convenient parameterizations of the real field: there is no reason to expect that the actual stellar field corresponds exactly to either of these structures (except, of course, that any cylindrically symmetric field must have a multipolar expansion, and that an arbitrary field must be expandable in spherical harmonics).

Smith: With regard to spectral line profile transients in Be stars, you were careful to point out that any fluctuation in **B** may be smaller than the strength of a background global field. But if the energy of this fluctuating field structure is dissipated on a short timescale, it may not matter that $|\mathbf{B}_{\mathrm{fluctuation}}| < |\mathbf{B}_{\mathrm{background}}|$.

Moss: I agree, if it is the energy dissipated by the fluctuation field that is important. It might be, however, that the relevant mechanism is an additional modulation of the wind by the fluctuation field.

Cassinelli: In your discussion of the core dynamo, you considered the effects of buoyancy and diffusion. Couldn't the circulation currents affect the rate at which the field is brought to the surface?

Moss: Their timescale is a diluted global thermal timescale, and they are thus much too slow to be important.

Vakili: If we were able to resolve angularly the photosphere of Be stars, we could then 'see' dipolar or quadrupolar magnetic structures. This would need two orders of magnitude improvement on currently operational stellar interferometers. If so, how would the 'magnetic image' of these stars appear?

Moss: I would like very much to know! Detailed field geometries of evolving CP star models depend to some degree on the initial flux distribution in the star. If we assume that the large scale surface fields of the Be stars are analogous to those of the Bp stars, ie approximately displaced dipoles, but that there is a strong wind, then the answer might depend somewhat on the height above the surface to which the observations referred.

Dudurov: What can you say from a theoretical viewpoint about the difference between the magnetic and normal hot stars?

Moss: From the fossil viewpoint one can imagine that stars reach the ZAMS with a range of magnetic fluxes, depending on local conditions in the interstellar medium where they formed, and their detailed history. Plausibly this initial flux distribution is peaked at relatively small fluxes, with a high flux tail (see Fig. 2). Then it is the members of this tail that are subsequently the observably magnetic stars.

Owocki: When thinking about magnetic structures on stars, I think it is important to keep in mind some properties of the solar magnetic field. Near solar minimum, the global coronal and interplanetary magnetic field is quite well described by a tilted dipole (much as you and the previous speaker described for magnetic stars), which has a magnitude of only about 1 gauss. Nonetheless, this can have a dominant influence on the solar wind, e.g. inducing high and low speed streams. In addition, there are small scale fields of much greater magnitude ($\sim 10^4$ G) that play an important role in solar activity. Although I understand that much of the physics of hot stars is different, it does not seem impossible that similar ranges of field scales could exist in them also.

Sareyan: Given that there is no direct detection of magnetic fields in Be stars, can you make a comment about correlations between magnetic field variations, photometry variability and Doppler imaging in some stars?

Moss: No!

FOSSIL MAGNETIC FIELDS AND ROTATION OF
EARLY-TYPE STARS

A.E. DUDOROV
Chelyabinsk State University, Russia

1. Theory of fossil magnetic fields

Observational data of the last 10 years allow two main conclusions:

a) Main sequence stars can be separated in two classes: – magnetic (Bp) stars with surface strengths of a dipole or quadrupole magnetic field of $B_s \approx n \cdot (10^2 - 10^3)$ G, $n = 2, 3, 4...7$, and – normal main sequence stars (F–O) with magnetic fields $B_s \approx 1 - 100$ G (< 300 G);

b) Typical star formation takes place in interstellar molecular clouds with magnetic field strengths $B \approx 10^{-5}$ G (See Dudorov 1990).

These observations form the base of the theory of fossil magnetic fields. The main goals are the study of the evolution of magnetic flux in the case of ambipolar and Ohmic diffusion and its interaction with rotation and turbulence, where various MHD instabilities may develop during the various stages of the protostellar-cloud collapse and (proto)star contraction.

For these investigations we use the MHD equations with the "diffusional" variables with additional equations (Dudorov 1990) for nonstationary equilibrium of ionization by cosmic and X rays, and the equation for nonstationary magnetic ambipolar and Ohmic diffusion (MAD). We used the weak-magnetic approach (Dudorov and Sazonov 1981, 1986). The numerical simulations are carried out with the help of a modified Lax-Vendroff scheme.

The calculation shows that the magnetic field is frozen in the collapsing gas in case of a transparent collapse and reaches after some time a quasi-radial geometry outside the core. If the ionization state is determined by cosmic rays and radioactive elements, the magnetic field in the opaque protostellar MAD decreases the magnetic flux in the opaque protostellar core if the central density $n_c \approx [10^5 \cdot n_0, 10^9 \cdot n_0]$, with $n_0 = 10^4 - 10^5$ cm^{-3} the initial density. Adiabatic heating of the opaque core switches on the thermal evaporation of grains and the thermal ionization of lighter elements like K, Na, Al etc. (Dudorov 1976), and in regions with temperature $T \approx 4000 - 5000$ K the magnetic field will be immersed into the whole gas again. A zone of powerful MAD moves to the surface in the case of protostars and young-star evolution, coinciding with the region of minimal ionization degree. The attenuation of the frozen-in magnetic field is $\approx 10^{-2}$ for a $5M_\odot$ star on the stellar birth line. The surface magnetic field (before interaction with convection) is

$$B_s \approx B_{s0} \cdot (M/M_\odot)^{0.25-0.35}, B_{s0} \approx F(\tau_{CR}, Z_{RE}, Z_q).$$

L. A. Balona et al. (eds.), Pulsation, Rotation and Mass Loss in Early-Type Stars, 184–185.

For typical values for the "optical" depth due to cosmic rays τ_{CR}, the abundance of radioactive elements Z_{RE} and radius and abundance of grains a and q, $B_{s0} \approx 1 - 100$ G for normal stars and $\leq 2000 - 3000$ G for the magnetic Bp stars. The magnetic field strength increases towards the center of the star and is in the core $\approx (0.1 - 10) \cdot 10^6$ G depending on the stellar mass.

2. Rotation

Be stars rotate very fast, near to their limit of centrifugal equilibrium. Variable Be stars rotate more slowly. These stars have surrounding disks and possibly a magnetic field, which may connect with the disk. The angular momentum evolution of such a system depends on the existence of a magnetosphere and various phenomena similar to the case of accretion from a stellar wind.

We consider the evolution of angular momentum of a magnetic star with a surrounding thin Keplerian disk. The magnetic field of the star may have three types of geometry: an open fossil magnetic field, a dipole or a quadrupole field. Accretion from the disk feeds angular momentum to the star. The magnetic coupling of the star with the disk diminishes the accretion acceleration of the stellar rotation and may decelerate the star if the magnetospheric radius R_m exceeds the radius of centrifugal equilibrium R_{co}. Therefore, the possibility of a so-called equilibrium rotation occurs when $R_m = R_{co}$. Determining the magnetospheric radius from the balance of magnetic and kinetic energy densities and the corotation radius from balancing the gravitational and centrifugal accelerations, we obtain:

$$V_r = V_{r0} \cdot (B_s/10^3)^{-6q_m} \cdot (\dot{M}/10^{-7})^{3q_m} \cdot (M/M_\odot)^{q_{Mv}} \cdot (R/3R_\odot)^{-q_{Rv}},$$

where

$$V_{r0} = (A_m^{-2q_m} \cdot A_{co})^{\frac{2}{3}}(km/s), q_{Mv} = \tfrac{3}{2}(q_m + \tfrac{1}{3}), q_{Rv} = \tfrac{3}{2}(5q_m + \tfrac{1}{3}).$$

For a multipole field with $B \approx r^k, q_m = (4k-5)^{-1}, A_m \approx 92, A_{co} = 40$. From this we find for Be stars with $R/2R_\odot = (M/M_\odot)^{q_R}, q_R \approx 1/3, B_{s0} = 10^3$ G, $\dot{M} \approx 10^{-7}$ $M_\odot$/yr, $k = 2, V_{r0} = 9.3$ km/s and $q_M \approx 2/3$. For $k = 3$ and the same values for the other parameters, $V_{r0} = 69$ km/s and $q_M \approx 2/11$. For $k = 4, V_{r0} = 120$ km/s and $q_M \approx 2/33$.

This shows that the existence of disks surrounding Be stars and its magnetic field leads to a decrease of their rotational velocities. This conclusion is consistent with the formation of a closed magnetosphere and with the theory of fossil magnetic fields.

Dudurov, A.E. 1977a, in Early Stages of Stellar Evolution, ed. I.G.Kolesnik, Kiev: Naukova Dumka, **56** (In Russian).

Dudorov A.E. 1990, Magnetic Fields of Interstellar Clouds (Moscow: VINITI), **158** (In Russian).

Dudorov, A.E. and Sazonov, Yu.V. 1981, Nauchnye Informatsii Astronomicheskogo Soveta AN SSSR, **49**, 114 (In Russian).

Dudorov, A.E. and Sazonov, Yu.V. 1987, ibid, **63**, 68.

3. X-RAY OBSERVATIONS

ROSAT OBSERVATIONS OF B AND Be STARS

JOSEPH P. CASSINELLI AND DAVID H. COHEN
Department of Astronomy, The University of Wisconsin-Madison

Abstract. We present results from a survey of X-ray emission properties of near main-sequence B stars, including several Be and β Cephei stars. The main conclusions of our survey are: 1) The X-rays are soft, probably because the shock velocity jumps are small since the terminal wind speeds are small. 2) A major fraction of the wind emission measure is hot, assuming wind theory estimates for the density distribution. A large fraction of the wind is not expected to be hot in current wind shock models. 3) A hard component is found to be present in τ Sco; possible causes are discussed. 4) For the Be stars, the X-rays emission is from a normal B-star wind that is coming from the poles as in the WCD model of Be stars. 5) None of the stars, including the β Cep stars, show noticeable variability in their X-rays. For the normal B stars we conclude from the lack of variability that the shocks are in the form of fragments in the wind instead of spherical shells. 6) Our observations suggest that all B stars are X-ray sources and that there is a basal amount of X-ray luminosity of about $10^{-8.5} L_{bol}$. The hot component in τ Sco and the high X-ray luminosity of B stars detected in the all-sky survey suggests that there is a source of X-ray emission in addition to wind shocks in some B stars.

1. Introduction

This paper summarizes results of a four-year survey of "near main-sequence" B stars carried out using the *ROSAT* PSPC. By near main sequence we mean stars of luminosity classes V, IV and III, and we refer to these as "BV stars". Some of the material described here is presented in greater detail in Cassinelli *et al.* (1994), but we also describe some results obtained since that paper was submitted.

The B stars are of special interest for several reasons. 1) Relative to the O stars their winds are poorly understood. The terminal speeds of the winds are not known from the ultraviolet profiles because in most cases the wind lines are weak, and the wavelength of the maximum doppler displacement does not clearly correspond to the maximal wind speed. These stars do not have a strong radio free-free continuum like that of the O stars, so the mass-loss rates are not known. Perhaps X-ray observations will provide information regarding mass loss properties. For example, if the winds cease to be driven at some spectral type, the X-ray emission should also cease. 2) The winds should be optically thin to X-rays of most energies, so for BV stars we can study emission from all portions of the wind. We perform our analysis under the assumption that the radiation-instability wind shock model (Owocki, Castor & Rybicki 1988; Cooper 1994) is the correct explanation for O- and B-star X-ray emission. The X-ray emission properties of the BV stars will help us constrain the properties of the shock model as well as indicate whether other modes of X-ray production should be invoked. For example, the maximum X-ray temperature will be determined by the

189

L. A. Balona et al. (eds.), Pulsation, Rotation and Mass Loss in Early-Type Stars, 189–199.
© 1994 IAU. Printed in the Netherlands.

maximum velocity jump across shock fronts, so the X-ray spectral properties provide information about the strength of the shocks. 3) There is an
interesting variety of stars in the B spectral class which may exhibit X-ray
emission properties reflecting the underlying physical processes which define
these classes of stars. The β Cep variables exist in the spectral range B1–B2.
The radial pulsations of these stars could lead to periodic shocks as a fast
wind from the star in its compressed state collides with a slower wind when
the star is in a more extended state. The Be stars are rapidly-rotating stars
of special interest to us: very little is known about their X-ray properties.
At the conclusion of the Einstein satellite survey of hot stars, only one non-
binary Be star had been observed in X-rays. Now there are observations of
several others, and we can see if the presence of a disk around these stars
gives rise to additional X-ray emission.

Even before our *ROSAT* observations were made, it could be assumed
that the X-rays would be soft and of low luminosity, L_X. The softness is
expected because of the relatively small shock-jump velocities likely for the
slower winds of B stars. The total luminosity is expected to be small for
two reasons. Firstly, if the $L_X/L_{bol} \sim 10^{-7}$ relationship known to hold for
O stars continues into the B-star range, then the lower-luminosity B stars
will have correspondingly lower X-ray emission. Secondly, as the winds get
thinner with later spectral type, there is simply less material with which
to produce X-rays. Beyond some spectral type we expect the level of X-ray
emission to fall below our detection threshold.

With these characteristics in mind, we chose to observe B stars that are
nearby and have low interstellar column densities. This latter attribute is
crucial for soft X-ray sources because the opacity of interstellar matter varies
roughly as ν^{-3}. Estimates of the column densities in stars with negligible
extinction were, for the most part, taken from the Na I interstellar absorption
studies of our collaborators: Welsh, Vedder & Vallerga (1990). The column
densities are typically in the range $\log N_H = 18.0$–19.5. The detection limit
for our sample stars varies, but is typically 5×10^{27} erg s^{-1}. This detection
limit is more than an order of magnitude below the limit of the *ROSAT*
all-sky survey (RASS).

2. Results of the *ROSAT* Observations

Fig. 1a shows the energy distribution of the X-rays from a typical B star,
α Vir. As expected, the X-rays are very soft, with the count distribution
peaking at about 0.2 KeV. This is to be compared with a peak at 0.8–1.0
KeV in a typical O star and OB supergiant X-ray spectrum. The decrease
in the count rate towards lower energies is due to the sensitivity of the
ROSAT detectors, so these stars have spectra about as soft as is possibly observed. To determine the temperature of the emitting plasma and

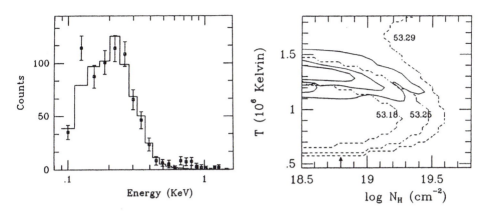

Fig. 1. The left panel shows the *ROSAT* spectrum of α Vir in PSPC counts versus energy. The right panel shows the results of a statistical analysis of the data, which provide information on the source temperature, source emission measure and the column density of the intervening interstellar material.

to assess the amount of absorption, we fit models of optically-thin thermal (line plus bremsstrahlung) emission (Raymond & Smith 1977). These models also include attenuation by neutral material (Morrison & McCammon 1983) with column densities equal to or greater than the known interstellar column densities. Fig. 1b illustrates the section of parameter space which is consistent with the data using χ^2 confidence limits defined for a joint probability distribution (Lampton, Margon & Bowyer 1976). The figure shows the results of calculations of a grid of models in $T - N_H$ space, with the emission measure of the best fit models at each point indicated by the dotted contours. The solid χ^2 contours correspond to confidence limits of 68, 95, and 99 per cent. As is typical for B stars in the sample, the column density we derive does not need to exceed the interstellar value (indicated by an arrow), although values somewhat higher than interstellar cannot be ruled out. Note that neutral gas with a column density of 10^{19} cm^{-2} has optical depth unity at the low energy end of the *ROSAT* PSPC detector. We are able to constrain the temperature to 1.3×10^6 within a factor of 25 per cent in either direction. The emission measure is a few times 10^{53}. The X-ray luminosity we derive for this source is 4×10^{30} erg s^{-1}.

Fig. 2 shows the emission measures and source temperatures of the stars in our sample to which we fit spectral models. The uncertainties correspond to the 68 per cent confidence limit as in Fig. 1b. With one exception, the temperatures are "low": about $1 - 4 \times 10^6$ K versus $5 - 8 \times 10^6$ K for typical early O-stars. The exception is the well studied "standard star" τ Sco. To fit its spectrum requires a two-component model with the high temperature being about 10^7 K as indicated in Fig. 2.

Fig. 3 shows the ratio of L_X/L_{bol} versus spectral type from O9.5–B7. For

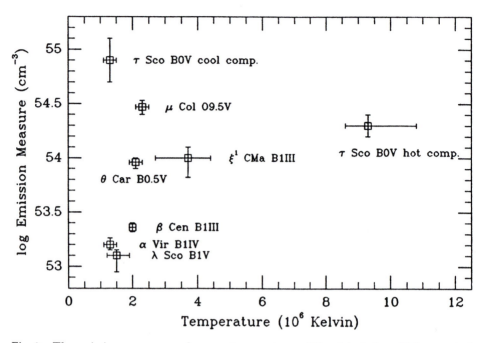

Fig. 2. The emission measures and source temperatures of the objects for which a spectral analysis could be made.

the earlier stars, the ratio is not much different from the value 10^{-7} in the O stars. However, in the B2–B3 spectral range there is a large decrease in the L_X/L_{bol} ratio to values near a few times 10^{-9}. Several classes of B stars are noted in the figure. The solid circles represent the β Cep variables. Also of special interest are observations of several Be stars. It is very interesting that there is essentially no difference between the X-ray luminosity of the Be stars and normal B stars of the same spectral class. The same can be said of the β Cep stars, but since there are no stars of the same spectral type which are not also variable, the comparison is more difficult.

These results for L_X/L_{bol} are quite different from those that have been reported from the RASS by Meurs *et al.* (1992) and Berghoeffer & Schmitt (1994). These authors find that there is a continuation of the 10^{-7} law throughout the B spectral range extending to late-B stars. However, the RASS observations were typically only about 400 s and only about 5 per cent of the B stars were detected. By contrast, we have a 2.5 σ level or higher detection in *all* our targets. Perhaps about 5 per cent of the stars are of a higher luminosity type similar to τ Sco. Alternatively, perhaps the ones that are seen in the RASS are just on the high X-ray luminosity tail of the distribution. It may be true that *all* B stars are X-ray sources at about

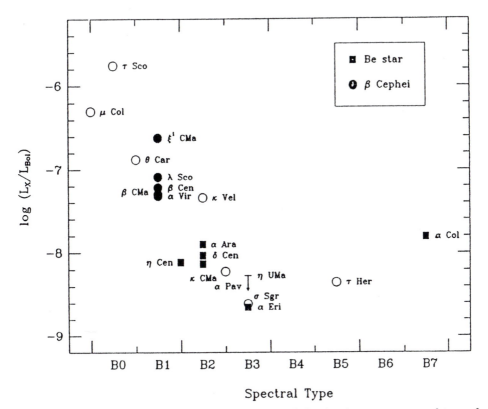

Fig. 3. The ratio of the X-ray luminosity to bolometric luminosity versus spectral type of our targets. All were detected at the 3 sigma level except for η UMa which was detected at the 2.5 sigma level.

the $L_X \sim 10^{-9} L_{bol}$ level or higher. This is probably the most surprising suggestion resulting from our survey thus far.

We also performed time variability analyses on those observations which had high enough count rates. This includes all those stars for which we fit spectral models. We observed most of these stars continuously for one or two kiloseconds. Several stars were observed two or more times, with the separate observations spread out over many hours or even months. The variability analysis was hampered by the presence of spacecraft wobble. This is employed to average obscuration by window support wires over a large area of the image plane (Downes *et al.* 1991). The wobble has a period of 400 s, making variability analysis on timescales of a few hundred seconds problematic. Variations of a *ROSAT* point source intensity up to about 25 per cent are known to occur (Dennerl & Kurster 1993). We found no evidence of variability above this level on timescales up to a few thousand seconds for any of the seven stars which we subjected to analysis, including the four β Cep variables. For two of the variable stars in the sample, α Vir and β Cen, and for τ Sco, we have several observations of each separated by days or months. For each object, the count rates of the different observations are consistent with a constant source at about 5–10 per cent. The hardness ratios of the different observation segments are also consistent with each other within the quoted uncertainties (of about 10 or 20 per cent). We caution that these results are for short segments of data or for poorly-sampled data. However, we can rule out large and rapid variability (say of a factor of one half over hundreds or thousands of seconds) in either the normal B stars or in the known variables. In a future paper, Finley, Cohen & Cassinelli (1994) will report on the final results of a *ROSAT* survey of β Cep variables which includes ten objects observed for up to 20 kiloseconds.

As mentioned above, we expected to find variability in the X-ray emission of the β Cep stars reflecting wind variability induced by photospheric pulsations. We thus far have no solid evidence for this. Furthermore, general variability is predicted in the hydrodynamical models of O- and B-star winds (Cooper 1994).

3. Interpretation of the L_X/L_{bol} behaviour

The softness of the spectra is not unexpected, but the fact that the X-rays persist through the full spectral range of our survey is surprising. The first question we should try to understand is: why is there a strong decrease in the L_X/L_{bol} relation at about B2–B3? Perhaps there is simply too little material in the wind to maintain the higher value of the ratio. To test this basic idea, we computed a plausible upper limit to the X-ray emitting emission measure in B stars. The upper limit represents the emission measure of the "entire wind". Our assumption is that the X-rays are produced by strong shocks which need a significant wind velocity to form. Thus we integrate the expected wind density structure from infinity down to the point in the wind where the velocity is 267 km s^{-1}, the jump speed required to produce gas at 10^6 K. We show three O stars for comparison and, due to their higher terminal velocities, we use 500 km s^{-1} for the integration limit. These upper limits are shown in Fig. 4.

To compute the emission measures, we need to know the density distribution in the wind. Unfortunately, observations provide almost no information about this, so we have chosen to follow Bjorkman & Cassinelli (1993) and use line-driven wind theory to provide the mass-loss rates and wind speeds that are expected for B stars. This requires that we specify the CAK line wind parameters k, α and δ—which we take from Abbott (1982)—and use the fitting formula of Kudritzki *et al.* (1989) to get the values for $\dot{M}$ and v_∞. We

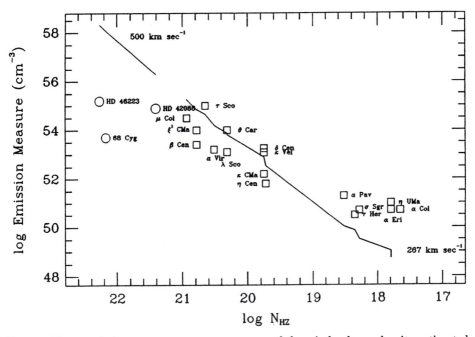

Fig. 4.   X-ray emission measure versus a measure of the wind column density estimated from wind theory. Also shown is a line of the estimated "maximal emission measure" taken to be the emission measure of the entire wind.

assume a smooth distribution of material in the wind for the calculation of the emission measure. Any clumping will increase the emission measure, but since the winds are relatively thin, the cooling lengths will be long and the shocks can be considered to be adiabatic. Hence the density enhancement will not be excessive. Finally, we use the ratio $N_{HZ} = \dot{M}/(v_\infty R_\star 4\pi\mu_H m_H)$ as an indicator of wind column density. This is a better quantity to use for comparison of the results than spectral type or T_{eff}, because we have stars with a range of luminosity classes (V–III) and wind properties depend also on luminosity and surface gravity, and their effects are included in N_{HZ}.

The trend we see in Fig. 4 is that both the observed and predicted emission measures fall with N_{HZ}, but that the observed values do not fall as fast as predicted by standard wind theory. By about spectral type B2.5, the observed emission measures exceed those predicted for the entire wind. In this instance we are learning something about the properties of the winds from the X-ray observations. The inconsistency could be resolved if we postulate that the B stars are better than expected in driving a stellar wind. The early-type B stars exhibit emission measures which, while consistent with the upper limit shown in the figure, are high compared to the predictions of hydrodynamical simulations (Cooper & Owocki 1994). These high emission measures pose a real problem for radiation-instability shock models which indicate that most shocks are reverse shocks with rarefied, fast gas accelerating into the shock zones. Only a small fraction of the wind material is hot at any given time in these models. Note that our calculations show that the very high X-ray luminosities found in a small portion of the RASS sample cannot be supported by any reasonable stellar wind. The source of their X-ray emission is a real puzzle.

4. X-rays from Be Stars

Since some of our targets are Be stars, it is interesting to draw conclusions about the nature and origin of their X-rays. With the Wind Compressed Disk (WCD) picture of Be stars in mind (see the paper by Jon Bjorkman in these Proceedings), there are several possible source locations. a) There should be shocks in radiatively-driven winds that are coming from the poles of the WCD envelope model. These should be very much like the X-ray emitting regions of ordinary B stars which do not have disks. b) There are shocks at the upper and lower boundaries of the WCD disk which are responsible for the high compression of the material in the disk around the stars. c) There is an infall shock at the surface of the star in the equatorial region as deduced from 2-D modelling by Owocki, Cranmer & Blondin (1994). They found that there should be a stagnation point in the disk, inwards of which there should be an infall toward the star and beyond which there should be an outflow. d) There could be hot gas confined to magnetic structures on the star. Smith *et al.* (1993) deduced from their X-ray observation of the Be star λ Eri that a flare of X-ray emission occurred. During the flare the X-ray emission changed from the soft X-ray spectrum that is seen in the other B stars, to a two-component picture with a 10^7 K component. Recall that in τ Sco we found a hot component that might perhaps be indirect evidence for the presence of hot, magnetically-confined gas. This may be related to the possible status of τ Sco as a pole-on Be star (Waters *et al.* 1993).

Since our observations of Be stars indicate that their X-rays have the same properties as those from normal B stars, we conclude that their X-rays are coming primarily from the polar wind component. The WCD boundaries are basically too cool to provide X-ray emission, although they are warm enough to provide the enhanced superionization that is detected as discrete absorption components in the UV spectra of Be stars (Grady *et al.*1989). There is also no clear evidence for X-rays from the infall shock. Perhaps the infall loss of material in the disk is occurring at a reduced rate. This could potentially explain one of the major problems with the WCD model: that the disk is "too leaky". Material flows out from the disk in two directions (J. Bjorkman, these Proceedings) and causes the mass in the disk to be too small to explain the IR excess and polarization observations.

Acknowledgements

We are grateful for useful discussions with Jon Bjorkman, Glenn Cooper, Dereck Massa, Geraldine Peters and Wayne Waldron. This research was supported by NASA grants NAG5-1579 and NAGW-2210.

References

Abbott, D.C.: 1982, *Astrophys. J.* **259**, 282.
Berghoeffer, T., Schmitt, J.H.M.M.: 1994, *Astron. Astrophys.* in press.
Bjorkman, J.E. and Cassinelli, J.P.: 1993, *Astrophys. J.* **409**, 429.
Cassinelli, J.P., Cohen, D.H., MacFarlane, J.J., Sanders, W.T. and Welsh, B.Y.: 1994, *Astrophys. J.* in press.
Cooper, R.G.: 1994, *Ph.D. Thesis*, University of Delaware.
Cooper, R.G. and Owocki, S.P.: 1994, in Moffat, A. *et al.* eds., *Instability and Variability in Hot Star Winds*, Kluwer: Dordrecht, in press.
Dennerl,K. and Kurster, M.: 1993, IAU Circular No. 5716.
Downes, R., White, R.E., Reichert, G., Pennerl, K., Engelhauser, J., Rosso, C. and Voges, W.: 1991, *The ROSAT Data Products Guide*, NASA: GSFC.
Grady, C.A., Bjorkman, K.S., Snow, T.P., Sonneborn, G., Shore,S.N. and Barker, P.K.: 1989, *Astrophys. J.* **339**, 403.
Kudritzki, R.P.Pauldrach, A., Puls, J. and Abbott, D.C.: 1989, *Astron. Astrophys.* **219**, 205.
Lampton, M., Margon, B. and Bowyer, S.: 1976, *Astrophys. J.* **208**, 177.
Meurs, E.J.A., Piters, A.J.M., Pols, O.R., Waters, L., Cote, J., van Kerkwijk, M.H., van Paradijs, J., Burki, G., Taylor, A.R. and de Martino, D.: 1992, *Astron. Astrophys.* **265**, L42.
Morrison, R. and McCammon, D.: 1983, *Astrophys. J.* **270**, 119.
Owocki, S.P., Castor, J.I. and Rybicki, G.B.: 1988, *Astrophys. J.* **335**, 914.
Owocki, S.P., Cranmer, S. and Blondin, J.: 1994, *Astrophys. J.* in press.
Raymond, J.C. and Smith, B.W.: 1977, *Astrophys. J. Suppl.* **35**, 419.
Smith, M.A., Grady, C.A., Peters, G.J. and Feigelson, E.D.: 1993, *Astrophys. J. Let.* **409**, L49.
Springmann, U.W.E. and Pauldrach, A.W.A.: 1992, *Astron. Astrophys.* **262**, 515.
Waters, L.B.F.M., Marlborough, J.M., Geballe, T.R., Oosterbroek, T. and Zaal, P.: 1993, *Astron. Astrophys.* **272**, L9.
Welsh, B.Y., Vedder, P.W. and Vallerga, J.V.: 1990, *Astrophys. J.* **358**, 473.

Discussion

Smith: Why are your L_X/L_{bol} values at spectral classes B2–B4 so much lower than is shown in papers in the literature and in posters at this meeting?

Cassinelli: The results that have been presented elsewhere are for the *ROSAT* all-sky survey observations, which were for durations of only about 400 s. Only about 5 per cent of the later-B stars were detected, whereas we detected all of our targets, and one of our 20 stars (τ Sco) is anomalous. So perhaps, the all-sky survey is picking up a "τ Sco class" of hard and luminous X-ray emitters. Perhaps there is a continuous distribution of B-star X-ray emitters, with the luminous ones being well out in the tail of the X-ray luminosity distribution. In either case, there is an indication that another cause of X-ray emission is present other than the wind shocks that we investigated. Even the relatively low X-ray luminosities that we are detecting seriously strains the shock model. Possibly the extra cause of emission is related to the magnetic phenomena that you observe in λ Eri (Smith *et al.* 1993).

Sofia: There has been a recent proposal that "dynamical friction" might heat the winds of early-type stars. This interaction is strong when the relative velocity between fast ions and wind is not too large, and stops for larger velocities. Perhaps the emission for B stars may be due to this effect.

Cassinelli: A dynamical friction or driving ion runaway model has recently been studied by Springmann & Pauldrach (1992). The drift speed between ions that are driven by the stellar radiation field and the wind material increases in the outward direction. This could in principle lead to the heating of the outermost regions of the wind and to a terminal wind speed that is less than expected without the runaway. I do not think the mechanism would lead to the production of X-ray emitting temperatures, however.

Ghosh: You have suggested a wind-shock model for the production of X-rays in Be stars. However, we should expect to find a frequency dependence of optical polarization from the envelope of Be stars, and that is not seen.

Cassinelli: To affect the polarization, a blob or shock fragment or whatever would need to have an optical depth near unity in electron scattering opacity. I am quite sure that the optical depths associated with the wind shocks of near main-sequence B stars are small. There can be large optical depths in the disks around Be stars. Karen Bjorkman (these Proceedings) has discussed the wavelength dependence of polarization in Be stars observed at ultraviolet wavelengths.

Owocki: When speaking about the canonical relation $L_X \sim 10^{-7} L_{bol}$, I think it is important to bear in mind that the *observed* X-rays could be a small fraction of the *intrinsic* X-ray emission produced by wind shocks. Theory suggests that the latter should perhaps be proportional to the mass-loss rate, which for the CAK model scales something like L_{bol}^2. For O-stars, it may be that true absorption in the wind effectively reduces this to $L_X \propto L_{bol}$. In this context it is actually somewhat encouraging to see a steeper L_X decline for the B stars, for which the winds are optically thin enough to observe even the soft X-rays.

Cassinelli: Yes, optical thinness is one of the real advantages for the B stars. We calculated the X-ray emission associated with just a single shock to find a plausible lower limit to the X-ray emission of these stars. The X-ray luminosities of BV stars of spectral class B0–B2 lie within the two limits derived from a single shock, and from the entire wind. In the case of the O stars, we find that the observed X-ray luminosity is roughly consistent with the idea that we see just the outermost shock. Nearly all of the X-ray emission from the inner shocks is attenuated by the overlying wind. The 10^{-7} law has been known since 1979, but it has not been adequately explained yet.

X-RAY PROPERTIES OF EARLY-TYPE STARS

THOMAS W. BERGHÖFER and JÜRGEN H. M. M. SCHMITT
Max-Planck-Institut für extraterrestrische Physik
Postfach 1603 D-85740 Garching

1. Introduction

Extensive stellar surveys with the *Einstein Observatory* (Chlebowski et al., 1989) and with ROSAT have clearly confirmed the presence of stellar X-ray emission over nearly the whole range of the HR diagram. In the ROSAT all-sky survey data approximately 20000 stellar X-ray sources were detected (Schmitt et al., 1992). Most of these stellar X-ray emitters are low mass late-type stars, the origin of their X-ray emission is thought to be coronal.

In the case of the early-type stars, the X-ray emission is thought to be generated by shock-heated gas (Lucy and White, 1980), present in the strong stellar winds of these stars. Model calculations for radiatively driven winds show that instabilities lead to the production of strong shocks and hence to the X-ray emission in the stellar wind (e.g. Owocki et al., 1988).

All the models for X-ray production in early-type star are based on assumptions for the underlying shock structures to heat the X-ray emitting gas. The search for variability in X-ray emission of early-type stars is a check for the so far assumed shock structures and is the only way to get access to parameters like the cooling time and the occurrence rate of shocks.

2. ROSAT observations and results

During the ROSAT all-sky survey (RASS) each star was observed for at least 2 days. We selected all detected early-type OB stars (**single stars, not X-ray binaries!**) with an X-ray count rate greater than 0.1 counts/s and listed in the Bright Star Catalog. For this magnitude and flux limited sample of early-type stars, (57 OB-type stars (O4-B3)), we generated the RASS X-ray light curves and tested them for variability. As an example we present in Fig. 1 the RASS light curves of the two O stars ζ Ori and ι Ori. The RASS X-ray light curves of our 57 OB stars show **no significant variations** above the 3σ level, and thus for 2 days these stars must have had an intrinsic memory of how much X-ray output to produce.

For a consistency check of the *Einstein* (Chlebowski et al., 1989, Grillo et al., 1992) and RASS data as well as a search for variability on a time scale of approximately 10 years we compared the *Einstein* and RASS X-ray count rates. For most of our OB stars the relation of the *Einstein* and RASS X-ray count rates is compatible with the relation of the effective area of the PSPC (ROSAT) and the IPC (*Einstein*). However, in the case of 4 B stars we observe, compared to the respective *Einstein* values, an overluminous RASS X-ray count rate.

L. A. Balona et al. (eds.), Pulsation, Rotation and Mass Loss in Early-Type Stars, 200–201.
© 1994 IAU. Printed in the Netherlands.

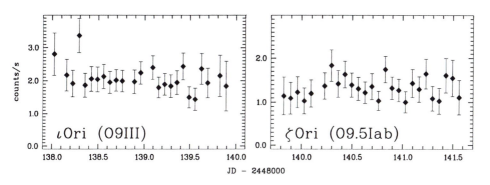

Fig. 1.   ROSAT all-sky survey light curves of the two O stars ι Ori and ζ Ori.

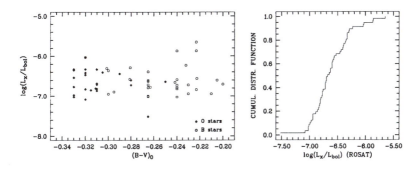

Fig. 2.   Plot of the computed L_x/L_{bol} values as a function of $(B - V)_0$ and the respective cumulative distribution function

As a result of the *Einstein* observations (Long and White, 1980), it is established that X-ray emission of OB stars obeys the relation $L_x/L_{bol} \approx 10^{-7}$. For stars later than B5 this relation breaks down and is definitely not valid for A-type stars (Schmitt et al., 1985). We have started to compute the relation L_x/L_{bol} for all detected OB stars. As a preliminary result we present for our sample of OB stars the L_x/L_{bol} values. For the bulk portion of our program stars (90%) we compute a $\log(L_x/L_{bol})$ value of -6.7 ± 0.4 dex. In the case of the O stars of our sample, the $\log(L_x/L_{bol})$ values agree with the respective *Einstein* values. However, some of the B stars appear to be overluminous compared to the observation with the *Einstein Observatory*.

References

Chlebowski, T., Harnden, F.R., Jr. and Sciortino, S.: 1989, *Astrophys. J.* **341**, 427

Grillo, F., et al.: 1992, *Astrophys. J. Suppl. Ser.* **81**, 795

Long, K.S., White, R.L.: 1980, *Astrophys. J.* **239**, L65

Lucy, L.B., White, R.L.: 1980, *Astrophys. J.* **241**, 300

Owocki, S.P., Castor, J.L., Rybicki, G.B.: 1988, *Astrophys. J.* **335**, 915

Schmitt J.H.M.M., et al.: 1985, *Astrophys. J.* **290**, 307

Schmitt, J.H.M.M., Kahabka, P., Stauffer, J., Piters, A.: 1993, *Astron. Astrophys.* **277**, 114

EFFECTS OF X-RAYS ON THE
IONIZATION STATE OF Be STAR WINDS

J. J. MACFARLANE

Dept. of Astronomy and Fusion Technology Institute
University of Wisconsin, Madison, WI 53706

It is important to understand the ionization state of Be star winds for several reasons, including: (1) if known, mass loss rates can be determined from UV P Cygni profiles; (2) the radiation line driving force which accelerates the winds depends on the ionization distribution; and (3) analysis of line profiles, in conjunction with polarization data, can help assess the credibility of various hypotheses for the assymetric nature of Be star winds.

Here, we descibe preliminary calculations to predict the ionization state in Be star winds. Calculations were performed for η Cen (B1.5 Ve), which has a mass loss rate constrained by UV observations of Si IV and Si III lines (Snow 1981), and an X-ray luminosity recently measured by ROSAT (Cassinelli *et al.* 1993). A list of stellar parameters used in our calculations is shown in Table I.

TABLE I
Adopted Parameters for η Cen

$\dot{M} = 3 \times 10^{-10}$ M$_\odot$/yr	$T_{eff} = 24,000$ K	$T_X = 2 \times 10^6$ K
$v_\infty = 770$ km/sec	$\log g = 4.0$	$L_X = 9.3 \times 10^{28}$ erg/sec
$T_{wind} = 24,000$ K	$R_* = 7.5\ R_\odot$	$L_X/L_{bol} = 7.8 \times 10^{-9}$

In our model, multilevel statistical equilibrium equations are solved self-consistently with the radiation field to determine atomic level populations (MacFarlane *et al.* 1993). Approximately 200 atomic levels were included in the multi-component plasma model consisting of H, He, C, N, O, and Si. The radiation field included contributions from the photosphere (Mihalas 1972), diffuse radiation from the wind, and X-rays from a high-temperature plasma source ($T = T_X$). Radiative transfer effects were computed assuming spherical symmetry using a multi-ray impact parameter model.

Results for the calculated ionization distributions for He and Si are shown as a function of scaled velocity ($\equiv v/v_\infty$) in Fig. 1. For both species, results are shown for 2 cases: one with an X-ray source distributed throughout the wind, and the other with no high-temperature X-ray source. Without X-rays, He II is predicted to be the dominant ionization stage throughout

L. A. Balona et al. (eds.), *Pulsation, Rotation and Mass Loss in Early-Type Stars*, 202–203.

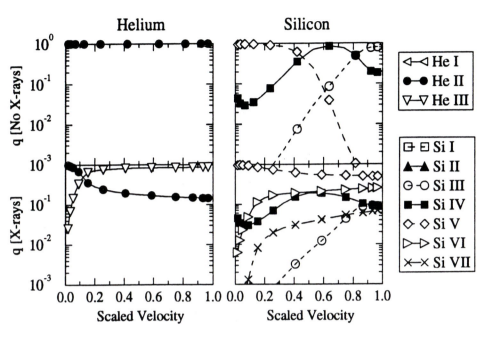

Fig. 1. Calculated ionization distributions for He and Si. The effect of X-rays can be seen by comparing the top and bottom figures for each species.

the wind, with the ionization fractions $q(\text{He I})$ and $q(\text{He III})$ being $< 10^{-4}$. However, when X-rays are included, He III is predicted to be the dominant ionization stage in the outer part of the wind ($v > 0.12\ v_\infty$; $r > 1.1\ R_*$).

Fig. 1 also shows that Si V is predicted to be the dominant ionization stage at $v < 0.45\ v_\infty$ when X-rays are neglected, and throughout the wind when X-rays are included. The Si V is produced by photoionization out of highly excited states of Si IV. With X-rays, $q(\text{Si IV})+q(\text{Si III}) \leq 0.2$ throughout the wind. This suggests that the mass loss rate determined by assuming that all Si resides in Si III or Si IV (Snow 1981) may be too low by roughly a factor of 5. It is important to note, however, that the above calculations do not include effects of density clumpling (due to shocks), non-spherical winds, and disks. Further investigations are in progress.

References

Cassinelli, J. P., Cohen, D. H., MacFarlane, J. J., Sanders, W. T., and Welsh, B. Y.: 1993, *Astrophys. J.*, in press.

MacFarlane, J. J., Waldron, W. L., Corcoran, M. F., Wolff, M. J., Wang, P., and Cassinelli, J. P.: 1993, *Astrophys. J.*, in press.

Mihalas, D.: 1972, *NCAR Tech. Note*, STR-76.

Snow, T. P.: 1981, *Astrophys. J.* **251**, 139.

INFRARED SPECTROSCOPY
OF BE/X-RAY BINARIES

C. EVERALL, M.J.COE and P. ROCHE
High Energy Astrophysics Group, Physics Dept., University of Southampton,
Southampton, SO9 5NH, U.K.

and

A. J. NORTON and S. J. UNGER
The Open University, Physics Dept., Walton Hall, Milton Keynes, MK7 6AA, U. K.

1. Introduction

We present infrared spectra of 4 Be/X-ray binaries in the K band, and 4 spectra in the J, H and K bands of 2 more sources. The HI IR emission lines are useful determinators of the conditions in the inner regions of the circumstellar disk about the Be star, due to optical depth effects. These are preliminary results, and hope to be followed up by high resolution echelle spectra, where we wish to estimate the velocity field, temperature and density structure of the circumstellar material.

2. Initial K band spectra

The spectra shown in Fig 1, of 4 Be/X-ray binaries (H0521+37, A0535+26, X Per and 4U0728-25) were take on January 13th 1992 in service time using the $75/\text{mm}^{-1}$ grating at short focal length with CGS4 on UKIRT. The resolution was $\lambda/\Delta\lambda{\sim}350$. Problems with tracking produced ripples that were difficult to correct. The spectra were combined to reduce this effect and are shown with line identifications in Fig. 2. The Pfund series was close to the case B ratios predicted by Hummer and Storey (1987), whereas Brγ must have a large optical depth. NaI emission could only arise in conditions cooler than the temperatures necessary for the free-free emission in the disk. Further details are available in Everall *et al.* (1993)

3. J,H,K spectra

The spectra shown in Figs 3- 6 were take on August 4-5th 1993 using the $75/\text{mm}^{-1}$ grating at long focal length with CGS4 on UKIRT. The resolution was $\lambda/\Delta\lambda{\sim}700$ at K. The targets were 2 Be/X-ray binaries, EXO2030+375 and 4U2204+56, and 4U1907+09, an OB supergiant X-ray binary. Absorption lines in the H band standard means the H band data is dubious.

L. A. Balona et al. (eds.), Pulsation, Rotation and Mass Loss in Early-Type Stars, 204–205.

Figure 1:

Figure 2:

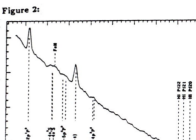

Figure 3:

Figure 4:

Figure 5:

Figure 6:

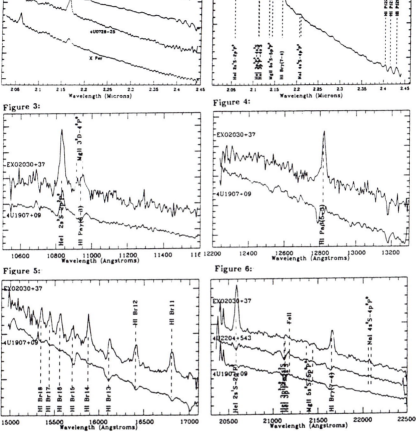

The spectra of 4U1907+09 varied greatly from the Be star's, with only the HeI 1.08μm ($2s^3S$-$2p^3P^0$) line in emission. 4U2204+56 had no emission lines, and therefore is likely to be in a diskless state. The spectra of EXO2030+375 had many HI emission lines, with a possible turnover in the Brackett series from optically thick to thin emission above Br14. The Paβ, Paγ and Brγ lines were all optically thick. These results will be discussed fully in a future publication (Everall *et al.* 1994).

References

Everall C. *et al.* 1993, MNRAS 262, 57-62
Everall C. *et al.* 1994 (in prep.)
Hummer D., Storey, P. 1987, MNRAS, 224, 801.
McGregor P. J., Hyland A. R., Hillier D. 1988, ApJ, 324, 1071

MULTIWAVELENGTH OBSERVATIONS OF THE BE-STAR / X-RAY BINARY EXO2030+375 DURING OUTBURST

A.J. NORTON

Dept. of Physics, The Open University, Walton Hall, Milton Keynes MK7 6AA, UK

M.J. COE, C. EVERALL and P. ROCHE

Astronomy & Space Physics Group, Univ. of Southampton, Southampton SO9 5NH, UK

L. BILDSTEN, D. CHAKRABARTY and T.A. PRINCE

Division of Physics, Mathematics & Astronomy, Caltech, Pasadena, CA 91125, USA

M.T. STOLLBERG

Department of Physics, University of Alabama, Huntsville, AL 35809, USA

and

R.B. WILSON

Space Sciences Lab., NASA Marshall Space Flight Center, Huntsville, AL 35812, USA

1. The EXO2030+375 System

EXO2030+375 consists of a neutron star in an eccentric 46 day orbit around a 20th magnitude Be-star companion (Coe et al., 1988; Parmar et al., 1989; Stollberg et al., 1993). The Be-star is thought to be surrounded by a shell/disc of material which is responsible for the infrared excess and Balmer emission lines which are characteristic of Be-stars in general. At periastron, the neutron star passes through this circumstellar material, giving rise to enhanced accretion onto the neutron star surface. As a result of this, the X-ray emission (pulsed at the neutron star spin period of 41.8s) increases dramatically, so producing the transient, outburst behaviour which is commonly seen in Be-star / X-ray binaries.

2. Observations

During the periastron passage of June 28th – July 7th 1993, we observed EXO2030+375 using the Palomar 200″ telescope, the William Herschel Telescope, the UK Infrared Telescope and the BATSE all sky monitor on the Compton Gamma-Ray Observatory. Hard X-ray measurements with BATSE span the entire period of the outburst, as do the infrared photometric observations, carried out in J, H, K and L′ bands. The optical spectroscopic observations were typically at < 1Å resolution, centred at the wavelengths of Hα and Hβ. These were obtained during the three nights covering the rise of the X-ray outburst.

L. A. Balona et al. (eds.), Pulsation, Rotation and Mass Loss in Early-Type Stars, 206–207.
© *1994 IAU. Printed in the Netherlands.*

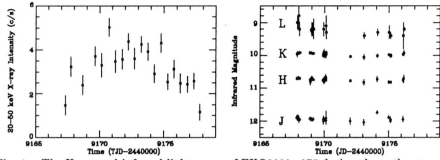

Fig. 1. The X-ray and infrared lightcurves of EXO2030+375 during the outburst

3. Results

As can be seen above, the X-ray intensity of EXO2030+375 increased by an order of magnitude over the course of about one week, followed by a decline over a similar timescale. Prior to JD2449167 and after JD2449178, the pulsed X-ray emission from the source was below the limits of detectability. It is clear that the scale of any correlated brightening or fading of the source at infrared wavelengths is less than one-tenth of a magnitude. Similarly the Hα spectroscopy shows no significant changes in either the equivalent width, the intensity or the profile of the Hα emission line. Hence, contrary to intuitive belief, we observed *no change* in the emission properties of the circumstellar material during this cataclysmic event. That is to say, the passage of the neutron star through the circumstellar material around the Be-star has *no detectable effect* on the structure or properties of that material.

4. Why no change ?

There are two ways in which the passage of the neutron star through the circumstellar material of the Be-star may be expected to influence its emission properties. Firstly, the intense flux of X-rays which is intercepted by the circumstellar material may be reprocessed into optical or infrared emission. Secondly, the passage of the neutron star itself may gravitationally disrupt the structure of the circumstellar disc/shell, so giving rise to changes in its emission characteristics. The fact that no such changes are observed implies that neither of these effects are significant.

References

Coe, M.J., Longmore, A., Payne, B.J. and Hanson, C.G.: 1988, *MNRAS* **232**, 865
Parmar, A.N., White, N.E., Stella, L., Izzo, C. & Ferri, P.: 1989, *ApJ* **338**, 359
Stollberg, M.T. et al.: 1993, 'BATSE Observations of EXO2030+375' in Friedlander, Gehrels & Macomb, ed(s)., *CGRO Symposium, St. Louis*, AIP Conf. Proc. 280, 371

TABLE II

Results of the Spectral Analysis of flares for 1983 August 8.

Fl.	Start Time (UT)	End Time (UT)	Δt (sec)	2–10 [a] keV flux	α [b]	N_H 10^{22} H cm^{-2}	L_X 10^{34} erg s^{-1}	E_{tot} 10^{36} erg	χ^2
E	18.75	18.94	700	20.277	$1.370^{+0.21}_{-0.20}$	$1.707^{+0.88}_{-0.61}$	14.333	100.331	1.162
B	17.21	17.36	550	19.092	$1.434^{+0.26}_{-0.24}$	$2.284^{+1.11}_{-0.99}$	13.495	74.224	0.975
C	17.36	17.78	1500	17.375	$1.643^{+0.17}_{-0.16}$	$2.239^{+0.65}_{-0.60}$	12.282	184.225	1.093
D	18.50	18.64	500	13.782	$1.690^{+0.37}_{-0.33}$	$3.141^{+1.62}_{-1.37}$	9.742	48.710	1.186
A	16.38	16.44	250	13.025	$1.673^{+0.56}_{-0.48}$	$3.096^{+2.75}_{-2.08}$	9.207	23.017	1.339

[a] In units of 10^{-11} ergs cm^{-2} s^{-1}.
[b] The errors in α and N_H are quoted at the 90% confidence level, that is, $\chi^2_{min} + 4.61$ for the two free parameters.

along with absorption in the line of sight give acceptable fits. There is a significant level of spectral variability in the source during the 1983 and the 1985 outburst. It was found that a simultaneous fit to the PH data from the different flares gave unacceptable values of χ^2. This implies that a single power law is not a good fit to the average spectrum of all the flares. The spectrum thus appears to vary from flare to flare. We examined the data for the evidence of line emission by adding to the continuum models, a Gaussian with a fixed width of 0.1 keV and a variable line energy. No improvement was found in the fits and we conclude that there is no evidence for the presence of a line in the spectrum. The upper limit to the iron K line equivalent width is 0.194 keV for 1983 and 0.103 keV for 1985 at the 90% confidence level.

To date nothing has been reported about the orbital period or the spin period of the compact object. We have searched for coherent pulsations indicating the signature of a pulsar from the two outburst data in a broad range in time. No periodicity was detected in the 50ms to 20s range at the 95% confidence level. Search for pulsations in the range 20s to 1000s resulted in a periodicity at 392.2s in the 1983 data and at 397.8s in the 1985 data. The above result, however, needs confirmation from further independent observations. If the spin period of the compact object is assumed to be 392s, a rough estimate of the orbital period of the binary can be made, using the P_{orb} vs. P_s relation of Corbet (1984); this gives P_{orb} between 150–250 days. It is strongly urged, that this source be monitored over a broad wavelength range to search for accurate pulse and orbital periods.

Saraswat,P. 1992, Ph.D. thesis and references therein
Apparao, K.M.V., Bisht, P., & Singh, K.P. 1991, ApJ **371**, 772
Saraswat, P. & Apparao, K.M.V. 1992, ApJ **401**, 678

PHOTOMETRIC AND H$_\alpha$ OBSERVATIONS OF LSI+61°303

J. M. PAREDES, J. MARTí, F. FIGUERAS, C. JORDI, G. ROSSELLÓ,
and J. TORRA
*Dep. d'Astronomia i Meteorologia, Universitat de Barcelona, Av. Diagonal 647, E-08028
Barcelona, Spain*

P. MARZIANI
*Dep. of Physics and Astronomy, University of Alabama, Tuscalosa AL 35487-0324,
U.S.A.*

J. FABREGAT, and V. REGLERO
*Dep. de Matemática Aplicada y Astronomía, Universidad de Valencia, 46100 Burjassot,
Valencia, Spain*

M. J. COE, C. EVERALL, P. ROCHE, and S. J. UNGER
Physics Department, University of Southampton, Southampton SO9 5NH, U.K.

J. M. GRUNSFELD
Caltech, Pasadena, CA 91125, U.S.A.

A. J. NORTON
Dep. of Physics, The Open University, Walton Hall, Milton Keynes MK7 6AA, U.K.

and

R. ZAMANOV
National Astronomical Observatory Rozhen, POB 136, 4700 Smoljan, Bulgaria

1. Introduction

The Be massive X-ray binary LSI+61°303 is a 26.5 days periodic radiosource
(Taylor & Gregory, 1984), exhibiting radio outbursts maxima between phases
0.6-0.8. Evidence of a photometric period of similar value has also been reported (Paredes & Figueras, 1986; Mendelson & Mazeh, 1989). The previous
spectroscopic radial velocity observations of Hutchings & Crampton (1981)
are in agreement with the radio period, and give support to the presence of
a companion. We present new optical and infrared photometric observations
and high resolution H$_\alpha$ spectra of LSI+61°303.

2. Observations and results

The photometric and spectroscopic observations are presented in Fig. 1.
The JHK data show a ~0.2 mag modulation with a deep minimum, which
is reminiscent of eclipsing binaries. A detailed model involving attenuation
and eclipsing of an emitting source associated to the secondary will be presented in Martí & Paredes (1993). A periodicity analysis applied to our V-

L. A. Balona et al. (eds.), Pulsation, Rotation and Mass Loss in Early-Type Stars, 211–212.
© 1994 IAU. Printed in the Netherlands.

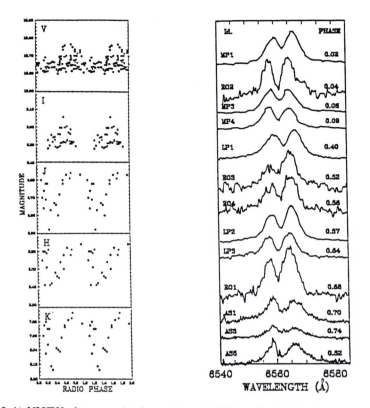

Fig. 1. *Left*) VIJHK photometric observations of LSI+61°303 folded on the 26.496 radio period. Phase zero has been set at JD 2443366.775. *Right*) Normalized H$_\alpha$ line profiles.

band data indicates 25.8±0.3 d as most significant period. A similar analysis merging our JHK data, after substracting their respective mean and dividing by the rms, gives 27.0±0.3. Both values are close to the radio period. On the other hand, our H$_\alpha$ spectroscopic observations show evidence of line parameter variability with radio phase. In particular, the FWHM of the H$_\alpha$ red hump increase significately during the phases of radio maximum, while the minimum value of the H$_\alpha$ EW and the maximum value of the B/R peak ratio are observed betwen radio phases 0.7-0.8. Further details are reported in Paredes et al. (1993).

References

Hutchings J.B., Crampton D. 1981, PASP, 93, 486
Martí J., Paredes J.M., 1993 (in preparation)
Mendelson H., Mazeh T. 1989, MNRAS, 239, 733
Paredes J.M., Figueras, F. 1986, A&A, 154, L30
Paredes et al., 1993 (in preparation)
Taylor A.R., Gregory P.C. 1984, ApJ, 283, 273

Be STAR DISK STRUCTURE AS IMPLIED BY X-RAY OBSERVATIONS

PRIYAMVADA SARASWAT

Indian Institute of Astrophysics, Bangalore, India

and

KRISHNA M.V.APPARAO

Tata Institute of Fundamental Research, Bombay, India

1. Modelling the Circumstellar Envelope of the Be/X-ray Binary 4U1907+09

1.1. OBJECTIVES OF THE STUDY

Compared to several other Be/X-ray binaries, 4U1907+09 has been observed more frequently due to the fact that it is found in an 'on' state more often. It also has a short orbital period of $\sim$ 8 days as compared to the long orbital periods commonly found in these binaries. But despite the attention it has received, the exact nature of the primary remains elusive. While some observers maintain it to be a Be/X-ray binary, others prefer to put it into the class of OB supergiants.

During January 1980, the X-ray intensity of 4U1907+09 flared up to 20 times its normal value over a period of $\sim$ 10 days as observed with the SSI on Ariel V. The subsequent decline was also continuously monitored by the SSI for a further 20 days. We have attempted to model this light curve of the outburst and to make an estimate of the number density and the velocity of the gas in the circumstellar envelope (CE) of the Be star.

1.2. THE ROTATIONALLY ENHANCED STELLAR WIND MODEL AS APPLIED TO 4U1907+09

The currently understood picture of Be/X-ray binaries is one in which the compact object is in an eccentric orbit around the Be star. The orbital plane of the binary may be inclined to the equatorial plane of the primary star. The high X-ray luminosities observed from these sources suggest that the X-ray source is fueled by a dense, slowly moving and probably equatorially condensed envelope. The Be/X-ray binary 4U1907+09 provides a unique opportunity to test the predictions of this model due to its short orbital period of $\sim$ 8.4 days. The variations in the X-ray flux give information on the densities and velocities in the outer regions of the extended envelope of the Be primary traced by the orbital motion of the neutron star. The neutron star acts as a probe of the outer wind regions of the Be star. Using

L. A. Balona et al. (eds.), Pulsation, Rotation and Mass Loss in Early-Type Stars, 213–215.
© 1994 IAU. Printed in the Netherlands.

this model we demonstrate semi–quantitatively that the observed shape of the X-ray flare from 4U1907+09 during the January 1980 outburst can be reproduced fairly satisfactorily.

In order to obtain the X-ray emission at a given time, the rate of accretion of matter onto the neutron star during its passage through the ring has to be calculated. This is given by

$$\dot{M}(r) = Av_{rel}^{-3}(r)\rho(r) \tag{1}$$

To find ρ and v_{rel} at any given time, the neutron star has to be located with respect to the Be star and the ring/disk. A numerical code was written for this purpose. The Be primary was assumed to have emitted a circular gas ring in the equatorial plane with an initial density ρ_{oi}. The eccentric neutron star orbit was determined using the parameters given for 4U1907+09. It lies in the equatorial plane of the Be star. In order to obtain v_{rel} at any point, we assume that the matter in the ring at any point has a Keplerian velocity, v_g, and a constant outflow velocity, v_o. The neutron star velocity, v_n, is also calculated at that point from the orbital parameters and the coordinates at that point. Then,

$$v_{rel} = [(v_g - v_{nt})^2 + (v_o - v_{nr})^2]^{\frac{1}{2}} \tag{2}$$

where v_{nt} and v_{nr} are the tangential and radial components of the neutron star orbital velocity. For the density of the matter in the ring, we find that a constant value throughout the ring does not yield the observed light curves, and a density distribution described mathematically by the form

$$\rho(r,t) = \rho_o(t) \left[\frac{r_o}{r}\right]^\alpha \quad \text{for } r \geq r_o$$
$$= \rho_o(t) \left[\frac{r_o}{(2r_o - r)}\right]^\alpha \text{ for } r \leq r_o \tag{3}$$

has to be prescribed. Here r is the radial distance of the matter in the ring from the center of the star, r_o is the radial distance of the center of the ring from the Be star where the density is $\rho_o(t)$ at that instant of time. We calculated the intensity of the X-ray emission from the neutron star at different positions of its orbit by obtaining $v_{rel}(t)$ and $\rho(t)$ at that position and time. The X-ray emission from the neutron star is attenuated by the matter in the portion of the ring between the observer and the neutron star. We found that to match the observed light curve, we have to postulate the existence of multiple rings ejected by the Be star. These rings have different values of density and propagate outwards at slightly different constant out-flow velocities. The shape of the X-ray light curve strongly depends on the outflow velocity of the gas ejected by the Be star. The outflow velocities in the case of 4U1907+09 lie in the range 100–250 km s^{-1} while the densities

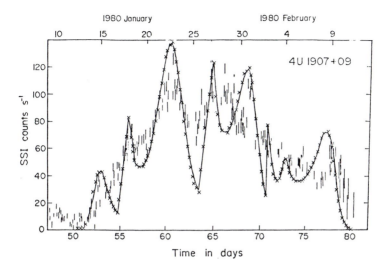

Fig. 1. Modelled light curve of 4U1907+09 with the Ariel V outburst superposed on it as determined on the basis of the disk model

in the rings vary between 10^9 to 10^{11} atoms cm^{-3}. The Ariel V light curve together with the modelled light curve superposed on it is shown in figure 1.

At certain phases in the light curve we see dips which are not present in the observed light curve. We postulate that the filling in of these dips in the observed light curve is due to the temporary formation of an accretion disk around the neutron star. This disk acts as a reservoir of material and results in a smoothening of the light curve at these points.

1.3. CONCLUSIONS FROM THE PRESENT STUDY

We can draw a few general conclusions from the above study. We find that if the source 4U1907+09 is a Be/X-ray binary (for a justification of this claim, see Saraswat, 1992) then the disk model is successful in explaining the X-ray light curve of this source to a reasonable degree of accuracy. We also see that the disk around the Be star is not a continuous one with a uniform density, but that the star ejects multiple rings, with different initial density values and gradients. Hence the circumstellar envelope is not a homogeneous medium but is segmented into rings, related to 'puffs' ejected from the Be star. The densities in these rings are $\sim 10^9$ to 10^{11} atoms cm^{-3}. Secondly, in addition to a Keplerian motion around the primary star, these rings also have a constant outflow velocity. However different rings have different outflow velocities. The shape of the X-ray light curve strongly depends on the outflow velocity of the gas ejected by the Be star. These velocities lie in the range 100–250 km s^{-1}.

Saraswat, P., 1992, Ph.D. thesis and references therein.

4. POLARIZATION

RECENT RESULTS FROM ULTRAVIOLET AND OPTICAL SPECTROPOLARIMETRY OF HOT STARS

K.S. BJORKMAN

University of Wisconsin, Space Astronomy Laboratory
1150 University Avenue, Madison, WI 53706 USA

Abstract. The first comprehensive linear polarization data on hot stars covering the spectral range from 1500 to 7600Å are presented. These results are based on recent observations made with the Wisconsin Ultraviolet Photo-Polarimeter Experiment (WUPPE), combined with ground-based observations from the Pine Bluff Observatory. Implications of the data for models of the circumstellar envelopes of hot stars are discussed, with particular emphasis on the surprising results found for the rapidly rotating Be stars. In particular, WUPPE discovered that the continuum polarization in Be stars decreases into the ultraviolet, which was not predicted by models prior to the observations. Time variability in the optical data is also discussed. Possible interpretations of these results are examined in the light of recent new models for Be star disks.

1. Introduction

The first flight of the Wisconsin Ultraviolet Photo-Polarimeter Experiment (WUPPE) aboard the Astro-1 mission in December 1990 provided an opportunity to obtain the first spectropolarimetry of hot stars in the ultraviolet. Ongoing ground-based support observations for this mission are being obtained at the University of Wisconsin's Pine Bluff Observatory (PBO), and have produced a valuable data set of optical spectropolarimetry for a broad range of hot stars. The combination of WUPPE and PBO data gives the first comprehensive linear polarization data covering the spectral range from 1500 to 7600Å.

2. Observations

2.1. OPTICAL SPECTROPOLARIMETRY

The optical observations were made with the PBO 0.9-m telescope and a dedicated spectropolarimeter. The PBO instrument is designed to be the complementary optical counterpart of WUPPE, measuring spectra and polarization from 3200 to 7600Å, with a spectral resolution of about 25Å. For details on the design of the instrument and the data reduction process, see Nordsieck *et al.* (1992). Observations of specific WUPPE targets were made contemporaneously at PBO. In addition, survey observations of a sample of Be stars have been made over the past 4 years, permitting a study of spectropolarimetric variability in Be stars.

L. A. Balona et al. (eds.), Pulsation, Rotation and Mass Loss in Early-Type Stars, 219–229.

2.2. Ultraviolet Spectropolarimetry

The ultraviolet observations were made in December 1990 using WUPPE, a 0.5-m f/10 Cassegrain telescope and spectropolarimeter; the design permits two orthogonally polarized beams to be recorded simultaneously on a dual intensified Reticon array. WUPPE obtains simultaneous spectra and polarization measurements, with a spectral resolution of about 12Å, from 1500 to 3200Å. For details about the instrumentation and design of WUPPE, see Nordsieck *et al.* (1993).

Several types of hot stars were observed with WUPPE, including OB supergiants (Taylor *et al.* 1991), Wolf-Rayet stars (Schulte-Ladbeck *et al.* 1992), OB main sequence stars (which were observed primarily as probes of interstellar polarization; Clayton *et al.* 1992), and Be stars (Bjorkman *et al.* 1991, 1993). Due to space limitations, only the results from the Be stars will be discussed in detail in these proceedings. The interested reader is referred to the above references for details on results of other types of hot stars.

Three Be stars were observed with WUPPE: ζ Tau, π Aqr, and PP Car. Contemporaneous PBO observations were made for ζ Tau and π Aqr (PP Car is a southern object). Figs. 1 and 2 show the combined WUPPE and PBO data for ζ Tau and π Aqr, respectively. Details regarding the data reduction process and interstellar polarization removal are found in Bjorkman *et al.* (1991). No interstellar removal was done for ζ Tau, since its interstellar polarization is small.

At the time of the Astro-1 mission, ζ Tau was in a typical high-polarization state, with a large polarization Balmer jump and strong line depolarization effects in the optical data. ζ Tau had shown this level of polarization over the previous two years, with only small variations at the level of about 0.1 per cent. π Aqr, on the other hand, was in a very low polarization state to which it had declined over the previous 18 months, from a peak of 1.5 per cent polarization (at around 4000Å) to the low of about 0.6 per cent. The optical data show that no polarization Balmer jump was present in π Aqr at the time of the WUPPE observations. The variability of π Aqr is discussed further in a later section.

3. UV Continuum Polarization of Be Stars

The typical models used to make predictions of the shape of polarization vs. wavelength assume that the polarization is a result of electron scattering of stellar flux within the circumstellar disk, reduced by the competing effects of hydrogen opacity within the disk material and dilution by unpolarized emission from the disk. Electron scattering alone produces a per cent polarization %P, vs. wavelength λ, curve which is flat, but the hydrogen bound-free opacity produces the characteristic "sawtooth" shape of the standard models. These models assume single scattering, i.e. that the elec-

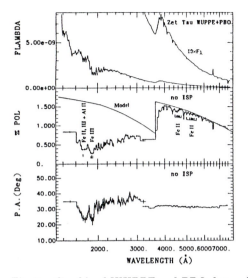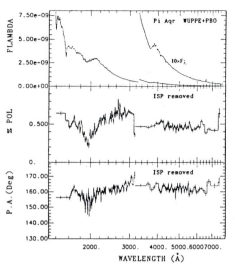

Fig. 1. Combined WUPPE and PBO data for ζ Tau. The upper panel shows the flux, the middle panel shows the % polarization, and the lower panel shows the polarization position angle. The WUPPE data are binned to a constant error of 0.02% and the PBO data to 0.01%. The solid line in the center panel indicates typical model predictions of the UV polarization continuum. (Figure from Bjorkman et al. 1991.)

Fig. 2. Combined WUPPE and PBO data for π Aqr. The panels are the same as shown for ζ Tau. Both the WUPPE and PBO data have been binned to a constant error of 0.02%. An estimate of the interstellar polarization has been removed from these data. Note the lack of a polarization Balmer jump, as compared with ζ Tau. (Figure from Bjorkman et al. 1991.)

tron scattering optical depth $\tau_e < 1$. Models prior to the flight of WUPPE predicted that the polarization would rise dramatically into the UV (e.g. Poeckert & Marlborough 1978; Cassinelli, Nordsieck & Murison 1987).

Several key features of the polarization data for ζ Tau and π Aqr are apparent, especially when contrasted with model predictions as in Fig. 1, which shows a comparison of the data for ζ Tau with a model prediction from Cassinelli et al. (1987). The "sawtooth" shape of the predicted polarization vs. wavelength in Fig. 1 is characteristic of all the existing models, and serves to illustrate the difference between the data and the models.

Firstly, although the general shape of the continuum polarization vs. wavelength curve matches fairly well with the model predictions in the optical, the UV polarization disagrees with the models in the case where a large polarization Balmer jump is present. While the models predicted there should be a strong increase in polarization toward shorter wavelengths in the UV, the data actually show a constant or slightly declining continuum polarization with decreasing wavelength.

Secondly, there are strong, broad UV polarization dips around 1700 and 1900 Å, corresponding to the location of numerous Fe lines in the spectra of both stars. Note that the 1700 Å dip is more pronounced in ζ Tau than in π Aqr, which may be an indication of differences in the density or temperature of the circumstellar material around the two stars. There is also a rotation of the polarization position angle across the UV polarization dips, but not across the polarization Balmer jump.

Thirdly, the optical data show depolarization effects across the Balmer hydrogen lines, as has been reported by others, as well as depolarization of Fe II lines in ζ Tau (Fig. 1). Since the Fe lines are much weaker than the hydrogen lines in the spectrum, the dilution effect from disk emission in the Fe lines is smaller, and hence the reduction in polarization may be instead an "attenuation" effect arising from the actual removal of polarized light from the line of sight. This implies that the Fe absorption occurs primarily in the disk, close to the star.

3.1. Implications for Models of Be Star Disks

The differences between the observations and the predictions point out several changes which must be made in the next generation of models for Be star polarization. Clearly the metal line opacities in the polarizing region must play an important role in UV (and also optical) %P vs. λ characteristics, suggesting that, in particular, Fe line opacities must be accounted for in any model for the UV polarization of Be stars. Line blanketing by Fe II and Fe III lines, as well as other metal lines, is apparently quite important in determining the wavelength dependence of polarization, especially in the UV. Initial simple models incorporating Fe line opacities show that this approach can probably come close to matching the observed UV polarization (A.D. Code, private communication).

Several other factors which could affect the wavelength dependence of the continuum polarization may also have to be considered—for example, the effect of gravity darkening. The von Zeipel theorem predicts that the equatorial regions of a rapidly-rotating star will have a cooler effective temperature than the polar regions. If the scattering region producing the polarization is located primarily in an equatorial disk, then the scattered radiation originates primarily in the cooler equatorial regions of the star, while the direct (unscattered) radiation originates primarily in the polar regions. Hence, this gravity darkening could change the wavelength dependence of the measured polarization, which is just the ratio of the cooler scattered flux to the hotter direct flux. However, while it does decrease the continuum polarization in the UV, gravity darkening alone does not appear to be adequate to explain the discrepancies between the observations and the model predictions (Bjorkman & Bjorkman 1994).

A potential difficulty for new models will be to explain the rotation in position angle across the Fe lines in the UV, which was seen in all three Be stars observed with WUPPE. This may be a result of optical depth effects in the Fe lines, but the details of how this translates into the disk geometry and density distribution must be worked out.

4. Optical Polarimetric Variability

Examination of the PBO optical data set provides information on the polarimetric variability of Be stars. The best studied case, with 40 observations from August 1989 through December 1992, is the Be star π Aqr. Fig. 3 shows the average polarization of π Aqr as a function of time, along with the variations in the equivalent width of the Hα emission line. Note that at least two, and possibly three, polarimetric "outbursts" were observed in π Aqr over the span of about 4 years. When the polarization was high, π Aqr showed a strong polarization Balmer jump similar to that seen in ζ Tau (see Fig. 1). When the polarization was low, π Aqr showed no polarization Balmer jump, just as is seen in Fig. 2.

The 1991 outburst (centered on JD 2448477), which was particularly well covered temporally, gives a good indication of typical time scales for such variations. The rise time of the outburst was about 3 months, with a comparable falloff time after the peak polarization was reached. Note also that the Hα equivalent width showed very little change during the time that the polarization level was changing dramatically. This implies that most of the variability is occurring at small radii, and not throughout the large Hα emitting region.

4.1. CONTEMPORANEOUS UV WIND LINE VARIABILITY

Examination of UV spectra of π Aqr from the archives of the International Ultraviolet Explorer (IUE) satellite as well as IUE data taken contemporaneously with the Astro-1 mission show an interesting change in the UV wind line profiles just prior to one of the polarimetric outbursts. Fig. 4 shows the IUE spectra around the N V ($\lambda\lambda$1238, 1242Å) region. As shown in this figure, on 21 November 1990 N V was in absorption with weak high velocity components. (This observation was taken in support of the ROSAT X-ray satellite All-Sky Survey. Note that Meurs *et al.* (1992) reported from the X-ray survey data that π Aqr had a value of $\log L_x/L_{bol} \approx -6.5$, which is strong for a B star, and that the spectrum was fairly hard.) Fifteen days later (during the Astro-1 mission) on 6 December 1990, the IUE data show that N V had developed a strong, low velocity ($\approx$ 200 km s^{-1}) discrete absorption component (DAC) in both components of the doublet line. Ten days later the DAC's in the N V had weakened considerably.

The development and disappearance of the low velocity DAC's in N V corresponded with the onset of a polarimetric outburst, which began some-

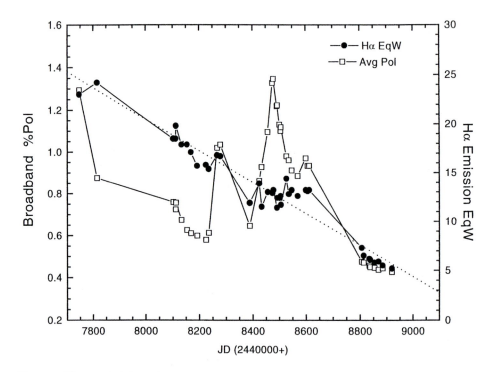

Fig. 3.   Time variability of π Aqr. The open squares are the polarization level averaged over the entire wavelength range of the PBO observations; the filled circles are the Hα line emission equivalent width in Å. Note the major polarimetric outburst around JD 2448477, as well as possible outbursts around JD 2448260 and JD 2447750. The Hα equivalent width changed little during these times.

time between 30 November 1990 and 10 December 1990. While there is not sufficient temporal coverage of IUE archival data to say whether such developments always occur prior to a polarimetric outburst, it is certainly a tantalizing hint that there may be a connection between changes in the wind and changes in the disk. The fact that the DAC developed at low velocities, more typical of disk velocities, also supports the idea that these wind changes may be associated more closely with the disk than with the polar wind. The WCD model of disk formation around rotating stars (see J.E. Bjorkman, these proceedings) would predict such a connection, especially in superionized lines such as N V.

4.2. INTERPRETING THE VARIABILITY OF π AQR

The observed timescale for polarimetric outbursts of π Aqr provides some potentially difficult constraints on models which might be proposed as the underlying cause of the variability. Since the outburst is fairly slow develop-

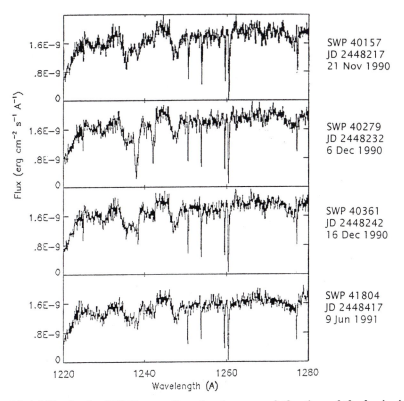

Fig. 4. Variability in the N V line profiles of π Aqr around the time of the beginning of a polarimetric outburst. The data are from the IUE satellite. The polarimetric outburst began between 30 November and 10 December 1990, or around the time of the development of the low-velocity discrete absorption component seen in the second panel. The IUE image number, Julian Date, and calendar date of the IUE observations are listed beside each panel.

ing, whatever produces the changing polarization must be stable over times of several months.

Assuming typical velocities of 100 km s^{-1} in the disk, the flow time for a density perturbation in the disk to move a distance of 1 stellar radius is only about 1 day. In the wind, where velocities are more typically 1000 km s^{-1}, the equivalent flow time would only be about 2 hours. This rules out any changes related to flow times in the disk or in the wind. Also, for the spectral type of π Aqr, the rotation timescale is only about 2 days, assuming an inclination angle of 90°.

Clearly, none of these time scales are long enough to explain the sustained and apparently smooth polarimetric changes over time scales of months. This would seem to rule out anything as simple as density perturbations (blobs) moving through the wind or disk. Also, the lack of much change in the Hα equivalent width suggests that the polarimetric changes are primarily due to changes in the inner part of the circumstellar envelope and that these changes do not affect the much larger region that produces the Hα emission.

Other suggestions to explain the polarimetric variations include increased mass-loss rate from the star (perhaps driven by an increased UV continuum flux, evidence for which is seen in the IUE data), which increases the mass in the disk due to the WCD mechanism, or a one-armed spiral density wave in the disk (which has also been proposed to explain V/R variations in Be stars). However, details of such pictures are not well enough developed yet to determine whether they might work in the case of π Aqr. There are also some potential problems with explaining the variability seen in the UV wind line profiles near the times of the polarimetric outbursts.

5. Future Work

The sample of hot stars with UV spectropolarimetric measurements will be increased by the flight of the Astro-2 mission, currently scheduled for launch in December 1994. For the Be stars, the effects of spectral type, rotation rate, and shell vs. non-shell classification on the UV polarization will be investigated. New predictions of the UV continuum polarization will be developed to include the effects of iron line opacities as well as to test predictions of new models for Be stars, such as the WCD model.

In the optical, the complete sample of Be stars observed spectropolarimet-rically at PBO will be analyzed to investigate the prevalence of polarimetric outbursts and to determine whether the time scales observed in π Aqr are typical for such outbursts. Analysis of the optical and UV polarization characteristics can give useful information about the nature of the circumstellar envelopes which is complementary to what can be learned from spectroscopy and photometry. Ideally, one would prefer to have a complete set of contemporaneous spectroscopy, photometry and spectropolarimetry to analyze concurrently.

Acknowledgements

I thank the WUPPE team, and the crew and NASA support personnel for Astro-1, for their hard work and dedication, without which these observations could never have been obtained. I thank the PBO team for assistance with ground-based observations and data reduction. Special thanks are due to Ken Nordsieck, Marilyn Meade, Brian Babler, and Art Code. Carol Grady provided much of the IUE data. Jon Bjorkman and Joe Cassinelli provided useful comments and discussion. This work has been supported by NASA grants NAS5-26777 and NAG5-1740 to the University of Wisconsin.

References

Bjorkman, J.E. and Cassinelli, J.P.: 1993, *Astrophys. J.* **409**, 429.

Bjorkman, K.S. and Bjorkman, J.E.: 1994, in preparation.

Bjorkman, K.S., Nordsieck, K.H., Code, A.D., Anderson, C.M., Babler, B.L., Clayton, G.C., Magalhhães, A.M., Meade, M.R., Nook, M.A., Schulte-Ladbeck, R.E., Taylor, M. and Whitney, B.A.: 1991, *Astrophys. J. Let.* **383**, 67.

Bjorkman, K.S., Meade, M.R., Nordsieck, K.H., Anderson, C.M., Babler, B.L., Clayton, G.C., Code, A.D., Magalhães, A.M., Schulte-Ladbeck, R.E., Taylor, M. and Whitney, B.A.: 1993, *Astrophys. J.* **412**, 810.

Cassinelli, J.P., Nordsieck, K.H., and Murison, M.A.: 1987, *Astrophys. J.* **317**, 290.

Clayton, G.C., Anderson, C.M., Magalhães, A.M., Code, A.D., Nordsieck, K.H., Meade, M.R., Wolff, M.J., Babler, B., Bjorkman, K.S., Schulte-Ladbeck, R.E., Taylor, M. and Whitney, B.A.: 1992, *Astrophys. J. Let.* **385**, 53.

Meurs, E.J.A., Piters, A.J.M., Pols, O.R., Waters, L.B.F.M., Coté, J., van Kerkwijk, M.H., van Paradijs, J., Burki, G., Taylor, A.R. and de Martino D.: 1992, *Astron. Astrophys.* **265**, L41.

Nordsieck, K.H., Babler, B., Bjorkman, K.S., Meade, M.R., Schulte-Ladbeck, R.E. and Taylor, M.J.: 1992, in Drissen, L., Leitherer, C. and Nota, A., eds., *Nonisotropic and Variable Outflows from Stars*, Atron. Soc. Pacific:San Francisco, 114.

Nordsieck, K.H. *et al.*: 1993, in Fineschi, S., ed., *SPIE Proceedings, Vol 2010, X-Ray and Ultraviolet Polarimetry*, SPIE Press, in press.

Poeckert, R. and Marlborough, J.M.: 1978, *Astrophys. J. Suppl.* **38**, 229.

Schulte-Ladbeck, R.E., Nordsieck, K.H., Code, A.D., Anderson, C.M., Babler, B.L., Bjorkman, K.S., Clayton, G.C., Magalhães, A.M., Meade, M.R., Sheperd, D., Taylor, M. and Whitney, B.A.: 1992, *Astrophys. J. Let.* **391**, 37.

Taylor, M., Code, A.D., Nordsieck, K.H., Anderson, C.M., Babler, B.L., Bjorkman, K.S., Clayton, G.C., Magalhães, A.M., Meade, M.R., Schulte-Ladbeck, R.E. and Whitney, B.A.: 1991, *Astrophys. J. Let.* **382**, 85.

Discussion

Smith: I would like to add one more logical possibility to your list of speculations for π Aqr's high-energy behavior in 1991. Imagine that you had a distribution of hot spots on the surface of π Aqr at that time that locally mimic an O-star's atmosphere. Then, locally, these spots would be expected to set up O-star-like winds, and therefore DAC's in the UV resonance lines. They would also tend to give you (according to J. Cassinelli's paper) slightly enhanced X-rays than those expected for a B2 star.

Bjorkman: Yes, that might be a possibility. I think the details of whether such a picture could explain the observations of DAC's in conjunction with polarization outbursts would have to be worked out, but that is true for all of these speculations.

Stee: Have you used single scattering for your computation? Multi-scattering close to the star (where the density is higher) may be a way to decrease polarization. Also, did you find any correlation between $v \sin i$ and polarization?

Bjorkman: Our calculations so far are quite simple and only assume single scattering. However, Barbara Whitney at CfA has done some preliminary

Monte-Carlo simulations which account for multi-scattering in an optically thick disk. She finds that although the optically thick case does produce somewhat lower overall polarization, the wavelength dependence doesn't change very much. So while the multi-scattering may help to lower the polarization some, the preliminary indications are that it will not explain the non-rising UV continuum polarization.

Hubert: Did you observe any important variability in ζ Tau? This star has begun a new sequence of activity (V/R variations and RV variations) since 1991.

Bjorkman: We have been monitoring ζ Tau for the past 4 years. During that time we have seen small-scale (at about the 0.1 per cent level) variability in the polarization level, but nothing comparable to the outburst changes we have seen in π Aqr. We are continuing to monitor it, however.

Ghosh: You have observed an increase of the equivalent width of N V during the outburst of an edge-on Be star. We have also observed an increase in the equivalent width of the absorption lines of Si II, Fe II, and N II (in the optical) in the pole-on Be star, μ Cen. Do you think that matter exists in the polar region, which we have proposed, based on our observations?

Bjorkman: If these lines are seen in absorption and if the star is truly pole-on, then obviously there must be some material in the polar region. Since the equivalent width increased, the amount of material seen through the column to the star must have increased. This does not necessarily mean, however, that there is more material in the polar regions than in the disk.

Peters: I would not expect the von Zeipel effect to be very important. According to current thought the Be-shell stars are viewed equator-on. In the case of ζ Tau, $v \sin i = 220 \approx v_{eq}$, which implies that $v_{eq}/v_{crit} \approx 0.4$. Therefore, the decreases in the temperature of the equator would be extremely small.

Bjorkman: You are correct that the temperature difference is probably small for ζ Tau, and thus unlikely to produce a large effect. The model results which I showed here were just intended to illustrate that gravity darkening can have an effect on the UV continuum polarization, and that it goes in the right direction to help with the discrepancy with the models. However, as I said, we do not think that gravity darkening alone will provide nearly enough of a change to explain the WUPPE observations, and the Fe line opacities will definitely play the dominant role.

Hummel: I have two questions. First, why did you start your observations with shell-type Be stars?

Bjorkman: The fact that WUPPE observed 3 Be shell stars and no non-shell stars was purely a coincidence. We had originally planned to observe

8 Be stars, of which some were Be and some were Be-shell. However, due to some problems during the Astro-1 mission, many planned observations were lost, and unfortunately the small number of observations that we did get included only shell stars. We hope to remedy that on Astro-2, which will fly in December 1994 if the current schedule holds.

Hummel: Second, what are the problems with associating the UV data to the density wave model?

Bjorkman: While the density wave model has potentially the right timescale for the polarization changes, the DAC's in the wind lines were observed at blue-shifted velocities of about 200 km s^{-1}. The spiral density wave model requires a Keplerian disk, which implies that the disk is not expanding. If these DAC's were caused by such a spiral density wave in the disk, they should not be blueshifted, but should occur close to zero velocity.

Baade: If the detection of variability is a stated goal, it may be advantageous to look at stars with weak Hα emission. For instance, there are long series of spectra of μ Cen which show that it underwent little Hα outbursts at a rate of up to one per 10–20 days. This is in line with the polarimetric and photometric monitoring of ω Ori by Hayes & Guinan, who found numerous polarization outbursts with a rise time of 1–2 days (as in μ Cen). I happened to obtain one Hα profile of ω Ori during such a high polarization state, and it had anomalously broad wings, which, if attributed to scattering, are consistent with the polarization data. Hα outbursts of μ Cen also seem to start with very broad Hα emission.

Bjorkman: It would be very interesting to try and correlate the polarimetric behavior with detailed line profile observations. Our current instrument does not provide the necessary spectral resolution to see line profile changes in any detail. I think it would be quite useful to coordinate observations of spectropolarimetry with high resolution Hα observations.

Spectropolarimetric determination of circumstellar disc inclinations

KENNETH WOOD

Department of Physics and Astronomy, The University, Glasgow, Scotland

1. Thomson scattered spectropolarimetric line profiles

Inference of the density and velocity structure of rotating/expanding circumstellar discs/winds is of considerable interest in the understanding of stellar mass loss. High resolution line spectropolarimetry creates the possibility of diagnosing such envelope structure much more fully than broad band polarimetry or high resolution spectrometry alone since each element of the scattered spectropolarimetric profile picks out the element of the envelope with the appropriate Doppler shift and provides orientation information on it. This problem has been formulated in detail by Wood, et al (1993) for scattering of a finite width line in a flattened envelope – the spectral shape of the scattered Stokes fluxes being determined by *isowavelength–shift* contours or surfaces which give the relative wavelength shift of the scattered radiation at different regions in the disc. It was also shown how, in the case of a narrow stellar line scattered in a rotating or expanding flat disc with a simply parametrised density and velocity structure (ignoring the smearing effect of electron thermal motions), it is possible to infer the system inclination and structure model parameters from the resulting spectropolarimetric line profile. This poster presents a method for determining the disc inclination from analysis of the scattered profiles – a parameter which cannot be determined uniquely from spectrometry alone – and thus illustrates the powerful diagnostic potential of high resolution line spectropolarimetry.

2. Determination of disc inclination

It was shown by Wood & Brown (1993) that the inclination of a rotating disc, in which the electron thermal velocity was negligible, could be determined from the ratio of the scattered polarised to total fluxes close to the centre of the scattered line, viz,

$$\frac{F_\nu^Q(\alpha \approx 0)}{F_\nu^I(\alpha \approx 0)} = \frac{-\cos^2 i}{1 + \sin^2 i} , \tag{1}$$

L. A. Balona et al. (eds.), Pulsation, Rotation and Mass Loss in Early-Type Stars, 230–231.
© 1994 IAU. Printed in the Netherlands.

where $\alpha = (\lambda - \lambda_0)/\lambda_0$ is the dimensionless wavelength shift from line centre, λ_0. The fact that i can be determined results from the fact that use of a scattered spectral line profile, as opposed to broad–band measurements – cf Brown & McLean (1977), allows different parts of the disc to be picked out. Setting $\alpha \approx 0$ picks out a line in the disc in the plane containing the observer and the disc axis along which the scattering angle is either $\pi/2 + i$ or $\pi/2 - i$, giving different values of the scattered total to polarised fluxes.

For a rotating and expanding disc the $\alpha = 0$ contour depends on the disc velocity in a complex way prohibiting a simple expression for determining the disc inclination. The inclusion of random thermal velocities is such that the regions of the disc contributing to the scattered flux at a given wavelength are no longer discrete contours and it would appear that the disc inclination may only be inferred for a rotating cold disc according to Eq. 1. However, by considering the total fluxes under the scattered profiles it is possible to determine the inclination uniquely from,

$$\frac{F^Q}{F^I} = \frac{\sin^2 i}{2 + \sin^2 i} , \qquad (2)$$

where F^I and F^Q are the integrated scattered Stokes fluxes with the scattered U flux averaging to zero across the line – as would be expected from the polarimetric cancellation properties of an axisymmetric disc. Since only Doppler wavelength redistribution was considered in this scattering analysis the result of Eq. 2 is an obvious one since it is precisely the ratio of the scattered total to polarised broad–band fluxes for any disc as derived by Brown & McLean.

So, providing the stellar line is narrow so that, apart from the region within the direct line core, the direct and scattered fluxes can be separated then Eq. 2 provides a means of determining disc inclinations irrespective of their velocity profiles. The beauty of this result is that it holds even when the Doppler redistribution due to scattering off thermal electrons is included.

As it stands the analysis presented in this poster for the determination of stellar inclinations is directly applicable to the scattering of *narrow* photospheric lines in hot, moving circumstellar discs. Forthcoming advances in high resolution CCD spectropolarimetry and the further development of this theoretical framework for interpreting spectropolarimetric line profiles will yield a novel method for determining the structure of stellar winds.

References

Brown, J.C., McLean, I.S. : 1977, *Astron. Astrophys.*, **57**, 141
Wood, K., Brown, J.C., Fox, G.K. : 1993, *Astron. Astrophys.*, **271**, 492
Wood, K., Brown, J.C. : 1993, *Astron. Astrophys.*, in press

THE POLARIZATION OF Hβ IN γ CAS

Y.M. JIANG, G.A.H. WALKER, N. DINSHAW and J.M. MATTHEWS
Department of Geophysics and Astronomy, University of British Columbia,
Vancouver, B.C., V6T 1Z4, Canada

High resolution (0.15 Å/pixel) spectropolarimetry across the Hβ emission line of the Be star γ Cas is presented (solid line in Figure B), as well as the intensity profile (Figure A). Earlier observations of the linear polarization across the emission lines (P_L) in Be stars (Coyne 1975, Poeckert 1975) found that P_L is inversely proportional to the total intensity of the emission line. However, our data clearly show that the width of the polarization profile is much wider than that of the emission line. The dotted line in Figure B gives P_L, based on the relationship of:

$$P_L = \frac{I_A}{I_{total}} P_C \tag{1}$$

which follows directly from Coyne (1975), where I_A is the intensity of the underlying Hβ absorption feature, I_{total} the observed intensity profile, and P_C the polarization in the adjacent continuum. The underlying stellar profile of Hβ (dotted line in Figure C), adopted from models by Hubeny (1988), gives a good match to the wings of the depolarization profile.

We have modified the Be star model developed by Poeckert and Marlborough (1978) for γ Cas, by adding the 6th level for hydrogen atoms in the envelope, representing the photosphere of the central star by a Hubeny atmosphere model (1988), and correcting the geometry of the calculation of polarization to properly take into account the finite size of the central star.

We find
- Linear polarization profiles across emission lines are not only affected by the emission from the disk, but also by the profile of the underlying stellar absorption feature, which in turn provides information about the central star.
- The modified Poeckert/Marlborough model gives a good fit to the Hβ emission line profile, but only a poor fit to the polarization profile, probably because of their assumption that polarization across the emission lines is only affected by the additional emission.

L. A. Balona et al. (eds.), Pulsation, Rotation and Mass Loss in Early-Type Stars, 232–233.
© 1994 *IAU. Printed in the Netherlands.*

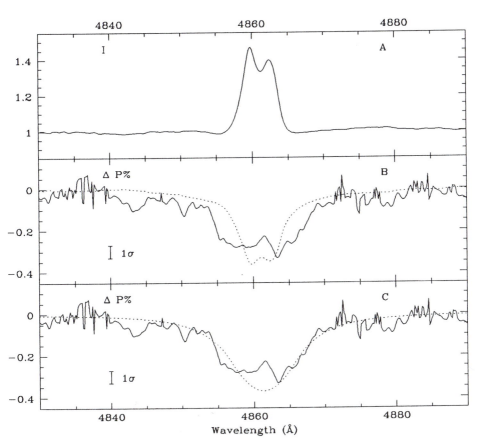

Figure A The emission line intensity profile of Hβ on γ Cas, Sep. 23, 1990.

Figure B The solid line is the difference between the polarization within Hβ and the polarization in the adjacent continuum. The dotted line is the polarization predicted by Equation (1), with the stellar Hβ absorption feature given by Hubeny NLTE atmosphere model with T_{eff} = 30000 K, log g = 4.0, $v \sin i$ = 325 km/s. P_C is taken to be constant, P_C = 0.75%.

Figure C The adopted stellar Hβ profile (see B) scaled to 1.45% superimposed on the differential polarization profile.

References

Coyne,G.V.: 1975, *I.A.U. Symp.* **70**, Be and shell stars, 233
Hubeny,I.: 1988, *Computer Physics Communications* **52**, 103
Poeckert,R.: 1975, *Astrophysical Journal* **196**, 777
Poeckert,R. and Marlborough,J.M.: 1978, *Astrophysical Journal* **220**, 940

LONG-TERM SPECTROSCOPIC AND POLARIMETRIC VARIATIONS OF THE Be STAR κ Dra

J. ARSENIJEVIC, S. JANKOV, S. KRSLJANIN
Astronomical Observatory, Belgrade, Yugoslavia

and

A.M. HUBERT, H. HUBERT, M.T. CHAMBON, and
J. CHAUVILLE, M. FLOQUET, A. MEKKAS
Observatoire de Paris, section d'Astrophysique de Meudon,
URA 335 du CNRS, F-92195 Meudon Cedex, France

1. Introduction

The bright Be star κ Dra (HD 109387, B5 III, v.sini=249 km/s) is one of the rare mid B type stars which have presented FeII emission lines during strong emission phases. Spectroscopic variations were previously reported by many authors over different time scales: years, days and hours. Juza et al. (1991) proposed that κ Dra was the primary component of a binary system with Porb. = 61.55 days. Short-term and long-term variations in the linear optical polarization were also reported by Arsenijevic et al. (1986). We have tried, with our extensive joint material, to search for correlation between long-term spectroscopic and polarimetric variations with the aim of obtaining information on the behaviour of the circumstellar envelope.

2. Data and results

This star has been regularly monitored in Hα at the Haute-Provence Observatory since 1953 at low, and since 1960 at high dispersion, and for its optical polarization, at the Belgrade Observatory, since 1979.

2.1. SPECTROSCOPY

The Hα emission line has presented a long term variation (24-25 years) of its equivalent width (defined here as the surface of emission above the stellar continuum); this time scale is an intermediate value between those obtained respectively by Jessup (1932) and McLaughlin (1949). For this determination additional data found in the literature were used.

Its profile has generally a rapidly changing complex structure which can be described as an asymmetric triple or quadruple emission with two prominent peaks. The variation of the separation of these prominent peaks shows an inverse correlation with the variation of the equivalent width (Δv(peaks)

L. A. Balona et al. (eds.), Pulsation, Rotation and Mass Loss in Early-Type Stars, 234–235.

= 206 km/s at the minimum of emission around 1980, and Δv(peaks) = 69 km/s at the maxima in 1961-1962, and in 1985-1986).

The V/R variation relative to the two prominent peaks was analysed in term of an orbital motion effect, using a programme written by the author J.C. based on Stellingwerf's PDM technique. The most probable period of 61.5 days has been found, in a very good agreement with the determination P = 61.55 days obtained by Juza et al. (1991) from numerous RV data.

2.2. POLARIMETRY

The daily mean value of the intrinsic optical polarization in percent obtained from 1979, exhibits a close correlation with the equivalent width of the $H\alpha$ emission line. As there is a lack of simultaneous observations we have to therefore assume that the minimum of polarization occured during the minimum of emission in 1979-1980, and the maximum of polarization during the probable maximum of emission in 1985-1986. As for ω Ori (Barker, 1986) and π Aqr (McLean, 1979) there is evidence for κ Dra of an increase of polarization associated with the development of a strong Be phase.

The position angle of polarization which increased from 5° to 22°, during the 1979-1982 period, has remained roughly constant since 1982. The polarization variation in the U,Q plane seems collinear, indicating an axisymmetric distribution of scattering.

3. Conclusion

Over the interval of time 1979-1986, the equivalent width of the $H\alpha$ emission line, after correction for underlying photospheric profile, increased by about a factor 3 and the percentage of the intrinsic polarization by about a factor 5-6. Using the statistical relation of Dachs et al. (1992), the effective emitting envelope of κ Dra is found to have increased from 4 to 6.5 stellar radii. According to polarization results, the internal layers of the envelope have an axial symmetry, but the outer $H\alpha$ emitting region, affected by the presence of the companion (V/R variation versus the orbital period), probably is associated with nonaxisymmetric external layers.

References

Arsenijevic J., Jankov S., Djurasevic G., Vince I.: 1986, *Bull. Obs. Astr. Belgrade* **136**, 6

Barker P.K.: 1986, *Publications of the ASP* **98**, 44

Dachs J., Hummel W., Hanuschik R.W.: 1992, *Astronomy and Astrophysics, Supplement Series* **95**, 437

Jessup M.K.: 1932, *Astrophysical Journal* **76**, 75

Juza K., Harmanec P., Hill G.M., Tarasov A.E., Matthews J.M., Tuominen I., Yang S.: 1991, *Bull. Astron. Inst. Czechosl.* **42**, 39

McLaughlin D.B.: 1949, *Pub. Astron. Obs. Univ. Michigan* **4**, 175

McLean I.S.: 1979, *Memoirs of the RAS* **186**, 265

COMMON PROPERTIES OF SOME Be STARS OPTICAL
POLARIZATION PARAMETERS

J. ARSENIJEVIĆ, S. MARKOVIĆ-KRŠLJANIN, A. KUBIČELA and
S. JANKOV

Astronomical Observatory, Volgina 7, 11050 Beograd, Yugoslavia

Abstract. The long-term changes of the intrinsic polarization percentage are evident in the stars o And, γ Cas, 88 Her (during the period 1974-1992), κ Dra (1979-1992) and BU Tau (1986-1992). The amplitudes of the polarization percentage variations are not more than a half of the percent. The changes of the position angles are within an interval of about 30 degrees.

It is generally accepted that the presence of intinsic polarization of Be stars confirms the basic assumption of an extended non-spherically symmetric envelope, as well as the origin of polarization in scattering of non-polarized stellar radiation on free electrons in the envelope.

The long-term polarization variations, for years to decades have been examined in a very limited number of stars. The behaviour of polarization parameters during all activity phases (B, Be, shell ...) is not known with certainty.

Having in mind the importance of polarization data of Be stars, on one side, and ability to observe in a very long series with the same telescope available at Belgrade Observatory, on the other, a program of stydying long-term optical polarization changes has been set up in 1974. The aim was to obtain reliable long-term data on polarization changes in V spectral region and examine these changes during different activity phases of Be stars.

Polarimetric observations at Belgrade Observatory were carried out with the 65-cm Zeiss refractor and the stellar polarimeter (Kubičela *et al.* 1976) which was modified in 1979 to enable one to obtain digital magnetic records suitable for further computer processing. The measurements were done in the V spectral region. Under "one measurement" we understand up to 8 one-minute polarimetric sine-wave signals phase-averaged. The typical standard deviation of one 8 – minute measurement is 0.07% for Stokes parameters. The interstellar components were estimated or used from the literature. For the star o And the observed polarization is presented.

In Fig. 1 the individual measurements of the observed polarization percentage (a) and corresponding position angles (b) of the star o And are shown. Figure 2a and b displays the intrinsic polarization percentage and position angles of the star 88 Her. The data for star κ Dra are presented in this Proceedings. The part of γ Cas data are published in the paper of Arsenijević *et al.* (1990). The BU Tau data are published in the article of Arsenijević *et al.* (1993).

From the sample of our polarimetric data of five stars and published data of V magnitude and H_α emission strength existing in the literature for the same star, one can come to the followinig conclusion:

The intrinsic visual polarization percentage for all five observed stars was smaller than 1 percent. The long-term variations of both intrinsic polarization parameters exsist for all five stars. The amplitude of the polarization percentage

L. A. Balona et al. (eds.), Pulsation, Rotation and Mass Loss in Early-Type Stars, 236–237.

changes are not higher than 0.5 percent. The amplitude of the position angle variations are under 30 degrees. The anticorrelation between polarization percentage and visual brightness was noticed always, when data existed. The correlation between polarization percentage and H_α emission strength is firmly confirmed for some of our program stars. More data are needed for the final conclusion.

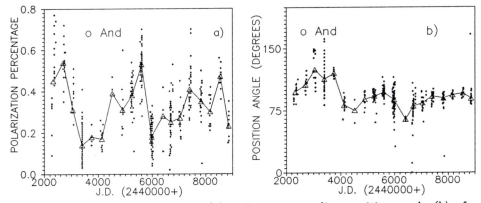

Fig. 1. *Polarization pertencage (a) and corrensponding position angle (b) of the star o And during the interval of time 1974 – 1992. Annual mean values are denoted by triangles.*

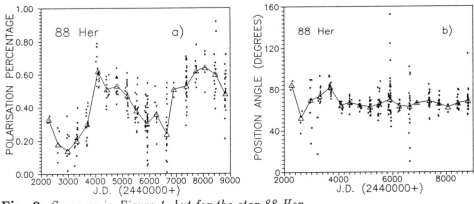

Fig. 2. *Same as in Figure 1, but for the star 88 Her*

REFERENCES

Arsenijević, J., Jankov, S., Vince, I., Djurašević, G.: 1987, *Bull. Obs. Astron. Belgrade* **137**, 49.

Arsenijević, J., Jankov, S., Djurašević, G.: 1990, in *XV SPIG*, ed. D.Veža, Inst. Phys. Univ. Zagreb, p.374

Arsenijević, J., Marković-Kršljanin, S., Jankov, S., Kršljanin, V. and Popović, L.: 1993, Publ. Obs. Astron. Belgrade **44**, 77.

Kubičela, A., Arsenijević, J., Vince, I.: 1976, *Publ. Dept. Astr. Univ. Belgrade* **6**, 25.

5. Be STARS:

SPECTROSCOPY AND PHOTOMETRY

PHOTOSPHERIC ACTIVITY IN SELECTED Be STARS:

λ Eri AND γ Cas

MYRON A. SMITH

IUE/CSC Observatory, 10000A Aerospace Rd., Lanham-Seabrook MD 20706, USA.

Abstract. Recent observations of rapid variations in optical He I lines, X-rays, and FUV wavelengths in the prototypical classical Be stars λ Eri and star γ Cas hint that the violent processes occur on the surfaces of these stars almost all the time. We suggest that of these phenomena show greater similarities with magnetic flaring than any other process thought to occur on stars.

1. Introduction

In retrospect, Peters' (1986) observation of the development of a high velocity feature in the λ6678 profile of μ Cen (B2e) was a call for the Be community to recognize that mass ejections of the surfaces of Be stars can occur discretely and violently over very rapid timescales. The extension of monitoring observations over several wavelength regimes, together with the development of multi-line spectroscopic detectors has brought necessary new tools to the search for the instability mechanism(s) in the atmospheres of classical Be stars. A reasonable strategy for investigating the local sites of mass ejections is to focus on a mild Be star the surface of which is visible most of the time because the outburst duty cycle is low, and a very active Be star seen at intermediate inclination. Examples of each of these categories are λ Eri (near edge-on, B2e; outburst duty cycle ~20 per cent) and γ Cas (B0.5e; in emission for 50 years), and I will confine my discussion to them. In addition to work by our group at Goddard, other teams have initiated *rapid* spectroscopic variability programs on these stars, e.g. in the U.S. (Gies, Peters), Canada (Kambe, Walker), Japan (Hirata), and India (Anandarao, Ghosh).

2. λ Eri's He I Line "Dimples"

In addition to regular line profile variations (*lpvs*) generally attributed to NRP, the He I λ6678 line of λ Eri shows an almost constant erratic activity which can be fit into a small group of patterns known as spectral transients (e.g. Smith 1989). At least 60 per cent of these events appear as central absorptions and weak flanking emissions known as "dimples" along the λ6678 profile. An additional ~10 per cent are "type d" or pure absorption events which are probably related to dimples. These features have a frequency of 0.2±0.05 events hr^{-1} and a duration of 2-4 hrs. They show a slow drift to the red during their lifetime. Occasionally, dimples have been observed to "come and go" over several hours, all the while moving along

L. A. Balona et al. (eds.), Pulsation, Rotation and Mass Loss in Early-Type Stars, 241–253.

the profile consistent with the stellar rotation rate, as if caused by a rooted active spot on the surface. With the exception of two possible simultaneous dimples in the $\lambda4922$ line, dimples have been observed only in $\lambda6678$. Smith & Polidan (1993; SP) note that the $\lambda6678$ line in five other Be/Bn stars with $v\sin i \geq 250$ km s^{-1} show dimples. In a simultaneous optical/IUE campaign, SP discovered that 10–20 per cent weakenings of the C IV and N V resonance doublets were correlated with the appearance of $\lambda6678$ dimples. This discovery suggested plasma had been added along the line-of-sight with a density in between those in which the C IV and $\lambda6678$ lines normally form. This indication, along with the observation that the $\lambda6678$ line conserves its EW during a dimple appearance, led SP to investigate an *ad hoc* model for a dimple consisting of an opaque (in $\lambda6678$) stationary, detached, intermediate density "slab" over the surface of λ Eri that scatters line photons back toward the star. Photons scattered a second time in a "penumbra" surrounding the slab's projection on the surface acquire a doppler shift from the local projected rotational velocity, thereby redistributing photons from the central absorption part of the feature into wings with slightly enhanced emission. The model predicts the slabs' areas, elevations over the star, vertical velocity, and also permits estimates of slab densities. Typical slabs show areas of $\sim$3 per cent of the star's area, an elevation of 0.1R$_*$, and a density of 10^{11-12} cm^{-3}. A slab cannot be associated with material moving with a vertical velocity exceeding the line width, for otherwise the slab will become transparent to $\lambda6678$ flux and the dimple will disappear. Typically, $V_{vert} \leq 30$ km s^{-1}. SP show that even some bizarre-shaped dimples with "inverse-P Cygni" profiles can be modeled with a slab falling at $\sim$35 km s^{-1} and also having an azimuthal component. If one assumes that 1.5 slabs are present somewhere over λ Eri's surface at any one time, they must comprise $\geq 3\times10^{-14}$ M$_\odot$. If dimples are visible in $\lambda4922$ the lower limit becomes 2×10^{-13} M$_\odot$. Given these masses, slabs should be just detectable over λ Eri as Hα emission (Marlborough, priv. comm.).

Dimples become optically thick within 15 mins. of onset (SP). This fact and the derived area means that the disturbance responsible for them propagates at $\sim$800 km s^{-1}, or about V_{esc}. One may discount formation scenarios requiring dimple-slabs to arise either from stellar ballistic ejections (dimples last too long) or from orbiting circumstellar debris (not long enough). Perhaps an Alfvenic disturbance propagates at this velocity, but this would not explain how plasma material can be transported at many times the sound speed to form a condensation. A paradigm that does account for many of the properties of slabs is the impulsive solar flare, which ejects a high energy plasma/electron beam to the chromo-/photosphere and heats it to a high temperature. The heated material evaporates from the surface at a velocity close to V_{esc} and encounters coronal magnetic loops, causing it to pile up there. The material cools to become a visible condensed prominence and eventually slowly falls back to the surface along field lines. The decay timescale for prominences is slow, $\sim$2 days, compared to dimple lifetimes, but the formation timescale, density, and elevation over the Sun are similar to SP's estimates from modeling λ Eri's dimples.

Although the doppler integration models of dimples fit their shapes, we are unable to reproduce intensity line profiles from NLTE models with black line cores, as implicitly required by SP's backscattering model. This failure exemplifies the present lack of understanding of the physics of formation of these He I lines.

3. Multi Line Observations of Rapid Emission Activity in λ Eri

One way to study the line formation process is to compare and model the simultaneous response of several lines arising from the same atom. Toward this end we obtained (Smith *et al.* 1994) three nights of time series of KPNO fiber-echelle spectra of the first two members of the He I singlet/triplet 2P–nD series, viz. λ6678, λ4922, λ5876, and λ4471 when λ Eri happened to be in emission. Time serial observations of λ Eri, μ Cen, γ Cas and other stars in their active phases have demonstrated that emission features in He I lines (and sometimes even Hα) can show substantial variability in much less than the star's rotation period. λ Eri's λ6678 rapid emission variability often falls into one of several patterns that suggest the occurrence of failed ejections (see e.g. Smith 1989). This includes high velocity blue absorption components, emission shifting continuously from the V to the R wing on a timescale consistent with the ejection/infall velocities, and a correlation of blue absorptions followed shortly by red emissions. R emissions can even evolve into higher velocity absorptions as infalling material crosses the line of sight to the stellar disk (Smith *et al.* 1991; "SPG91"). Thin rings of material orbiting within 1R$_*$ of the surface have also been noted (SPG91). It should therefore not be surprising that much of the V/R emission comes from matter projected over the limb matter. However, in general it is difficult to distinguish between emission produced by over-the-limb material and foreground matter with an enhanced source function. In the case of λ Eri, the potential exists for this ambiguity to be broken because R emissions tend to be stronger and because they are more prevalent than V emissions (Smith 1989).

The means to break the geometrical ambiguity was provided by simultaneous observations of several He I lines using Kitt Peak's 2.1-m echelle spectrograph on 1991 November 3–5 (Smith *et al.* 1994). We monitored the first two members of the singlet and triplet 2P–nD series, viz. λ6678, λ4922, λ5876, and λ4471. Our data showed contrasting behaviors of the V, R emission in the line wings and inner profile in this way: V emissions tend to scale with the line's log gf, with the blue lines showing small or no emission as

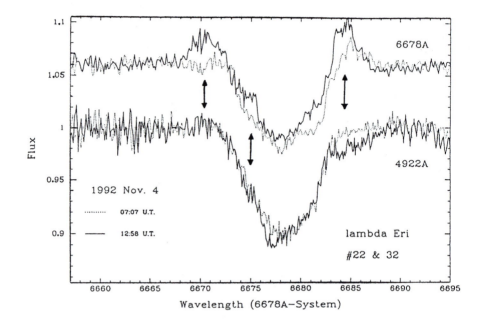

Fig. 1. λ Eri's R-wing fluctuations in emission (λ6678) and absorption (λ4922). Note V-wing/core emission in λ6678 compared to no change in λ4922.

the red lines varied from large to moderate emission strengths. Increases in R-wing emission correlated with definite weak *absorptions* in the V-wing of λ4922. When the "excess" emission wanes in λ6678 so does the absorption in λ4922. Fig. 1 shows an example of this contrasting behavior. Fig. 2 shows emission/absorption in the red cores of the two lines.

This behavior must necessarily be interpreted through NLTE processes affecting atoms in *foreground* material. Using the sophisticated NLTE code TLUSTY written by Hubeny (1988), Smith *et al.* (1994) were able to understand the simultaneous emission and absorption of these two lines by an NLTE effect first noted by Auer & Mihalas (1972) in hot atmospheres ($T_{eff} \geq 45000K$). At these temperatures, He I becomes a trace ion, causing the λ584 resonance line to become transparent and its radiation to leave the star. This drains electrons from the lower state of λ6678, 2^1P, thereby enhancing the source function of all transitions from this level. The effect is augmented for λ6678 by stimulated emission, driving the line almost preferentially into emission. Further study shows that this mechanism is confined to plasmas with at least a photospheric density. This is because at lower densities corresponding to the same ionization state the temperature and UV radiation field are too low to populate the upper atomic levels substantial-

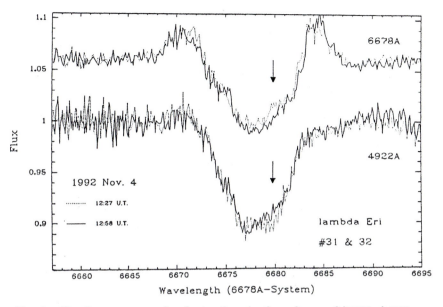

Fig. 2.   Simultaneous em., abs. fluctuations in the red core of λ6678, λ4922.

ly, and this decreases the source function of all these lines. As Fig. 3 shows, the combination of λ6678 excess emission and λ4922 absorption confines the region of the $(T_{eff}, \log g)$ (i.e. T, N_e) domain further, requiring temperatures of at least 40000 K and densities of $\geq 10^{14}$ cm^{-3}. Another implication of this diagram is that the region where the required emission/absorption occurs does not coincide with the requirements for greatest efficiency of λ6678 emission. This implies that a much larger area of the star is responsible for λ6678 emission than if there were no λ4922 absorption. We plan to estimate typical hot spot areas in the near future.

The TLUSTY results imply that in contrast to the V-wing, the emission from the R-wing during λ Eri's outburst phase probably arise from downward-moving plasma within the atmosphere itself. It is easy to show from Virial Theorem arguments that the liberation of gravitational potential energy by infalling matter is too weak by $\geq 10^{4-5} \times$ to heat the implied hot spots on the star down to a level $\tau_c \sim 10^{-3}$. In contrast, there is enough CS mass around Herbig Be stars like HR 5999 to power the emission of lines by an infall mechanism (Blondel *et al.* 1993). In the energy hierarchy of astrophysical energy mechanisms, gravity is rather powerful. It dwarfs almost all other energy sources operating on stellar surfaces. We are able to think of only one other source (excepting nucleosynthesis), *viz.* flaring, that can produce more energy. Once again magnetic energy dissipation seems implicated as a destabilizing mechanism in the atmosphere of λ Eri.

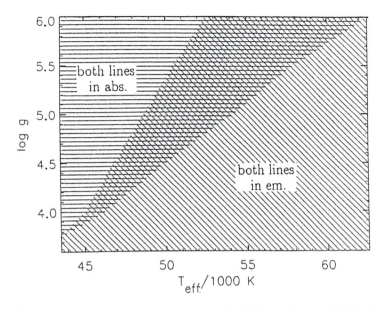

Fig. 3.   T$_{\text{eff}}$, log g diagram, showing range of λ6678 emission, λ4922 absorption (cross-hatched region).

4. An X-Ray Flare on λ Eri

All this talk about possible flaring on λ Eri led us to request and be granted 29 Ksec of pointed (PSPC) observations with *Rosat*, an instrument capable of photon counting and spectral binning in the range 0.2–2 keV. Our observations were conducted on 1991 February 21–22 at the very beginning of the AO1/GO period with little notice to arrange simultaneous observations at other wavelengths. However, we were able to obtain an SWP camera *IUE* spectrum within several hours of the completion of the X-ray observations.

As detailed in Smith *et al.* (1993), our observations were distributed over 13 orbits spanning 38 hours. During the first two and last six orbits λ Eri was detected and showed a L_x/L_{bol} ∼2×10^{-7} typical for an early B star. During the middle five orbits the X-ray flux rose to seven times its initial value; the e-folding timescale was ≤2 orbits. After reaching a peak $L_x = 4 \times 10^{31}$ ergs s^{-1}, the emission began declining, though with a slower decay rate. No significant fluctuations in emission were found on a shorter timescale. The final orbits of our timeline coincide with one rotational cycle after the maximum. Since the flux had returned to its initial state, one can rule out a rotationally modulated hot spot and characterize the brightening as an extended (∼50,000 sec) flare, and a giant flare at that. The spectral analysis of our data shows the flare energy comes exclusively from photons having energies ≥ 0.7 keV and is characterized by a Raymond-Smith temperature of

1.4×10^7K and an Emission Measure of 3×10^{53} cm^{-3}. Both the temperature and EM of the softer component, presumably arising from the basal flux entirely, are each seven times smaller than the flare values.

The *IUE* spectrum showed a typically low wind flux for λ Eri, so that even if there was a hypothetical neutron star or white dwarf orbiting around this star too little X-ray flux would be emitted to account for the flare. As well as showing no detectable radial velocity variations that would betray binary motion, λ Eri's spectrum does *not* show traces of strong metal lines expected from a G–K dwarf capable of producing RS CVn-like or T Tauri flares. Smith *et al.* (1994) concluded that the best hypothesis was that this X-ray activity came from the Be star itself and probably from a flare-related process.

λ Eri is the second star to have been reported as an X-ray Be star; the first is γ Cas. The latter is well known to undergo rapid, chaotic X-ray fluctuations with a Raymond-Smith temperature of $\sim 1.5 \times 10^8$K (Horaguchi *et al.* 1993, Parmar *et al.* 1993), if a thermal description is appropriate. Peters (1982) noted the first X-ray flare, an event on 1977 January 28 that was recorded as emission in several UV metallic lines as well as Hα (Slettebak & Snow 1978). Murakami *et al.* (1986) recorded a flare spectrum lasting several minutes and having a high temperature. Coincidentally or not, Yang et al. (1988) noted λ6678 line profile transients in observations a few days before this flare. Whereas the prevailing sentiment among the X-ray community is that γ Cas is a member of the group of X-ray Be binaries, the evidence for binarity is weak (no detectable RV variations; also, it has a symmetrical disk resolved by optical interferometry (e.g. Quirrenbach, these Proceedings.). Thus it does not fit into that group easily. Perhaps the best argument that γ Cas has a close degenerate companion is the high temperature of its X-ray spectrum. In most other respects the case for flares arising in Be stars, in our opinion, equals or outweighs the mass-accretion/neutron star scenario.

It is difficult to prove that γ Cas and λ Eri are typical Be stars. Why, for example, do surveys show early Be stars to be normal, even subnormal X-ray emitters, (Berghofer & Schmitt, these Proceedings.), when these two stars are so active? A possible explanation is that the X-ray emission occurs very close to the star and is attenuated by well developed Be disks. If so, the two stars we are discussing could be coincidentally good X-ray monitoring candidates either because the star's surface is usually visible or because X-ray flux is not absorbed by disks owing to a low aspect to our line of sight. We suggest that other low-sin$i$ Be stars, e.g. μ Cen, should be good candidates for X-ray monitoring.

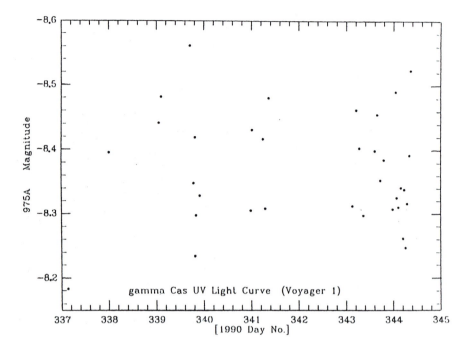

Fig. 4.   FUV rapid flux variations in γ Cas during 1990-1.

5. Far-UV and X-ray Activity Correlations in γ Cas

Our suspicion that the X-ray variability in γ Cas is produced near the Be star motivated us to request *Voyager 1* observations. Our request was granted, and we have obtained 1990-91 UVS data using the reduction program of Holberg & Watkins (1992) for calibration and descattering. The result is a series of 9.3Å-resolution spectra over the range λλ930–1650 (except for poor sensitivity in λλ1200–1360). We grouped our data into 61 useable 0.5–1 hr. bins of spectra. A temporal plot of monochromatic FUV fluxes (subset shown in Fig. 4) shows that variability over the shortest timescales we could compare binned spectra to, 0.5–2 hrs., dominates any possible longer timescale variations. These variations can exceed 0.2 mags.

To understand the nature of these rapid fluctuations we performed two analyses. First, we used all the spectra to synthesize the spectrum of the fluctuating FUV component. This was constructed at each wavelength by subtracting the flux of the third brightest spectrum from the third faintest. The result, shown in Fig. 5, is an almost white difference spectrum, except for a marginal enhancement at the Lyβ and Lyγ lines. Because this component has the same spectrum, the origin of the FUV variations is likely to be close to or even in the photosphere, not a hot site somewhere else. Second, we

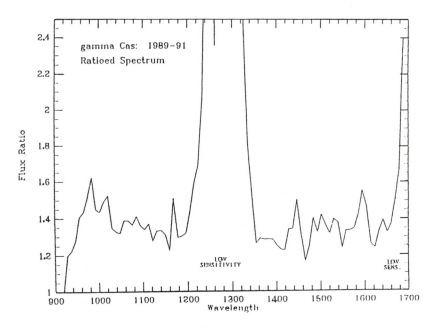

Fig. 5. Derived spectrum of FUV fluctuating component of γ Cas.

constructed a histogram of monochromatic flux differences (absolute value) between consecutive time binned observations ~1 hr. apart. This is shown in Fig. 6a. Note the flat-topped distribution out to a limit of 0.2 mags., beyond which there are few observations. This suggests that the fluctuating FUV component comes from several independent sources at any time.

To compare the FUV with the X-ray variations, we performed the same histogram analysis on the mean fluxes of consecutive orbits from the Parmar et al. (1993; Exosat; 5–9 keV) dataset in Fig. 6b and the Horaguchi et al. (1993; Ginga; 1–9 keV) data (not shown). The FUV and X-ray distributions are nearly identical in shape and magnitude threshold. Further investigation shows that the histogram form holds for data grouped even 0.5 hr apart. In view of these similarities, we suggest that the FUV and X-ray variabilities have the same origin.

What could this origin be? The energy of the fluctuating X-rays is too high for them to arise from gravitational potential energy on the surface of the Be star. The energy in the FUV component is actually about 20× that of the X-rays. This suggests that it is unlikely to originate from the steep potential well near a collapsed object either because the emitted radiation would be nearly monoenergetic. On the other hand, the three spectral characteristics we have noted, the panchromatism, dominance in the FUV,

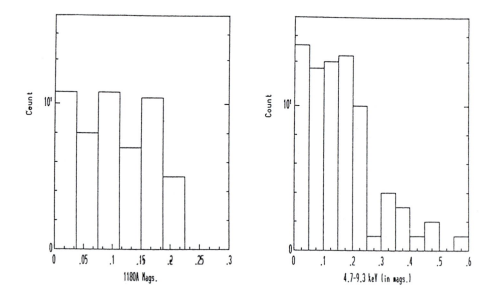

Fig. 6.   Histogram comparison of rapid FUV and X-ray flux variations of γ Cas.

and possible emission in hydrogen lines are all characteristics of White Light
Flares on the Sun. These flares are thought to be the reaction of the chromo-
/photosphere by magnetically guided electron/ion beams, causing marginal
heating of the irradiated photosphere. From the standpoint of X-ray flaring
and panchromatic rapid flux variability, magnetic flaring must once again
be considered as a possible mechanism. A final, important argument for the
Be star origin of this activity is the observation of rapid variations along
the V, R emission components (including fine structure) of the λ6678 line
of γ Cas. This suggests that different sectors of the overlying plasma are
excited by several transient hot sources *close* to the Be star. In our view it
becomes increasingly difficult to maintain the degenerate binary hypothesis
in light of these new observations.

Finally, we may turn to a number of optical photometric reports of sudden brightenings on a 1–2 day timescale in other Be stars, most recently
κ CMa (Balona 1990) and ε Cap (Balona 1993). These reports tend to cor-
roborate earlier visual reports of sudden, several-minute optical brightenings
of HD160202 (Bakos 1970) and 66 Oph (Page & Page 1970). It would appear
that magnetic flaring should be considered as the most likely single mechanism responsible for each of the aperiodic activity outlined in the foregoing
(cf. Underhill & Fahey 1984).

Acknowledgements

This work was supported by NASA Contracts NAS5-31221 and P.O. S-97229-E.

References

Auer, L.H. and Mihalas, D.M.: 1972, *Astrophys. J. Suppl.* **24**, 293.
Bakos, G.A.: 1970, *Sky and Telescope* **40**, 214.
Balona, L.A.: 1990, *Mon. Not. Roy. Astr. Soc.* **245**, 92.
Balona, L.A.: 1993, in Peters, G., ed., *Be Star Newsletter* **26**, 5.
Blondel, P.F., Televera, A. and Djie, H.R.: 1993, *Astron. Astrophys.* **268**, 640.
Holberg, J. and Watkins, R.: 1992, *Voyager Data Analysis Hdbk.* vers. *1.2*, Univ. Arizona.
Horaguchi, T. *et al.*: 1993, *Publ. Astr. Soc. Japan* **46**, in press.
Hubeny, I.: 1988, *Comp. Phys. Commun.* **52**, 103.
Murakami, T., Koyama, K., Inque, H. and Agrawal, P.C.: 1986, *Astrophys. J.* **310**, L31.
Page, A.A. and Page, B.: 1970, *Proc. Astr. Soc. Aust.* **1**, 324.
Parmar, A., Israel, G., Stella, L. and White, N.: 1993, *Astron. Astrophys.* **275**, 227.
Peters, G.J.: 1982, *Publ. Astr. Soc. Pacific* **94**, 157.
Peters, G.J.: 1986, *Astrophys. J.* **301**, L61.
Slettebak, A. and Snow, T.P.: 1978, *Astrophys. J.* **224**, L127.
Smith, M.A.: 1989, *Astrophys. J. Suppl.* **71**, 357.
Smith, M.A. and Polidan, R.S.: 1993, *Astrophys. J.* **408**, 323 (SP).
Smith, M.A., Peters, G.J. and Grady, C.A.: 1991, *Astrophys. J.* **367**, 302 (SPG91).
Smith, M.A., Peters, G.J., Grady, C.A. and Feigelson, E.D.: 1993, *Astrophys. J.* **409**, L49.
Smith, M., Hubeny, I, Lanz, T. and Meylan, T.: 1994, *Astrophys. J.* submitted.
Underhill, A.B. and Fahey, R.P.: 1984, *Astrophys. J.* **280**, 712.
Yang, S., Ninkov, Z. and Walker, G.: 1988, *Publ. Astr. Soc. Pacific* **100**, 233.

Discussion

Balona: Supposing your slabs are concentrated over a much wider area, or you have a grand slab covering 20–30 per cent of the stellar disk. Then you will not observe a dimple but rather a much broader line profile variation like that hypothesized for an $l = 2$ NRP mode. Can you comment?

Smith: Dimples have a smaller characteristic width than the global lpv's from an NRP $l = 2$ mode. It is unlikely that several slabs could "conspire" to emulate a profile shape from this mode, and far less likely still that they would do so over the many profiles that the global $l = 2$ distortion has been documented. However, it is possible at times that a few dimples could be confused with bumps from Penrod's $l = 8$ mode.

Owocki: You mentioned that dimples appear mostly at line center, but wouldn't your "slabs" also appear off disk center and thus away from line center. The one episode you did mention that appeared from line center you interpreted in terms of a falling slab, but couldn't this just be a slab off disk center?

Smith: The preference for observing dimples near line center is probably

RAPID VARIABILITY OF H α EMISSION LINE IN Be STARS

B. G. ANANDARAO and A. CHAKRABORTY

Physical Research Laboratory, Ahmedabad - 380009, India

and

R. SWAMINATHAN and B. LOKANADHAM

Osmania University, Hyderabad - 500007, India

We have initiated an observational campaign on some bright Be stars in order to investigate the rapid variability in emission lines using a Fabry-Perot spectrometer($\lambda/\delta\lambda = 10^4$; $FSR = 21.3\overset{\circ}{A}$) at the Nasmyth focus of the 1.22 m JRO telescope at Hyderabad, India. The PMT dark counts were $1 - 2sec^{-1}$. Here we report our first observations on four stars.

1. γ Cassiopieae

Within the observation time of about 1 hour on 27/12/92, the $H\alpha$ line profile from this star has undergone variability from a well-defined asymmetric shape(Fig. 1a) into disappearance and reappearance sequences with the changes occurring in a few minutes. On 29/12/92, the profile was broader and more spiky and probably there was continuum variation within the scan time($\approx 1.5min$).

2. λ Eridani

The $H\alpha$ emission profiles observed on 29/12/92 show a strong triple- spiked structure(Fig. 1b) which disappeared in the very next scan taken 2 minutes later. On 14/02/93 the $H\alpha$ emission was substantially weaker and broader than in Dec '92. The transient-like episodes were not observed in Feb '93.

3. κ Draconis

This star exhibited rapid variability in the $H\alpha$ emission profile structure on time scales of $\leq 2min$ on 14/02/93 as well as on 15/02/93. The coadded and averaged profile on 14/02/93 shows a double-humped structure with the V component narrower than the R(Fig. 2). The $FWHM$ of the V component is close to the stellar $vsini$ indicating that the shell is in corotation while the R component could be due to earlier ejections circularised after the redistribution of the angular momentum(Hanuschik et al 1993). The absence of these features 24 hours later shows that the corotating and out-going shell has attained circularisation phase during the rotation period viz. 0.8 day, becoming broader and falling back on to the star.

L. A. Balona et al. (eds.), Pulsation, Rotation and Mass Loss in Early-Type Stars, 254–256.
© 1994 IAU. Printed in the Netherlands.

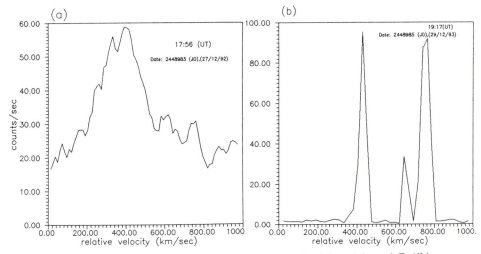

Fig. 1.   Profiles of $H\alpha$ emission line from γCas(a) and from λEri(b)

4. 27 Canis Majoris

Essentially this star showed a similar behavior as κDra with a double-humped profiles on both the nights(14 & 15 Feb '93)(Fig. 3). The V component is corotating with the star and the R component is broader. The presence of the narrower V component on the second night could be attributed to the longer rotation period for $27CMa$(cf. κDra) and perhaps the circularisation process was not completed and/or there had been fresh ejecta. The presence of broad HeI emission line indicated that it is associated with the thermalised region. The minute-scale variations are present in this case also.

5. Conclusions

Clearly we need to establish the nature of the minute-scale variability in all these stars. These rapid variations cannot be due to radial pulsations and need either a binary hypothesis or flares due to localised magnetic fields(Smith 1991). The day-to-day variability observed in the unique double-humped emission line profile in κDra and $27CMa$ may be understood in terms of episodic mass ejection followed by redistribution of angular momentum.

6. References

D. Baade, Vol.36, 59.
Hanuschik, R.W. et al, 1993, A&A, 274, 356.
Smith, M.A., 1991, in ESO Workshop on Rapid Variability of OB stars, Ed.

but the length of C curves varies from a relative maximum at the centre of the stellar disc (which produces the central emission), passes through an absolute maximum (which produces the maximum absorption), and then, decreases toward zero where $[(\Omega(\varpi)\eta \sin i)]_{Max} = V \sin i$. An example of rotationally broadened profile of HeI $\lambda 4471$ for $V \sin i = 100$ km s^{-1} ($T_{eff} = 20000$ K, log g $= 4.0$) is shown in Fig. 1b. Allowing for the gravity darkening effect, the emission-like reversal can be more or less contrasted depending on the sensitivity of the line to the temperature and to the gravity. It is expected that time dependent $\Omega(\varpi, t)$ laws will introduce variable emission-like components. Such an effect may also be due to superimposed horizontal currents in the stellar surface. When the ratio $\tau = K/|W|$ (K : rotational kinetic energy; W: gravitational potential energy) is $\tau_c^R \leq \tau \leq 0.10$ (τ_c^R : ratio for critical rigid rotation ≤ 0.008) the strong polar deformation of the star introduces a double-valued problem for the V sin i parameter (Zorec et al. 1988).

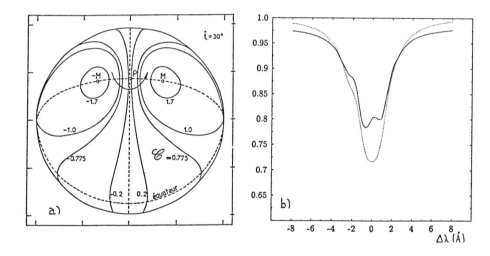

Fig. 1.   a) Curves C of equal radial velocity at $i = 30^\circ$. **P** : stellar pole; $\pm$ **M** points of radial velocities $\pm V_{Max} \sin i$. b) (—): broadened profile of the HeI $\lambda 4471$ line for $V_{Max} \sin i = 100$ km s^{-1}; (....): profile broadened by a rigid rotation and same V sin i.

References

Baade, D.: 1990,   in L.A. Wilson and R. Stalio, ed(s)., *Angular Momentum and Mass Loss for Hot Stars*, Kluwer Acad. Publ., 177
Jeffery, C.S.: 1991, *Mon. Not. R. astr. Soc.* **249**, 327
Zahn, J.-P.: 1983, *Astroph. Processes in Upper Main Seq. Stars*, Obs. Genève, 253
Zorec, J., Mochkovitch, R., Garcia, A.: 1988, *C.R. Acad. Sci. Paris* **306**, 1225

STUDY OF THE HeI λ4471 LINE-PROFILE OF Be STARS(*)

D. BALLEREAU, J. CHAUVILLE

Observatoire de Paris-Meudon - 92195 Meudon, France

and

A. GARCIA, J. ZOREC

Institut d'Astrophysique de Paris - 98bis, Bd. Arago, 75014 Paris, France

Abstract. In this paper we study the veiling effect and the strength of an extra emission-like component filling partially the HeI λ4471 line of Be stars. It is shown that this component is roughly correlated with the emission in the continuum and in the Hγ line.

1. The HeI λ4471 line and the emission in the continuum and in the Hγ line

For a set of 36 Be stars observed on february 1990 at the ESO (La Silla-Chile) with the Echelle Spectrograph at the 1.52 m telescope (resolution $\simeq$ 35000, S/N $\simeq$ 150-200), and using the Didelon's (1982) calibration of photospheric lines, we compare the measured equivalent width W_{HeI}^{obs} of the observed HeI λ4471 line with those corresponding to the spectral type of the studied stars $W_{HeI}^{Sp.Type}$. We can see that $W_{HeI}^{obs} < W_{HeI}^{Sp.Type}$ and that the stronger deviations are seen for the hottest Be stars, where also the emission phenomenon is generally stronger. This deviation may be due: a) to a veiling produced by the emission excess in the continuum, b) to some emission-like phenomenon in the line: this phenomenon can be a real emission, a perturbing effect due to some velocity field on the stellar surface (Zorec, this issue), or perturbations due to non-radial pulsations (Waelkens 1990), etc. Within this picture we have : $|W_{HeI\ 4471}^{th}| = (1+r) \times (|W_{HeI\ 4471}^{obs}| + |W_e|)$, where $r = \Delta F_c / F_c$ is the emission excess in the continuum and $|W_e|$ is the absolute value of the equivalent width due to an extra component filling the photospheric line. To obtain the veiling factor and the V sin i parameter, we look for the best fit between the wings of the observed HeI λ4471 line and the theoretical one (Stoeckley and Mihalas 1973). In Fig. 1 is shown the relation between W_e and veiling factor 1+r for 16 over 36 studied stars where $W_e \neq 0$. The error bars correspond to imposed uncertainties of 15% in the adopted T_{eff} and log g parameters.

Using simultaneous observations of the Hγ line we can prove that there is a correlation between $|W_e|$ and the equivalent width of the emission in the Hγ line. When $W_e \neq 0$, the V sin i parameters we obtained are systematically smaller than those derived using the FWHI-method or by fitting the full line profile: $(\Delta V\ sin\ i / V\ sin\ i)\% \sim 143 \times W_e \leq 30\%$.

L. A. Balona et al. (eds.), Pulsation, Rotation and Mass Loss in Early-Type Stars, 259–260.
© 1994 *IAU. Printed in the Netherlands.*

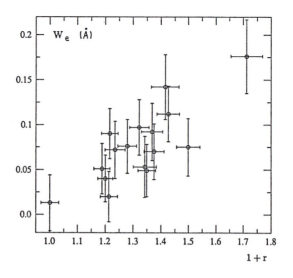

Fig. 1. Relation between W_e and the veiling factor $1 + r$

2. Conclusion

The extra component producing the W_e value seems to show a kind of statistical correlation both with the emission in the Balmer $H\gamma$ line and with the emission excess in the continuum. Only the hottest stars of our sample ($T_{eff} \geq 21000$ K) have $W_e \neq 0$. This may imply that the observed HeI $\lambda 4471$ line profile of Be stars are somewhat filled up by a real emission. It can also be due to some effect (photospheric velocity fields, or other activity on the stellar surface) which has to be correlated with the emission power of Be stars. Our results also indicate that the V sin i of Be stars may be systematically overestimated up to 30 % due to the perturbing effects acting on the photospheric HeI $\lambda 4471$ line.

(*) *Observations obtained at the ESO, Chile and at the OHP, France*

References

Didelon, P.: 1982, *Astron. Astrophys Suppl.* **50**, 199
Stoeckley, T.R., Mihalas, D.: 1973, *Limb Darkening and Rotation Broadening of HeI and MgII Line Profiles in Early-Type Stars*, NCAR-TN/STR-84
Waelkens, C.: 1990, in L.A. Wilson and R. Stalio, ed(s)., *Angular Momentum and Mass Loss for Hot Stars*, Kluwer Acad. Publ., 235

THE HeI λ4471 LINE BEHAVIOUR OF 31 PEG(*)

D. BALLEREAU, J. CHAUVILLE

Observatoire de Paris-Meudon - 92195 Meudon, France

and

J. ZOREC

Institut d'Astrophysique de Paris - 98bis, Bd. Arago, 75014 Paris, France

Abstract. In this paper we report the spectroscopic behaviour of 31 Peg in the HeI λ4471, MgII λ4481 and the Hγ lines at four observing dates. The behaviour of the HeI line indicates that perhaps there exists a photospheric activity and/or some exophotospheric perturbing effect acting on the stellar photosphere.

1. Introduction

Long-term spectroscopic observations of μ Cen (HD 120324) showed a significant anticorrelation between the equivalent width of the HeI λ4471 and MgII λ4481 absorption lines (Barrera and Vogt 1987). It was proposed that this could be due to some atmospheric activity, possibly related to non-radial pulsations. In this paper we report observations of 31 Peg (HD 212076), where contrary to μ Cen, the MgII λ4481 and Hγ lines are in emission.

2. The observations

The spectra were obtained at the ESO (La Silla, Chile) on 27 and 31 Aug. 1991 with the Echelle Spectrograph at the 1.52 m telescope (resolution $\simeq$ 35000; S/N = 150) and at the OHP (France) on 3 and 6 Sep. 1992 with the Aurelie Spectrograph (resolution $\simeq$ 30000; S/N = 300). The profiles (I_λ/I_c) obtained for the Hγ, the HeI λ4471 and MgII λ4481 lines of 31 Peg at the four observing dates are shown in Fig. 1.

3. The star and the spectroscopic variations

For 31 Peg, a B2 IV-Ve star, the expected equivalent width of the HeI λ4471 line is $W^{Sp.\ Type}_{HeI\ \lambda4471} = 1.35$ Å. The equivalent widths obtained from the spectra shown in Fig. 1 are $W^{Observed}_{HeI\ \lambda4471} = 0.91$ to 1.1 Å, which are smaller than $W^{Sp.\ Type}_{HeI\ \lambda4471}$. As in a previous paper (Ballereau et al. BCGZ, this issue), we may argue that the difference between the observed and the expected equivalent widths of the HeI line is due at least to: a) some photospheric activity; b) a veiling produced by the continuum emission excess; c) some emission-like component partially filling the photospheric line. This last effect may be related to the non-radial pulsation phenomenon as suggested by Barrera

L. A. Balona et al. (eds.), Pulsation, Rotation and Mass Loss in Early-Type Stars, 261–262.

and Vogt (1987) and Waelkens (1990), but also to some real emission in the
HeI line (BCGZ) or to some velocity field on the stellar surface as reported
by Zorec (this issue).

The emission strengths in the Hγ and in the MgII $\lambda4481$ lines are well
correlated, due perhaps to a growing common formation region in the circumstellar envelope. However, for the HeI $\lambda4471$ line, two behaviours are
distinguished: 1) in 1991 (low emission in both Hγ and MgII lines) the
absorption in the HeI line is smaller when the emission in Hγ and MgII
lines is stronger; 2) in 1992 (stronger emission in Hγ and MgII lines) the
absorption in the HeI line is enhanced but always $W_{HeI}^{Observed} < W_{HeI}^{Sp.\ Type}$.

The behaviour of 31 Peg seem to be different to that described in BCGZ,
where a kind of correlation was found between the emission-like feature
in the HeI $\lambda4471$ line and the emission in Hγ and with the continuum.
Both studies show however that even the exophotospheric perturbation on
the photospheric HeI line cannot be denied, the detailed description of the
phenomenon needs to take into account the physical characteristics of the
circumstellar envelope and the photospheric activity.

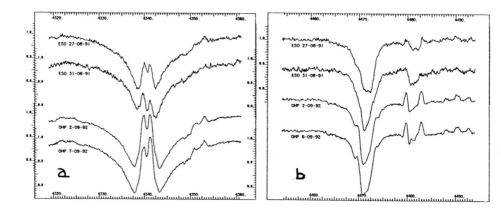

Fig. 1.   Line profiles of 31 Peg: a) Hγ, b) HeI $\lambda4471$ and MgII $\lambda4481$

(*)*Observations obtained at the ESO, Chile and at the OHP, France*

References

Barrera, L.H., Vogt, N.: 1987, *Rev. Mexicana Astron. Astrof.* **14**, 323
Waelkens, C.: 1990,   in L.A. Wilson and R. Stalio, ed(s)., *Angular Momentum and Mass
    Loss for Hot Stars*, Kluwer Acad. Publ., 235

THE FeII λ5317 LINE PROFILE OF NINE Be STARS(*)

D. BALLEREAU, J. CHAUVILLE

Observatoire de Paris-Meudon - 92195 Meudon Cedex, France

and

J. ZOREC

Institut d'Astrophysique - 98bis, Bd. Arago, 75014 Paris, France

Abstract. Our aim is to analyze the FeII λ5317 emission line of nine Be stars recorded in 1990, to deduce some constraints on the modeling of their envelope and to compare them to the results obtained in 1985 and 1987 by other authors.

1. Introduction and observations

Be stars are known to present emission lines on the first Balmer terms of their spectrum, episodically or permanently. Some early (<B5) Be stars show in addition FeII emission lines, when the emission on Balmer lines reaches a threshold of strength. These lines are produced by a moving circumstellar gas supposed to be more or less concentrated in the equatorial plane.

The FeII line profiles displayed in Figure 1 (α Col, HR2142, β^1 Mon, κ CMa, NV Pup, o Pup, PP Car, 48 Lib and χ Oph) were observed in Feb. 1990 at the La Silla Observatory (T1.52m, ECHELEC spectrograph, R = 32000, S/N $\simeq$ 200). The Hγ line of seven of nine of the program stars were also recorded.

2. Discussion and conclusions

Two of our sample stars (β^1 Mon and 48 Lib) present a shell reversal. The one of 48 Lib has two components, which are explained by Hubert et al. (1989) by the presence of two layers with different RVs. Seven stars display Class 1 profiles, one a Class 2 profile (κ CMa) and one an intermediate Class profile (χ Oph), according to the classification of Hanuschik (1987). We measured the widths of each emission line at three levels (base, half-height, peaks) and found that a clear linear correlation exists between them and V sin i, which confirms the results of Hanuschik (1987), Slettebak et al. (1992) and Dachs et al. (1992). If we take into account the peak separations for FeII and Hγ, the regression lines are respectively represented by ΔV^{FeII}_{peaks} = V sin i - 7 and $\Delta V^{H\gamma}_{peaks}$ = 0.95 V sin i - 63 ; the plot of the peak separations of Hγ versus FeII's ones leads to a regression line $\Delta V^{H\gamma}_{peaks} = \Delta V^{FeII}_{peaks}$ - 67 , which indicates that the FeII emitting region is inside the Hγ one.

If we use the relation $\Delta V_{peaks} = 2$ V sin i r^{-j} (Huang 1972), r being the outer radius of the FeII and Hγ emitting disks (case of a keplerian movement

L. A. Balona et al. (eds.), Pulsation, Rotation and Mass Loss in Early-Type Stars, 263–264.

264

with j = 1/2), we obtain the r values given in the Figure 1, for the epochs 1985, 1987 and 1990. We note that the FeII emitting disk of six stars vary in a monotonic way, while two show an extremum.

In conclusion, we may emphasize the following points :

- the width parameters of the FeII emission lines are correlated with the V sin i of the central star

- for several stars, the total width of the FeII emission lines is larger than 2 V sin i, which implies additional broadening phenomena

- the FeII disk is closer to the star than the Hγ disk by a factor of 0.75

- an increase of the equivalent width of FeII emission lines is not always correlated with an increase of the outer radius of the disk

- the long term modeling constraints obtained from different sets of spectroscopic data are sometimes in contradiction.

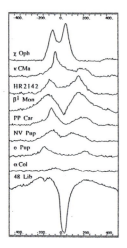

star	r_{FeII}/r_* 1985(a)	r_{FeII}/r_* 1987(b)	r_{FeII}/r_* 1990(c)	$r_{H\gamma}/r_*$ 1990(c)
α Col	3.7	5.1	5.1	-
HR2142	9.5	8.4	7.8	9.1
β^1 Mon	3.8	4.3	5.2	5.5
NV Pup	6.7	6.6	6.9	-
o Pup	3.1	4.2	3.3	5.0
PP Car	4.6	3.4	3.3	3.7
48 Lib	-	8.9	5.0	5.9
χ Oph	8.2	7.5	6.3	1 4

Remarks :
(a)... from Hanuschik (1987)
(b)... from Dachs et al. (1992)
(c)... this work

Fig. 1. Intensity tracings of the FeII λ5317 lines of our sample stars. The intensity of the emission decreases from the top to the bottom. The Table gives the outer radii of the FeII emitting disks at three epochs (with the Hγ emitting disk in 1990).

(*)Observations obtained at the ESO (La Silla, Chile)

References

Dachs, J., Hummel, W., Hanuschik, R.W.: 1992, Astron. Astrophys. Suppl. 95, 437
Hanuschik, R.W.: 1987, Astron. Astrophys. 173, 299
Huang, S.-S.: 1972, Astrophys. J. 171, 549
Hubert, A.-M., Hubert, H., Floquet, M. et al.: 1989, in P. Delache, S. Laloe, C. Magnan and J. Tran Thanh Van, ed(s)., Modeling the Stellar Environment : How and Why ?, Editions Frontières, 263
Slettebak, A., Collins II, G.W., Truax, R.: 1992, Astrophys. J. Suppl. 81, 335

FE II LINE WIDTHS AS TRACERS FOR THE
GEOMETRY OF Be STAR ENVELOPES

R. W. HANUSCHIK*
Astronomical Institute, Ruhr-Universität, D-44780 Bochum, Germany

1. Motivation

The geometry of Be star envelopes is not directly observable, apart from those very few cases where interferometry has been successful. This is even more true for the kinematical conditions in these envelopes. An indirect measure of kinematics, density law and geometry can be achieved by comparing line widths of *photospheric absorption lines* ($\Longleftrightarrow v_* \sin i$) and *circumstellar emission lines* ($\Longleftrightarrow v(r) \sin i$). Hitherto existing determinations of line widths have been, however, quite unsatisfactory. The reason is that in these studies *Balmer line* parameters were used [see Hanuschik (1989) and references therein] which are strongly broadened by radiative transfer and Thomson scattering in addition to kinematic broadening. Because the question of geometry and kinematics is of crucial importance for understanding the Be phenomenon, I have started a new study, using measurements of the Fe II $\lambda5317$ line. This line shows almost optically thin emission and is broadened primarily by kinematics.

2. Fe II line width

In Fig. 1 (top), total width of this line, Δv_{tot}, is plotted vs. stellar $v_* \sin i$ (taken from Slettebak 1982). There exists an unambiguous correlation,

$$0.5\Delta v_{\text{tot}} \approx 1.2 v_* \sin i . \tag{1}$$

This correlation proves axisymmetry of the envelope and gravitationally dominated gas motion (*rotation* or *infall*). Observations of shell lines can help to further differentiate between these two cases. Shell line widths essentially indicate the range of radial velocities, v_{rad}, with disappearing derivative dv_{rad}/dx in front of the star (x being the coordinate towards the observer). For a Keplerian disk, this spread is only $\approx -\Delta v_{\text{D}} ... + \Delta v_{\text{D}}$ or about 4 km s^{-1} in case of iron at 10^4 K. For gravitational infall, this spread is larger, and the line becomes strongly redshifted. Figure 1 (bottom) shows the Fe II

---

* Now at: Astronomical Institute, University of Tübingen, D-72076 Tübingen, Germany

L. A. Balona et al. (eds.), Pulsation, Rotation and Mass Loss in Early-Type Stars, 265–266.
© 1994 IAU. Printed in the Netherlands.

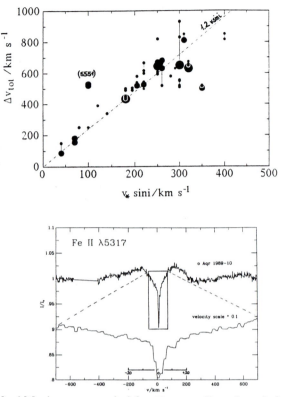

Fig. 1. Top: Total width Δv_{tot} vs. $v_* \sin i$ for 27 stars. Size of symbols corresponds to number of measurements. Bottom: Fe II $\lambda 5317$ in o Aqr.

shell line in o Aqr in Oct. 1989: it is symmetric around zero, and unresolved at resolution 6 km s^{-1}! This line is clearly originating in a perfect Keplerian disk, putting a very strict upper limit on *radial* motions (less than about 3 km s^{-1}), thus contradicting any model with wind-like outflow.

Other stars (e.g. 48 Lib, ζ Tau) show broader, but asymmetric shell lines which are cyclically variable. These are presumably due to perturbed disks, cf. Hanuschik et al. and Hummel & Hanuschik (these proceedings).

Acknowledgements

I would like to thank Wolfgang Hummel for providing unpublished measurements and for valuable discussions.

References

Hanuschik R.W.: 1989, *Ap&SS* **161**, 61.
Slettebak, A.: 1982, *ApJS* **50**, 55.

FeII, FeIII and MgII spectra of Be stars

L. H. Barrera and J. Dachs
Astronomisches Institut, Ruhr-Universität, Postfach 102148, D-44780 Bochum , Germany.

and

T. W. Berghöfer
Max-Planck-Institut für Extraterrestrische Physik, Giessenbachstraße, D-85748 Garching, Germany.

For a total of 33 bright Be stars, archival high-dispersion spectrograms obtained by the International Ultraviolet Explorer (IUE) satellite were analyzed together with optical spectrograms in order to study the FeII, FeIII and MgII spectra in the entire 1600 Å to 6500 Å wavelength range. Ultraviolet FeII resonance lines are always found to be in absorption which usually is of interstellar origin. By applying curve-of-growth methods to equivalent widths measured for FeII multiplets UV-1 through UV-8, column densities of interstellar FeII are derived ranging typically between about 10^{14} and 10^{15} cm^{-2}. For a few program stars showing shell-type spectra in the optical region, additional FeII resonance absorption lines are detected starting from excited fine-structure levels above the ground state, pointing to FeII absorption occurring in the dense circumstellar envelope in front of these star. Analysis of their equivalent widths is used to infer typical column densities of circumstellar ground state FeII ions of order 10^{15} cm^{-2}.

FeII emission lines are only observed in the optical part of Be star spectra, involving transitions connecting both quartet and sextet levels with minimum excitation energies of about 2.5 eV. The relatively large strengths of optical FeII emission lines (with equivalent widths up to 0.7 Å) are compared to upper limits for undetected FeII resonance line emission in the circumstellar disks. This leads to the conclusion that FeII emission in Be star envelopes is mainly caused by radiative recombination of FeIII. This conclusion is also supported by the fact that in variable Be star spectra, equivalent widths of FeII emission lines are correlated with optical depths for electron scattering as inferred from variable extensions of Hα emission line wings. For about one quarter of program stars unusually strong FeIII absorption is observed in front of the stars in multiplets UV-34 and UV-48, arising from metastable lower levels about 5 eV above the FeIII ground state. These stars either show strong CIV stellar wind characteristics in their ultraviolet spectra (HR 2855, HR 4140, HR 5440, HR 8539), or shell-type features in their optical spectra (HR 5941, HR 8260, HR 8402).

MgII optical (multiplet 4) and ultraviolet (UV-1) lines contain photospheric (absorption) as well as circumstellar (emission or shell absorption) components or both and for multiplet UV-1 also strong interstellar absorption. Intensities of MgII emission are correlated to intensities of FeII emission lines. Four different cases of the combination of photospheric, interstellar and variable circumstellar

L. A. Balona et al. (eds.), Pulsation, Rotation and Mass Loss in Early-Type Stars, 267–268.
© 1994 IAU. *Printed in the Netherlands.*

absorption and emission features of MgII UV-1 are illustrated in Fig. 1 (HR 3498, HR 4140, HR 5440 and HR 6118). Partly, our conclusions are supported by data collected in the Be Star Atlas of High-Resolution Spectra by Doazan et al. (ESA SP-1147, 1991).

This work has been supported by a grant from the Deutscher Akademischer Austauschdienst (DAAD) to L.H. Barrera.

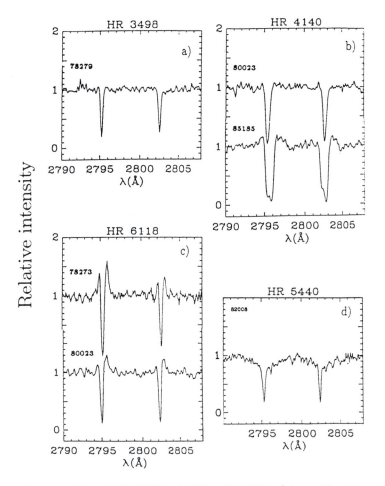

Fig. 1.   Resonance lines of MgII UV-1, λ 2795 and λ 2802; a) interstellar absorption; b) interstellar absorption superimposed by variable circumstellar absorption and emission; c) variable circumstellar emission superimposed by interstellar absorption; d) photospheric absorption superimposed by interstellar absorption. Dates are indicated as (year - 1900)*1000 + day number.

SPECTROSCOPY OF Be STARS IN NGC 330

PAOLO A. MAZZALI and F. PASIAN
Osservatorio Astronomico, Via G.B.Tiepolo, 11, Trieste, Italy

D.J. LENNON
Universitäts Sternwarte, München, Germany

P. BONIFACIO
SISSA, Trieste, Italy

and

V. CASTELLANI
Istituto di Astronomia, Universitá di Pisa, Italy

Medium resolution (2Å/px) but high s/n spectra of approximately twenty of the brightest blue stars in the young open cluster NGC 330 in the SMC have been obtained with EFOSC1 on the ESO 3.6m telescope, and analyzed in order to determine the atmospheric parameters and the evolutionary status of the stars. LTE and NLTE model atmosphere calculations were used to determine the stellar parameters. The T_{eff} values were derived from fits of the UV continua for all stars where these were available, using Robertson's (1974) B and V photometry to scale the Kurucz model fluxes for metallicity $Z = 0.1Z_\odot$. Luminosities of the sample stars lie in the range $4.0 < \log(L_*/L_\odot) < 5.0$ and spectral types between B0 and late-B.

We find that all but one of the stars in our sample with spectral types in the range B0–B3 V–II show emission in H_α, confirming previous findings based on low resolution spectroscopy (Feast 1972) and photometry (Grebel et al. 1992) that the incidence of Be stars in this metal-poor cluster is very high (60-70%, Grebel et al. 1992). This is a peculiar feature requiring explanation. Stars are classified as Be if they show emission in H_α, but in some the emission is observed also in H_β, H_γ and H_δ. The resolution is too small ($\sim 100 \, \mathrm{km/s}$) to resolve any V/R components in the emission lines.

For the strongest Be stars, where emission was present even in the cores of H_γ and H_δ, we tried to determine T_{eff} and $\log g$ by comparing the profiles of their Balmer lines with those of stars with less emission, and by fitting the wings of the lines, which were assumed to be unaffected by the Be emission. A comparison with evolutionary tracks can be found in Lennon et al. 1993. We measured $v_{rot} \sin i$ values from the widths of the H_α emission, when present, and the emission intensity from the equivalent width of the emission itself. The various parameters of the stars in our sample are summarized in Table 1. The stars are referred to by their Robertson (1974) number.

The H_α intensity is strongly correlated with the value of $v_{rot} \sin i$. This is in some sense an unexpected result, since if the H_α emission arises from

269

L. A. Balona et al. (eds.), Pulsation, Rotation and Mass Loss in Early-Type Stars, 269–270.
© 1994 IAU. Printed in the Netherlands.

TABLE I

Parameters for the stars in NGC 330.

STAR	T_{eff} K	$\log g$	$\log(L/L_\odot)$	Type	EW(H_α) Å	$v \sin i$ km/s
A01	29000 ± 1000	4.25 ± 0.20	4.74	B0.5 Ve	-0.4	125
A02	16000 ± 1000	2.50 ± 0.20	4.86	B4 Iab/b	–	< 100
B04	25000 ± 1000	3.90 ± 0.20	4.23	B1.5 IVe	-2.4	180
B05	22000 ± 1000	3.50 ± 0.25	4.16	B2 IIIe	-38.7	300
B06	22000 ± 1000	3.50 ± 0.30	4.21	B2 IIIe	-26.9	300
B07	22000 ± 3000	3.70 ± 0.30	4.09	B2 III/IVe	-1.6	160
B11	12000 ± 2000	3.25 ± 0.50	3.56	B7 II/III	–	< 100
B12	22000 ± 2000	3.40 ± 0.30	4.22	B2 IIIe	-37.5	300
B13	22000 ± 2000	3.60 ± 0.40	4.07	B2 III/IVe	-5.6	160
B16	10000 ± 1500	2.60 ± 0.50	4.02	B9.5 Ib/II	–	< 100
B18	32000 ± 1000	4.50 ± 0.20	4.49	B0 Ve	-9.7	180
B21	22000 ± 1000	3.00 ± 0.25	4.59	B1.5 II/IIIe	-36.6	320
B22	20000 ± 1000	3.20 ± 0.20	4.53	B2 IIe	-1.9	150
B24	25000 ± 3000	3.90 ± 0.50	4.42	B1 IVe	-1.6	150
B28	30000 ± 3000	4.40 ± 0.40	4.34	B0 Ve	-0.6	125
B30	20500 ± 1000	3.25 ± 0.25	4.57	B2 II	–	< 100
B35	25000 ± 5000	3.50 ± 0.50	4.38	B1.5 III/IVe	-49.2	400
B37	18000 ± 1000	2.60 ± 0.20	4.81	B3 Ib	–	< 100

discs surrounding the Be stars, as it is normally supposed, then for a given Be star the observed H_α emission should be higher for smaller $\sin i$ values. Since we observe the opposite correlation, we suggest that we are seeing the Be stars in NGC 330 under roughly the same inclination angle, so that the H_α emission correlates directly with v_{rot}.

Two interesting conclusions can therefore be drawn: first, that the rotational axes of all stars in NGC 330 are aligned, which has implications on the process of star formation in the cluster, and, secondly, that the extension of the discs surrounding Be stars depends on the stars' rotational velocities.

References

Feast, M.W.: 1972, Monthly Notices of the RAS **159**, 113

Grebel, E.K., Richtler, T., de Boer, K.S.: 1992, Astronomy and Astrophysics **254**, L5

Lennon, D.J., Mazzali, P.A., Pasian,F., Bonifacio, P., Castellani, V.: 1993 Space Science Reviews, in press

Robertson, J.W.: 1974, Astronomy and Astrophysics, Supplement Series **15**, 261

Be STARS IN YOUNG LMC CLUSTERS *

H. KJELDSEN, D. BAADE

European Southern Observatory

Karl-Schwarzschild-Str. 2, D-85748 Garching bei München, Germany

Abstract. Slitless field Hα spectroscopy with a resolution of 0.2 nm has detected numerous Be stars in 3 young open star clusters in the LMC but only a few each in one Galactic and one LMC cluster. The line widths indicate rapid rotation as is typical of Galactic field Be stars. In NGC 2004 (LMC) their distribution seems inconsistent with the assumption of either very similar equatorial rotation velocities or random orientation of the rotation axes or both, which are believed to be valid for Galactic field Be stars.

1. Introduction and Observations

There is no widely accepted model which predicts either which rapidly rotating B stars (Bn stars) become Be stars or, if all Bn stars transform into Be stars, when this happens. In the Galaxy (Bright Star Catalog), an average ∼10% of the B-type stars show line emission and roughly twice as many between B1 and B3. The open star cluster NGC 330 in the SMC has become known for its unusually high content of Be stars (e.g., Grebel et al. 1992). Since NGC 330 possibly was the first Magellanic Cloud cluster which has been looked at in this regard, other clusters may be expected to show a similar excess of Be stars. In fact, in the LMC clusters NGC 2004 (Grebel et al. 1993) and 2100 (Bessell and Wood 1993) the same symptoms have been found since.

We have obtained slitless grism spectra with ESO's 3.5-m NTT at a dispersion of 0.2 nm/mm and over a field of about 7×7 arcmin². Each cluster was observed twice, with the grism being rotated through 90 degrees for the second exposure in order to reduce crowding problems. For the same reason the length of the spectra was limited to 7.2 nm (FWHM) with an interference filter roughly centered on Hα. In each 15 minute exposure, stars down to m_V between 20 and 21 were detected which at a distance of 55 kpc (LMC) covers all B-type stars on the main sequence and beyond.

2. Detections

From a preliminary analysis, we have so far found 17 Be stars in NGC 1818 (LMC), 43 in NGC 2004 (LMC), 30 in NGC 2100, 3 in NGC 2122, and 2 in NGC 3293 (Galaxy). Since only limited photometry and spectral classification is available from the literature, it is not presently possible to reliably estimate the fraction of B-type stars with emission lines. However, the following conclusions are justified:

* Based on observations obtained at the European Southern Observatory, La Silla, Chile

L. A. Balona et al. (eds.), Pulsation, Rotation and Mass Loss in Early-Type Stars, 271–272.
© *1994 IAU. Printed in the Netherlands.*

1. The widths of the Hα emission lines detected by us rule out unresolved HII regions as their origin. They strongly support the assumption that these Be stars belong to only one population, which furthermore is very similar to the Galactic one, as is also suggested by their homogeneous photometric variability (Balona 1992).
2. Both high and low proportions of Be stars occur in low- (SMC, LMC) as well as solar-metallicity environments (for instance, 50% of all early-B type stars in χ Per are Be stars, Schild 1965).
3. Clusters rich in Be stars may be more common in the Magellanic Clouds.
4. The Be stars in the LMC clusters observed by us are spread over the full field of view and appear less concentrated to the nucleus than the other cluster members.

3. NGC 2004

Age determinations for this LMC aggregate cluster around 10 Myr but also reach 40 Myr. The metallicity [Fe/H] = -0.7 ± 0.1. Grebel et al. (1993) find ~55 photometric Be star candidates. This is consistent with our census of 43 Be stars. Bessell and Wood (1993) also report a large, but unspecified, proportion of Be stars. We estimate a mean fraction of 30% Be stars which may peak at 45% roughly around B3-B4. Our numbers refer to the full field and, because of contamination by non-cluster members, only mark a lower limit. Equivalent widths (uncorrected for underlying Hα absorption) range from 0.2 to 11.5 nm with the peak of the line emission reaching up to ten times the continuum level.

After coarse correction for instrumental broadening, the line widths (FWHM) were found to lie between 100 and 500 km s^{-1}. A histogram (bin width 50 km s^{-1}) shows a broad plateau between 225 and 475 km s^{-1} which appears irreconcilable with the assumption (found to be acceptable for Galactic field Be stars, Lucy 1974) of roughly identical equatorial velocities and random orientation of the rotation axes. One may wonder whether in a young cluster the latter assumption is well satisfied by single stars.

References

Balona L.A.: 1992, *MNRAS* **256**, 425
Bessell M.S., Wood, P.R.: 1993, *in New Aspects of Magellanic Cloud Research, B. Baschek, G. Klare, J. Lequeux (eds.), Springer, Lect. Notes in Phys.* **416**, 271
Grebel E.K., Richtler T., de Boer K.S.: 1992, *A&A* **254**, L5
Grebel E.K., Richtler T., de Boer K.S.: 1993, *in New Aspects of Magellanic Cloud Research, B. Baschek, G. Klare, J. Lequeux (eds.), Springer, Lect. Notes in Phys.* **416**, 366
Lucy L.B.: 1974, *AJ* **79**, 745
Schild R.E.: 1965, *ApJ* **142**, 979

(The full results will be submitted to Astronomy & Astrophysics when the analysis has been completed.)

PUZZLING PROBLEMS OF He I LINE FORMATION IN EARLY B STARS

MYRON A. SMITH,
IUE/CSC Observatory, 10000A Aerospace Rd., Lanham-Seabrook MD 20706, USA

and

IVAN HUBENY, AND THIERRY LANZ
Goddard Space Flight Center, Greenbelt, MD 20771, USA

1. Introduction

Although NLTE model atmospheres have been shown to resolve most of the equivalent width (EW) discrepancies for blue He I lines (Auer and Mihalas 1972, 1973), Wolff and Heasley (1984, 1985) have demonstrated that discrepancies remain for the leading members of the singlet/triplet 2P − nD series, *viz.* λ6678 and λ5876. These two lines are the strongest nonresonance He I transitions and are important because they respond to thermal changes in the superficial atmosphere ($\tau \sim 10^{-3}$) of early B stars. In order to understand the observed rapid variations of the λ6678 line in mild Be stars, we undertook a survey of EWs of λ6678 and λ4388, namely the first and third member of the same series. These two lines have a log gf ratio of 15 but have similar EWs in B star spectra. Our new observations confirm the red line discrepancy noted by WH85 and point to additional EW differences among various groups of B stars not noted hitherto.

2. Observations and Models

Observations were conducted at the McMath Solar telescope using a resolution of 50,000 and 30,000, respectively, for the red and blue line. We observed 100 chemically normal B0.5-B5 stars known not to have obvious secondary contamination. We converted their published $uvbyH\beta$ colors to (T_{eff}, log g) from WH85's calibration, and when necessary the WH85 Hγ profile criterion to determine log g's.

We used the TLUSTY code (Hubeny 1988) to compute pure H/He NLTE model atmospheres and line profiles. These models include 14 discrete singlet and triplet He I levels plus one for the He II ground state; additional He II states are unimportant for T_{eff}'s < 30,000K. Profiles computed with various ξ_t values showed negligible difference in EW. *Figs. 1* and *2* show our observations against the models of AM73 and TLUSTY. Because Be stars with emission have contaminated photospheric EWs, these stars are omitted in the following discussion.

L. A. Balona et al. (eds.), Pulsation, Rotation and Mass Loss in Early-Type Stars, 273–274.
© 1994 IAU. Printed in the Netherlands.

3. Results

λ4922 shows good agreement between EWs predicted by AM73, TLUSTY, and our data – nor do λ4922 EW differences exist among subgroups, except that giants are predicted/observed to show 250mÅ EWs than dwarfs. Yet as *Fig. 1* shows, while there is agreement between predictions from the two codes, their predictions fall short by ∼100mÅ of the observations. Also, contrary to theory, the EWs of giants are stronger than B dwarfs. Finally, EWs of known pulsating B stars, and Be stars without strong emission (at obsn.) are all larger than for B dwarfs.

The intergroup EW differences for λ6678, but not for λ4922, is a new result. We have tried to model these differences with a variety of toy model atmospheres with modified ρ, T distributions, including dense slabs. None of these can enhance the λ6678 EW without also influencing λ4922 and disturbing its agreement. To resolve this conflict, we are currently building a new generation of model atmospheres with blanketing by $\sim 10^6$ lines. This will include investigation of the influence of raised microturbulence in pulsating and Be star atmospheres.

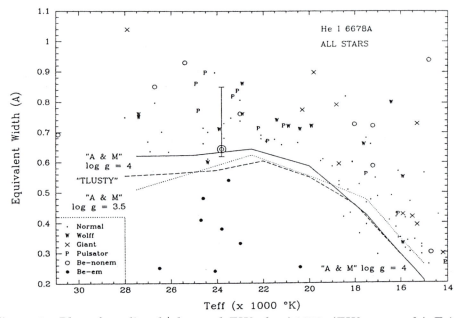

Figure 1 – Plot of predicted/observed EWs for λ6678. (EW range of λ Eri noted.)

References

Auer, L. H., and Mihalas, D. M. 1972, ApJS, 24, 293.
Auer, L. H., and Mihalas, D. M. 1973, ApJS, 25, 433, ("AM73").
Heasley, J., Wolff, S, and Timothy, G. 1982, ApJ, 262, 663.
Hubeny, I. 1988, Comp. Phys. Commun., 52, 103.
Wolff, S., and Heasley, J. 1985, ApJ, 292, 589, ("WH85").

Be STARS IN THE ULTRAVIOLET SPECTRAL CLASSIFICATION SYSTEM

JANET ROUNTREE

Science Applications International Corporation, Tucson Arizona

and

GEORGE SONNEBORN

NASA Goddard Space Flight Center, Greenbelt Maryland

Rountree and Sonneborn (1991) developed a system for the classification of ultraviolet B-star spectra, using MK standards drawn from the optical region. The observational material consisted of high-dispersion spectra obtained with the Short Wavelength Prime (SWP) camera on the International Ultraviolet Explorer (IUE) spacecraft. The classification criteria were based exclusively on photospheric absorption lines, primarily lines of C III, Si II, and Si III. The stellar wind lines of Si IV, C IV, and N V were not used in the classification.

Once the spectral type and luminosity of a program star were determined from the photospheric lines, the wind lines were compared with those in the appropriate standard star. If the program star appeared to have anomalous wind lines (usually stronger than in the standard), a suffix "w" was appended to its ultraviolet spectral type.

All of the program stars were known to have well-determined optical MK types in the range B0 - B8 III - V. Stars with a suffix "e," "n," "nn," or "p" in their MK types were explicitly excluded from the observing list. Therefore, it may be assumed that none of the program stars exhibited emission at Hβ, at least at the epoch of their MK classification. However, an investigation of the literature concerning those stars which had been given a "w" suffix in the ultraviolet classification system revealed that many of them had a history of reported emission at Hα. Two such stars are HD 180968 (2 Vul, B1 IVw) and HD 192685 (B2.5 Vw), which were studied by Grady, Bjorkman and Snow (1987). Several examples of this phenomenon are shown in the ultraviolet spectral atlas of Rountree and Sonneborn (1993).

In conclusion, it appears that the extended atmosphere effects that manifest themselves as a Be spectrum in the optical region can also be recognized as anomalies in the ultraviolet stellar wind lines. There is not a one-to-one correspondence, but an enhanced stellar wind may be an "early warning" or a residual sign of hydrogen-line emission.

L. A. Balona et al. (eds.), Pulsation, Rotation and Mass Loss in Early-Type Stars, 275–276.
© *1994 IAU. Printed in the Netherlands.*

References

Grady, C. A., Bjorkman, K. S., and Snow, T. P.: 1987, *Astrophysical Journal* **320**, 376

Rountree, J., and Sonneborn, G.: 1991, *Astrophysical Journal* **369**, 515

Rountree, J., and Sonneborn, G.: 1993, *"Spectral Classification with the International Ultraviolet Explorer: An Atlas of B-Type Spectra" (NASA Reference Publication)* ,

UV SPECTRAL CLASSIFICATION AND STELLAR WINDS
IN A SAMPLE OF Be AND STANDARD STARS

ARNE SLETTEBAK
Department of Astronomy, Ohio State University
174 W. 18th Ave., Columbus, Ohio 43210, USA

Equivalent widths of 16 lines of CI, CII, CIII, CIV, SiII, SiIII, SiIV, AlII, AlIII, FeII, and FeIII, plus centroid and edge velocities of the SiIV and CIV lines, were measured in IUE spectra of 39 B1e–B8e and 18 B1–B8 standard stars. These suggest the following:

1. Certain line ratios of SiII/III, CII/III, AlII/III, and FeII/III are very sensitive to spectral type and represent excellent UV criteria for spectral classification.

2. UV line strengths and line ratios show that there are no significant differences between the photospheric line spectra of Be and normal, non-emission stars of corresponding type.

3. Despite the fact that the SiIV and CIV wind lines are variable in the Be stars, certain conclusions can be drawn from a statistical "snapshot" study such as this:

   a) The SiIV and CIV wind lines in the Be stars are correlated with both spectral type and luminosity class in the sense that the hottest stars have the strongest lines, and the giants and subgiants (at least for the B1–B3 stars) have stronger lines than the main-sequence stars. (see Fig. 1 and 2).

   b) The SiIV wind lines persist to spectral type B8 in both the Be stars and the standard stars but are stronger in the Be stars than in the standards for the earlier types (B1–B3). (See Fig. 1)

   c) The CIV wind lines persist to spectral type B8 in the Be stars but only to B3 in the standard stars. They are stronger in the Be stars than in the standards at all spectral types. (see Fig. 2).

   d) The equivalent widths of the SiIV and CIV wind lines are only very weakly correlated with $v\sin i$, if at all, but a threshold in $v\sin i$ near 150 km s^{-1} (as found earlier by Grady et al.) exists below which no large equivalent widths of SiIV or CIV may be seen. Assuming that the Be stars are all rapid rotators, such a correlation is essentially a correlation with i and suggests that the winds from Be stars arise preferentially from the equatorial regions. The aforementioned conclusion is supported by a plot of SiIV 1394 centroid velocities versus $v\sin i$, which shows that stars with large velocity shifts also have large $v\sin i$, while those with unshifted lines all have $v\sin i$ less

L. A. Balona et al. (eds.), Pulsation, Rotation and Mass Loss in Early-Type Stars, 277–279.

    than about 150 km s^{-1}. Again, it seems that strong stellar winds are more likely to arise from equatorial than polar regions.

e) Additional evidence that the winds are stronger for Be stars of early type comes from the SiIV and CIV centroid velocities, which are much larger for the hotter Be stars. The stars with the strongest lines also tend to have the largest velocity shifts, suggesting that winds with more mass are also faster moving.

f) The SiIV and CIV lines in the standard stars, while they may be asymmetric, never show displaced line centers.

g) The edge velocities for both the SiIV 1394 and the CIV 1548 lines increase with earlier spectral type in the Be stars (therefore, stronger winds in the hotter stars), and are considerably higher in the Be stars than in the standard stars, suggesting again that the winds are stronger in the Be stars.

h) Shell stars have weaker CIV absorption and smaller edge and centroid velocities than other Be stars, suggesting that they have weaker winds. Since there is considerable evidence that these are stars with cool, low-velocity disks which are being viewed edge-on or nearly edge-on, the winds may be inhibited and modified by the denser material in the equatorial regions.

4. Mg II emission is detected in about half of the program Be stars with long wavelength IUE spectra, and seems not to be correlated with spectral type, $v\sin i$, or strength of the SiIV wind lines. Since the MgII emission presumably originates in the cool, low velocity envelope and since MgII emission also correlates with hydrogen Balmer emission in the Be stars, this suggests that there is no strong physical relationship between the stellar winds and the cool disk.

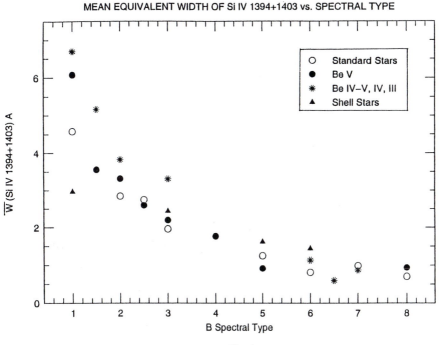

Fig. 1.

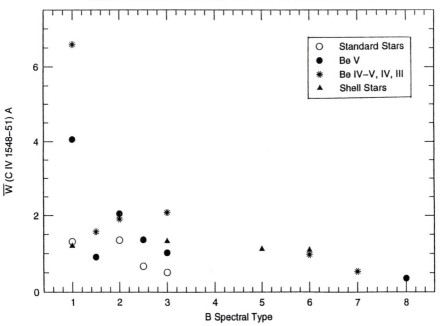

Fig. 2.

Analysis of IUE spectrograms for Be stars

T. Eversberg[1], J. Dachs[1], T. W. Berghöfer[1,2], C. Huilai[1,3], U. Lemmer[1,4]
[1] Astronomisches Institut der Ruhr-Universität Bochum, D-44780 Bochum, Germany
[2] Max-Planck-Institut für Extraterrestrische Physik, Giessenbachstrasse, D-85748 Garching, Germany
[3] Beijing Astronomical Observatory, Chinese Academy of Sciences, Beijing 100 080, China
[4] Fischbachauer Str. 8, D-81539 München, Germany

Archival high-dispersion spectrograms obtained by the International Ultraviolet Explorer satellite in the short-wavelength ($\lambda\lambda$1150-2000Å) region were inspected and analyzed for a total of 33 Be stars, including one Oe star (HR 6397) and three stars showing shell-type spectra in the optical region (48 Lib, ε Cap, o Aqr). The following atomic and ionic transitions were investigated: HI-Lyα, CII-UV1, CIII-UV4, CIV-UV1, NV-UV1, SiII-UV1...-UV4, SiIII-UV2, SiIV-UV1.

Measured line parameters include absorption equivalent widths, central line depths, full widths at half-maximum intensity, and short-wavelength edges for asymmetric profiles indicating the presence of a stellar wind. As far as possible, photospheric, circumstellar, and interstellar contributions to line absorption in these multiplets were identified according to their different line widths and evaluated separately. By means of curve-of-growth methods, column densities were derived for interstellar HI and SiII in the line of sight to program stars. HI column densities were then used to determine interstellar contributions to $(B - V)$ colour excesses measured for these stars.

For Be stars showing shell-type spectra in the optical region, a strong circumstellar contribution to HI and SiII column densities is revealed by the appearence of SiII resonance absorption line transitions arising from excited fine-structure levels of its ground state (e.g., SiII λ1265, 1533, 1814 Å). Shell absorption in these SiII lines is also distinctly visible in one spectrum obtained for HR 4140 on 1980 January 23, but not in another one measured for the same star 1983 April 02. All these shell stars, including HR 4140, also show strong narrow absorption lines in the FeIII-UV34 multiplet (λ1895Å, 1914Å and 1926Å).

Photospheric absorption lines in the spectra of Be stars are characterized by strong symmetric broadening due to rapid rotation of the star. Equivalent widths measured for photospheric components of SiII , SiIII and SiIV lines were compared to results of theoretical line strength predictions found in the literature (e.g., by Kamp (ApJ Suppl. **36**, 143, 1978) and Lennon et al. (Mon. Not. R. Astr. Soc. **222**, 719, 1986)). For SiII λ1533Å(UV2) we generally observe higher values of equivalent widths as calculated for photospheric lines at an adopted Si abundance of $N(Si)/N(H) = 3 \cdot 10^{-5}$ (Fig. 1). The largest values of SiII absorption are found for stars also showing strongest, usually asymmetric SiIV absorption lines (with their asymmetry due to stellar wind).

L. A. Balona et al. (eds.), Pulsation, Rotation and Mass Loss in Early-Type Stars, 280–281.

Photospheric and interstellar components in CII and SiII resonance lines are separated and discussed, and the spectral type dependence of CIII-UV4 and SiIII-UV2 multiplets is investigated.

Stellar wind was studied as indicated in the spectra of Be stars by the presence of extended blue wings in the resonance lines of SiIV-UV1 and CIV-UV1. Our data clearly show the CIV resonance doublet equivalent widths to increase both with increasing T_{eff} and with increasing $v \sin i$ of the stars. For the equivalent widths of SiIV-UV1 absorption, we find a similar correlation with T_{eff} but no firm correlation between terminal velocity and stellar latitude (Fig. 2).

Equivalent widths and terminal velocities measured for the SiIV $\lambda 1393$Å(UV1) line were used to determine mass loss rates $\dot{M}$, starting from equation (37) of Dachs and Hanuschik (A&A **138**, 140, 1984). Optical depths in the wind flow at wind velocity $V_\infty/2$ were derived by fitting observed line profiles to theoretical P Cygni-type profiles calculated by Castor & Lamers (ApJ Suppl. **39**, 481, 1979), with resulting values of optical depth $T_{1/2}$ ranging between 0.2 and 0.8. For the ion density fraction, $N_{SiIV} / N_{Si,total}$, the maximum value of 0.35 obtained in the calculations by Arnaud & Rothenflug (A&A Suppl. **60**, 425, 1985) was adopted as a uniform ratio, while the relative Si abundance in the wind, N_{Si}/N_H was taken to equal $4 \cdot 10^{-5}$.

Resulting mass loss rates derived from SiIV profiles range between 10^{-11} and 10^{-9} solar masses per year and, on the average, distinctly increase with increasing effective temperature of the star. Evidence for extended blue wings due to stellar winds is also detected in the photospheric components of SiII λ 1526 Å resonance lines. For several Be stars, time variations of the stellar wind were studied. Spectacular changes were noted in particular for HR 2855 between 1981 and 1987 in the NV and CIV doublet resonance absorption profiles.

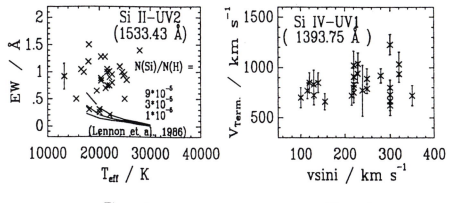

Figure 1   Figure 2

FIRST RESULTS OF AN INTERNATIONAL MULTISITE
MULTITECHNIQUE CAMPAIGN ON omicron AND

J.P. SAREYAN[1], J. CHAUVILLE[2], D. BRIOT[3],
S.J. ADELMAN, M. ALVAREZ, I. BALEGA, Y. BALEGA,
D. BARDIN, I. BELKIN, D. BONNEAU, M. BOSSI, V. DESNOUX,
M. ESPOTO, R. FRIED, S. GONZALEZ-BEDOLLA,
X.Z. GUO, Y.L. GUO, J.X. HAO, L. HUANG, A. KLOTZ,
N. LEISTER, F. MORAND, D. MOURARD, F. VAKILI and
F. VALERA, F.Y. ZHAO

1 Observatoire de la Cote d'Azur, 06304 Nice Cedex 04, France
2 Observatoire de Meudon, 92195 Meudon Cedex, France
3 Observatoire de Paris, Av.de l'Observatoire, 75014 Paris, France

Omicron Andromedae is a multiple system of at least four stars : a B ↔Be star (component A), a spectroscopic binary (components B1-B2) and a close companion (component a). According to several studies (see Hill et al. 1988, 1989) :
  - the distance between A and the B1-B2 system decreased from 0.39" in 1975 to 0.25" in 1987 (McAlister and Hartkopf 1988)
  - the few previous speckle measurements of component a have shown the possibility of a 3.7 years orbit around A, according to the 1975 to 1984 observations (mean distance 0.05"). The calculations with this 3.7 yr orbit lead to the prediction of a maximum distance of 0.77" at 1992.738, i.e. at the end of september 1992, with a North-South orientation.

But a strong contradiction appears : given the spectroscopic distance (188 parsecs), this orbit would lead to a mass of 180 M⊙ for the A-a system; or if we adopt a mass of 7 M⊙ for A, we deduce an orbital period of 14 years for A-a, which cannot fit the speckle positions.

From a photometric point of view, the B ↔ Be (with an eventual shell phase) ↔ B switching occurs every 4 to 9 years (with a 8.5 years average), and can be described as follows :
  - when the star is at its light minimum (Be phase, lasting from 2 to 7 years), it displays a double-wave light curve of 1.6 day period, with a 0.1 mag amplitude in the visible.
  - it has been recently shown (Sareyan et al., 1992) that this 1.6 day period still exists when the star is in its "normal" B phase (the star being then bluer), but with a 0.01 mag amplitude, i.e. very difficult to detect with such a long period.

We decided of a campaign on this star with several techniques, in order to solve the contradictions mentioned above, and also to get a better understanding of this rather complex system, i.e. investigating the relationship

L. A. Balona et al. (eds.), Pulsation, Rotation and Mass Loss in Early-Type Stars, 282–283.
© 1994 IAU. Printed in the Netherlands.

between the data obtained by photometry (activity + rotation or pulsation) and spectroscopy (photospheric line profiles, activity in the Balmer series), and trying to determine if any coincidence (coupling) exists between the Be ↔ Be shell phase of the star "A" and the proximity of "a" (which also means a better orbit determination for the latter).

Observations have been planned in China, France, Italy, Mexico, Russia, Switzerland, and USA. Bad weather conditions did not allow photometric observations in Europe. A few preliminary results can be pointed out :

- strong night-to-night variations occur in the HeI λ4471 and MgII λ4481 line profiles, probably not entirely due to the B1-B2 binarity.

- the center of the Hα line undergoes some small variations with a time constant of the same order that the 1.6 day photometric period. It is important to stress here that spectroscopic observations made two months before show a similar aspect, while in august 1991, Hα appeared on interferometric data with a strong emission on both sides of the central absorption.

- the speckle interferometry carried out in Russia with the 6 meter telescope (Aug.10th and Oct.6th) shows that the B1-B2 distance to component A has decreased to 0.187 - 0.186" in 1992, while the component a - provided its detection is confirmed - could be at a distance around 0.02 to 0.04" only. So, we already know that the 3.7 yr period is very probably wrong, as no close companion has been detected at the predicted distance of 0.077".

Provisional conclusions :

Due to the large amount of photometric data collected, we will probably be able to derive our own ephemeris on which the photospheric lines behavior will be "phased", i.e. giving a strong constraint on any future explanation and model. Particularly, the small variations which affect the Hα line center have to be interpreted (in terms of variable activity or rotation of an active feature ?)

The speckle observations show that the whole system, as a multiple star, has to be entirely reconsidered (masses and orbital periods).

As the campaign did not coincide with a B to Be transition phase (still unpredictable), we have no idea about the time constants involved in such a process, and we have no measurements of the phase lag between the beginning of the larger photometric variations and the beginning of the Be phase, when emission appears. This will deserve further studies.

References

Hill G.M., Walker G.A.H., Dinshaw N., Yang S., Harmanec P.: 1988, *PASP* **100**, 243
Hill G.M., Walker G.A.H., Yang S., Harmanec P.: 1989, *PASP* **101**, 258
Mac Alister H.A., Hartkopf W.I.: 1988, *2nd Catalogue of Interferometric measurements of binary stars (Georgia State Univ. Atlanta, 30303 USA)* ,
Sareyan J.P., Gonzalez-Bedolla S., Chauville J., Morel P.J., Alvarez M.: 1992, *Astron.Astrophys.* **257**, 567

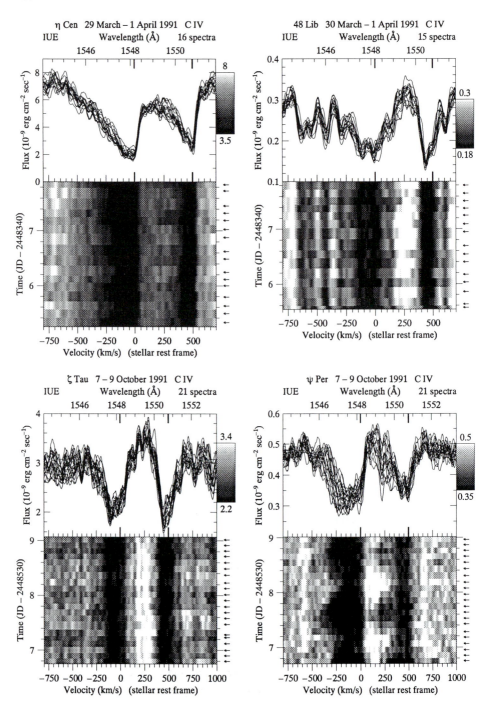

PERIODIC VARIATIONS AND MASS LOSS IN Be STARS

HIDEYUKI SAIO

Astronomical Institute, Faculty of Science, Tohoku University, Sendai 980, Japan

Abstract. We discuss the connection between the periodic light variations and the equatorial mass loss of Be stars. The observed properties of the short period ($\sim$ day) variations seem to indicate that they arise in the photosphere. An upper limit for the surface magnetic field of Be stars is derived from the rate of angular momentum loss expected from the typical mass loss-rate in Be stars. The upper limit suggests that surface magnetic fields of Be stars are too weak to make a spot. We argue that the periodic variations of Be stars are explained by nonradial pulsations whose periods on the stellar surface are much longer than the rotation period. They transport angular momentum from the core to the envelope to accelerate the surface regions. If this mechanism works sufficiently well, the rotation speed near the surface will reach to the critical velocity and an excretion disk will be formed around the star. A simple model for a steady-state excretion disk around a Be star is found to be consistent with the density structure inferred from the IR fluxes.

1. Introduction

Be stars are B-type stars which have or had emission in the Balmer lines (see e.g., Slettebak 1988, Bjorkman & Cassinelli 1993 for recent reviews). The emission line profile and polarization of light are consistent with the existence of an equatorial disk around a Be star.

Be stars, which tend to have large projected rotational velocities, show various light and spectral variations with timescales ranging from a fraction of a day to years. Among the variations, many Be stars show short period ($\sim$ day) light variations (e.g., Balona 1990), which seem to be associated with low-order line profile (asymmetry) variations (e.g., Baade 1987). According to Balona (1990), we call these stars λ Eri stars.

Another important property of Be stars is that the mass-loss rate determined from the IR excess is much higher than the rate from UV resonance lines, which is interpreted as that the mass-loss rate is much higher in the equatorial disk compared to the radiation driven wind in the polar region (Waters *et al.* 1987).

In this paper we regard rapid rotation and short-period light (and low-order line profile) variation as necessary conditions for a star to become a Be star, and discuss the relation between the short-period variation and mass loss.

2. Observed Properties of the Periodic Variations of Be Stars

Fig. 1 shows the position of λ Eri stars on the HR diagram together with β Cep stars (Lesh & Aizenman 1978) and slowly pulsating B (SPB) stars (Waelkens 1991; sometimes called 53 Per stars). As seen in this figure, λ Eri

L. A. Balona et al. (eds.), Pulsation, Rotation and Mass Loss in Early-Type Stars, 287–298.

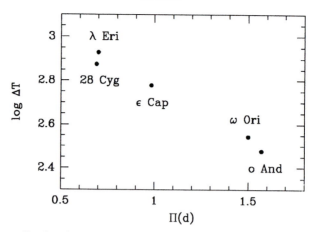

Fig. 3. The amplitude of temperature variation derived from FUV-optical observation (Percy & Peters 1991) versus period.

ation in EW Lac. Since this phenomenon was accompanied by an abrupt change in the phase of its periodic variation, it is likely that the persistent periodic light variation was interrupted by another phenomenon. This phenomenon is probably similar to the sudden appearance of periodic variations accompanied by a large increase in the mean brightness observed in the Be star κ CMa by Balona (1990). These phenomena could be related to the transition phases of μ Cen observed by Hanuschik *et al.* (1993), in which a circumstellar disk is being formed. In such transition phases periodic light variation can be produced by the existence of inhomogeneities in the newly-born circumstellar envelope. (Hanuschik *et al.*, 1993, found that such an inhomogeneity was erased in a few revolutions.) In the remainder of this paper, we will consider persistent periodic variations only.

In Fig. 3 the amplitude of the photospheric temperature variation ΔT determined from FUV-optical data (Percy & Peters 1991, Peters 1991) is plotted against the period of variation. It is apparent that the temperature variation is larger for stars with shorter periods. Although the number of stars for which ΔT is determined is still small, if this correlation is confirmed by other λ Eri stars, it would give an important constraint for the model of the light variations.

We consider in the following sections two possibilities for the photospheric variations which cause light and low-order line profile variations in λ Eri stars. One is the photospheric temperature inhomogeneity (spots) possibly generated by a dipole magnetic field. The other possibility is a nonradial pulsation which has a low azimuthal degree (m) and a small oscillation frequency in the co-rotating frame of the envelope.

3. Magnetic Fields

A strong magnetic field could make spots on the photosphere, but such a field has not been detected in λ Eri stars. The observationally estimated upper limit of the surface magnetic field is $\sim 10^2$ gauss (Baker 1987, Bohlender, these Proceedings). Smith & Polidan (1993) inferred the existence of $\sim 10^2$ gauss magnetic fields in λ Eri from its peculiar line profile variations.

An upper limit of the surface magnetic field may be estimated theoretically in a way similar to that employed by Washimi *et al.* (1993). The total angular momentum J of a uniformly rotating star may be written as

$$J = fMR^2\Omega, \tag{3.1}$$

where M, R, and Ω are the mass, radius, and the angular frequency of rotation respectively and $f \sim 0.1$ for a massive main-sequence star (e.g., Eriguchi *et al.* 1992). The rate of angular momentum loss from a magnetic star $\dot{J}$ is estimated as

$$\dot{J} \simeq \dot{M}\Omega R_A^2, \tag{3.2}$$

where $\dot{M}$ and R_A are mass loss rate and the Alfven radius respectively. Since Be stars have no preferred evolutionary phase in the main-sequence band (see e.g., Mermilliod 1982, Grebel *et al.* 1992), the timescale of spin down $(J/\dot{J})$ should be longer than the main-sequence life time τ_{MS}. This condition may be written as

$$f\frac{M}{\dot{M}}\left(\frac{R}{R_A}\right)^2 > \tau_{MS}. \tag{3.3}$$

The relation between the strength of a dipole magnetic field on the surface B_0 and at the Alfven radius gives the following relation

$$\left(\frac{R_A}{R}\right)^2 = \left(\frac{B_0}{2}\right)^{2/3}(4\pi\rho_A v_A^2)^{-1/3}, \tag{3.4}$$

where ρ_A and v_A are, respectively, the gas density and outward velocity at the Alfven radius. Substituting Eq. 3.4 into inequality 3.3, we obtain a constraint on B_0;

$$B_0 < 4\sqrt{\pi\rho_A}v_A\left(f\frac{M}{\dot{M}\tau_{MS}}\right)^{3/2}. \tag{3.5}$$

If we adopt the following typical values $\rho_A \sim 10^{-13}$g cm^{-3}, $v_A \sim 20$km s^{-1}, $M/\dot{M} \sim 10^9$yr, and $\tau_{MS} \sim 10^7$yr, we have a constraint of $B_0 \lesssim 100$ gauss, which is consistent with the observationally determined upper limit.

In order for a spot to be formed on the photosphere, the magnetic pressure should be comparable to (or larger than) the photospheric pressure. The upper limits of surface magnetic pressure corresponding to the upper limit of

Next we consider the relation between $\mathcal{F}_J$ and the energy flux of the nonradial pulsation $\mathcal{F}_E$. The energy flux may be evaluated as the rate of work done by the nonradial pulsation: i.e.,

$$\mathcal{F}_E = <v_r p'> = \frac{\omega}{2\pi} \int_0^{2\pi/\omega} v_r p' dt, \qquad (5.6)$$

where p' is the Eulerian perturbation of pressure. The pressure perturbation is related to the pulsation velocity through the momentum equation in the ϕ direction (on the equatorial plane) by

$$-\frac{1}{\rho r}\frac{\partial p'}{\partial \phi} \simeq \partial v_\phi/\partial t = -mAk\omega \sin(\omega t + m\phi + \delta). \qquad (5.7)$$

Integrating this equation with respect to ϕ, we obtain

$$p' \simeq -Ak\omega\rho r \cos(\omega t + m\phi + \delta) = -v_\phi \omega \rho r/m. \qquad (5.8)$$

Substituting Eq. 5.8 into Eq. 5.6 and using Eq. 5.5, we obtain the relation between $\mathcal{F}_E$ and $\mathcal{F}_J$,

$$\mathcal{F}_E = -\frac{\omega^2 \rho r}{2\pi m} \int_0^{\frac{2\pi}{\omega}} v_r v_\phi dt = -\omega \mathcal{F}_J/m = \frac{1}{2}\omega r \rho A^2 k \sin\delta. \qquad (5.9)$$

It is apparent that a prograde mode ($\omega/m < 0$) transports angular momentum in the same direction as the energy flow.

The low-frequency nonradial pulsations which are overstable in a rotating massive main-sequence star are likely to be prograde (see §4). For these oscillations energy flows from the convective core to the damping zone near the surface (Lee & Saio 1989). Therefore angular momentum is transported outward and the surface region is accelerated. Although the photospheric rotation speeds of Be stars are constrained observationally to be less than 90 per cent of the critical velocities (Collins & Sonneborn 1977), there is some indication that the rotational velocity increases outward above the photosphere (Chen et al. 1989). Therefore it is possible that the surface region is accelerated by nonradial pulsations and that the rotational velocity reaches the critical velocity at a layer above the photosphere.

6. Excretion Disk around a Be Star

If angular momentum is supplied steadily at the surface of a Be star, an excretion disk is formed around the star. In the excretion disk, the gravitational force in the radial direction is balanced by the centrifugal force of the Keplerian motion as in an accretion disk, but the direction of the drift is outward. Hanuschik et al. (1993) call it a decretion disk. Lee et al. (1991)

investigated the vertical structure of excretion disks. Here we discuss simple properties of a steady excretion disk analytically.

In a steady state Keplerian disk, the conservation of angular momentum may be written as

$$4\pi \frac{d}{dr}(r^2 W) + \dot{M}\sqrt{GM/r} = 0, \tag{6.1}$$

where $\dot{M} = 2\pi r \Sigma v_r$ represents the mass-loss rate (which is constant throughout the disk), M is the mass of the central star, and W is the vertically integrated viscous stress. In order to evaluate W we use the α description for the viscous stress (Shakura & Sunyaev 1973) and assume that the disk is vertically isothermal. Then we obtain

$$W = \alpha \int_{-\infty}^{\infty} p\,dz = \alpha c_s^2 \int_{-\infty}^{\infty} \rho\,dz = \alpha c_s^2 \Sigma, \tag{6.2}$$

where $c_s(r) = \sqrt{p/\rho}$ is the isothermal sound speed, Σ is the surface density, and α is a free parameter of $O(1)$. Integrating Eq. 6.1 yields the relation

$$2\pi \alpha c_s^2 r^2 \Sigma = \dot{M}\sqrt{GMr_1}(1 - \sqrt{r/r_1}), \tag{6.3}$$

where the subscript 1 indicates a quantity at the outer boundary of the disk ($\Sigma_1 \sim 0$). From Eq. 6.3 we obtain

$$\Sigma \propto r^{-2}c_s^{-2} \quad \text{and} \quad v_r = \dot{M}/(2\pi r\Sigma) \propto rc_s^2 \quad \text{for} \quad r \ll r_1. \tag{6.4}$$

In order to convert the surface density to the density on the equatorial plane, $\rho_e(r)$, we consider the vertical structure of the disk. Since $p = c_s(r)^2 \rho$, the equation for vertical hydrostatic equilibrium, $dp/dz = -GMz\rho/r^3$ may be integrated to yield $\rho(z,r) = \rho_c(r)\exp[-GMz^2/(2r^3 c_s^2)]$. Then we have

$$\Sigma = \int_{-\infty}^{\infty} \rho\,dz = \rho_e r^{3/2} c_s \sqrt{\frac{2\pi}{GM}}. \tag{6.5}$$

If we assume that the temperature of the excretion disk varies as $r^{-\beta}$ with $\beta > 0$ (i.e., $c_s \propto r^{-\beta/2}$), we obtain

$$\rho_e \propto \Sigma r^{-3/2} c_s^{-1} \propto r^{-3.5+1.5\beta}. \tag{6.6}$$

Steady excretion models of Lee $et\ al.$ (1991) indicate $\beta \sim 1 - 1.5$. If a shock wave is formed at the surface of the excretion disk due to ram pressure of the polar wind (Bjorkman & Cassinelli 1993), β would have a smaller value. In spite of our simple analysis, Eq. 6.6 is consistent with the observational relation $\rho \propto r^{-n}$ with $2 < n < 3.5$ obtained from IR data by Waters $et\ al.$ (1987).

Acknowledgements

I thank Shinpei Shibata and Osamu Kaburaki for helpful conversations on the effect of magnetic fields.

References

Ando, H.: 1986, *Astron. Astrophys.* **163**, 97.
Baade, D.: 1982, *Astron. Astrophys.* **105**, 65.
Baade, D.: 1984, *Astron. Astrophys.* **134**, 105.
Baade, D.: 1987, in Slettebak, A. and Snow, T.P., eds., *Physics of Be Stars*, Cambridge Univ. Press: Cambridge, 361.
Baker: 1987, in Slettebak, A. and Snow, T.P., eds., *Physics of Be Stars*, Cambridge Univ. Press: Cambridge, 38.
Balona, L.A.: 1990, *Mon. Not. Roy. Astr. Soc.* **245**, 92.
Balona, L.A.: 1992, *Mon. Not. Roy. Astr. Soc.* **256**, 425.
Balona, L.A.: 1993, *Mon. Not. Roy. Astr. Soc.* **260**, 795.
Balona, L.A., Sterken, C. and Manfroid, J.: 1991, *Mon. Not. Roy. Astr. Soc.* **252**, 93.
Bjorkman, J.E. and Cassinelli, J.P.: 1993, *Astrophys. J.* **409**, 429.
Bolton, C.T., Fullerton, A.W., Bohlender, D., Landstreet, J.D. and Gies, D.R.: 1987, in Slettebak, A. and Snow, T.P., eds., *Physics of Be Stars*, Cambridge Univ. Press: Cambridge, 82.
Chen, H., Riguelet, A., Sahade, J. and Kondo, Y.: 1989, *Astrophys. J.* **347**, 1082.
Collins, G.W.,II and Sonneborn, G.H.: 1977, *Astrophys. J. Suppl.* **34**, 41.
Cox, J.P.: 1974, *Rep. Prog. Phys.* **37**, 563.
Dziembowski, W.A. and Pamyatnykh, A.A.: 1993, *Mon. Not. Roy. Astr. Soc.* **262**, 204.
Eriguchi, Y., Yamaoka, H., Nomoto, K. Hashimoto, M.: 1992, *Astrophys. J.* **392**, 243.
Gautschy, A. and Saio, H.: 1993, *Mon. Not. Roy. Astr. Soc.* **262**, 213.
Grebel, E.K., Richtler, T. and de Boer, K.S.: 1992, *Astron. Astrophys.* **254**, L5.
Hanuschik, R.W., Dachs, J., Baudzus, M. and Thimm, G.: 1993, *Astron. Astrophys.* **274**, 356.
Kiriakidis, M., El Eid, M. and Glatzel, W.: 1992, *Mon. Not. Roy. Astr. Soc.* **255**, 1p.
Lee, U.: 1988, *Mon. Not. Roy. Astr. Soc.* **232**, 711.
Lee, U., Jeffery, C.S. and Saio, H.: 1992, *Mon. Not. Roy. Astr. Soc.* **254**, 185.
Lee, U. and Saio, H.: 1986, *Mon. Not. Roy. Astr. Soc.* **221**, 365.
Lee, U. and Saio, H.: 1987, *Mon. Not. Roy. Astr. Soc.* **225**, 643.
Lee, U. and Saio, H.: 1989, *Astrophys. J.* **360**, 590.
Lee, U. and Saio, H.: 1993, *Mon. Not. Roy. Astr. Soc.* **261**, 415.
Lee, U., Saio, H. and Osaki, Y.: 1991, *Mon. Not. Roy. Astr. Soc.* **250**, 432.
Lesh, J.R. and Aizenman M.L.: 1978, *Ann. Rev. Astron. Astrophys.* **16**, 215.
Mermilliod, J.-C.: 1982, *Astron. Astrophys.* **109**, 48.
Moskalik, P. and Dziembowski, W.A.: 1992, *Astron. Astrophys.* **256**, L5.
Osaki, Y.: 1986, *Publ. Astr. Soc. Pacific* **98**, 30.
Percy, J.R. and Peters, G.J.: 1991, in Baade, D., ed., *Rapid Variability of OB-stars: Nature and Diagnostic Value*, ESO: Garching, 97.
Peters, G.J.: 1991, in Baade, D., ed., *Rapid Variability of OB-stars: Nature and Diagnostic Value*, ESO: Garching, 171.
Shakura, N.I. and Sunyaev, R.A.: 1973, *Astron. Astrophys.* **24**, 337.
Slettebak, A.: 1988, *Publ. Astr. Soc. Pacific* **100**, 770.
Slettebak, A., Collins, G.W. II and Truax, R.: 1992, *Astrophys. J. Suppl.* **81**, 335.
Smith, M.A. and Polidan, R.S.: 1993, *Astrophys. J.* **408**, 323.
Stagg, C.R., Božić, H., Fullerton, A.W., Gao, W.S., Guo, Z.H., Harmanec, P., Horn, J., Huang, L., Iliev, L.H., Koubský, P., Kovachev, B.Z., Pavlovski, K., Percy, J.R., Schmidt, F., Štefl, S., Tomov, N.A. and Žižňovský, J.: 1988, *Mon. Not. Roy. Astr.*

Soc. **234**, 1021.
Unno, W, Osaki, Y., Ando, Y., Saio, H. and Shibahashi, H.: 1989, *Nonradial Oscillations of Stars*, 2nd ed., Univ. Tokyo Press: Tokyo
Washimi, H., Mori, M. and Shibata, S.: 1993, *Nature* submitted.
Waters, L.B.F.M., Coté, J. and Lamers, H.J.G.L.M.: 1987, *Astron. Astrophys.* **185**, 206.
Waelkens, C.: 1991, *Astron. Astrophys.* **246**, 453.
Waelkens, C., Van den Abeele, K. and Van Winckel, H.: 1991, *Astron. Astrophys.* **251**, 69.

Discussion

Smith: I have two comments to make.

(1) It may be possible to produce an excretion disk by hydrodynamical interaction of particles initially ejected from a star by a *deus ex machina*. In such a picture, many particles would lose angular momentum and return to the star, leaving the other particles angular-momentum rich and able to remain in orbit. Although no one has explored the validity of this model yet, it should be considered for inclusion in your list of mechanisms (even in the absence of radiation pressure).

(2) Your J-transfer mechanism from NRP (the Ando mechanism) relies on $m < 0$. Actually, NRP models seem to always have $m > 0$ and therefore the J-transfer will be inward, not outward! (Note that even Harmanec's lower mass-radius relation will change the sign of m.)

Lafon: My question is concerned with dynamics. You have a star rotating at a given (even if perturbed) velocity, but you say that the excretion disk rotates with Keplerian velocity. There are too many dynamical conditions in the star and the disk to be matched without some *boundary layer* between them. Can you comment about this problem in your model?

Saio: I am afraid that my model is too simple to answer your question properly. I just speculate that the rotational velocity increases outward and attains Keplerian velocity at some height above the photosphere.

Owocki: I didn't quite understand the mechanism for transporting angular momentum outward. But assuming it is true, I still don't understand why it wouldn't make the surface layers spin up past critical velocity, in contradiction to observations of widths of photospheric lines. Also, even if your angular momentum is contained in the pulsations and not simply in the systematic rotation of the atmosphere, wouldn't it still impart a centrifugal effect that would expel the pulsating photospheric material?

Saio: I speculate that the rotational velocity becomes critical above the photosphere. The perturbation of angular momentum caused by nonradial pulsations is assumed to be small, but on a timescale very long compared to the period of nonradial pulsation, angular momentum can be accumulated in the surface regions where the wave dissipates.

Henrichs: What is the effect of Eddington-Sweet currents on the rate of outward angular momentum transport?

Saio: I did not take into account the effect of Eddington-Sweet currents, but I think that the timescale involved is much longer than the timescale of angular momentum transport by nonradial pulsation.

Hanuschik: In Hanuschik *et al.* (1993), we published a series of Hα observations of μ Cen (HR 5193) where we observed isolated outbursts, and a decay of emission thereafter, with a typical decay timescale of about 50–100 days. We proposed the decretion disk model in order to explain these findings. Do you have any ideas on the origin of the decay timescale and, if it is due to viscosity, about its magnitude?

Saio: The timescale of radial drift of gas in a Keplerian disk is determined by viscosity. Since we use the α-disk assumption, the timescale depends on the parameter α. If we assume $\alpha = 1$, the drift timescale for an excretion disk around a typical Be star is a few hundred days, which is more or less consistent with your results on the decay timescale of μ Cen.

Harmanec: How can you explain the observed periodic light curves of Be stars by NRP given the fact that they are (i) singly-periodic, (ii) systematically change their amplitudes and shapes from cycle to cycle, and (iii) their periods remain constant over years?

Saio: Amplitude modulation is possible when at least two pulsation modes are excited simultaneously. If the periods of these modes are very close to each other, only one period is detected observationally unless a light curve with a very long baseline is analyzed. Since the pulsation periods are much longer than the rotation period in the co-rotating frame of the convective core, the periods in the observer's frame should be very close to the rotation period of the core divided by $|m|$ and hence should be stable.

PHOTOMETRIC STUDIES OF Be STAR VARIABILITY

JOHN R. PERCY

Erindale Campus, University of Toronto
Mississauga, Ontario, Canada L5L 1C6

Be stars are hot stars which have shown emission in at least one Balmer line on at least one occasion. As the definition implies, the Be phenomenon can be variable with time: on time scales of days to decades as the circumstellar disc develops and disperses; on time scales of days to months in a few Be stars which are interacting binaries; on time scales of 0.2 to 2 days due to non-radial pulsation or possibly rotation. The Be stars are worthy of photometric study because they are bright and numerous; the nature of the short-term variability is not yet agreed upon; the cause of the development of the disc – and its relationship to the short-term variability – is also not yet known.

1. Long-Term Photometric Monitoring

For over a decade, as part of an international UBV photometric campaign proposed and organized by P. Harmanec and his colleagues, we have been photometrically monitoring several bright, active Be stars using two facilities: the 0.4m "teaching telescope" on the main campus of the University of Toronto (Percy et al. 1988), and the 0.25m Automatic Photoelectric Telescope in Arizona (Percy & Attard 1992). The program currently includes θ CrB, 4 Her, 88 Her, 66 Oph, MWC 601, CX Dra, 2 Vul, V923 Aql, V1294 Aql, 12 Vul, 25 Cyg, 28 Cyg, QR Vul, o And, KX And and KY And. Other stars are occasionally observed as part of special projects or campaigns. Many of these stars show short-term variability; some of them (MWC 601 and QR Vul, for instance) deserve further study using a multi-longitude approach. I would be happy to send copies of the observations made since Percy and Attard's (1992), or to collaborate in multi-longitude studies of some of these stars.

2. Short-Term Photometric Variability

The majority of Be stars vary on time scales of 0.2 to 2 days. This variability is difficult to study because: the time scale (about a day) is very inconvenient for study from a single site; the short-term variability is not always strictly periodic; there is often additional variability from day to day because of changes in the emission or absorption by the disc; and the amplitude of the variability is a few hundredths of a magnitude at the most. We have therefore made many of our observations as part of multi-longitude "cam-

L. A. Balona et al. (eds.), Pulsation, Rotation and Mass Loss in Early-Type Stars, 299–300.
© 1994 *IAU. Printed in the Netherlands.*

paigns" - often as part of multi-wavelength campaigns which use optical and UV photometry and spectroscopy, along with polarimetry to understand the short-term variability (Peters et al., this meeting). The following are preliminary results of the photometric parts of some of these campaigns.

25 Cyg (HR 7647) has a suspected short period of 0.21 day (Percy et al. 1981), which is too short to be easily explained by rotation. From observations from Hungary, Toronto and Arizona, we have derived a "best" period of 0.226 day, with a V range of 0.01, though lesser peaks are present in the power spectrum.

ψ Per (HR 1087) has a suspected short-term variability of 0.05 in V (Percy et al. 1981); the period is unknown. From observations from Hungary and Arizona, we have derived a "best" period of 1.021 days. The ranges in V, B and U are 0.019, 0.016 and 0.065. The period is consistent with that derived from UV data (Peters et al., this meeting).

ζ Tau (HR 1910) has a variety of suspected short periods determined by various observers by various techniques. From observations from Hungary and Arizona, we have obtained a very complex power spectrum, with peaks around 0.3, 0.67 and 1.8 days. The period of 0.67 day fits the data well; it agrees with a period determined spectroscopically by Yang et al. (1990); and is reasonably consistent with that derived from UV data (Peters et al., this meeting). It should, however, be regarded as tentative. The ranges in V, B and U are 0.071, 0.064 and 0.106.

2 Vul (HR 7318) has a suspected period of 0.6096 day (Lynds 1959). From observations from Arizona only, we have obtained a power spectrum with the highest peak at 0.615 day. The ranges in V, B and U are 0.10, 0.11 and 0.15.

Note that, in each case, the B and V amplitudes are comparable, and the U amplitude is larger. The UV amplitude is larger still. This is consistent with a model in which the variability is due to temperature fluctuations in a non-radially pulsating star (Peters et al., this meeting).

Acknowledgements. I thank my campaign collaborators R.E. Fried, K. Hayhoe, I.I. Ivans, M. Paparo and S. Roy Choudhury, as well as D. Gies, H. Henrichs, D. McDavid and G. Peters for their substantial contributions to this work, and the Natural Sciences and Engineering Research Council of Canada for support.

3. References

Lynds, C.R. 1959, ApJ, **130**, 577

Percy, J.R., Jakate, S.M. and Matthews, J.M. 1991, AJ, **86**, 53

Percy, J.R., Coffin, B.L., Drukier, G.A., Ford, R.P., Plume, R., Richer, M.G. and Spalding, R. 1988, PASP **100**, 1555

Percy, J.R. and Attard, A. 1992, PASP **104**, 1160

Yang, S., Walker, G.A.H., Hill, G.M. and Harmanec, P. 1990 ApJS **74**, 595

PHOTOMETRIC CHARACTERISTICS OF THE Be STARS: ALMOST TWENTY YEARS OF UBV MONITORING AT THE HVAR OBSERVATORY

K. PAVLOVSKI, H. BOŽIĆ, Ž. RUŽIĆ,
Hvar Observatory, Zagreb University, 41000 Zagreb, Croatia

and

P. HARMANEC, J. HORN, P. KOUBSKÝ, K. JUZA, and
P. HADRAVA, S. KŘÍŽ, S. ŠTEFL, F. ŽĎÁRSKÝ
Astronomical Institute, 251 65 Ondřejov, Czech Republic

1. The scope of the observational project

Since their discovery (by Father Secchi in 1866) until the end of sixties, Be stars were not a subject of any systematic studies of their possible light and colour variations. Already at that time, the astronomical literature contained ample evidence showing that a number of Be stars *were* light variables. However, almost all such findings resulted as by-products of studies of different or wider groups of objects. Feinstein (1968) was probably the first who pointed out explicitly that many Be stars are light variables. A pioneering study which was aimed at the detection of light variations of a large group of Be stars *by means of differential photoelectric photometry* was carried out by Haupt & Schroll (1974).

In 1972, our group has started systematic photoelectric observations of bright Be stars at the jointly built observing station which belongs to the Hvar Observatory in Croatia. Our original goal was to search for new eclipsing binaries among Be stars. However, already the first results arose our interest also in their long-term and rapid variations.

2. Observations and reductions

All observations were obtained with the Ondřejov 0.65-m Cassegrain telescope installed at the Hvar Observatory. A single-channel photoelectric photometer with an unrefrigerated EMI 6256S photomultiplier and UBV (Schott glass) filters has been used. During the first epoch (1972 - 1978), the data were recorded with a strip-chart recorder. In 1979, the recorder was replaced by a VF digital converter. After 1988, some observations were secured with a new, computer-controlled UBV photometer, which was tested there. The project was temporarily terminated after January 1991.

L. A. Balona et al. (eds.), Pulsation, Rotation and Mass Loss in Early-Type Stars, 301–302.
© 1994 *IAU. Printed in the Netherlands.*

The final reduction was carried out with a FORTRAN program HEC22 (see Harmanec et al. 1993 for the details of reductions and software). HEC22 ensures a stable and accurate transformation into the standard Johnson's UBV system. Much more accurate UBV magnitudes of all comparison, check and standard stars were also derived from about 46 000 all-sky observations secured between 1972 and 1988.

3. The principal results

The principal finding of our and other studies is that the long-term light and colour variations of Be stars are usually the most pronounced ones and that they are somehow related to the long-term spectral variations of respective stars. While we failed to discover many new eclipsing binaries among the Be stars, we *did find* a half-dozen of phase-locked periodic light variations for some known Be binaries, e. g., CX Dra, KX And, ζ Tau, or LQ And.

Soon after the observing program at Hvar was started, Percy and his collaborators begun a search for new β Cep variables among the Be stars (Percy et al. 1981). Although they were also unsuccesful in achieving their original goal, they succeeded in demonstrating that a large fraction of Be stars are low-amplitude ($<0^m.1$) light variables, varying on a time scale of 10^{-1} - 10^0 d.

We found several well-documented cases of a clear correlation between the amplitudes of periodic short-term variations and the phase of the long-term cycle (e.g. EW Lac or o And). This fact was used as a diagnostic tool to predict new shell episodes (Harmanec et al. 1988, Pavlovski & Ružić 1991).

In passing, we wish to mention that at least in one case our original goal was finally achieved: A recent analysis of our archive data of V360 Lac (14 Lac, HD 216200) indicated that this Be star is a new peculiar eclipsing binary (P = 10.085 d). A detailed analysis, which combines our and other available photometry, and the Dominion Astrophysical Observatory spectroscopy by Dr. G. Hill, is currently under way (Hill et al., to be published).

References

Feinstein, A.: 1968, *Z. Astrophys.* **68**, 29

Harmanec, P., Hadrava, P., Ružić, Ž., Pavlovski, K., Božić, H., Horn, J., Koubský, P.: 1988, *IBVS, No. 3263* ,

Harmanec, P., Horn, J., Juza, K.: 1993, *A&AS, in press* ,

Haupt, H., Schroll, A.: 1974, *A&AS* **15**, 311

Pavlovski, K., Ružić, Ž.: 1991, in *Rapid Variability of OB-stars: Nature and Diagnostic Value*, Ed. D. Baade, ESO Conf. Proc. No. 36, 191

Percy, J.R., Jakate, S.M., Matthews, J.M.: 1981, *AJ* **86**, 53

PHOTOMETRY OF EARLY TYPE STARS IN OPEN CLUSTERS

CLUSTERS

(NGC 1444, NGC 1662, NGC 2129, NGC 2169 AND NGC 7209)

J. H. PEÑA and R. PENICHE
Instituto de Astronomía, UNAM

This is part of a series which has the purpose of examining the nature of the stars belonging to open clusters. The aim of this series is, among others things, to study short period pulsating stars, mainly of the Delta Scuti type, by first establishing the membership of each star to the cluster, to determine the abundance of the Be and Ap phenomena and blue stragglers in open clusters for clusters of different ages and metalicities and, eventually, to study the chemical enrichment of the galaxy when age, dynamics and metalicity are known for a fair number of clusters.

In the present study, an analysis of the open clusters NGC 1444, NGC 1662, NGC 2129, NGC 2169 and NGC 7209 is presented. These clusters were selected because they seem to have a relatively large number of early type stars. The observations were carried out at the National Astronomical Observatory of the UNAM. The 1.5 m telescope with a multichannel uvby-β spectrophotometer was utilized for the acquisition of the data. A compilation of the observations and results are presented in Table 1.

Since the aim of the present paper is to establish physical and geometric characteristics of the cluster stars, the first step was to determine membership of the observed stars to each of the clusters. Then we defined which stars were Main Sequence stars and the broad spectral regions to which they belonged by constructing a $[m_1] - [c_1]$ diagram. It defines three main spectral regions: early type stars of class B and early A; A and F stars, and late type stars. The distance for each group has been calculated separately.

The calibration of the B and early A type stars follows a method proposed by Shobbrook in 1984, and by Balona and Shobbrook (1984) and for the A and F stars a procedure proposed by Nissen in 1988 was employed. The membership probability has been defined by adjusting a gaussian distribution to the histogram of the distances in parsecs.

In order to estimate the age of the cluster the following procedures were undertaken: An estimation of the turn-off point was carried out by means of the grids of Relyea and Kurucz (1978). Then a direct comparison was

L. A. Balona et al. (eds.), Pulsation, Rotation and Mass Loss in Early-Type Stars, 303–304.

made with the theoretical models of VandenBerg (1985) or, for early type stars, with those of Maeder (1991) or Meynet et al. (1991) which consider overshooting.

The determination of the Ap abundance was carried out through uvby-β photometry since it is well-known that the Ap stars lie in a specifically defined zone in the $[m_1] - [c_1]$ diagram. From a compilation of several articles on Ap stars in open clusters, a ratio of 0.051 for the frequency of Ap stars in the clusters and of 0.070 for field stars was determined. From the results obtained in the present paper, a higher value is determined, see Table 1.

Two new blue stragglers that belong to the clusters have been found in the present study, see Table 1.

TABLE I

Compilation of Results.

cluster NGC	obs all	stars mbr	Ap all	Ap mbr	dist (pc)	E(b-y)	age $\times 10^6$	binaries all	binaries mbr	blue strg	Ap/BAF
1662	42	28	7	7	390	0.231	490	4	2	0	0.24
1444	25	6	1	0	920	0.568	170	0	0	1	0.00
2169	20	8	2	0	860	0.147	50	0	0	1	0.00
2129	37	-	1	-	—	—	–	-	-	-	-
7209a	54	16	7	1	760	0.108	500	1	0	0	0.06
7209b	54	17	7	2	1190	0.145	710	1	1	0	0.12

References

Balona, L. and Shobbrook, R. R., 1984, MNRAS, 11, 375.

Maeder, A., 1991 Astrophysical Age and Dating Methods, 71, Ed. Frontiers, Paris.

Meynet, G., Mermillod, J. C., and Maeder, A., 1991, Astrophysical Age and Dating Methods, 91, ED. Frontiers, Paris.

Nissen P. E., 1988, A&A, 199, 146.

Reylea, L. J., and Kurucz, R. L., 1978, ApJS, 37,45.

Shobbrook, R. R., and Balona, L., 1984, MNRAS, 11, 659.

RAPID MULTICOLOUR VARIATIONS OF SELECTED
SOUTHERN BE STARS *

S. ŠTEFL, P. HARMANEC
Astronomical Institute, Academy of Sciences of the Czech Republic,
CZ-251 65 Ondřejov, Czech Republic

and

D. BAADE
European Southern Observatory, Karl-Schwarzschild-Str. 2,
D-85748 Garching bei München, Germany

1. Introduction and observational database

Many previous surveys of the rapid variability of Be stars (e.g., Balona et al. 1987, Cuypers et al. 1989, Balona et al 1992) were limited to the Strömgren b-band only. We have undertaken a new observational effort with the specific aim of studying colour variations. A search for new rapidly variable stars should provide us with statistical material needed to generalize properties of the phenomenon with respect to the stellar parameters. Some preliminary results of these items are presented here.

$uvby$ and Hβ measurements were obtained in three observing runs at the European Southern Observatory with the Danish automatic 0.50-m telescope in 1991 November, 1992 May/June, and 1992 December/1993 January. Additional photometric observations were secured from the South African Astronomical Observatory in parallel to the first two runs. Thirty stars covering all B spectral sub-classes were observed.

2. Results

The amplitudes of the variations usually stay below $0.^m12$ and do not differ much between the four passbands. But generally they are the largest in the u band and decrease towards longer wavelengths. The variations in the $b-y$ and m_1 indices only in some cases reach a few hundredths of a magnitude but mostly remain within the scatter of the observations of the check stars ($0.^m01$ - $0.^m02$). By contrast, variations in the $u-b$ or c_1 indices are well detectable for 85 % of the variable stars observed.

We can distinguish 3 types of variations which often occur in various combinations:

---

* Based on observations obtained at the European Southern Observatory, La Silla, Chile, and the South African Astronomical Observatory.

L. A. Balona et al. (eds.), Pulsation, Rotation and Mass Loss in Early-Type Stars, 305–306.

1. (Quasi-) periodic variations with periods of the order of one day which, at least for early-B spectral sub-classes, are characteristic of Be stars.
2. Ephemeral brightness changes by up to $0.^m2$ on a typical time scale of days. We observed both a drop and an increase in brightness. A decrease by $0.^m05$ in u and $0.^m02$ in b accompanied by a reddening of $0.^m03$ in $u-b$ and no significant change in $b-y$ was observed for HD 98922. The event was almost symmetric in time, but the star was fainter by about $0.^m01$ after it returned to its quiescent phase. The pole-on star μ Cen became brighter by $0.^m26$ in u and $0.^m16$ in b during 4 days in January 1993. The star got bluer by $0.^m1$ in $u-b$ but redder by $0.^m05$ in $b-y$. The observed part of the event was nearly symmetric, too. It may correspond to the "outbursts" in Hα equivalent width reported for μ Cen by Baade et al. (1988) and Hanuschik et al. (1993).
3. Smooth medium-term variations on a time scale of up to tens of days and amplitudes of a few hundredths of a magnitude (e.g. DX Eri).

From a comparison of the type-1 variabilities in $u-b$ and u, we find that most stars are bluer when brighter. Only for one-third of our sample such a relation is not well pronounced, and no star was found that showed the opposite trend.

No variations above the noise limit ($\approx 0.^m01$) were detected in 5 of 7 program stars with spectral types later than B7, in agreement with previous photometric and spectroscopic searches (Balona et al. 1992, Baade 1989). The full amplitude of the B8-9 IIIe star HR 4221 was $0.^m025$. Balona et al. (1992) derived a period of 0.870 d for variations of the same amplitude observed in January 1988. The star HD 98922 (B9Ve), which showed the large-amplitude event described above, may be the most active known late B-type star if future observations prove the photospheric origin of its activity and thereby exclude a binary model.

In addition to HD 98922 also HR 4009 was identified as a new rapidly variable Be star. Our observations of HR 4009 span only 20 nights but indicate 2 candidate periods near 0.8 d in all four Strömgren bands and the $u-b$ and c_1 indices.

References

Baade D.: 1989, *A&A* **222**, 200
Baade D., Dachs J.,v.d. Weygart R., Steeman F.: 1988, *A&A* **198**, 211
Balona L.A., Marang F., Monderen P., Reiterman A., Zickgraf F.-J.: 1989, *A&AS* **71**, 11
Balona L.A., Cuypers J., Marang F.: 1992, *A&AS* **92**, 533
Cuypers J., Balona L.A., Marang F.: 1989, *A&AS* **81**, 151
Hanuschik R.W., Dachs J., Baudzus M., Thimm G.: 1993, *A&A* **274**, 356

SIMULTANEOUS PHOTOMETRY AND SPECTROSCOPY
OF THE RAPID VARIABILITY OF THE BE STAR η CEN *

S. ŠTEFL
Astronomical Institute, Academy of Sciences of the Czech Republic,
CZ-251 65 Ondřejov, Czech Republic

D. BAADE
European Southern Observatory
Karl-Schwarzschild-Str. 2, D-85748 Garching bei München, Germany

P. HARMANEC
Astronomical Institute, Academy of Sciences of the Czech Republic,
CZ-251 65 Ondřejov, Czech Republic

and

L. A. BALONA
South African Astronomical Observatory
P.O. Box 9, Observatory 7935, Cape Town, South Africa

1. Introduction and observational results

The reality of the previously reported triple-wave light curve (Cuypers et al. 1989, A&A Suppl. 81, 151), and relations between rapid line profile and brightness variations of the B1-2III-Ve star η Cen (HD 127972) were investigated by means of simultaneous high-resolution spectroscopy (4 nights, 86 Si III 455.2622 nm profiles) and two-station *uvby* photometry (14 nights) secured in 1992. To this database were added the *uvby* photometry by Cuypers et al. and unpublished *b* observations by L.A.B. from 1988-1991.

After prewhitening for seasonal and medium cyclic ($\approx$ 30 days) changes, all 887 *b* data of 1987–1992 can be reconciled with a period of 0.642424 d according to the ephemeris:

$$T_{max.light} = (\text{HJD}2448767.582 \pm 0.012) + (0^d.6424241 \pm 0^d.0000027) \times E$$

The scatter *is not* reduced for either a double- or triple-wave period. There are systematic cycle-to-cycle variations of the light curve but no other coherent periodicity or "flickering" with an amplitude larger than about $0.^m005$. The full amplitude of rapid brightness variations is $\approx 0.^m10$ in u and $0.^m06$ in v, b and y; $b - y$ varies only slightly while the $c1$ index varies for $0.^m04$. The star is bluest in $u - b$ when brightest.

---

* Based on observations obtained at the European Southern Observatory, La Silla, Chile and the South African Astronomical Observatory

307

L. A. Balona et al. (eds.), Pulsation, Rotation and Mass Loss in Early-Type Stars, 307–308.
© 1994 IAU. Printed in the Netherlands.

The RV of the outer wings of the Si III line varies nearly sinusoidally with the 0.6424 d period and peak-to-peak amplitude of 62 ± 5 km s^{-1}. Similar periodicities can also be detected in W_λ, 1st and 3rd moment, I_c, and FWHM. The object is brighter and bluer close to the RV maximum, when also W_λ, FWHM, and I_c attain their maxima.

The Si III line profiles vary significantly on a time-scale of minutes. Moving bumps, carried across the line profile from blue to red, obviously cause the large part of the observed variability. They are barely visible near the line center where most of them disappear or weaken substantially. The appearance of the strong bumps apparently correlates with the phase of light minimum. Free fits of the composite velocity curves of individual travelling bumps led to periods of 0.683 ± 0.017 and 0.654 ± 0.042 days for the absorption and emission bumps, respectively, similar to the photometric period.

2. Confrontation with current models

Corotating surface or circumstellar structures in a single star can hardly be reconciled with periodic RV variations. Neither do they explain the observed profile modulation which is stronger in the wings than at the centre.

The nonradial pulsation (NRP) model explains the moving bumps by analogy to other early-type stars. It can also address the light and RV variations and make plausible the observed increased strength of the moving bumps in the line wings if the horizontal-to-vertical velocity amplitude ratio is large. The enormous difference between the amplitude of the T_{eff} variations as determined from $u - b$ and W_λ of Si III 455.3 nm (1200 K vs. 6000 K) can be understood in terms of NRP if effects of varying microturbulence and electron pressure are included. If the pulsation period is of the order of several days in the corotating frame, the amplitude of g-modes cannot exceed a few km s^{-1} in order to be consistent with the observed amplitudes of the photometric and spectroscopic variations. It is furthermore not clear that a pure NRP model can explain the large RV variations observed or the presence of higher-order bumps in just one half of the profile at a time.

Orbital motion in a binary system is the only straightforward interpretation for the sinusoidal RV curve observed. An orbital solution leads to a contact binary with Roche lobe radii of about 4 R$_\odot$ and 1.3 R$_\odot$, respectively. The model does not offer any simple explanation for the relative phasing of the RV and light curves observed and no direct spectroscopic evidence of the secondary was found. – The available data cannot discriminate between the pulsating-star and binary model which are based on almost disjunct sub-volumes of the observational parameter space.

(*The full version of the paper will be submitted to Astron. & Astrophys.*)

SIMULTANEOUS *uvbyβ* PHOTOMETRY AND Hα
SPECTROSCOPY OF Be STARS IN OPEN CLUSTERS

J. FABREGAT, J.M. TORREJÓN and P. REIG

Dep. de Matemática Aplicada y Astronomía, Universidad de Valencia, 46100 Burjassot, Valencia, Spain

and

G. BERNABEU

Dep. de Ingeniería de Sistemas y Comunicaciones, Universidad de Alicante, Apdo. 99, 03080 Alicante, Spain

1. Introduction

The usual methods of spectral clasification, equivalent widths of Balmer lines or photometric calibrations are not suitable for the determination of the astrophysical parameters of the underlying star in Be-type objects. The spectrum is distorted by the circumstellar envelope lines, while the contribution of the envelope continuum radiation contaminates the photometric indices.

The aim of the present work is to develop a method to determine such intrinsic parameters from *uvbyβ* photometry and Hα equivalent widths. We are currently carrying out simultaneous photometric and spectroscopic observations of Be stars in open clusters. The main pourpose is the decoupling of the underlying star contribution to the photometric indices from the circumstellar disc contribution, and the elaboration of a photometric calibration for the physical parameters of the underlying star. A first exposition of this method has been already published (Fabregat & Reglero 1990).

As a preliminary result, in this paper we present the analysis of the *uvbyβ* photometric diagrams for the h and χ Persei clusters (NGC 869 and NGC 884).

2. Discussion and conclusions.

The photometric HR diagrams are presented in Fig 1. We have represented with an open square the Be stars observed in an active fase, i.e. with the Hα line observed in emission or with the photometric index β lower than 2.55, which implies that the Hβ line is in emission. With a filled square we represent the stars previously classified as Be but which have been observed whit Hα in absorption, indicating the occurrence of a non-active phase. Finally, we represent with asterisks the "normal" absorption-line B stars. We

L. A. Balona et al. (eds.), Pulsation, Rotation and Mass Loss in Early-Type Stars, 309–310.

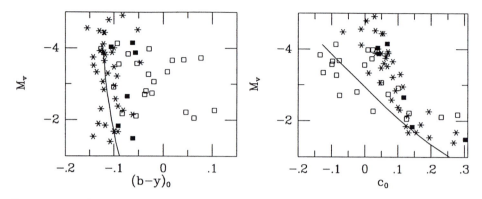

Fig. 1. M_V–$(b-y)_0$ and M_V–c_0 photometric diagrams for the h and χ Persei clusters. Open squares represent active Be stars, filled squares non-active Be stars and asterisks absorption-line B stars. Lines represent the ZAMS.

have included, as well as our own data, stars non classified as Be observed by Crawford *et al.* (1970), in order to obtain a better definition of the locus of the normal stars in the photometric diagrams.

In both diagrams Be stars in a non-active phase are placed in the same position than the normal B stars. This implies that the usual anomalous position of Be stars in the photometric diagrams affects only the Be stars in an active phase, and then it is an effect of the circumstellar emission, and not due to the underlying star itself. In consecuence, such anomalous positions should not be considered as differencies in evolutionary status between Be and absorption-line B stars.

Be stars in active phase deviate systematicaly in both diagrams. In the M_V–$(b-y)_0$ plane they deviate towards redder $(b-y)$ values, showing a significant continuum contribution of the circumstellar emission to the photometric indices. In the M_V–c_0 plane active Be stars deviate to lower "bluer" c_0 values. As the c_0 index is a measure of the Balmer discontinuity strength, this deviation is produced by the Balmer discontinuity emission, the so-called "Balmer jump" described, for instance, by Kaiser (1990).

We can conclude that the anomalous position usualy showed by Be stars in the photometric diagrams is produced by the contribution of the circumstellar continuum emission, an does not imply any particular evolutionary status.

References

Crawford, D.L., Glaspey, J.W. and Perry, C.L.: 1970, *AJ* **75**, 822
Fabregat, J. and Reglero, V:: 1990, *MNRAS* **247**, 407
Kaiser, D: 1990, *A&A* **222**, 187

PERIODIC VARIABILITY OF Be STARS:
NONRADIAL PULSATION OR ROTATIONAL
MODULATION?

D. BAADE

European Southern Observatory
Karl-Schwarzschild-Str. 2, D-85748 Garching bei München, Germany

and

L.A. BALONA

South African Astronomical Observatory
P.O. Box 9, Observatory 7935, Cape Town, South Africa

Abstract. A large fraction of Be stars show periodic light and line profile variations with a timescale of about one day. The mechanism which causes these periodic variations has been attributed to nonradial pulsation (NRP) or rotational modulation (RM). The authors present arguments supporting the two opposing points of view with the purpose of stimulating subsequent discussion by the Symposium participants.

1. The Case for Nonradial Pulsation (D. Baade)

1.1. WEAKNESSES OF THE ROTATIONAL MODULATION HYPOTHESIS

The primary reason, which has enabled the RM hypothesis to become a contender for the explanation of the periodic variability of Be stars, is that the observed periods are *statistically* indistinguishable from the best currently possible guesses for the rotation periods. This is certainly suggestive, but it must be kept in mind that presently rotation periods of *individual* stars cannot nearly be determined with the required accuracy. Increasing the sample further does *not* help because the *systematic* uncertainties are much larger.

More important is another objection. Recall that (i) only single-channel photometry has been performed, (ii) periods have been derived which broadly overlap with the expected rotation periods, and (iii) only these periods are interpreted: Is it, then, possible to draw any other conclusion than that the observed variability is due to some surface inhomogeneity which is carried around the star by rotation? The answer can only be a clear no. However, such a hypothesis is *observing strategy-limited* and should, therefore, be received with utmost skepticism so long as it has not been demonstrated that the RM hypothesis can accommodate more observational facts.

Many such facts (mainly supplied by spectroscopy) exist already (for a more complete description see the reviews by Gies IAU162*, Gies 1991, and Baade 1987):

* References which in the text are marked 'IAU162' concern papers included in these Proceedings.

L. A. Balona et al. (eds.), Pulsation, Rotation and Mass Loss in Early-Type Stars, 311–324.
© 1994 IAU. Printed in the Netherlands.

- The line profile variability (*lpv*) extends virtually always over the full width of the lines: Why would putative corotating *surface* structures in virtually all stars be concentrated on the equator? In known magnetic stars there does not seem to be a preferred latitude for spots (except, perhaps, polar caps).

- *Circumstellar* structures are difficult to reconcile with both of the following two pairs of observations:

  1. Line profile variability is seen also in pole-on stars whereas ordinary shell spectra are confined to equator-on stars.

  2. The amplitude of the *lpv* is quite similar for many photospheric lines of very different strengths. This would require an amazingly precise tuning of the circumstellar physics including abundances. By contrast, conventional shell spectra are not nearly as exotic.

- If corotating inhomogeneities do not have a pronounced vector-like component, they cannot explain the increase in amplitude of the profile variability from the line centre to the wings (cf. Gies IAU162).

- Kambe *et al.* (1993) report that in ζ Oph the range in radial velocity, over which moving bumps can be observed, is different during different phases of mass loss episodes. The authors attribute the additional mass loss to accelerated rotation at the equator and the latter to the NRP-supported transfer of angular momentum from the stellar core. However, at the same time the observed periods do not change. Such a behavior would imply that the surface inhomogeneities assumed by the RM hypothesis would have to be rooted in very deep layers. It appears, therefore, very important to obtain an independent confirmation of this observational result.

- The periodic component of the light variability often only accounts for the lesser part of the total power in the statistics used for the time series analysis (e.g., Cuypers, Balona & Marang 1989). The explanatory power of a purely rotational model is accordingly reduced.

- What would cause the multi-pole symmetry which is required to explain the high-order line profile variability which is periodic (but probably only with limited phase coherence so) in both space and time?

Finally, it is disturbing that qualitatively indistinguishable phenomena, especially the high-order variability, should require different explanations in Be and δ Scuti stars (cf. Kennelly *et al.* IAU162). This is all the more so since advocates of the RM hypothesis have carefully avoided elaborating on a physical model of the inhomogeneities invoked by them. A reminder that weak magnetic fields cannot easily be sustained in a radiative atmosphere has been provided by Saio (IAU162).

My personal expectation is that spectroscopic observations will in the not too distant future prove beyond doubt that the line profile variability of some Be star is multi-periodic, i.e., Be stars *do* pulsate.

1.2. Ways to Further Consolidate the NRP Model

1.2.1. Observations

In spite of what I just said, it appears essential not to get obsessed with the hunt for multi-periodicity even though a positive result would be extremely important. However, I suspect that it will still not terminate the dispute because some people will argue that of n periods found in some star only $n-1$ are due to pulsation and the nth one is caused by rotational modulation.

Therefore, it is essential to devise observational experiments which can give conclusive results. This includes covering as large an observational parameter space as possible and requires that *all* information is exploited in a synoptic way:

o   One high-priority goal must be to study the atmospheric response to the variability. This is especially important since in the corotating frame periods are very long so that non-adiabaticity will be prominent:

   – From the work on β Cephei and δ Scuti stars it is well known (cf. Jerzykiewicz IAU162; Cugier *et al.* IAU162 [two papers]) how much information is contained in color variations. Furthermore, the periods of Be stars are much longer than in the other stars named, and the amplitudes are conveniently large (Štefl *et al.* IAU162 [two papers]). Since virtually every reasonable photometer has a filter wheel or equivalent device, it should simply be used.

   – Technically more difficult to satisfy is the equivalent spectroscopic demand to observe more than one line simultaneously. Work by, e.g., Fullerton *et al.* (IAU162), Smith (IAU162), and Reid (IAU162) convincingly shows the value of studying lines which are formed at different atmospheric depths or stellar latitudes.

o   Long-period g-modes are characterized by large horizontal-to-vertical amplitude ratios of the pulsation amplitude. This provides a prime discriminant between a pulsation velocity field and most scalar models:

   – The change of the *lpv* between line wings and centre should be studied more systematically (cf. Gies IAU162), especially for low-order variations. The high-order variability presently gives a rather confusing picture: the phase coherence is low (Gies IAU162); temporary differences between the two halves of a spectral line have been reported (Štefl *et al.* IAU162); and the velocity range over which traveling bumps are seen may vary with phase of the mass loss cycle but the periods are maintained (Kambe *et al.* 1993).

   – Observations of low-$v \sin i$, i.e. pole-on, stars can reveal the vector character of the variability in a second dimension. Particularly attractive is the possibility to discriminate between sectorial ($\ell=m$) and tesseral ($\ell \neq m$) modes. Because of cancellation across the stellar disk, the difference between the two categories of modes is quite sub-

tle for equator-on situations so that observers generally still *assume*
sectorial modes. In pole-on stars this should not be necessary.

o   Since the κ-mechanism is sensitive to metallicity, it is desirable to:

   – observe Be stars in low-metallicity environments such as the Magel-
     lanic Clouds (e.g., Mazzali *et al.* IAU162; Kjeldsen and Baade IAU162).
   – determine chemical abundances of Be stars, especially in comparison
     with Bn stars (cf. Kolb and Baade IAU162).

o   Be, Bn, and 53 Per stars should be compared with respect to their line
   profile variability (e.g., Fieldus and Bolton IAU162). Bn stars with low-
   order *lpv* are interesting because it is thought that the incidence of such
   variations in Bn stars is much lower than in Be stars. Since this appears
   to be supported by photometry, photometric follow-up of such stars is
   important.

o   Attempts to verify suggestions that changes of the pulsation amplitude
   and mass loss events in Be stars may be correlated (Penrod 1986) have
   yielded negative results (Smith 1989; Bolton and Štefl 1990). However,
   the underlying speculation may be revived, if it is confirmed that the
   range in velocity, over which traveling bumps are seen, changes with
   phase of mass loss episodes. If $v \sin i$ varies while a low-order mode
   is excited, one may not notice this but rather diagnose a change in
   the pulsation amplitude. Within the picture sketched by Kambe *et al.*
   (1993), the primary role of the pulsation would be a cyclic increase of
   the equatorial rotation velocity by transfer of angular momentum from
   the stellar core. The effect of the pulsation on circumstellar emission
   and absorption lines is also worthwhile studying (cf. Gies IAU162).

o   Observed series of line profiles should always be presented also as gray-
   scale coded, so-called dynamical spectra (e.g., Gies IAU162).

o   Interferometric resolution of the central stars of Be systems will answer
   many of the questions that we have today. However, these answers may
   not come very soon as baselines of the order of a kilometer are required
   even for relatively nearby stars.

Of course, any of the above observational approaches can be combined into
even more powerful strategies. Especially simultaneous spectroscopy and
photometry bear many promises (cf. Štefl *et al.* IAU162).

1.2.2. *Theory*

It is encouraging that now two driving mechanisms, namely the κ-mechanism
(Dziembowski IAU162) and the core convection (cf. Saio IAU162), are 'competing' for the explanation of the pulsations of Be stars. Current observations, that may help to discriminate between the two and in any case require
some guidance by theory for their analysis and the definition of future work,
include:

- the existence of pulsating Oe- and other O-stars (where, however, the concept of luminosity classes and related evolutionary stages gets increasingly blurred towards higher T_{eff}),
- the high proportions of Be stars in metal poor environments (cf. Balona 1992, see also Kjeldsen and Baade IAU162),
- the dominance of very few modes although at long periods the g-mode spectrum should be dense,
- the preference for very low frequencies in the co-rotating frame,
- the pronounced amplitude variability (cf. Dziembowski IAU162),
- the constancy of pulsation periods during phases of apparently accelerated surface rotation (Kambe *et al.* 1993).

It might be that some of these points are presently more readily accommodated by the overstable core convection model. But the rôle of rapid rotation in many of these areas probably is very important and may easily invalidate premature conclusions.

For the modeling of observations, two domains appear particularly important:

- the atmospheric response to the pulsation (Cugier *et al.* IAU162; Lee *et al.* 1991); this should also include model atmospheres for extremely rapidly rotating stars,
- eigenfunctions of rapidly rotating stars (Clement IAU162, see also discussion thereafter; Aerts IAU162; Lee and Saio 1990).

1.2.3. Observations and Theory

Although I have discussed observational and theoretical efforts separately, it is obvious that, for maximum efficiency, problems should be tackled jointly.

2. The Case for Rotational Modulation (L.A. Balona)

2.1. THE DISTINGUISHING CHARACTERISTICS OF BE STARS

For many decades the only distinguishing characteristic of Be stars, apart from the emission lines, is that as a group they rotate more rapidly than non-emission B stars of the same temperature and luminosity class. The discovery that more than half of the Be stars show periodic light variations (λ Eri variables), and that this characteristic is unique to Be stars, has added another distinguishing feature. Clearly, these are two powerful clues to the mechanism which causes enhanced mass loss in Be stars. It is important to bear in mind that only *low degree* spherical harmonic variations can lead to observable light variations. Several instances are known of non-Be stars in which line-profile variations of high degree ("moving bumps") are found, but rapidly-rotating stars with variations of low-degree and periods between 0.5 and 2 d are invariably Be stars. It follows that the enhanced mass loss in Be stars is directly connected in some unknown way with rapid rotation

an extremely constant period, is unknown in any pulsating star; certainly our current understanding of pulsation theory is unable to accommodate these facts.

2.4. Physical Limitations of NRP

If it is accepted that the photometric periods of Be stars are precisely equal to the periods of rotation, it follows that the period of NRP in the corotating frame is infinite. This is equivalent to a stationary geometrical distortion of the star (the pulsation velocity is zero) and a stationary temperature distribution of the photosphere. This is nothing more than a special case of RM. The greater the difference between the photometric and rotation period, the shorter the pulsation period in the corotating frame and the larger the pulsational velocity for the same displacement amplitude. Proponents of NRP wish to claim that there is a difference between the two periods. As we have seen, observational uncertainties allow a difference of about 20 per cent between these two periods at the most. If we assume a typical pulsation amplitude of about 10 per cent, it is a simple matter to show that the horizontal pulsational velocity amplitude cannot exceed 15 km s^{-1} in these stars. Radial velocity amplitudes of many Be stars, such as λ Eri, η Cen and others, are often much larger than this. This can only be accommodated in the NRP hypothesis if it is assumed that the line profile variations are dominated by temperature perturbations.

For g-mode pulsations the radial displacement is very small. The geometric distortion is negligible and cannot account for light amplitudes of 0.1 mag. or more seen in some λ Eri stars. In any case, for $l = 1$ sectorial modes seen equator-on the amplitude due to the geometrical effect should be zero. In the NRP hypothesis, it is very clear that practically all the light and line profile variation must be caused by the temperature perturbation. The periodic variation of temperature across the photosphere, in combination with rapid rotation, leads to periodic line profile variations. Additional observational evidence to show that the temperature effect dominates in λ Eri stars comes from multicolour observations (Cuypers et al. 1989): the $u - b$ colour is bluest when the star is brightest. The far-UV light amplitudes are greater at shorter wavelengths as would be expected if the variations were due to temperature perturbations (Percy & Peters 1990).

From these simple and rather general deductions, we are forced to conclude that the NRP model is in reality practically identical to the starspot model. The "starspot" in this case consists of a spatial distribution of temperature described by the spherical harmonic degree of the pulsation. Because the period of pulsation in the corotating frame is very long, this "starspot" is almost stationary. In spite of this quite general conclusion, the line profile variations are still being interpreted in terms of a pure velocity field and not an almost scalar temperature field as demanded by the above arguments.

The fact that low-degree NRP in λ Eri stars is observationaly indistinguishable from a smooth, stationary temperature distribution over the photosphere is inconsistent with the detailed light curves which clearly imply the presence of many small-scale variable features. It is probably inconsistent with the line profiles. Indeed, NRP advocates have always argued against such a model themselves (failing to realize that their representation of the profile variations by a pure velocity field is physically impossible). They also fail to grasp the difference between the low-order and the high-order *lpv*, mistakenly attributing both to the same cause. This, in particular, has led to much confusion.

2.5. CONCLUDING REMARKS

The mechanism which is responsible for the enhanced mass loss in Be stars is, of course, crucial to our understanding of these stars. It is well established that the majority of Be stars rotate with an equatorial velocity which is not much larger than half of the critical rotation speed. A velocity component of several hundred km s^{-1} is required to accelerate the equatorial layers to escape velocity. As we have seen, the largest pulsational amplitude that can be accommodated is no more than 10–20 km s^{-1}. Under the circumstances, it is difficult to see how NRP can play any role in the process.

I feel that the mechanism must be sought elsewhere. There are indirect but rather persuasive observations that suggest the presence of magnetic fields in these stars (Smith IAU162). Even a weak magnetic field can play a role in enhancing the mass loss by forcing material into corotation. At this stage this is mere speculation, but the idea would certainly gain strength if observations of corotating material could be found. Indeed, Peters (1991) shows that the C IV wind lines are modulated with the phase of the light curve, indicating corotation above the photosphere. I believe that the evidence has been there all along, but we have been blinded by trying to force every unknown periodic phenomenon into the NRP mould. Be stars are very complex objects and there is no reason why they should fit into this mould.

References

Baade, D.: 1987, in Slettebak, A. and Snow, T.P., eds., *IAU Coll. 92: Physics of Be Stars*, Cambridge Univ. Press: Cambridge, 361.

Balona, L.A.: 1975, *Mon. Not. Roy. Astr. Soc.* **173**, 449.

Balona, L.A.: 1990, *Mon. Not. Roy. Astr. Soc.* **245**, 92.

Balona, L.A., Sterken, C. and Manfroid, J.: 1991, *Mon. Not. Roy. Astr. Soc.* **252**, 93.

Balona, L.A.: 1992, *Mon. Not. Roy. Astr. Soc.* **256**, 425.

Balona, L.A.: 1994, *Mon. Not. Roy. Astr. Soc.* in press.

Balona, L.A. and Koen, C.: 1994, *Mon. Not. Roy. Astr. Soc.* in press.

Bolton, C.T.: 1981, in Jaschek, M. and Groth, H-G., eds., *Be Stars: IAU Symposium 98*, Reidel: Dordrecht, 181.

Bolton, C.T., Štefl, S.: 1990, in Willson, L.A. and Stalio, R., eds., *Angular Momentum and Mass Loss for Hot Stars*, Kluwer: Dordrecht, 191.

Cuypers, J., Balona, L.A. and Marang, F.: 1989, *Astron. Astrophys. Suppl.* **81**, 151.

Gies, D.R.: 1991, in Baade, D., eds., *Rapid Variability of OB-Stars: Nature and Diagnostic Value*, ESO: Garching, 229.

Kambe, E., Ando, H., Hirata, R., Walker, G.A.H., Kennelly, E.J. and Matthews, J.M.: 1993, *Publ. Astr. Soc. Pacific* submitted.

Lee, U., Jeffery, C.S. and Saio, H.: 1991, in Baade, D., ed., *Rapid Variability of OB-Stars: Nature and Diagnostic Value*, ESO: Garching, 245.

Lee, U. and Saio, H.: 1990, *Mon. Not. Roy. Astr. Soc.* **349**, 570.

Penrod, G.D.: 1986, *Publ. Astr. Soc. Pacific* **98**, 35.

Percy, J.R. and Zsoldos, E.: 1992, *Astron. Astrophys.* **263**, 123.

Percy, J.R. and Peters, G.J.: 1991, in Baade, D., ed., *Rapid Variability of OB-Stars: Nature and Diagnostic Value*, ESO: Garching, 97.

Peters, G.J.: 1991, in Baade, D., ed., *Rapid Variability of OB-Stars: Nature and Diagnostic Value*, ESO: Garching, 171.

Smith, M.A.: 1989, *Astrophys. J. Suppl.* **71**, 357.

Underhill, A. and Doazan, V.: 1982, *B Stars with and without Emission Lines*, NASA SP-456.

Zsoldos, E. and Percy, J.T.: 1991, *Astron. Astrophys.* **246**, 441.

Discussion

Le Contel: Maybe comparison with other groups of stars could enlighten the discussion:

– β Cep stars: short periods and fundamental mode.
–   53 Per stars: slowly rotating stars (1–3 day periods).

In between there are a few stars where the two time scales are present. As an example that the two phenomena (RM and NRP) could be present in a single star, I would like to emphasize the case of ET And (see Kuschnig *et al.* IAU162) which is a Bp star where the rotational period is well known and a slow pulsation is also present.

Balona: Yes, it is important to place λ Eri stars within the context of the pulsating B stars. We can distinguish several kinds of intrinsic variables based on their observed characteristics:

–   The β Cep stars: periods characteristic of p-modes.
–   The 53 Per stars: slowly-rotating stars with periods in the g-mode range.
– ζ Oph stars: generally rapid rotators showing line profile variations of high degree. In the corotating frame these have very long periods (g-modes).

According to current thinking, the λ Eri stars may be the rapidly-rotating counterparts of the 53 Per stars. The discovery of λ Eri stars in a metal-poor system (Balona 1992) is a problem since pulsational driving is not expected in stars with low metal abundance. Moreover, it seems that rapid rotation will inhibit pulsation as there are no 53 Per stars in NGC 3293 and NGC 4755 — two young open clusters containing rapidly-rotating stars (Balona 1994, Balona & Koen 1994). These observations do not favour the NRP interpretation for λ Eri stars. However, I do believe that NRP might be the correct explanation for the high-degree *lpv* in Be stars, so I think RM and NRP can indeed co-exist in these stars. The interesting case of ET And needs further study: perhaps it is a late-type 53 Per star.

Percy: In other (though different) stars, e.g., ρ Cas and HR 8752, pulsation *seems* to be able to trigger a brightening of the star followed by a slow decline (Percy & Zsoldos 1992, Zsoldos & Percy 1991). Why can pulsation *not* trigger an outburst in κ CMa and ϵ Cap and *how* can the rotational modulation theory explain it?

Balona: According to the references you quote, the light curves of these hypergiants are complicated by episodes of shell ejection. This is hardly a case of pure pulsation and cannot be taken as an example of how pulsation can lead to rapid fluctuations. We know that there *is* photospheric activity in Be stars. All that I am suggesting to account for observations in κ CMa and ϵ Cap is a sudden localized outburst of one of these active regions. Rotational modulation is a direct result of such an outburst. What causes the outburst is not known, though one may speculate on flaring or other magnetically-related mechanism.

Le Contel: How can you be sure you do not have shorter periods in your observations? In fact, the sharp changes in the light curve of stars such as κ CMa may mask the presence of other periods.

Balona: The observations certainly cannot be represented by a single sinusoidal period. On the other hand, there is only one *coherent* period. You show this by calculating a periodogram of the data. You then find only one significant peak in the periodogram. The Fourier decomposition of the light curves require very many frequency components to describe them, but only one component is coherent.

Gies: J-P. Zahn argues that the differential rotation rate of the core differs from the surface rate by no more than 20 per cent. If NRP is driven by convective motions in the core, then the superperiods would be similar to the surface rotation periods.

Balona: If NRP at the surface is driven by the rotating core and if, as you say, the period that would be observed differs from the rotational period by no more than 20 per cent, it would be consistent with observations in this one respect. However, the core is very small indeed for giants at the end of core hydrogen burning and for late B-type stars. I cannot see how the near-absence of a convective core could drive pulsations, but it would be important to predict the instability strip for such a mechanism. The Be stars do not appear to be confined to early main sequence dwarfs. I therefore think that this is not the explanation for the λ Eri stars. Besides, how can one explain sudden increases and decreases in amplitude in such a model?

Harmanec: I think that what is important to the whole problem is the observed co-existence of slow light and line-profile variations and travelling sub-features which often move with the *same* acceleration across the line profile. Free NRP should not have such commensurable periods. Possible explanations can perhaps be found in the theoretical work of Dr Aerts who showed us that travelling bumps can be produced by *low-order* nonradial pulsation.

Baade: It should be kept in mind, though, that genuine commensurability has not been demonstrated in any star with any certainty. Often, the phase coherence of the high-order variability is not very pronounced. My intensive search for one common superperiod in 2 weeks worth of high-quality spectra of the Be star μ Cen firmly excludes any superperiod between one and a few days.

Gies: In Balona's talk he mentions that the expected velocity amplitudes for periodic Be stars do not exceed 20 km s^{-1}. Model NRP profiles show shape changes that are very significant! Velocity changes in the line centroid can be large. I will measure the velocity variations in these model profiles and send him my results.

Baade: We must also remember the effect of rotation which strongly amplify any "real" radial velocity amplitude.

Balona: The point I want to make is that if the photometric period and the rotational period differ by no more than 20 per cent, as the observations indicate, then the contribution of the pulsational velocity to the radial velocity is very low. The changes in the line profile are dominated by temperature variations which in a rotating star produces a shift in the line centroid (radial velocity), as Dietrich has just mentioned. If the dominating effect of NRP is to produce a variation of temperature in the photosphere, this is in reality equivalent to a washed-out star spot which the NRP proponents assure me does not fit the line profiles. I do not believe in such a spot model: the rotational modulation must be due to a more complex phenomenon such as a corotating obscuration or localized hot spot at or close to the photosphere.

Smith: Actually, I believe there are two separate mechanisms on the surfaces of many Be stars, but that is not the question. The question concerns the *periodic* variations. I would like to make the following points in defense of NRP as opposed to RM:

(1) As Gies and I have both been attempting to show over a few years, *both* the footprints (and their associated wings) move back and forth in agreement with the line profile variations, radial velocity and photometric periods of several Be stars. I can't see how anything other than a velocity field can cause this!

(2) I have done the disk-wind modelling of the *lpv* variations in λ Eri (e.g., Smith 1989) and only small velocity amplitudes are necessary to fit them, values that do not violate any fluid flow conservation conditions.

(3) I think we should establish a picture of these RM entities and call them "spots". Thanks to Holberg *et al.* IAU 162, we now have a colorimetric paradigm of what one of these double-wave photometric variables, α Eri, is doing and we see from the UV-optical color that the star is blue when it is bright. Therefore $T_{spot} < T_{eff}$ and we can say that the immaculate phase of the star's light curve should also be its maximum. If this is true, then we expect light curves from several stars, statistically, to show a flat light maximum for 50–60 per cent of the period during the time that the spot is occulted behind the visible disk. I've looked for this simple signature among λ Eri light curves and I'm not sure that I can find an extended light maximum even once! Therefore, I conclude there could be a fundamental problem with the RM-spot hypothesis.

Balona: (1) I do not think that the variation of the footprints is explicable only if there is a velocity field. The radiation flux which causes the extreme ends of the line profiles originates at the equatorial limbs of the star. As one limb darkens and the opposite limb brightens (by a large-scale obscuration, say), the intensities of the footprints and wings will change accordingly and give rise to the effect you mention. Also, we know that moving bumps are present in some Be stars on many occasions. The nature of these high-degree line profile variations is not understood but is probably NRP. The footprints can be severely distorted by these features, but this is not relevant to the present discussion which is confined to the *low-degree* line profile variations unique to Be stars.

(2) In the first place, it is not really correct to model the line profiles using the eigenfunction of a non-rotating star for the Be stars which are very rapid rotators. Even worse, the temperature perturbation is ignored. The periodic Be stars do, of course, change amplitude quite markedly from time to time. Over the last few years λ Eri has had a very low amplitude, so I can quite believe that you can model the line profiles with small pulsational amplitudes and no temperature perturbation as there are many free parameters. Bear in mind however, that many years ago the situation was quite different. When λ Eri was at high amplitude, Bolton (1981) obtained a radial velocity amplitude of 60 km s^{-1}. As I point out, this can only be understood if the temperature variation dominates the line profile.

(3) I agree that the simple picture of a starspot is not adequate to explain all the complex line profile and light variations in the λ Eri stars. Note, however, that the spot model is almost indistinguishable from the NRP model (because temperature is so dominant) and in arguing against the spot model you argue against NRP. I cannot pretend to have a full picture

of what is causing the rotational modulation, but I suspect it is probably a corotating obscuration or hot area rather than a star spot as in the Sun.

Saio: Balona, Sterken & Manfroid (1991) show that the light curves of the λ Eri stars are not correlated with the Hα emission line intensity variations. It seems to me that this observational fact is inconsistent with the rotational modulation model.

Balona: The Hα emission strength is determined by the envelope, not by what is happening in the photosphere. What these authors show is that the intensity of Hα emission does not seem to depend on the amplitude of the periodic light variations. This seems to rule out a direct connection between the mass-loss rate and the amplitude of the periodic variations, but it has no bearing on what is causing the periodic variations itself. One can use the same argument against NRP of course.

Peters: To Petr Harmanec: Have you attempted to reconcile your conclusions on η Cen (see Štefl *et al.* IAU162) with the wind variability that we observed during our campaign?

Harmanec: The seemingly periodic RV variation of η Cen is a real phenomenon which affects the whole line. The centroid, bisector and outer wings move in phase and should be taken into account by any model attempting an interpretation. My feeling is that the periodic variations are very complex and that both RM and NRP might be occurring. In η Cen large bumps reappear exactly at the phase of light minimum which seems to speak in favour of some projection effect.

Peters: To Luis Balona: (1) How can you explain the observed wind behaviour with the spot model?

(2) Comment on periods: For all the stars we studied in the campaigns (except φ Per), the observed FUV/optical photometric and spectroscopic periods are systematically smaller than those expected from rotation, assuming that the objects are not precisely all zero-age main sequence stars.

Balona: (1) Without an understanding of what is the actual cause of the enhanced mass loss (which not even proponents of NRP claim to know), it is impossible to answer your question.

(2) Calibrations of the radii of B stars differ and it is not surprising that systematic errors exist. You have to look at the statistics. *Within the errors of these calibrations*, the photometric and spectroscopic periods are the same.

LINE-PROFILE VARIABLE ε PER:

SPECTROSCOPIC BINARY AND (?) MILD Be STAR

A.E. TARASOV, V.V. KOSTUNIN
Crimean Astrophysical Observatory, Nauchnyj, Crimea, 334413 Ukraine

P. HARMANEC, J. HORN, P. KOUBSKÝ
*Astronomical Institute, Academy of Sciences of the Czech Republic,
251 65 Ondřejov, Czech Republic*

C. BLAKE
*Department of Astronomy, University of Western Ontario,
London, Ontario, N6A 5B9 Canada*

and

G.A.H. WALKER, S. YANG
*Department of Geophysics and Astronomy, University of British Columbia,
129-2219 Main Mall, Vancouver, B.C., V6T 1Z4 Canada*

ε Per (45 Per, HR 1220, HD 24760, ADS 2888A) is a star which was used for years as a spectrophotometric and MKK classification standard. Nowadays it is known as an archetype of early-type line-profile variables (LPV). The variations are so pronounced that they were misinterpreted for a composite spectrum (B0.5+A2) and the star was even reported to be a double-lined spectroscopic binary (Petrie 1958). Since 1983, the characteristics of line-profile variability in the spectrum of ε Per have been intensively studied (Bolton 1983, Smith 1985, Gies & Kullavanijaya 1988). The recent investigations of RV variations of ε Per led to the conclusion that ε Per is a spectroscopic binary with a period of about 14.05 days and a rather eccentric orbit (Harmanec & Tarasov 1990). The issue of the binary nature (with a very eccentric orbit) is of utmost importance since it could be causally related to the extremely large observed line-profile variations (e.g., Polfliet & Smeyers 1990, Tassoul & Tassoul 1992).

To derive a reliable binary orbit of ε Per, to investigate the characteristics of the rapid line-profile variability and to derive possible relationship between rapid line-profile changes and orbital motion we obtained several new series of spectroscopic observations of ε Per at four observatories over a period of ten years. This study is devoted to a preliminary analysis.

Principal results

The principal results of this study are:

a) Confirmation of the binary nature of ε Per.

From the analysis of all high-dispersion data the following elements were

L. A. Balona et al. (eds.), Pulsation, Rotation and Mass Loss in Early-Type Stars, 325–326.
© 1994 IAU. Printed in the Netherlands.

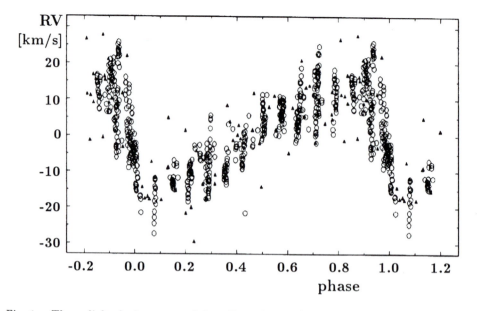

Fig. 1. The radial-velocity curve of the ε Per primary (all data). Allegheny velocities of 1907–1912 are shown by triangles while all RVs obtained since 1983 are shown by open circles.

derived:

P = 14.075978 days, $T_{periastr.}$ = 47767.49, e = 0.503, ω = 109.7°, K = 16.8 km s^{-1}, γ = 5.85 km s^{-1}.

Phase plot for all available RVs of ε Per is shown in Fig. 1.

b) Discovery that ε Per may be a mild Be star.

c) Further evidence that the line profile variations of the ε Per primary may be controlled by only one physical period.

d) There may be a 1:2 resonance between the synchronization in periastron and the rotational period of the primary.

In any case, the analyses of the (multi)periodicity of line-profile variations of ε Per should be re-considered in the light of the fact that the RV of the whole profile varies with the binary orbital period for about 30 km s^{-1}.

References

Bolton C.T.: 1983, *Hvar Obs. Bull.* **7**, 141
Gies D.R., Kullavanijaya A.: 1988, *ApJ* **326**, 813
Harmanec P., Tarasov A.E.: 1990, *Bull.Astron.Inst.Czechosl.* **41**, 273
Petrie R.M.: 1958, *MNRAS* **118**, 80
Polfliet R., Smeyers P.: 1990, *A&A* **237**, 110
Smith M.A.: 1985, *ApJ* **288**, 266
Tassoul J.-L., Tassoul M.: 1992, *ApJ* **395**, 259

AN IMPROVED MODEL OF

THE B0.5e + Be BINARY SYSTEM φ Per

P. HARMANEC, J. HORN, P. KOUBSKÝ, K. JUZA

Astronomical Institute, Academy of Sciences of the Czech Republic
251 65 Ondřejov, Czech Republic

H. BOŽIĆ, K. PAVLOVSKI

Hvar Observatory, Faculty of Geodesy, Zagreb University,
Kačićeva 26, 41000 Zagreb, Croatia

and

A.-M. HUBERT, H. HUBERT

Observatoire de Paris, Section d'Astrophysique de Meudon,
URA 335 du CNRS, F-92195 Meudon Cedex, France

1. Current knowledge

φ Per (HD 10516) is a spectroscopic binary with a 126.699-day period (Ludendorff 1910, Cannon 1910, Harmanec 1985). However, most of the published RV curves of the primary are based on H I shell lines and exhibit a typical distortion with a sharp maximum, shallow minimum and a bump at $0.^P4$ after the RV maximum (c.f., e.g., Harmanec 1985). There has been controversy on the nature of the secondary. Hynek (1940) and Hendry (1976) concluded that the binary was composed of two B stars. Peters (1976) suggested that the secondary of φ Per is a Roche-lobe filling K giant. Poeckert (1981) reported the discovery of a weak He II 4686 emission which moved in antiphase to the Be primary and suggested that it originated in the disk around the secondary. In his interpretation, the secondary is a helium star, a remnant of an originally more massive star which in the past transferred its mass to the present Be star. There is now no mass tranfer in the system according to Poeckert. Using RVs of the broad absorptions for primary, and of the He II 4686 emission for the secondary, Poeckert obtained two roughly sinusoidal RV curves and
$$M_1\sin^3i = 21.1\ m_\odot\ \text{and}\ M_2\sin^3i = 3.4\ m_\odot.$$
The primary mass seems too high for a B0.5III-V star.

Harmanec (1985) advocated a model in which the secondary is evolutionary shrinking towards the helium main sequence *and loses mass* towards the Be primary due to the rotational instability at its equator. Recently, Gies et al. (1993) discovered two He I 6678 emissions moving in antiphase to the primary and argued in favour of Poeckert's (1981) orbital elements but Harmanec's interpretation of the system dynamics.

L. A. Balona et al. (eds.), Pulsation, Rotation and Mass Loss in Early-Type Stars, 327–328.
© 1994 IAU. Printed in the Netherlands.

2. New preliminary orbital elements

To obtain better orbital elements, we adopted *emission-wing velocities for both components* of φ Per.

The finding that the hydrogen emission lines move in orbit with the primary of φ Per is not new. It was first established by Jordan (1913). We have verified that the Balmer emission-wing RVs published by various authors and our own lead to virtually identical sinusoidal RV curves - in spite of secular changes in the emission strength of Balmer emission.

We concluded that what Gies et al. (1993) describe as the main and secondary peaks of the emission (seen in both elongations) may in fact be **a double emission** from the disk around the secondary. We measured RVs of these emission wings of He I 6678 line on all Ondřejov Reticon and OHP ISIS and Aurelie CCD spectra we secured. Separate solutions for these data led to a sinusoidal RV curve in an exact antiphase to that of the Balmer emission, with a semiamplitude of 98 ± 5 km s^{-1}, i.e. *lower* than that obtained by Poeckert (1981) and Gies et al. (1993) but higher than that found by Hendry (1976). Solution for both components leads to

$$\text{Period} = 126.699 \text{ days} \quad T_{\text{prim.ecl.}} = \text{HJD } 2448226.8 \pm 1.3$$
$$e = 0 \text{ (assumed)} \quad K_1 = 10.12 \pm 0.54 \text{ km s}^{-1} \quad K_2 = 98 \pm 5 \text{ km s}^{-1}$$
$$M_1 \sin^3 i = 15.0 \text{ m}_\odot \quad M_2 \sin^3 i = 1.55 \text{ m}_\odot \quad A \sin i = 270.7 \text{ R}_\odot.$$

Since the orbit is very nearly circular, our orbital elements define – for the first time – the true phases of elongations and conjunctions of the binary. There is a hint of possible phase-locked light minima around phases 0.6 and possibly 0.05 in our Hvar and published UBV photometry. If real, they must be associated with projection effects of some circumstellar material within the system.

Dereddening of our standard mean UBV values of φ Per from Hvar V = $4^m.019$, B-V = $-0^m.081$, U-B = $-0^m.909$ leads to values which correspond to a spectral type B0.5V or slightly earlier, in a very good agreement with the primary mass of 15 m$_\odot$ which follows from our orbital solution. The mass of the secondary, 1.5 m$_\odot$, seems reasonable for a helium star.

References

Cannon J.B.: 1910, *J.R.Astron.Soc.Can.* **4**, 195

Gies D.R., Willis C.Y., Penny L.R., McDavid D.: 1993, *PASP* **105**, 281

Harmanec P.: 1985, *Bull.Astron.Inst.Czechosl.* **36**, 327

Hendry E.M.: 1976, in A. Slettebak, ed., *Be and Shell Stars*, *IAU Symp. 70* Reidel: Dordrecht, 429

Hynek J.A.: 1940, *Contr. Perkins Obs.* **No. 14**, 1

Jordan F.C.: 1913, *Publ. Allegheny Obs.* **3**, 31

Ludendorff H.: 1910, *Astron. Nachr.* **186**, 17

Peters G.J.: 1976, in A. Slettebak, ed., *Be and Shell Stars*, *IAU Symp. 70* Reidel: Dordrecht, 417

Poeckert R.: 1981, *PASP* **93**, 297

HdBe STARS: HYDROGEN-DEFICIENT SUPERGIANTS
WITH EMISSION LINES

C.S. JEFFERY

Dept of Physics and Astronomy, University of St Andrews, North Haugh, St Andrews,
FIFE KY16 9SS, Scotland

1. The hydrogen-deficient Be stars

The hydrogen-deficient B stars (often known as extreme-helium stars: EHes) show weak or absent Balmer lines and the following properties (Heber 1986a): $T_{eff} \sim 12 - 25\mathrm{kK}, \log g \sim 2 - 3, \log n_H \sim -3 - -4, n_{He} \approx 0.99, n_C \sim 0.01$. Their galactic positions and apparent magnitudes indicate that they are low-mass supergiants belonging to the Galactic bulge (Jeffery et al. 1987).

All EHes have been observed using high-resolution échelle spectrographs (at ESO and the AAT). A small number show an emission-line spectrum, usually consisting of He $_I$, C $_{II}$ and Si $_{II}$ lines; other lines including Balmer lines are occasionally seen. Whilst the spectra vary from object to object, the general characteristics (see table) and specific behaviour (Jeffery & Heber 1992) resemble that observed in normal Be stars (Balmer lines excepted). The term 'Hydrogen-deficient Be stars' refers to these objects.

2. Origin of the emission lines

From the type of ions detected, the stellar T_{eff} and the relative strengths in the He $_I$ lines, the emission lines are deemed to arise in an extended atmosphere (shell) with a temperature similar to or slightly below that of the photosphere.

One object, BD$-9°4395$, has a low surface gravity close to the Eddington limit, a relatively high rotational velocity ($v_{eq}/v_{crit} > 0.36$), pulsations (Jeffery et al. 1985, Jeffery & Heber 1992) and a radiatively driven wind (Hamman et al. 1981). Although less well studied, LSE 78 and DY Cen show similar properties. It is reasonable to suppose that the combination of high luminosity, rapid rotation and pulsations supports an extended atmosphere.

3. Related objects

Although extremely rare, the EHes may be related to other hydrogen-deficient stars, including RCrB stars, late-type Wolf-Rayets (WC10,11), H-deficient subdwarf O stars (HdsdO) and planetary nebula central stars

L. A. Balona et al. (eds.), Pulsation, Rotation and Mass Loss in Early-Type Stars, 329–330.

(HdCPN), and PG1159 stars. Some of these (see table) have similar T_{eff} to the EHes. They differ in having stronger emission spectra with P Cygni profiles (e.g. V348 Sgr: Leuenhagen & Hamman 1993). It will be proposed elsewhere (Jeffery, in preparation) that although qualitatively different, HdBes, WC11s, RCrBs, HdsdOs, HdCPNs and PG1159s share a common evolutionary origin, probably as a white dwarf (possibly binary) which experiences helium-shell reignition. Diversification depends critically on the structure of the progenitor(s) and the process leading to helium-shell ignition.

Star	T_{eff}/K	n_H/n_{He}	Type	Reference
Emitting ions				
MV Sgr	15 400	low	EHe / hot RCrB	Jeffery et al. 1988
Hα HeI OI MgII CaII SiI TiII FeII				
BD −9°4395	22 700	0.0015	EHe	Jeffery & Heber 1992
LSE 78	18 000	< 0.0001	EHe	Jeffery 1993
DY Cen	19 500	0.1	EHe / hot R CrB	Jeffery & Heber 1993
Hαβ HeI CII SiII				
V348 Sgr	19 000	0.05	hot RCrB / WC11	Leuenhagen et al. 1993
Hαβγδ HeI CI,II NI,II OI MgII NeI AlII,III SiII,III,IV PII,III SI,II FeIII				
CPD −56°8032	∼ 25 000	low	WC10 / HdCPN	in preparation
He 2-113	∼ 25 000	low	WC10 / HdCPN	in preparation
Hβγδ HeI CII,III NII OII ...				
HD 160641	32 000	low	EHe	Heber 1986b
LSS 5121	28 000	low	EHe	Heber 1986b
448.6, 450.4 nm id ?				

References

Hamann W.-R., Heber U., Schönberner D.: 1981, *AA* **116**, 273
Heber U.: 1986a, '' in Hunger, Schönberner & Rao, ed(s)., *Hydrogen-deficient Stars and Related Objects, IAU Coll. 87*, Reidel:Dordrecht, 33
Heber U.: 1986b, '' in Hunger, Schönberner & Rao, ed(s)., *Hydrogen-deficient Stars and Related Objects, IAU Coll. 87*, Reidel:Dordrecht, 73
Jeffery C.S.: 1993, *AA* , in press
Jeffery C.S., Heber,U.: 1992, *AA* **260**, 133
Jeffery C.S., Heber U.: 1993, *AA* **270**, 167
Jeffery C.S., Heber U., Hill P.W., Pollacco D.: 1988, *MNRAS* **231**, 175
Jeffery C.S., Skillen I., Hill P.W., Kilkenny D., Malaney R.A., Morrison K.: 1985, *MNRAS* **217**, 701
Leuenhagen U., Hamann W.-R.: 1993, *AA* , submitted
Leuenhagen U., Heber U., Jeffery C.S.: 1993, *AAS* , in press

SPECTROSCOPIC ANALYSES OF HdBe STARS

C.S. JEFFERY
Dept of Physics and Astronomy, University of St Andrews, North Haugh, St Andrews, FIFE KY16 9SS, Scotland

1. The hydrogen-deficient Be stars

A few hydrogen-deficient B stars (or extreme helium stars: EHes) exhibit Be-type behaviour (the HdBe stars). The principal class members are BD $-9°4395$, LSE 78 and DY Cen. The hot RCrB stars MV Sgr and V348 Sgr and the hot EHes HD 160641 and LSS 5121 are related to these objects. Since there are wide abundance variations amongst the EHes, as well as wide variations in Be-type behaviour, the principal HdBes have been analysed for atmospheric parameters, including T_{eff}, $\log g$ and composition.

2. Atmospheric parameters

Improved model atmosphere techniques, including line blanketing, have been adopted, together with high resolution high S/N spectra (Jeffery & Heber 1992,1993, Jeffery 1993). The H abundance is found to vary from less than 0.01% up to 10%. The C abundance is uniformly about 1 dex above solar, but the O abundance varies between a subsolar value up to about 1/3 of the C abundance. This very high value of the O/C ratio is similar to that found at the end of core He burning and has important consequences for the evolution of HdBe and EHe stars. In summary the HdBe (EHe) surface compositions reflect the history of the star, including primordial abundance variations (Fe), the residual component of the H-rich envelope (H), CNO (N,He) and 3α (C) and α-capture (O,Ne) processed material. The results for all analysed EHes and HdBes are shown in the table.

3. Related Objects

It is proposed that EHes share a common evolutionary origin with several other H-deficient objects (Jeffery 1994, in preparation). Whilst most analyses of EHes have been carried out using consistent methods (plane-parallel static atmospheres in LTE), successive improvements have modified the results. This process is incomplete, since line-blanketing and non-LTE effects have been shown to be severe. For example, new non-LTE calculations for C II in EHe atmospheres indicate that LTE methods considerably underestimate the C abundance (Dundas 1993, priv. comm.). These processes must be correctly treated before EHes can be reliably compared with other objects.

331

L. A. Balona et al. (eds.), Pulsation, Rotation and Mass Loss in Early-Type Stars, 331–332.
© 1994 IAU. Printed in the Netherlands.

Atmospheric parameters for HdBes (normalized to $\log \Sigma_i \mu_i \nu_i = 12.15$)

Star	HD 168476	MV Sgr	HD 124448	BD+10 2179	DY Cen	BD−9 4395	LSE 78	V348 Sgr*	γ Peg
Class	EHe	RCrB HdBe	EHe	EHe	RCrB HdBe	EHe HdBe	EHe HdBe	RCrB WC11	B
$T_{\rm eff}/{\rm kK}$	13.7	15.4	15.5	16.8	19.5	22.7	18.0	19.0	
$\log g$	1.35	2.5	2.5	2.55	2.15	2.55	2.00	2	
H	< 7.8		< 7.5	8.5	10.76	8.74	< 7.54	0.20	12.00
He	11.54	11.6	11.53	11.53	11.52	11.54	11.54	0.50	10.96
C	9.4	7.8	9.46	9.54	9.51	9.17	9.54	0.25	8.45
N	8.9	8.0	8.83	8.11	8.01	7.97	8.33		7.82
O	8.4		8.5	8.1	8.85	7.90	9.06	0.05	8.66
Ne	9.3:				9.6	8.8			8.5
Mg	7.7		8.2	8.0	7.3	7.3	7.2		7.5
Al	7.2:		6.2	6.3	5.9	5.6	5.8		6.5
Si	7.7	6.9	7.5	7.3	8.1	7.8	7.1		7.4
P	6.3		5.6	5.5	5.8	6.2	6.3		5.2
S	7.0	6.5	7.2	7.12	7.1	7.8	7.1		7.2
Ar			6.6	6.4	6.1	7.2	6.6		6.7
Ca	7.0		< 6.9	< 5.9 :			6.3		6.4
Fe	7.5		7.4	6.49	5.0	6.6	6.8		7.4

* V348 Sgr – relative abundances by number   : value uncertain

HD 168476	Walker & Schönberner 1981
MV Sgr	Jeffery et al. 1988
HD 124448	Schönberner & Wolf 1974
BD +10°2179	Heber 1983
γ Peg	Peters 1976
BD −9°4395	Jeffery & Heber 1992
DY Cen	Jeffery & Heber 1993
LSE 78	Jeffery 1993
V348 Sgr	Leuenhagen et al. 1993a,b

References

Heber U.: 1983, *AA* **118**, 39

Jeffery C.S.: 1993, *AA* , in press

Jeffery C.S., Heber,U.: 1992, *AA* **260**, 133

Jeffery C.S., Heber U.: 1993, *AA* **270**, 167

Jeffery C.S., Heber U., Hill P.W., Pollacco D.: 1988, *MNRAS* **231**, 175

Leuenhagen U., Hamann W.-R.: 1993a, *AA* , submitted

Leuenhagen U., Heber U., Jeffery C.S.: 1993b, *AAS* , in press

Peters G.J.: 1976, *ApJS* **65**, 95

Schönberner D., Wolf R.E.A.: 1974, *AA* **37**, 87

Walker H.J., Schönberner D.: 1981, *AA* **97**, 291

Detection of calcium abundance stratification in Ap stars

J. BABEL

Service d'Astrophysique, Centre d'Etudes de Saclay, 91191 Gif-sur-Yvette, France

Abstract. We report the discovery of a systematic and large calcium abundance stratification in cold Ap stars. These detections are in very good agreement with diffusion theory and set stringent upper limits on turbulent processes.

1. Context and observations

The peculiarity of the Ca II K line at 3933 Å (see Fig.1 of Babel 1993b) is a well-known but unexplained feature of Ap stars (since Babcock 1958). On the theoretical side, abundance stratification is a major prediction of radiative diffusion (e.g. Michaud 1970) and has to be tested.

We made a high resolution spectroscopic survey of the Ca II K and H lines at 152cm of OHP. It includes 28 Ap stars with $7500 \lesssim T_{eff} \lesssim 11000$ K (3 with $vsini \simeq 100$ km/s, 2 rapidly oscillating Ap (roAp)). The K line profile was parametrized to allow quantitative study of the K line shape.

Our goal was to discriminate spotted-nonstratified models from stratified models on a statistical ground as any peculiar K line can be reproduced either by abundance stratification or by abundance spots (Babel 1993a)

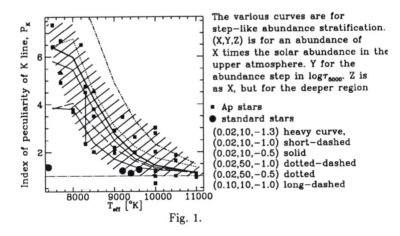

The various curves are for step–like abundance stratification. (X,Y,Z) is for an abundance of X times the solar abundance in the upper atmosphere. Y for the abundance step in $\log\tau_{5000}$. Z is as X, but for the deeper region

■ Ap stars
● standard stars

(0.02,10,−1.3) heavy curve,
(0.02,10,−1.0) short–dashed
(0.02,10,−0.5) solid
(0.02,50,−1.0) dotted–dashed
(0.02,50,−0.5) dotted
(0.10,10,−1.0) long–dashed

Fig. 1.

2. Results and Discussion

In various diagrams relative to the shape of the K line (see Babel 1993b), Ap stars follow a very different trend than normal stars. In particular, we did not find Ap stars with nonpeculiar profiles, $P_K \simeq 1$ for $T_{eff} < 9000$ K

333

L. A. Balona et al. (eds.), Pulsation, Rotation and Mass Loss in Early-Type Stars, 333–334.
© 1994 IAU. Printed in the Netherlands.

334

(Fig. 1). Our results exclude statistically nonstratified-spotted models as an explanation of the peculiar shape of the K line. NLTE effects can also be excluded (Babel 1993b). In contrast, the observations are well explained by a large Ca stratification with decreasing abundance towards the surface.

Stringent additional test comes from the study of the blend H_ϵ-CaII H and gives another proof of calcium abundance stratification (see Fig. 2.a).

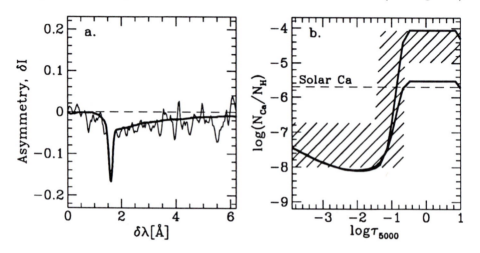

Fig. 2. a.Asymmetry of H_ϵ in HD 204411. Heavy line: stratified model (0.005,10,-1.3) (see Babel 1993b). Thin line: observation at DAO by Adelman (private comm.)b. Abundance of Ca as a function of optical depth. The heavy curves are for the diffusion-mass loss model of 53 Cam with $\dot{M} = 3\ 10^{-15}\ M_\odot yr^{-1}$ (upper curve) and $\dot{M} = 0$ (lower curve) (Babel 1992). The hatched zone is the range of stratification of Ca (step-functions) obtained from the Ca II K and H line for Ap stars with $T_{eff} \le 9000$ K.

The results on Ca abundance stratification deduced from the Ca II K and H lines (for an assumed step-function) for Ap stars with $T_{eff} \lesssim 9000$ K are summarized by the shaded area in Fig. 3. We obtain that a large Ca stratification, with a variation by 2 dex of the Ca abundance in the line-formation region, seems very common in Ap with $T_{eff} \lesssim 9000$ K, without effects to first order from rotational velocity or from pulsation for the roAp.

We obtain a very good agreement with equilibrium abundance distributions from the diffusion model (Babel 1992). These results indicate a very large stability of the photospheric regions.

References

Babel, J.:1992, A&A 258, 449
Babel, J.:1993a, in Peculiar versus normal phenomena in A-type and related stars, IAU Coll. 138, Eds. Dworetsky et al., Astron. Soc. Pacific, Conf. Series, 44, p. 458
Babel, J.:1993b, A&A, in press
Babcock H. W.:1958, ApJS 3, 141
Michaud, G.:1970, ApJ 160, 641

STARK WIDTHS OF ASTROPHYSICALLY IMPORTANT
FOUR- AND FIVE-TIMES CHARGED ION LINES

MILAN S. DIMITRIJEVIĆ

Astronomical Observatory, Volgina 7, 11050 Belgrade, Yugoslavia

Important astrophysical applications of Stark broadening of spectral lines of multiply charged ions are in the physics of stellar interiors (Seaton 1987). In subphotospheric layers, the modelling of energy transport requires radiative opacities and thus, certain atomic processes must be known accurately. At these high temperatures (10^5 K or more) and densities ($10^{17} - 10^{22} \mathrm{cm}^{-3}$) Stark broadening of strong multicharged ionic lines plays a non-negligible role in the calculation of the opacities, especially in the UV. Moreover, with the development of spectroscopic investigations from space, UV and extreme UV spectral line research has been further stimulated.

In order to provide such data for four- and five-times charged ions, comprehensive studies of electron-, proton- and ionized helium-impact broadening parameters for 30 N V (Dimitrijević and Sahal-Bréchot 1992a), 30 O VI (Dimitrijević and Sahal-Bréchot 1992b) and 21 S VI (Dimitrijević and Sahal-Bréchot 1993) multiplets have been made recently, by using the semiclassical perturbation approach (Sahal-Bréchot 1969ab). In the case of C V, O V and P V lines, there exist sufficient atomic data for sophisticated semiclassical calculations for some or all astrophysically interesting lines. But, for other four- and five-times charged ions, the atomic data set is not sufficiently complete.

In order to complete the Stark broadening data for four- and five-time charged ions, Stark widths of astrophysically important spectral lines within 3 C V, 50 O V, 12 F V, 9 Ne V, 3 Al V, 6 Si V, 11 N VI, 28 F VI, 8 Ne VI, 7 Na VI, 15 Si VI, 6 P VI and 1 Cl VI multiplets, have been calculated by using the modified semi-empirical approach (Dimitrijević, Konjević, 1980). Results for 159 Stark line widths (FWHM) calculated using the modified semi-empirical approach (Dimitrijević and Konjević 1980) - (WMSE) will be published in Dimitrijević (1993a). Moreover, in order to compare the different theoretical methods, for 88 of the above mentioned multiplets calculations were performed by using the symplified semiclassical approach (Griem, 1974) as well (Dimitrijević, 1993b).

Comparison of the present values with values calculated by using Eq. (526) in Griem (1974) have been performed, and the obtained agreement is satisfactory. As an example, a comparison for the C V $3s^1S - 3p^1P$, N VI $2s^1S - 2p^1P$ and O V $4p^1P - 4d^1D$ cases is presented in Table 1. In comparison with the experiment of Purić et al. (1988) for two O V lines, both approaches give about two times smaller values.

L. A. Balona et al. (eds.), Pulsation, Rotation and Mass Loss in Early-Type Stars, 335–336.
© 1994 IAU. Printed in the Netherlands.

TABLE 1

Comparison of present results for Stark broadening full half width (WMSE) with values obtained by using Eq. (526) in Griem (1974) (WG). The electron density is 10^{17} cm^{-3}.

Transition	λ(Å)	χ(eV)	T(K)	WMSE(Å)	WG(Å)
C V $3s^1S - 3p^1P$	12202.6	56.6	50000	1.79	1.63
			100000	1.57	1.36
			200000	1.46	1.18
			400000	1.22	1.05
			800000	1.21	0.949
N VI $2s^1S - 2p^1P$	2833.7	2.95	50000	0.700E-02	0.812E-02
			100000	0.516E-02	0.607E-02
			200000	0.420E-02	0.475E-02
			400000	0.357E-02	0.395E-02
			800000	0.305E-02	0.349E-02
O V $4p^1P - 4d^1D$	11913.1	29.2	50000	4.35	4.16
			100000	3.79	3.43
			200000	3.30	2.96
			400000	2.98	2.64
			800000	2.81	2.41

References

Dimitrijević, M.S.: 1993a, *Astronomy and Astrophysics, Supplement Series* **100**, 237

Dimitrijević, M.S.: 1993b, *Astrophys. Lett. Commun.* , in press

Dworetsky, M.M., Konjević, N.: 1980, *JQSRT* **24**, 451

Dimitrijević, M.S. and Sahal-Bréchot, S.: 1992a, *Astronomy and Astrophysics, Supplement Series* **95**, 109

Dimitrijević, M.S. and Sahal-Bréchot, S.: 1992b, *Astronomy and Astrophysics, Supplement Series* **93**, 359

Dimitrijević, M.S. and Sahal-Bréchot, S.: 1993, *Astronomy and Astrophysics, Supplement Series* **100**, 91

Griem, H.R.: 1974, *Spectral Line Broadening by Plasmas*, Academic Press, New York ,

Purić, J., Djeniže, S., Srećković, A., Platiša, M., Labat, J.: 1988, *Phys. Rev. A* **37**, 498

Sahal-Bréchot, S.: 1969a, *Astronomy and Astrophysics* **1**, 91

Sahal-Bréchot, S.: 1969b, *Astronomy and Astrophysics* **2**, 322

Seaton, M.J.: 1987, *J. Phys. B* **20**, 6363

STARK BROADENING OF STELLAR Pt II LINES

MILAN S. DIMITRIJEVIĆ

Astronomical Observatory, Volgina 7, 11050 Belgrade, Yugoslavia

Lines of Pt II have been discovered in Hg Mn stars by Dworetsky (1969). The analysis of a few strong Pt II transitions, which are also observed in IUE spectra of stars, has shown (Dworetsky et al., 1984) "that Pt is, like Hg, among the most overabundant elements in the atmospheres of Hg Mn stars, with enhancements of the order of 10^4 to 10^5 over the solar system abundances". Dworetsky et al. (1984) selected also the four Pt II lines which might be used for astrophysical applications. Moreover, they determined the corresponding theoretical gf values. The aim of this contribution is to investigate the Stark broadening of these Pt II lines and to provide the corresponding Stark widths.

In the case of more complex atoms or multiply charged ions the lack of accurate atomic data needed for more sophisticated calculations diminishes the reliability of the semiclassical results. In such cases approximate methods might be very interesting. Good possibilities provide e.g. the modified semi-empirical method (Dimitrijević and Konjević, 1980), which have been used here for these calculations. Our results for four Pt II lines selected by Dworetsky et al. (1984) as the most interesting ones from an astrophysical point of view, are presented in Table 1. In order to see the influence of the differences in oscillator strengths, results of calculations with gf values obtained by using the Coulomb approximation are presented as well. The differences might give an impression of the error bars in the obtained results.

We can also see from Table 1 that all lines belong to the same supermultiplet and that their widths are not very different. Using the analysis of Stark broadening parameters within a supermultiplet (Dimitrijević, 1982) we can estimate the Stark widths of other members within multiplets and the supermultiplet using $W_1 = (\lambda^2_1/\lambda^2_2) W_2$, taking for W_2 the most appropriate value for the considered case (i.e. e.g. the nearest available member of the same multiplet).

For an order of magnitude estimate, we might use the above mentioned equation, taking for W_2 the data for the transition with the same upper level (or the nearest available member of the same multiplet). For optical Pt II lines given in Dworetsky et al. (1984) and Dworetsky and Vaughan (1973), we might scale by this relation the $\lambda = 2245.5$ Å data for the $\lambda = 4148.30$ Å, 4061.66 Å, 4034.17 Å, 4023.81 Å and 3447.78 Å lines; the $\lambda = 1781.9$ Å data for the 3806.91 Å line and the $\lambda = 2144.2$ Å data for the $\lambda = 4514.17$ Å, 4288.40 Å, 4046.45 Å, 3766.40 Å, and 3577.20 Å lines.

L. A. Balona et al. (eds.), Pulsation, Rotation and Mass Loss in Early-Type Stars, 337–338.

TABLE 1

Full Stark widths in Å of astrophysically important Pt II lines as a function of temperature T in K. The electron density is 10^{17} cm^{-3}. The Stark width W_1 has been calculated by using oscillator strengths of Dworetsky et al. (1984) and W_2 with oscillator strengths calculated with the Coulomb approximation.

Transition	$\lambda(\text{Å})$	$T(\text{K})$	$W_1(\text{Å})$	$W_2(\text{Å})$
Pt II $6s^4F_{9/2} - 6p^4G_{11/2}$	1777.1	5000	0.0496	0.0353
		10000	0.0351	0.0249
		20000	0.0248	0.0176
		40000	0.0176	0.0125
		80000	0.0124	0.00882
Pt II $6s^4F_{7/2} - 6p^4G_{9/2}$	2245.5	5000	0.0529	0.0557
		10000	0.0374	0.0394
		20000	0.0264	0.0279
		40000	0.0187	0.0197
		80000	0.0132	0.0139
Pt II $6s^4F_{9/2} - 6p^4F_{9/2}$	1781.9	5000	0.0502	0.0358
		10000	0.0355	0.0253
		20000	0.0251	0.0179
		40000	0.0177	0.0127
		80000	0.0125	0.00895
Pt II $6s^4F_{9/2} - 6p^4D_{7/2}$	2144.2	5000	0.0692	0.0484
		10000	0.0489	0.0342
		20000	0.0346	0.0242
		40000	0.0245	0.0171
		80000	0.0174	0.0122

References

Dimitrijević, M.S.: 1982, *Astronomy and Astrophysics* **112**, 251
Dimitrijević, M.S., Konjević, N.: 1980, *JQSRT* **24**, 451
Dworetsky, M.M.: 1969, *Astrophysical Journal* **156**, L101
Dworetsky, M.M., Storey, P.J., Jacobs, J.M.: 1984, *Physica Scripta* **T8**, 39
Dworetsky, M.M., and Vaughan, Jr.H.: 1973, *Astrophysical Journal* **181**, 811

6. Be STARS: CIRCUMSTELLAR ENVIRONMENT

VARIABILITY IN THE CIRCUMSTELLAR ENVELOPE

OF Be STARS

A.M. HUBERT

Observatoire de Paris, section d'Astrophysique de Meudon
URA 335 du CNRS, F-92195 Meudon Cedex, France

Abstract. A progress report will be given on recent studies of circumstellar variability of Be stars, their implications on the formation process, the dynamics, the structure and the morphology of the envelope. The following points are discussed: recurrence of outbursts, simultaneous infall and outflow events, variability in hot regions versus variability in the cool envelope, the establishment and vanishing of Be and Be-shell phases, periodic medium-term and quasi-periodic long-term phenomena, and constraints on models.

1. Introduction

The main difficulty in establishing envelope models for Be stars lies in the complexity of the observed phenomena, in their wide time-scale variability (days, months, years), and in the fact that they are superimposed on short-term variations of the central star. Rapid variability can be generated by shocks, local density enhancements and discrete mass-loss events. Long-term variations are a sign of global change in the circumstellar envelope, but their connection with the properties of the central star is still very difficult to establish. Intensive observations of a few stars over several months or years and using various techniques, have improved our understanding about the nature and time scale of circumstellar variability.

In view of the great number of publications in the field and of the length allowed for this review, it is of course incomplete and I apologize in advance for omission of any important points.

2. Outbursts - Simultaneous Infall and Outflow

In the star μ Cen the birth of the envelope is not the result of a continuous process, but of separate individual events or outbursts (Hanuschik *et al.* 1993 and references therein). The stronger the Hα emission, the longer the Be phase (Peters 1984, Baade *et al.* 1988). C II line profile changes, as well as dramatic V/R variations, are present during the rising phase of outburst, but disappear at maximum and thereafter. Hanuschik *et al.* (1993) analyzed the development of major and minor outbursts over a 200-d interval in 1987 and proposed a decretion-accretion model with magnetic activity producing the ejection process.

The geometry and the kinematics of the circumstellar material in λ Eri during the outbursts of 1987–1988 have been investigated by Smith *et al.*

L. A. Balona et al. (eds.), Pulsation, Rotation and Mass Loss in Early-Type Stars, 341–350.
© 1994 *IAU. Printed in the Netherlands.*

(1991). High-velocity emission and absorption features in the He I λ 6678 line have been interpreted as the signature of vertical motions of ejected blobs returning to the star. The variation of the separation of the Hα emission line peaks was explained by the slow expansion of a detached, thin, quasi-Keplerian disk. The complex structure of the Hα emission line profile during minor outbursts could be interpreted as the result of a "continuous" rain onto the star of low-angular momentum material from an unstable ring in low orbit.

Recurring shell-infall events in FY CMa were detected in the UV by Grady *et al.* (1988) between 1979 and 1988. Peters (1988) has noted the coexistence of red-shifted circumstellar lines of low- and moderately-ionized species and narrow violet-shifted components in the N V lines. Similar observations have been made for the Be-binary systems HR 2142 and ϕ Per. A possible explanation for these simultaneous violet- and red-shifted components could be a rapidly decelerated gas colliding with the stellar wind. The resulting shock produces a high-temperature region where the N V resonance lines and the He I emission component could be formed.

Short-lived outbursts are usually detected in early-type Be stars, but they can also be present in late-type Be stars. Indeed, Ghosh *et al.* (1989) reported a short-lived outburst in Hα and in the C II λ 6578 emission lines for the B9e star HR 4123. They estimated that the Lyman continuum radiation has to be enhanced by a factor of 10^4 to explain the strong temporary Hα emission, though they did not include ionization from the second level of hydrogen.

3. Hot Regions—Cool Circumstellar Envelope

From a multi-technique analysis of the ω Ori outburst during 1982–1983, Sonneborn *et al.* (1988) at first explained the lack of correlation between the highly-ionized UV lines, the polarized continuum and the visual flux by a spatial separation between the wind acceleration and the inner part of the disc. However, Brown & Henrichs (1987) made a more quantitative analysis of the same data, and argued that two types of mass loss could be present: an isotropic density enhancement effect, and an episodic highly-flattened mass ejection.

Coordinated far-UV and optical observations of 59 Cyg from 1978–1987 allowed Doazan *et al.* (1989) to describe the formation and evolution of a new, cool, Hα emitting envelope and to discuss modelling implications. They concluded from their data that the mass outflow is highly time-dependent and interacts strongly with the cool envelope.

A Be star atlas of far-UV and optical high-resolution spectra was published by Doazan *et al.* (1991). It contains selected spectral regions in the UV and visual wavelengths for 166 Be stars.

4. Establishment and Vanishing of Shell Phases

Changes in the mass-loss rate and in the radiative flux are thought to induce variability in the opacity, the size, and the geometry of the envelope and to give rise to an alternation between a quasi-normal B phase, and a Be or a Be-shell phase, as is commonly observed for many stars.

An alternation of Be and Be-shell phases in 27 CMa from 1985–1989 was reported by Ghosh (1990). P Cyg profiles in Hα appeared several months before a new shell phase, indicating a strong episodic mass loss.

Having been observed for more than one century, Pleione is well-known for its alternation of B, Be and Be-shell phases. The light variations associated with phase changes in its envelope has often been described in the literature. Radiative energy flux changes in the far UV through the Be-shell and the Be transitions were considered by Doazan et al. (1993). The apparent continuum level increases in the UV at the time when the strong shell phase is vanishing. A decrease of the UV flux during a shell phase was reported, for the first time, by Beeckmans (1976) for 59 Cyg. It was explained by the large number of absorption lines and enhanced continuum and line opacity (see also Slettebak & Carpenter 1983).

The last Be-shell phase of Pleione, observed from 1973–1986, has been extensively described in the visual range by many authors. Ballereau et al. (1988) have interpreted the increasing negative radial velocity of the shell lines associated with a decrease of the emission-line peak separation as being due to an expanding phase of the envelope. According to Doazan et al. (1988), the fading of the shell lines is accompanied by the appearance of C IV and Si IV resonance lines. Strong features in the Mg II resonance lines are present at the maximum of the shell phase. This is similar to the strong feature in the Ca II resonance lines discovered by Hirata & Kogure (1976) at the beginning of the same shell phase and interpreted as due to the ejection of a dense, rapidly-rotating ring.

According to Koubský et al. (1993), quasi-emission central bumps have been detected in the metallic shell lines of 4 Her: they are associated with the establishment of a new Be-shell phase. Up to now, such bumps have been observed only in the photospheric lines of B and Be stars (see e.g. Porri & Stalio 1988, Baade 1989). They were at first interpreted as an effect of equator-to-pole variations of photospheric parameters (Baade 1990), then of cool polar stellar caps (Jeffery 1991). Is there a physical connection between the quasi-emission central bumps in photospheric and shell lines?

A recent phase change in X Per has been reported by Norton et al. (1991): it has reverted to a "normal" O-B absorption-line star. A decrease in the IR flux by over 1 mag accompanied the disappearance of Hα emission, indicating that the same regions are responsible for these variations.

5. Medium-term Periodic Variations

In this section we consider only medium-term periodic variations observed in binary Be stars.

Periodicities in the variation of line profiles (V/R ratio, radial velocity), in light, and in the level of polarization—as well as the presence of additional absorptions due to gas streams at certain orbital phases—provide very interesting information on the interaction between the gas and the stellar components and on the geometry of the envelope in Be binary systems. However, in such binary systems the photospheric lines of the primary, and sometimes also of the secondary star, are strongly affected by the circumstellar features. Even if the spectral lines of the companion are detected, it is often difficult to determine accurately the true radial velocity and the mass function. The following examples are illustrative.

The radial velocity curve of 4 Her based on photospheric lines has lower amplitude and eccentricity than the curve derived from shell lines (Koubský et al., these Proceedings). This difference can be interpreted as a consequence of a mass transfer effect.

The puzzle of KX And (Štefl et al. 1990) has to be recalled: the photospheric He I radial velocity of the presumed primary star is in phase with the radial velocity of the shell lines, but shifted by about -100 km s^{-1}!

In ϕ Per there is no evidence for the photospheric lines of the secondary, but only lines which originate from the circumstellar gas around this star have been detected (Poeckert 1981; Gies et al. 1993). Furthermore, there is an important circumstellar contamination of the photospheric lines of the primary. Depending on the lines selected and the method of measurement, a difference in the amplitude of the radial velocity of the primary and the secondary is obtained. Compare the values obtained by Gies et al. (1993), and on the other hand those determined by Harmanec et al. (these Proceedings). It is clear that in such systems the determination of the mass of each component is very difficult.

An important point concerns the number of Be stars in which a cool companion overflows its Roche lobe. The number of such systems is in fact very small. In the past, several Be binaries have been proposed as Be stars associated with cool giant companions, but this has never been really established. The presence of a cool companion was not confirmed in ζ Tau (Floquet et al. 1989) and in HR 2142 (Waters et al. 1991). Baade (1992) did not succeed in finding new candidates in a survey in the southern hemisphere. However, some detections were nevertheless confirmed, as in the cases of KX And (Floquet & Hubert 1991) and CX Dra. This last example has been revealed as an interacting system. There is evidence for a gas stream and the satellites of UV resonance lines depend on the orbital phase (Horn et al. 1992 and references therein). Disruption of the axial symmetry of the envelope has been detected from the variation in polarization (Huang *et al.* 1989), in agreement with the model of Brown & Fox (1989).

The possibility of a high mass-transfer rate was investigated in the case of several interacting binaries with Be primaries. The answer is negative for AX Mon, which has revealed no significant change of its orbital period over 20 years (Mastenova 1989), and positive for β Lyr (Harmanec 1990 and references therein). Furthermore, a flux excess in the mid-UV region attributed to the presence of an accretion disc, has been detected in some Be stars with giant companions (Parsons *et al.* 1988). While some Be binaries may have an accretion disc, it is clear that the circumstellar envelope of the majority of Be stars in binary systems is produced by ejection from the star itself. According to Pols *et al.* (1991), companions to Be stars should preferentially be helium stars and compact objects.

Newly-discovered spectroscopic Be binaries include 17 Tau, η Tau, 48 Per, β CMi (Jarad *et al.* 1989), κ Dra (Juza *et al.* 1991), V923 Aql (Koubský *et al.* 1989), o And B (Hill *et al.* 1988), HD 45677 (Halbedel 1989) and the X-ray transient source V568 Cyg (Blanco *et al.* 1990).

The periodic V/R variations may be explained either by obscuration due to a binary companion, or a hot spot created by the stream impact upon the accretion disc, or composite emission from an ionized region created by a compact object and from the Be star envelope (Apparao & Tarafdar 1989). Recently Chakrabarti & Wiita (1993) proposed that the V/R variations could also be the result of effects of spiral shocks on the disc.

6. Long-term Quasi-periodic Variations

The history of variable shell stars is mainly based on spectroscopic information. During the last few years, photometric and polarimetric data have been progressively obtained (e.g. Percy *et al.* 1992, McDavid 1990; see also Borkjman, Arsenijevic *et al.* and Pavlovski *et al.* in these Proceedings). The study of the correlations between long-term variations of light, colour, emission-line strengths, radial velocities of shell lines, and polarization levels will introduce new constraints on quantitative envelope models. In particular, polarization should be a very powerful tool for studying the morphology of envelopes. A partial illustration of such correlations is given in Figs. 1a and 1b.

6.1. LONG-TERM V/R VARIATIONS

The most important studies in recent years concern EW Lac (Hubert *et al.* 1987a), V1294 Aql (Ballereau & Chauville 1989), 59 Cyg (Doazan *et al.* 1989), γ Cas (Horaguchi *et al.* 1993, Telting *et al.* 1993), ζ Tau (Mon *et al.* 1992), and V923 Aql, the twin of ζ Tau (Koubský *et al.* 1989), whose observed radial-velocity variation arises from a superposition of cyclic long-

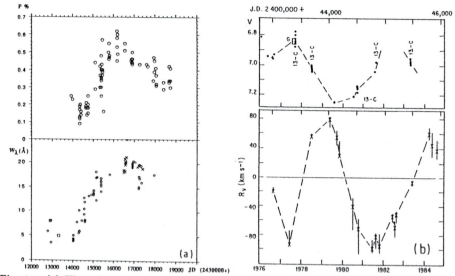

Fig. 1. (a) The relation between the equivalent width of the Hα emission line (OHP + literature) and the intrinsic polarization (Belgrade Obs.—Arsenijevic and coworkers) in κ Dra. (b) The relation between the magnitude variation and the radial velocities of shell lines in V1294 Aql (Alvarez *et al.* 1987).

term velocity variations of variable amplitude and with cycle length of an orbital period.

A summary of typical features of long-term V/R variations can be found in Okazaki (1991): cycles range from years to decades, are insensitive to spectral type, and tend to cluster around 7 years on average (e.g. Hirata & Hubert-Delplace 1981). There is a blueward shift of the whole profile when the red emission component is stronger, and vice versa. There are phase lags between V/R cycles of individual lines as in EW Lac (Kogure & Suzuki 1987), in ζ Tau (Hubert *et al.* 1987b) and in γ Cas (Horaguchi *et al.* 1993).

Information on the light variation associated with the radial velocity or the V/R variation is lacking, though it would be very useful for testing various models. There is only one example, that of V 1294 Aql, in which a phase lag of about 0.15–0.20 period is found between the minimum of the radial velocity and the maximum of the light curve (Fig. 1b). Up to the end of the eighties, the cyclic V/R variations were interpreted with qualitative models: a symmetric rotating-pulsating cool envelope or an elongated disc in apsidal rotation (e.g. Telting *et al.* 1993). Though still very simple, a quantitative model has been recently proposed by Kato (1989) and Okazaki (1991): the one-armed oscillating, quasi-Keplerian disc. These authors suggest that the possible global oscillations of a quasi-Keplerian disc are low-frequency and one-armed, since neither global axisymmetric oscillations, nor global non-axisymmetric oscillations are persistent except for these one-armed modes.

The elongated disc model is a geometrical version of the hydrodynamical one-armed oscillation model.

There are two variants of the one-armed oscillating disc: the first one by Okazaki (1991), considers that a modal precession results from pressure forces in the disc. The eigenmodes are strongly dependent on the outer edge of the disc and on the density gradient in the radial direction. Modes are generally retrograde, and periods range from years to decades. They are insensitive to the spectral type of the star, in agreement with observations. As the one-armed oscillation is multiperiodic, a new analysis of previous data is needed to verify this property.

The second variant is by Papaloizou et al. (1992) and Savonije & Heemskerk (1993) and takes into account the deviation from a point mass potential for the rotationally flattened star. This deviation forces elliptical particle orbits to precess, and the precession can manifest itself as a low-degree ($m = 1$) mode with a low pattern speed. In this frame, modes are found to be prograde, naturally confined to a few stellar radii, and rather insensitive to the size of the disc or to its density distribution. Observed periodicities of the V/R variations can be reproduced by this model for plausible values for the Mach number of the disc and for the stellar rotational deformation.

According to Okazaki (1991, 1992 and these Proceedings) and Hummel & Hanuschik (these Proceedings), the global one-armed oscillating disc model seems to be promising for explaining:
- the phase lag between the radial velocities of individual lines,
- the long-term variations of Balmer line profiles for both optically thin and optically thick discs and
- the phase lag between the light variations and the radial velocity variations of the shell lines and also the V/R ratio variation.
It must be emphasized that measurements of the Hα disc of γ Cas by the I2T, GI2T and Mark III interferometers (Thom et al. 1986, Mourard et al. 1989, Quirrenbach et al. 1993) seem to favour a deviation from angular momentum conservation in the envelope and support the quasi-Keplerian motion assumed in the theory of the one-armed oscillating model. However, this model does not take in account the effects of viscosity and envelope expansion which is often observed. Detailed comparison with observations must await calculations of the nonlinear evolution of the disc oscillations.

7. Conclusions

This review concerns only variability in the circumstellar envelope. Yet, one of the more crucial problems in Be-star research is the detection of a physical relation between the short-term stellar variations and long-term activity (or active and quiescent phases). This question remains open. A correlation between the amplitude of the line profile variations and Be episodes was found for ζ Oph and for λ Eri by Kambe *et al.* (1993a,b), but Smith (1989) failed to detect such a correlation in the later star. Another fundamental problem concerns the difference between the cool envelope of a Be star and of a Be-shell star. The conclusions given by different authors conflict on this point. A more extensive and denser envelope was found in Be-shell stars by Marlborough *et al.* (1993) in agreement with the work of Dougherty & Taylor (1992), while a lower electron density and a more compact envelope was derived for Be-shell stars by Slettebak *et al.* (1992). On the other hand, Kogure (1990) found a denser, disc-like envelope in Be-shell stars and a spheroidal, more extended envelope, in Be stars. We would like to stress the importance of long-term monitoring of some well-chosen objects using many different techniques in improving our knowledge of the physics of Be stars.

Acknowledgements

It is a pleasure for me to have stimulating discussions with Ryuko Hirata, Eduardo Janot-Pacheco, Michael Friedjung, Michele Floquet, Pavel Koubský and A.T. Okazaki. I am very grateful to my husband, Henri Hubert, for his invaluable assistance in preparing this review.

References

Alvarez, M., Ballereau, D., Sareyan, J.P., Chauville, J., Michel, R. and Le Contel, J.M.: 1987, *Rev. Mexicana Astron. Astrof.* **14**, 315.
Apparao, K.M.V. and Tarafdar, S.P.: 1989, *Astrophys. Space Sci.* **153**, 61.
Baade, D., Dachs, J., Van de Weygaert, R. and Steeman, F.: 1988, *Astron. Astrophys.* **198**, 211.
Baade, D.: 1989, *Astron. Astrophys.* **222**, 200.
Baade, D.: 1990, in Willson, L.A. and Stalio, R., eds., *Angular Momentum and Mass Loss for Hot Stars: NATO Advanced Research Workshop*, Kluwer: Dordrecht, 177.
Baade, D.: 1992, in Kondo, Y., Sistero, R. and Polidan, R.S., eds., *Evolutionary Processes in Interacting Binary Stars: IAU Symp. 151*, Kluwer: Dordrecht, 147.
Ballereau, D., Chauville, J. and Mekkas, A.: 1988, *Astron. Astrophys. Suppl.* **75**, 139.
Ballereau, D. and Chauville, J.: 1989, *Astron. Astrophys.* **214**, 285.
Beeckmans, F.: 1976, *Astron. Astrophys.* **52**, 465.
Blanco, C., Mammano, A., Margoni, R., Stagni, R., Munari, U., Baratta, G.B., Coluzzi, R. and Croce, V.: 1990, *Astron. Astrophys.* **230**, 307.
Brown, J.C. and Henrichs, H.F.: 1987, *Astron. Astrophys.* **182**, 107.
Brown, J.C. and Fox, G.K.: 1989, *Astrophys. J.* **347**, 468.
Chakrabarti, S.K. and Wiita, P.J.: 1993, *Astron. Astrophys.* **271**, 216.
Doazan, V., de la Fuente, A., Barylak, M., Cramer, N. and Mauron, N.: 1993, *Astron. Astrophys.* **269**, 415.
Doazan, V., Sedmak, G., Barylak, M. and Rusconi, L.: 1991, *A Be Star Atlas*, SP-1147 ESA: Noordwijk.
Doazan, V., Thomas, R.N. and Bourdonneau, B.: 1988, *Astron. Astrophys.* **205**, L11.
Doazan, V., Barylak, M., Rusconi, L., Sedmak, G., Thomas, R.N. and Bourdonneau, B.: 1989, *Astron. Astrophys.* **210**, 249.
Dougherty, S.M. and Taylor, A.R.: 1992, *Nature* **359**, 808.

Floquet, M. and Hubert, A.M.: 1991, in Jaschek, C. and Andrillat, Y., eds., *The Infrared Spectral Region of Stars*, Cambridge Univ. Press: Cambridge, 160.

Floquet, M., Hubert, A.M., Maillard, J.P., Chauville, J. and Chatzichristou, H.: 1989, *Astron. Astrophys.* **214**, 295.

Ghosh, K.K., Apparao, K.M.V. and Tarafdar, S.P.: 1989, *Astrophys. J.* **344**, 437.

Ghosh, K.K.: 1990, *Astrophys. J.* **360**, 239.

Gies, D.R., Willis, C.Y., Penny, L.R. and McDavid, D.: 1993, *Publ. Astr. Soc. Pacific* **105**, 281.

Grady, C.A., Bjorkman, K.S., Peters, G.J. and Henrichs, H.F.: 1988, in Rolfe, E.J., ed., *A Decade of UV Astronomy with the IUE Satellite*, Vol.1, SP-281 ESA: Paris, 257.

Halbedel, E.M.: 1989, *Publ. Astr. Soc. Pacific* **101**, 999.

Hanuschik, R.W., Dachs, J., Baudzus, M. and Thimm, G.: 1993, *Astron. Astrophys.* **274**, 356.

Harmanec, P.: 1990, *Astron. Astrophys.* **237**, 91.

Hill, G.M., Walker, G.A.H., Dinshaw, N. and Yang, S.: 1988, *Publ. Astr. Soc. Pacific* **100**, 243.

Hirata, R. and Hubert-Delplace, A.M.: 1981, in GEVON and Sterken, C., eds., *Workshop on Pulsating B Stars*, Obs. de Nice: Nice, 217.

Hirata, R. and Kogure, T.: 1976, *Publ. Astr. Soc. Japan* **28**, 509.

Horn, J., Hubert, A.M., Hubert, H., Koubský, P. and Bailloux, N.: 1992, *Astron. Astrophys.* **259**, L5.

Horaguchi, T., Kogure, T., Hirata, R., Kawai, N., Matsuoka, M., Murakami, T., Doazan, V., Slettebak, A., Huang, C.C., Cao, H., Guo, Z., Huang, L., Tsujita, J., Ohshima, O. and Ito, Y.: 1993, *Publ. Astr. Soc. Japan* in press.

Huang, L., Hsu, J.C. and Guo, Z.H.: 1989, *Astron. Astrophys. Suppl.* **78**, 431.

Hubert, A.M., Floquet, M., Chauville, J. and Chambon, M.T.: 1987a, *Astron. Astrophys. Suppl.* **70**, 443.

Hubert, A.M., Floquet, M. and Chambon, M.T.: 1987b, *Astron. Astrophys.* **186**, 213.

Jarad, M.M., Hilditch, R.W. and Skillen, I.: 1989, *Mon. Not. Roy. Astr. Soc.* **238**, 1085.

Jeffery, C.S.: 1991, *Mon. Not. Roy. Astr. Soc.* **249**, 327.

Juza, K., Harmanec, P., Hill, G.M., Tarasov, A.E., Matthews, J.M., Tuominen, I. and Yang, S.: 1991, *Bull. Astron. Inst. Czechosl.* **42**, 39.

Kambe, E., Ando, H. and Hirata, R.: 1993a, *Astron. Astrophys.* **273**, 435.

Kambe, E., Ando, H., Hirata, R., Walker, G.A.H., Kennelly, E.J. and Matthews, J.M.: 1993b, *Publ. Astr. Soc. Pacific* in press.

Kato, S.: 1989, in Meyer, F., Duschl, W.J., Frank, J. and Meyer-Hofmeister, E., eds., *Theory of Accretion Disks*, Kluwer: Dordrecht, 173.

Kogure, T. and Suzuki, M.: 1987, in Slettebak, A. and Snow, T.P., eds., *Physics of Be Stars: IAU Coll. 92*, Cambridge Univ. Press: Cambridge, 192.

Kogure, T.: 1990, *Astrophys. Space Sci.* **163**, 7.

Koubský, P., Gulliver, A.F., Harmanec, P., Ballereau, D., Chauville, J., Gráf, T., Horn, J., Iliev, L.H. and Lyons, R.W.: 1989, *Bull. Astron. Inst. Czechosl.* **40**, 31.

Koubský, P., Horn, J., Harmanec, P., Hubert, A.M., Hubert, H. and Floquet, M.: 1993, *Astron. Astrophys.* in press.

Marlborough, J.M., Chen, H. and Waters, L.B.F.M.: 1993, *Astrophys. J.* **408**, 646.

Mastenova, K.: 1989, *Bull. Astron. Inst. Czechosl.* **40**, 118.

McDavid, D.: 1990, *Publ. Astr. Soc. Pacific* **102**, 773.

Mon, M., Kogure, T., Suzuki, M. and Singh, M.: 1992, *Publ. Astr. Soc. Japan* **44**, 73.

Mourard, D., Bosc, I., Labeyrie, A., Koechlin, L. and Saha, S.: 1989, *Nature* **342**, 520.

Norton, A.J., Coe, M.J., Estela, A., Fabregat, J., Gorrod, M.J., Kastner, J., Payne, B.J., Reglero, V., Roche, P. and Unger, S.J.: 1991, *Mon. Not. Roy. Astr. Soc.* **253**, 579.

Okazaki, A.T.: 1991, *Publ. Astr. Soc. Japan* **43**, 75.

Okazaki, A.T.: 1992, in Takeuti, M. and Buchler, J.R., eds., *Nonlinear Phenomena in Stellar Variability: IAU Coll. 134*, Astrophys. Space Sci. series, in press.

Papaloizou, J.C., Savonije, G.J. and Henrichs, H.F.: 1992, *Astron. Astrophys.* **265**, L45.

Parsons, S.B., Dempsey, R.C. and Bopp, B.W.: 1988, in Rolfe, E.J., ed., *A Decade of UV Astronomy with the IUE Satellite*, Vol.1, SP-281 ESA: Paris, 225.

Percy, J.R. and Attard, A.: 1992, *Publ. Astr. Soc. Pacific* **104**, 1160.

Peters, G.J.: 1984, *Publ. Astr. Soc. Pacific* **96**, 960.

Peters, G.J.: 1988, *Astrophys. J. Let.* **331**, 33.

Poeckert, R.: 1981, *Publ. Astr. Soc. Pacific* **93**, 297.

Pols, O.R., Coté, J., Waters, L.B.F.M. and Heise, J.: 1991, *Astron. Astrophys.* **241**, 419.

Porri, A. and Stalio, R.: 1988, *Astron. Astrophys. Suppl.* **75**, 371.

Quirrenbach, A., Hummel, C.A., Buscher, D.F., Armstrong, J.T., Mozurkewich, D. and Elias, N.M.: 1993, *Astrophys. J. Let.* **416**, 25.

Savonije, G.J. and Heemskerk, M.H.M.: 1993, *Astron. Astrophys.* **276**, 409.

Slettebak, A. and Carpenter K.G.: 1983, *Astrophys. J. Suppl.* **53**, 869.

Slettebak, A., Collins, G.W. and Truax, R.: 1992, *Astrophys. J. Suppl.* **81**, 335.

Smith, M.A.: 1989, *Astrophys. J. Suppl.* **71**, 357.

Smith, M.A., Peters, G.J. and Grady, C.A.: 1991, *Astrophys. J.* **367**, 302.

Sonneborn, G., Grady, C.A., Wu, C.C., Hayes, D.P., Guinan, E.F., Barker, P.K. and Henrichs, H.F.: 1988, *Astrophys. J.* **325**, 784.

Štefl, S., Harmanec, P., Horn, J., Koubský, P., Kríz, S., Hadrava, P., Bozic, H. and Pavlovski, K.: 1990, *Bull. Astron. Inst. Czechosl.* **41**, 29.

Telting, J.H., Waters, L.B.F.M., Persi, P. and Dunlop, S.R.: 1993, *Astron. Astrophys.* **270**, 355.

Thom, C., Granes, P. and Vakili, F.: 1986, *Astron. Astrophys.* **165**, L13.

Waters, L.B.F.M., Coté, J. and Pols, O.R.: 1991, *Astron. Astrophys.* **250**, 437.

Discussion

Waters: I would like to point out that the fact that Be stars lose their Hα emission does not necessarily imply that these stars lose their disc altogether. It may be that these stars go through a phase of much lower density without noticeable Hα emission.

Peters: (1) I have been attempting for some time now to find any linkage between long- and short-term variability cycles. In general, it is my opinion that different physical phenomena are responsible for each. In most cases, if the star displays long-term variations, short-term variations are absent or very weak. One exception is λ Eri in which I have recently found that the outbursts recur with a quasi-period of 1.3 yr.

(2) The nature of the secondary in HR 2142 is still an open question. The phase-dependent variations in the strengths and velocities of the shell lines are exactly the same type one sees in conventional Algol systems. This implies the presence of a cool secondary.

Hubert: The energy distribution in the IR range does not reveal the presence of a cool giant companion for HR 2142.

Rountree: Some quiescent Be stars can be detected in the ultraviolet because they have stronger C IV (and sometimes Si IV) lines than "normal" stars of the same spectral type. This is possible only if the spectra are classified by comparing "photospheric" absorption lines with the standards. The wind lines can then be used to detect anomalous conditions.

PHYSICS OF THE PHASE CHANGES OF Be STARS

K.K. GHOSH

Indian Institute of Astrophysics, Vainu Bappu Observatory, Kavalur, India

Abstract. Based on spectroscopic observations of more than 100 Be stars, we present the results of the phase changes of Be stars.

1. Introduction

Be stars display B, Be and Be shell spectra and the cause of the phase changes is mostly unknown. Kogure (1990) has suggested a simple model of phase-changing variation between Be stars and shell stars from the viewpoint of the formation of the shell absorption lines in the envelopes of these stars. Here we present the observational results which are directly connected with the phase-change variations of Be stars.

2. Observations

We have started a spectroscopic program, since 1988, at the Cassegrain and Coudé foci of the 2.34 m and 1 m reflectors of VBO, for faint (brighter than 14 mag) and bright (brighter than 7 mag) Be stars, respectively. Using the Coudé Echelle spectrograph with CCD system we get the spectra, simultaneously, in the He I (4471 Å), Mg II (4481 Å), Hβ (4861 Å), He I (5876 Å), Si II (6374 & 6371 Å) and H regions. From the obtained spectra we have detected phase change variations in many Be stars (γ Cas, ϕ Per, HR 1423, 27 CMa, λ Eri, FY CMa, μ Cen, 48 Lib, etc.).

3. Results and discussion

We have found that the phase-change variations of Be stars from Be to Be-shell phase take place via a strong outburst with P Cygni profiles in their spectra (in Figs. 1–3, the P-Cygni profiles of FY CMa during February 1991 have just been mentioned (Peters 1991, Ghosh et al. 1993). Before the outburst the Hβ and He I (5876 Å) lines displayed inverse P Cygni profiles during March 1990. Also we have detected inverse P-Cygni profiles in the He I (5876 Å) and Hβ lines of many other Be stars before their outburst (Ghosh 1990, Ghosh et al. 1993). Inverse P-Cygni profiles indicate the infall of matter on the star and this may cause the instability of the star which leads to the outburst. The amount of material ejected due to the outburst is much larger then that ejected by the mass loss due to the non-radial pulsation of the star (Ghosh et al. 1993). After the outburst the star enters

351

L. A. Balona et al. (eds.), Pulsation, Rotation and Mass Loss in Early-Type Stars, 351–353.

into the Be shell phase (see Figs. 1–3, the shell profiles of FY CMa during 29 March 1991). After some time the envelope of the shelter star disperses into the interstellar medium (ISM) and the star enters either into a B phase (if there is no new mass loss between the outburst and the disappearance of the envelope into the ISM) or into Be phase (if there is new mass loss from the star).

4. Conclusion

Phase-change variations of Be stars from the B to Be phase take place via mass loss of the star due to its non-radial pulsations (Baade 1989) and from the Be to Be-shell phase is due to the disappearance of the shell envelope into the ISM when no new mass loss occurs between the outburst and the disappearance of the envelope. If there is new mass loss, the star will enter into Be phase from the Be-shell phase (see Fig. 4).

References

Baade, D.: 1989, *Astronomy and Astrophysics* **198**, 211
Ghosh, K.K.: 1990, *Astrophysical Journal* **360**, 239
Ghosh, K.K. et al., 1993, in preparation
Kogure, T.: 1990, *Astrophysics and Space Science* **193**, 7
Peters, G.J.: 1991, *Be Star News Lett.* **24**, 17

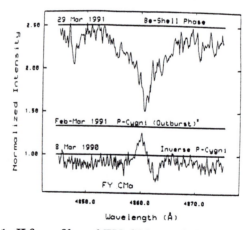

Fig. 1. Hβ profiles of FY CMa at different phases.

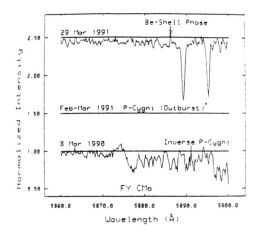

Fig. 2. Same as Fig. 1 but for He I profiles

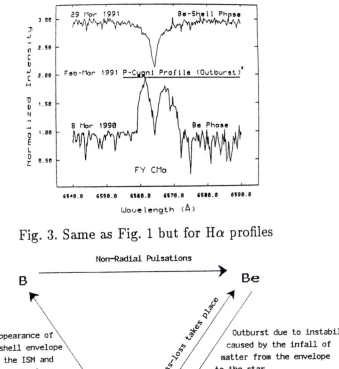

Fig. 3. Same as Fig. 1 but for Hα profiles

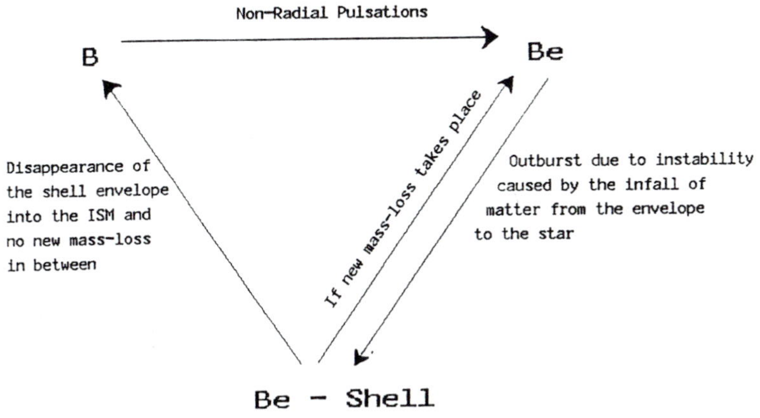

Fig. 4. Schematic diagram of the phase-change variations of Be stars.

DIFFERENCE BETWEEN Be STARS AND SHELL STARS FROM Hα EMISSION

H. CHEN and A. R. TAYLOR
Department of Physics and Astronomy
University of Calgary, Canada

Abstract. We examine Hα equivalent widths versus spectral-types for 41 Be stars. Although the W_α value is scattered for a given spectral-type, a well-defined upper limit exists. Most shell stars are located in the upper limit region. The shell stars all have $v \sin i/v_{cri}$ larger than the normal Be stars in the upper limit region. This strongly suggests that the distinction between shell stars and normal Be stars of high equivalent width is a result of variation in inclination angle i. Polarization data also support this hypothesis.

1. Introduction

What causes the difference between a "normal" Be star and a shell star is still not clearly understood. On one hand, Hutchings (1976) suggested that "A shell may however be simply a very extended Be star envelope." On the other hand, Slettebak et al. (1992) recently claim that shell stars have envelopes which are more compact and have lower electron densities than Be stars, based on their spectroscopic observations of 41 bright stars.

In this investigation we re-examine the data of Slettebak et al. (1992) to see if there is any difference between shell stars and "normal" Be stars in terms of their W_α values.

2. Results

In Fig. 1, we plot W_α versus spectral types for 41 stars using Slettebak et al. data. Although W_α value is scattered for a given spectral-type, a well-defined upper limit exists. This upper limit peaks at about B2 and decreases towards later-types. A similar plot for a larger number of Be stars suggests that this upper limit might also decrease towards earlier-types (Chen & Talyor 1993). With one exception, the shell stars are located in the upper limit region (above the dash line). As a group, they have higher equivalent widths than "normal" Be stars with same spectral type. The only exception is ζ Tau, which is in a binary system (Slettebak, 1982).

In the upper limit region, there are also stars which show "normal" Be star spectra. It is noteworthy that the shell stars in this region all have $v \sin i/v_{cri}$ larger than the normal Be stars (Fig. 2). This suggests that the distinction between shell stars and normal Be stars of high equivalent width is a result of variation in $\sin i$ or v/v_{cri}. If all these stars are rotating at a high speed, one can conclude normal Be stars have smaller inclination angles

L. A. Balona et al. (eds.), Pulsation, Rotation and Mass Loss in Early-Type Stars, 354–355.

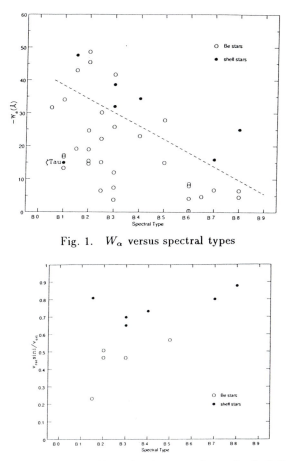

Fig. 1. W_α versus spectral types

Fig. 2. The values of $v \sin i/v_{cri}$ for the stars above the dash line in Fig. 1

than shell stars. Of seven shell stars and five normal Be stars above the dash line in Fig. 1, intrinsic polarization data are available for six shell stars and three normal Be stars (McLean & Brown 1978). All shell stars have higher intrinsic polarization than normal Be stars of the same spectral types. This again supports the hypothesis that the shell stars have higher inclination angles than the normal Be stars of high W_α.

3. The References

Chen, H. & Taylor, A. R., in preparation, (1993).

Huchings, J. B., in IAU Symp. 70, *Be and Shell Stars*, ed. A. Slettebak (Sordrecht:Reidel), 13 (1976).

McLean, I. S., & Brown, J. C., *Astr. Astrophys.* **69**, 291 (1978).

Slettebak, A., *Astrophys. J. Suppl.*, **50**, 55 (1982).

Slettebak, A., Collins II, G.W., &Truax, R., *Astrophys. J. Suppl.* **81**, 335 (1992).

THE NUMBER OF Be STARS COMPARED WITH THE NUMBER OF B STARS, TAKING INTO ACCOUNT THE SPECIFIC PHYSICAL CHARACTERISTICS OF Be STARS

D. BRIOT

Observatoire de Paris, 61 Avenue de l'Observatoire, 75014 Paris

and

J. ZOREC

Institut d'Astrophysique de Paris, 98bis Boulevard Arago, 75014 Paris

It is very important for the understanding of the Be phenomenon, and particularly for locating a possible Be phase in the evolutionary track of B stars, to accurately determine the proportion of Be stars among all B stars. This type of study was already made several times in the past. Results obtained generally show a maximum Be frequency around spectral type B2 then a decrease towards late spectral types. Actually Be stars do not have the same characteristics as "normal" B stars and we have to take this into account in the determination of the ratio : number of Be stars / number of B stars. We use the Bright Star Catalogue (Hoffleit & Jaschek 1982) and the Supplement to the Bright Star Catalogue (Hoffleit, Saladyga & Wlasuk 1983) containing stars V= 7.10 and brighter. This study needed to be made separately for the different spectral types because :

-Physical parameters of B stars are very different from B0 to B9 ;
-Emission characteristics of Be stars vary very much, with a decrease from B0e to B9e.

We successively consider three effects which can influence the frequency of Be stars :

- The over-luminosity of Be stars as compared with B stars ;
- Spectral type changes during constant mass evolution ;
- Spectral type changes due to the fast rotation of Be stars.

We know that Be stars are intrinsically brighter than B stars and more so when the stars are hotter (earlier type stars) (Zorec & Briot 1991). If the stars are counted, as currently done, up to a limiting magnitude, Be stars are counted in a larger volume than B stars, which artificially increases the ratio of Be stars to B stars.

Stars change their spectral type while they evolve with a constant mass from the main sequence (luminosity class V) up to the giant branch (luminosity class III). The determination of the ratio of Be stars for the various spectral types must be done by classifying the stars of luminosity classes III and IV according to their mass, that is to say not according to their present

356

L. A. Balona et al. (eds.), Pulsation, Rotation and Mass Loss in Early-Type Stars, 356–357.

spectral type but according to their spectral type they had when they were of luminosity class V.

Because of the fast rotation of Be stars, we took into account the change of spectral type for the case of rigid rotation (Collins & Sonneborn 1977). The distribution function of the angular velocity was calculated for each spectral type and luminosity class.

All results are shown on Fig. 1. We see that the ratios of Be stars to B stars change when specific physical characteristics of Be stars are taken into account. In particular, effects due to rotation shift the Be maximum towards early types. So we obtain the result that the hotter Be stars are, not only are the emission characteristics more pronounced, but also that the proportion of these stars is greater.

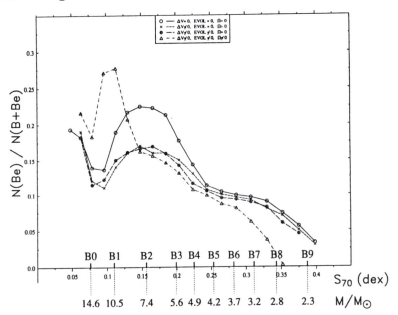

Fig. 1. Frequency of Be stars among B stars as a function of the spectral type. The results directly obtained from the Bright Star Catalogue and the Supplement to the Bright Star Catalogue are represented by open circles. The results are represented by crosses when the absolute magnitudes characteristic of Be stars are taken into account, by circled crosses when spectral type changes during stellar evolution are also taken into account and by triangles when spectral type changes due to fast rotation are in addition considered.

References

Collins, G.W., II and Sonneborn, G.H.: 1977, *Astrophys. J. Suppl. Series* **34**, 41

Hoffleit, D. and Jaschek, C.: 1982, *The Bright Star Catalogue*, Yale Univ. Obs.

Hoffleit, D., Saladyga, M. and Wlasuk P.: 1983, *A Supplement to the Bright Star Catalogue*, Yale Univ. Obs.

Zorec, J. and Briot, D.: 1991, *Astron. Astrophys.* **245**, 150

SHELL PROFILES IN Be STAR SPECTRA

R. W. HANUSCHIK*

Astronomical Institute, Ruhr-Universität, D-44780 Bochum, Germany

1. Problem

Although very different in shape, Be-type and Be-shell-type spectral lines are thought to arise from the same kind of circumstellar environment, a disk-like envelope. The only discrimination between these two types of spectra is the inclination angle: If sufficiently high, a large column depth of circumstellar material causes partial obscuration of the stellar disk, i.e. shell absorption; otherwise a pure emission line is observed. A shell line can be considered as equivalent to P Cygni-type absorption trough for different geometry (disk instead of sphere) and kinematics (rotation instead of outflow; see Hanuschik, these proceedings).

Obviously, this very pronounced difference in line shape can be used to determine the *average opening angle of a Be star disk*. This is, however, difficult in practice since there seems to be nothing like a unique definition of "shell profile" in the literature. The mostly used criterion – central depression (cd) in Balmer lines visible below (or even above) stellar continuum – depends on spectral resolution and is ambiguous because of other effects influencing this parameter like underlying stellar absorption profile, additional circumstellar emission, kinematical broadening. Clearly, a reliable and objective definition of "shell line" is desirable.

2. Fe II line profiles

Circumstellar Fe II lines are the key to the problem because their shape is not distorted by most of the above effects. The straightforward shell criterion for these lines is $I_{cd}(\text{FeII}) < I_*$. However, they are not always visible in Be star spectra, while Hα is (per definition). I have therefore searched for spectra with both Hα *and* Fe II lines, derived an Hα shell criterion from the above Fe II criterion, and applied this to a large sample of Be and shell-type stars. The Hα parameter is I_p/I_{cd} (I_p = mean double peak intensity). Figure 1 shows the relation of these two parameters. Clearly, the domain with

$$I_p/I_{cd}(H\alpha) \geq 1.5 \tag{1}$$

* Now at: Astronomical Institute, University of Tübingen, D-72076 Tübingen, Germany

L. A. Balona et al. (eds.), *Pulsation, Rotation and Mass Loss in Early-Type Stars*, 358–359.

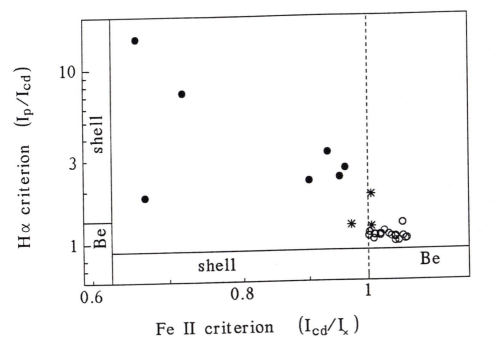

Fig. 1. Hα criterion I_p/I_{cd} vs. Fe II criterion I_{cd}/I_* for a sample of programme stars with both Hα and Fe II $\lambda 5317$ measurements available.

is exclusively occupied by shell stars, giving confidence to adopt this criterion in the following.

Applying it to a sample of 109 stars from our atlas (Hanuschik et al., 1994) and that of Doazan et al. (1991), I find that $p = 21\%$ of the spectra are shell spectra. Assuming random distribution of inclination angles i, the average half opening angle $\alpha = 90° - \arccos p$ comes out as 12°. This value is in approximate agreement with results from model line profile calculations assuming a Keplerian disk in hydrostatic equilibrium (Hummel 1993).

Acknowledgements

I would like to thank Wolfgang Hummel for many valuable discussions.

References

Doazan V., Sedmak G., Barylak M., Rusconi L.: 1991, ESA SP-1147.
Hanuschik R.W. et al.: 1994, in preparation.
Hummel W.: Ph. D. thesis (Bochum).

PHASE CHANGES IN Be STARS:

The Be-shell and Be phases of Pleione

V. DOAZAN
Observatoire de Paris, 61, Av. de l'Observatoire, F-75014 Paris

A. DE LA FUENTE
INSA-VILSPA, Apartado 50727, E-28080 Madrid

N. CRAMER
Observatoire de Genève, CH-1290 Sauverny, Suisse

and

M. BARYLAK
ESA IUE Observatory, Apartado 50727, E-28080 Madrid

Phase changes from Be to Be-shell and/or B normal, and conversely, presently remain unpredictable in Be stars (except in some binaries where the transition to a shell phase seems to be associated with the orbital period). Because of their unpredictable character, these phenomena have been monitored in very few cases and are still very poorly understood. However, the existence of phase transitions have strong modelling implications. Under the disk model, the inclination of the star's rotation axis determines the type of spectrum, Be or Be-shell, that a given Be star may exhibit. It is obvious that this picture is contradicted by phase transitions (Doazan 1982).

In view of investigating the phenomena occurring during phase transitions and the changes between the Be and Be-shell phases, we undertook systematic coordinated observations in both photometry and high dispersion spectroscopy, in the far UV and optical spectral regions, for the two phases, Be-shell and Be, of Pleione which is the only Be star for which such simultaneous observations exist.

Regular observations made in the Geneva photometric system showed that the excursion of Pleione in the HR diagram in 1960-1992 starts out from the vicinity of the main-sequence B6V–B8V stars (Be phase, 1960-1972), "evolves" up to the luminous supergiant branch among the A5Ia–A7Iab stars (maximum shell phase, 1983), and finally returns to its initial place among the late main sequence B stars in the well-developed Be phase (1992) (Cramer et al. 1993).

In this paper we summarize results obtained from our coordinated observations made in the time-interval 1979-1991 when the star exhibited a strong shell phase (1979) and a well developed Be phase (1991).

L. A. Balona et al. (eds.), Pulsation, Rotation and Mass Loss in Early-Type Stars, 360–361.

Between 1979, when Pleione exhibited a strong shell spectrum, and 1991, when it showed a Be-type spectrum:

1. The CIV and SiIV resonance lines, which were not present/detectable during the shell phase, appeared and strengthened as the shell spectrum vanished and the Be phase developed (Doazan et al. 1988,1991)
2. The absorption bump at 2200 Å showed large changes, clearly indicating that it cannot be used confidently for measuring the interstellar component of extinction for Be stars (Doazan et al. 1993a)
3. The observed far UV radiative flux increased by more than a factor of two (Doazan et al. 1993b)
4. The observed dereddened absolute energy distribution, from the far UV to the optical spectral regions increased in all the observed wavelengths between the Be-shell and Be phases
5. A fit of the Kurucz models (ATLAS 9) to the normalized absolute dereddened energy fluxes from the far UV and optical wavelength range gave the following results (Doazan et al. 1993c):
   - The 1979 data of the shell phase are best fitted with a model with $T_{eff} = 11,000°$ K and $\log g = 4$.
   - The 1991 data of the Be phase are best fitted with a model with $T_{eff} = 12,500°$ K and $\log g = 4$.

These data and analysis would imply that shell phases cannot be said to be characterized by a far UV deficiency, as is often stated. Rather, we observe, when coordinated far UV and optical data are used, a complete change in the energy distribution, from the far UV to the optical regions, suggesting/simulating a temperature change of the star.

References

Cramer, N., Doazan, V., de la Fuente, A., Nicollet, N., Barylak, M., 1993, to be published

Doazan, V., 1982, in Underhill, A. B., and Doazan, V. (eds), *The B stars with and without Emission Lines*, NASA SP-456

Doazan, V., Thomas, R. N., Bourdonneau, B., 1988, *Astronomy and Astrophysics*, **205**, L11

Doazan, V., Sedmak, G., Barylak, M., Rusconi, L., 1991, *A Be star Atlas of far UV and optical high resolution spectra*, ESA SP-1147

Doazan, V., de la Fuente, A., Barylak, M., Cramer, N., 1993a, in Weiss, W. W., Baglin, A. (eds), *Inside the stars*, IAU Colloq. 137, p 787

Doazan, V., de la Fuente, A., Barylak, M., Cramer, N., Mauron, N., 1993b, *Astronomy and Astrophysics*, **269**, 422

Doazan, V., de la Fuente, A., Barylak, M., Cramer, N., 1993c, to be published

THE PHOTOSPHERE OF PLEIONE(*)

J. ZOREC

Institut d'Astrophysique de Paris - 98bis, Bd. Arago, 75014 Paris

Abstract. In this paper we report results concerning the photospheric Balmer discontinuity of Pleione. They show that after the Be → Be-shell phase transitions, the stellar
photosphere keeps its characteristics unchanged.

1. Introduction

The energy distribution around the Balmer discontinuity (BD) gives us informations on the emission or absorption power of the Be phenomenon, and
on the photosphere of the underlying star. Two BD's are seen in Be stars:
one is variable, it may be in emission or in absorption and it is due to the
circumstellar envelope; the other one resembles that of a normal B star and
doesn't show any change within the observational uncertainties: $\delta D \leq 0.02$
dex (Divan 1979). The observed value of the stellar BD (T_{eff}-indicator) as
well as its mean position (log g-indicator) are the same after the characteristic Be "phase" variations (Zorec 1986).

2. The observations

Pleione (HD 23862) has been observed in the BCD (Barbier-Chalonge-
Divan) spectrophotometric system since 1951 (Zorec 1986). The BCD observations were unfortunately done only during Be-shell phases. Meanwhile, the
star has however shown at least two Be phases. The mean value of the photospheric BD and its mean position $\lambda_1 - 3700$ are: $D_* = 0.369 \pm 0.010$ dex
and $\lambda_1 - 3700 = 53 \pm 1$Å (observation dates: Nov. 51, Nov. 77, Jul. 80,
Nov. 80, Nov. 81). Note the constancy of D_* and of $\lambda_1 - 3700$! These values
give: spectral type : B7-8 IV-V; $T_{eff} = 11820 \pm 270$ K; log g $= 3.96 \pm 0.09$;
$R/R_\odot = 3.17 \pm 0.14$; $M/M_\odot = 3.35$ (Divan and Zorec 1982, Zorec 1986).
In Fig. 1 are shown the BCD spectrophotometric and some photometric
observations of Pleione [UBVRI system (Johnson et al. 1966), UBV system
(Sharov and Lyutiy 1972), Geneva system (Rufener priv. comm.), 13-colours
(Johnson et al. 1967)] transformed to the total BD: $D = D_* + d_{cs}$ (stellar
+ circumstellar components) and to the gradient Φ_{rb} ($\lambda\lambda 0.4 - 0.6\mu$). The
difference between the BCD gradients and those obtained from photometry
is due to the shell-line crowding which produces redder photometric colours.
Note the constancy of the V mag. and that of the Φ_{rb} gradient in spite of
the strong and variable absorption in the circumstellar component of the
BD.

L. A. Balona et al. (eds.), Pulsation, Rotation and Mass Loss in Early-Type Stars, 362–363.

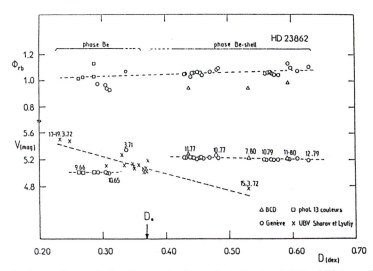

Fig. 1. Gradient Φ_{rb} and the V magnitude against the total BD of Pleione: $D = D_* + d_{cs}$. Points of 65-66 are for a Be phase. Points of 71-72 correspond to a Be → Be-shell transition. Since 77 points are for a Be-shell phase. D_* is the photospheric BD.

3. Conclusion

The photosphere of Pleione seems to remain the same after two phase changes of the type Be → Be-shell. The remarkable constancy of the stellar component of the BD as well as its position indicates that the phase variations leaves the stellar photosphere unchanged. Hence, the observed photometric changes have to be produced by a variable exophotospheric structure. It can be shown that changes of the geometry, size and temperature of the circumstellar envelope may explain the observed photometric variations during a phase transition (Zorec et al. 1989, Zorec and Briot 1991).

(*) *Observations obtained at the ESO, La Silla, Chile and at the OHP, France*

References

Divan, L.: 1979, in M.F. McCarty, G.V. Coyne, A.G.D. Philip, ed(s)., *Spectral Classification in the Future*, IAU Coll. **47**, 247

Divan, L., Zorec, J.: 1982, *The Hipparcos Space Mission*, ESA SP-177, 101

Johnson, H.L., Mitchel, R.I., Iriarte, B., Wisniewski, W.Z.: 1966, *Lun. Plan. Lab.* **4**, 99

Johnson, H.L., Mitchel, R.I., Latham, A.S.: 1967, *Lun. Plan. Lab.* **6**, 85

Sharov, A.S., Lyutiy, B.M.: 1972, *IAU Inf. Bull. Var. Stars* , 698

Zorec, J.: 1986, *Thèse d'Etat*, Universite Paris VII

Zorec, J., Höflich, P., Divan, L.: 1989, *Astron. Astrophys.* **210**, 279

Zorec, J., Briot, D.: 1991, *Astron. Astrophys.* **245**, 150

SHORT-TERM SPECTROSCOPIC VARIATIONS OF THE
SOUTHERN Be STAR 48 Lib

M. FLOQUET, A.M. HUBERT, H. HUBERT

Observatoire de Paris, Section d'Astrophysique de Meudon, URA 335 du CNRS,
F-92195 Meudon Cedex, France

and

E. JANOT-PACHECO, N.V. LEISTER

Instituto Astronomico e Geofisico, Sao Paulo University, Brazil

1. Introduction

48 Lib (B3IV, vsin i=395 km/s) is a well known Be shell star. Cuypers et al. (1989) found a period of 0.4017 d, the single wave light curve being asymmetric. Rapid variations in the radial velocity of metallic shell lines have been found by Ringuelet-Kaswalder (1963) with a period of 0.115 d.

For the first time, we have searched for multiperiodic spectroscopic variations in the HeI-MgII $\lambda\lambda$4471-4481 photospheric lines of 48 Lib.

2. Observational data and reduction

Observations were performed at La Silla (ESO) observatory from 10 to 14 June 1989 with the CAT+CES+CCD (R=50000 and S/N<250). The mean exposure time was 30 min. Reductions have been done with eVe package on the Vax 4500 of the Paris-Meudon observatory.

First, we scanned the profiles every 3px, i.e. 0.102Å (the element of spectral resolution). Fourier Transform and Clean algorithms were then applied on each time series data at each λ. Due to the observing window, frequencies higher than 12.0 c/d and lower than 0.3 c/d cannot be detected. Accuracy on frequency determination is 0.3 c/d.

Second, we have searched for moving bumps on residuals obtained by subtracting a nightly mean profile from each spectrum.

3. Results

3.1. PERIODOGRAM

On the HeI line we clearly see 3 frequencies: 10.6, 3.1 and 1.3 c/d. The frequency 10.6 c/d is present on a large part of the profile and the variation of the complexe phase of the power spectrum across the line profile gives a lower value of the modal number m=6.3. The frequency 1.3 c/d is less

L. A. Balona et al. (eds.), Pulsation, Rotation and Mass Loss in Early-Type Stars, 364–365.

important but is present on the two lines. The situation is less clear for the
MgII line and the periodogram is difficult to interpret. Different frequencies
appear on small parts of the line profile. On the blue part, 3 frequencies
are present: 6.5, 7.5 and 8.5 c/d. It would be reasonable to assume that the
true frequency is $\nu=7.5$ c/d and the two other ones are aliases at $^+_-1$ c/d.
On the red part, it appears $\nu=3.7$ c/d which is half the above period 7.5
c/d. But we do not find $\nu=10.6$ c/d as on the HeI line. It should be kept
in mind that the $\lambda4481$ MgII line is strongly influenced by NLTE (that is
envelope effects). This periodogram does not display the value $\nu=2.49$ c/d
found by Cuypers et al. (1989) in photometry, but $\nu=1.25$-1.30 c/d, which
was a secondary peak in their analysis, is present on the two lines.

3.2. RESIDUALS

Five bumps can be followed on the 2 lines, on the 13-14 June 1989, crossing
the profiles from blue to red. Bumps acceleration a_0 and intervals δt between
two consecutive bumps at the line center have been computed. In the frame
of Non Radial Pulsations, modal values m can be deduced from a_0 and δt.
For both lines the frequency associated to the mean interval $\delta t=0.0975$ d
($\nu=10.6$ c/d), that is the NRP frequency in the corotating frame, is very
close to one of the frequency found on the periodogram for the HeI line.
The m value computed with vsin i=395 km/s and the mean value of the
acceleration $a_0=3161$ km/s/d is m=8.

4. Conclusion

A frequency $\nu=10.6$ c/d is found either in the periodogram of HeI line and
in residuals analysis of the two lines. In the frame of NRP, it corresponds
to a sectorial mode m=8. The periodogram, which is different for HeI and
MgII, displays two other frequencies for the HeI line (3.1 and 1.36 c/d), and
three frequencies for MgII (7.5, 3.7 and 1.25 c/d; the first one could be an
harmonic of the second one). The photometric frequency found by Cuypers
et al. (1989) is apparent in the periodogram on MgII line only. This could
also be linked to some envelope phenomenon. 48 Lib is a good candidate for
multifrequency, but further multisites observations performed with higher
S/N at different phases of the envelope activity are needed to study the
frequency stability and to explain the variable mass loss through NRP.

References

Cuypers J., Balona L.A., Marang F.: 1989, *Astronomy and Astrophysics, Supplement Series* **81**, 151
Ringuelet-Kaswalder A.E.: 1963, *Astrophysical Journal* **137**, 1310

STRUCTURE OF THE ENVELOPE OF 48 LIB

A. CRUZADO, L. CIDALE

Observatorio de La Plata - Paseo del Bosque S/N - 1900 La Plata - Argentina

and

J. ZOREC

Institut d'Astrophysique de Paris - 98 ^{bis}, Bd. Arago - 75014 Paris - France

Abstract. From line and continuun data of the ultraviolet and visible spectrum of 48 Lib, the fundamental parameters of the circumstellar envelope are derived.

1. Introduction

48 Lib (HD 142983) is a $B3Ve$ star that presents cyclic V/R variations in the photographic region. In order to investigate the physical properties of its circumstellar envelope we have studied the continuum spectrum around the Balmer discontinuity and the UV spectral regions.

2. Parameters of the envelope of 48 Lib

Assuming that the regions contributing to the continuum and line spectrum can be represented by a thin slab-like envelope, we derived physical parameters such as continuum and line optical depths (τ_λ), excitation and electron temperatures, atom columns, and the distance from the central star (R_e/R_*) to the continuum and line forming-regions.

Taking into account the 13-color photometry from Schuster and Guichard (1984) and the calibration in absolute fluxes of Johnson and Mitchell (1975), we derived physical parameters assuming that the luminosity L_λ emitted by the star-envelope system may be represented by:

$$L_\lambda = 4\pi R_*^2 \pi F_\lambda^* exp(-\tau_\lambda) + \alpha B_\lambda(T_\epsilon)[1 - exp(-2\tau_\lambda)],$$

where $\alpha \propto (R_e/R_*)^2$ depends on the geometry of the envelope (Cidale and Ringuelet 1989); F_λ^* is the stellar flux, and B_λ is the Planck function at the excitation temperature $T_\epsilon = \epsilon T_{eff}$. The photosphere of the star was characterized with the BCD parameters; $T_{eff} = 17300$ K and log g = 3.8.

Table 1 gives the values of τ_u and τ_v ($\lambda_u = 0.37\mu$; $\lambda_v = 0.55\mu$), T_ϵ and the ratio R_e/R_* as it can be obtained when the envelope is supposed to be spherical. From this table, it can be seen that the changes responsible for the observed photometric variations may be due to variations in the value of T_ϵ and in the size of the envelope.

From IUE images of 48 Lib (SWP 3842, LWR 3425; January 1979, and SWP 14479, LWR 11065, July 1981) we selected Fe II lines with large gf

L. A. Balona et al. (eds.), Pulsation, Rotation and Mass Loss in Early-Type Stars, 366–367.

TABLE I

Parameters of the envelope derived from the continuum spectrum

Date	$F_{0.37\mu}^{obs}/F_{0.37\mu}^{*}$	τ_u	τ_v	$T_\epsilon(K)$	$(R_e/R_*)_{spher.}$
22.03.1981	0.604	1.04	0.89	8397	1.47
26.03.1981	0.622	0.71	0.56	7170	1.70
15.04.1983	0.473	0.87	0.44	5150	3.21
18.04.1983	0.447	0.95	0.59	5118	3.47

values and a collision-dominated line source function. The physical and geometrical parameters of the line-forming region were derived applying Cidale and Ringuelet (1989) method. We found that from 1979 to 1981 there was an enhancement of the line optical depths and of the atom columns, and the distance, R_e/R_*, of the Fe II line-forming region decreased from 14.5 to 2.4 stellar radii, respectively, keeping the same electron temperature at a value of 8200 K.

3. Conclusions

From the R_e/R_* ratios obtained from the line and continuum spectrum, we see that for 1981 it is roughly $(R_e/R_*)^{lines} \sim 2 (R_e/R_*)^{continuum}$. This should imply that the temperature of the continuum forming region is expected to be higher than that obtained here. However, the excitation temperature T_ϵ is related to the electronic temperature T_e by the relation $B_\lambda(T_\epsilon) = B_\lambda(T_e)/b$, where b is the non-LTE deviation coefficient. These results also show that the canonical value $T_\epsilon = 10^4$ K currenly used to study the circumstellar properties of Be stars may not always be the best approximation.

On the other hand, the distances of the regions of line-formation deduced from the UV data are consistent with the distances of regions of similar density obtained introducing the polarization value of 48 Lib (0.95 %) into Fox's (1993) results. Therefore, we may suggest that cyclic variations of a weak magnetic field could explain the variations in the V/R ratio, the radial velocity, and the geometric extension of the envelope.

References

Cidale, L. S., and Ringuelet, A. E.: 1989, *PASP* **101**, 417
Fox, G. K.: 1993, *MNRAS* **260**, 525
Johnson, H. L., and Mitchell, R. I.: 1975, *Rev. Mexicana Astron. Astrof.* **1**, 299
Schuster, W. J., and Guichard, J.: 1984, *Rev. Mexicana Astron. Astrof.* **9**, 141

GROWTH OF THE LINE-PROFILE VARIATION REGION DURING BE EPISODES

E. KAMBE*

Department of Geophysics and Astronomy, University of British Columbia,
Vancouver, British Columbia, Canada, V6T 1Z4

Abstract. We have found that the range in wavelength of the line-profile variations (*lpv*) in the Be star λ Eridani increases by some 20% during its emission phases. This growth is associated with a possible increase in the equatorial velocity of the star during emission. We discuss the significance of this in connection with other features of the *lpv* and our earlier discovery of a weak correlation between the amplitude of the *lpv* and Be outbursts in ζ Ophiuchi.

1. Introduction

In an attempt to examine features of *lpv* and their relation to the mass loss phenomena, we have monitored two Be stars, λ Eri and ζ Oph in particular, at Okayama Astrophysical Observatory in Japan since 1987. Dominion Astrophysical Observatory and other sites have joined the project since 1992 as part of an international campaign. In this paper, we discuss the features of the observed *lpv* especially in the context of mass loss episodes.

2. LPV during emission and quiescence

We have found that the *lpv* region of λ Eri is systematically extended during emission by ~ 100 kms^{-1} (380 kms^{-1} $\implies$ 480 kms^{-1} at the extreme edge of the profiles). The *lpv* during emission: 1) are well confined within 480 kms^{-1} of the line center; 2) appear similarly at both extreme wings; 3) have amplitudes comparable to those in the main profile; 4) are seen repeatedly at every emission epoch observed; and 5) in a least one case, have a period comparable to that of the *lpv* in the main profile (see Kambe et al. 1993b for details).

An increase of the *lpv* amplitude, which was detected in ζ Oph (Kambe et al 1993a), is not significant in λ Eri, which is consistent with earlier studies (Smith 1989, Bolton and Stefl 1990). However, this may be partly due to large variations of double peak emissions in the star. An accumulation of data and a monitoring of non-emission lines are necessary to examine such a slight change of *lpv* amplitudes.

* On leave from Department of Geoscience, National Defense Academy, Yokosuka, Kanagawa 239, Japan

L. A. Balona et al. (eds.), Pulsation, Rotation and Mass Loss in Early-Type Stars, 368–369.
© 1994 *IAU. Printed in the Netherlands.*

3. Discussion

The growth of the *lpv* region in λ Eri seems to suggest that the extension is connected with stellar rotation and we propose the existence of a rotationally accelerated equatorial region, in which *lpv* can propagate, during emission phases of the star (Kambe et al 1993b). If the acceleration occurs during emission, it could be connected with mass loss episodes. One of the possible mechanism of such a rotational acceleration is a redistribution of angular momentum by nonradial pulsations at the stellar surface.

However, our study of HeI λ 6678 alone can provide limited information about the proposed accelerated region and the related *lpv*. It is impossible to examine if the region has a property of a stellar photosphere and/or an inner disk. Multi-line spectroscopic observations are necessary for a detailed examination. A possible change of amplitudes of *lpv* should also be monitored in the context of an acceleration mechanism by nonradial pulsations (Ando 1991).

The confirmation of an existence of multiple modes, the specification of an origin of photometric variations could be other important aspects for examing a relation between short-term and long-term variations in Be stars and also for future asteroseismology (g-modes and/or r-modes). International campaigns have been completed and more are planned for the near future by our group (Hirata 1993) and others. We hope they will produce fruitful results soon.

References

Ando H., 1991, in Rapid Variability of OB-stars: Nature and Diagnostic Value, ESO Conference and Workshop Proceedings No. 36, ed.D. Baade (ESO-WS91), p. 303

Bolton C.T., Stefl S., 1990, in Angular Momentum and Mass Loss for Hot Stars, eds. L.A. Wilson and R. Stalio (Kluwer Academic Publishers, The Netherlands), p.191

Hirata R., 1993, Be Star Newsletter, No 26

Kambe E., Ando H., Hirata R., 1993a, AA, 273, 435

Kambe E., Ando H., Hirata R., Walker G.A.H., Kennelly, E. J., Matthews, J. M.,1993b, PASP, in press

Smith M.A., Peters G.J., Grady C.A., 1991, ApJ, 367, 302

Walker G.A.H., 1991, in ESO-WS91, p.27

70 YEARS OF OBSERVATIONS OF 4 HER:

CHANGES THROUGH THREE SHELL EPISODES

P. KOUBSKÝ, P. HARMANEC, J. HORN

Astronomical Institute, Academy of Sciences of the Czech Republic,
251 65 Ondřejov, Czech Republic

and

A.-M. HUBERT, H. HUBERT, M. FLOQUET

Observatoire de Paris, Section d'Astrophysique de Meudon, URA 335 du CNRS,
F-92195 Meudon Cedex, France

1. History

4 Her (HD 142926, HR 5938; V=5^m.75, v.sini=300 km s^{-1}) is a well known and rather frequently observed Be and shell star. It was recognized as a Be star by Heard (1939) and Mohler (1940). The estimates of its spectral type by different authors vary between B7 IV-V and B9e. Hubert(1971) reported remarkable spectral changes of 4 Her which occurred between 1953 and 1970. Harmanec et al.(1973) discovered periodic radial-velocity variations of the hydrogen shell lines with a period 46.023 days and suggested the object is a single-line spectroscopic binary. The system elements were later refined by Heard et al.(1975) to P=46.194 days, K=12 km s^{-1} and e=0.3. In a subsequent paper, Harmanec et al.(1976) studied the variations of emission and absorption components of the hydrogen lines and concluded that 4 Her is an interacting binary and that the observed eccentricity of the orbit is spurious, caused by the effects of circumstellar matter.

The long-term spectral variations have been monitored at Ondřejov since 1969 and at Haute Provence since 1953 at low, and since 1960 at high dispersion. The history of the recurrent shell changes of 4 Her can be reconstructed from data published by Hubert (1971) and Harmanec et al. (1976), and on new data from Ondřejov and Haute Provence Observatories. We have been able to identify two periods of different lengths when the Hα emission was absent (1948–1963, 1987–1991). We thus cannot estimate the lengths of the emission period which started after 1920 and was on in late 1930's. The present emission period re-appeared 28 years after the onset of a episode in 1960's which lasted about 18 years. The IUE spectra of 4 Her were taken in 1979 (i.e. at the end of the previous emission episode), in 1983 (normal B type spectrum) and in 1992 (the present shell phase). Most of the UV lines of 4 Her appear unchanged on all three spectra, i.e. not influenced by the presence of the shell. The important exception are the C IV lines 1548 and

370

L. A. Balona et al. (eds.), Pulsation, Rotation and Mass Loss in Early-Type Stars, 370–371.
© 1994 *IAU. Printed in the Netherlands.*

1551 Å which closely correlate with the changes of the envelope signalled by Hα.

2. Binary nature

All so far published orbital elements of 4 Her were based exclusively on the radial velocities from shell lines. A few attempts to measure the velocities of photospheric helium lines failed (c.f., e.g., Heard et al.1975). We measured RVs on the wings of the stellar hydrogen lines from the photographic spectra digitized with a microdensitometer. We also took the advantage to measure the stellar profiles from the two epochs without shell. First, we derived the new value of the period using all available H shell RVs. Then, we calculated another orbital solution based solely on the new H-wing RVs and keeping the period fixed from the previous solution for all shell RVs (P=46.192 days). The solution based on stellar lines leads to a circular orbit (e=0, K=7.5 km s^{-1}) while an eccentric orbit has resulted from RVs of the shell lines (e=0.330, K=11.3 km s^{-1}). We thus reinforce the conclusion by Harmanec et al.(1976) that the orbital eccentricity indicated by the shell lines is spurious, caused by velocity distortions due to circumstellar matter. We also tried to verify the model of 4 Her as an interacting binary advocated in the latter study. Regrettably, we have started the spectroscopic search for the lines of the secondary only after the onset of the new shell. It is therefore very difficult to distinguish faint and sharp shell lines from the lines of the secondary – unless a reasonable phase coverage by spectra can be achieved.

The phase distribution of available high-quality data from the Aurélie spectrograph at OHP and Ondřejov Reticon does not allow to check on the conclusion by Harmanec et al. (1976) that the V/R ratio of Balmer lines varies in phase with the RV curve of 4 Her.

Acknowledgements

We are grateful to the OHP and Ondřejov Observatory staff for their help during the observations. This study was partly supported from the internal grant No. 30318 of the Czechoslovak Academy of Sciences.

References

Harmanec P., Koubský P., Krpata J.: 1973, *A&A* **22**, 337
Harmanec P., Koubský P., Krpata J., Žďárský F.: 1976, *Bull. Astron.Inst.Czechosl.* **27**, 47
Heard J.F.: 1939, *J. Roy. Astron. Soc. Canada* **33**, 384
Heard J.F., Hurkens R., Harmanec P., Koubský P., Krpata J.: 1975, *A&A* **42**, 47
Hubert H.: 1971, *A&A* **11**, 100
Mohler O.: 1940, *ApJ* **92**, 315

Activity in the Circumstellar Envelope of the Be/Shell Star ζ Tau

Yulian Guo   Lin Huang   Jinxin Hao

Beijing Astronomical Observatory, Chinese Academy of Sciences

1. Introduction

ζ Tau is a well-known V/R-variable shell star. It is a single-lined spectroscopic binary with an orbital period of 133 days (Harmanec 1984, and Jarad 1987). Delplace (1970) found that the long-term radial velocity variations of the Balmer shell absorption lines are cyclic in 1960-67. Subsequently, similar behaviour has been observed and studied by several authors (Delplace and Chambon 1976, Hubert-Delplace *et al* 1983, Harmanec 1984, and Guo and Cao 1987). Mon *et al* (1992) showed that the cyclic variation had terminated and the star seems to have entered a new quiet phase around 1982.

2. Observations and Data Reduction

All observations were carried out at Xinglong station of Beijing Observatory. The spectroscopic data were obtained using the grating spectrograph attached to the 60/90 cm Schmidt telescope with plate during 1978 Dec.-1990 Feb.. After 1990 Dec., spectroscopic observations were made with the All-Fiber-Coupler grating spectrograph at the 216 cm telescope with a CCD detector with 576x384 pixels. The reciprocal liner dispersions of the spectrum were 17Å/mm, 50Å/mm and 86Å/mm at the blue region and 50Å/mm and 86Å/mm at the H_α region. Spectrograms have been digitized on the PDS microdensitometer of the Purple Mountain Observatory. All spectroscopic data are reduced with Starlink image processing software on the VAX 11/780 computer of Beijing Observatory. To analyse and study the shell activity, we measured the radial velocities of the stronger shell lines. The photometry was made on the 60cm reflector using the filters of the standard UBV system during 1984 Dec.-1992 Dec..

3. Results and Discussion

Fig.1a shows the variations of the mean velocity of H_γ, H_δ and H_ϵ shell lines with time. It seems that the cyclic variations which started in 1976 (Huber-Delplace et al. 1983) ceased around 1982. The velocity values still displayed an oscillation of the cycle in 1982-1985 but the cycle lasted shorter and the amlitude was less. During 1986-1990, the variations of the radial velocity conspicuously diminished and the cyclic variation disappeared too.

The variations of the V/R ratio of H_α is in phase with the radial velocity curve of the shell absorption lines before 1982. Between 1983 and 1987, the V/R ratios fluctuated in an irregular manner but remained generally close to unity. After 1987 the cyclic changes of the V/R ratio reappeared.(see Fig.1b).

The equivalent width of H_α emission declined gradually from 1978 to 1988, but afterwards, it began to rise rapidly, although there could be some fluctuations in-between.(see Fig.1c).

In addition to the above-mentioned long-term variations, we also observed some short-term activities on a time scale of night-to-night and even more rapid changes in H_α line. For example, the H_α profiles obtained on 1983 Dec. 10-14, appeared remarkable variations on line shape. The equivalent widths of the H_α emission increased from 10.0Åto 15.0Åin a 4-night interval.

during 1990-1992, the brightness of the star showed evident variation which was correlated to the V/R variation of H_α and the minimum of the brighness occurred between the end of 1991 and the beginning of 1992. In addition the rapid change of the UBV magitudes were reveated in 1984 Dec.-1985 Jan. and the beginning of 1989(see Fig.1d, e and f).

L. A. Balona et al. (eds.), Pulsation, Rotation and Mass Loss in Early-Type Stars, 372–373.

Our spectroscopic observations indicate that the V/R cyclic variations of H$_\alpha$ emission emerged once again at the end of 1990 after their disappearance for several years. We do not have the necessary information about the radial velocity after 1990 but, based on the correlation between the V/R and the radial velocity variation revealed in our observations in 1978–1982 and those of Hubert-Delplace *et al.* (1982) in 1960 -1981, we may suggest that the radial velocity could have entered new cyclic variation phase. To our best knowledge, there are two possible interpretations to the origin of the V/R cycle associated with the radial velocity curves: the rotating-pulsating envelope and the non-axisymmetric elliptical disk. We consider that an elliptical disk model in slow precession motion appears more convincing for explaining the various variations we observed in the cyclic variation phase. But other observed facts, such as that the periods and the amplitudes from one cycle to another were different and that the cyclic variations disappeared completely as in 1985-1987, can't be interpreted only in terms of a simple slowly rotating elliptical envelope. In modelling the envelope of the star we need to consider the stage which is without the presence of the cyclic change, namely, usually what is called a quiet stage and the process of the disappearance of the cyclic change. In addition, we observed important short-term variations. The rapid changes can be due to erratic activities of matter streams inside the envelope. Affected by such matter stream activities, the elliptical disk will vary with time. A sufficiently strong activity could even have destroyed elliptical disk. In consequence,the cyclic variation vanished. Of course owing to other violent activities of matter and accumulations of the outflow matter, a new elliptical disk could form again. In the meantime,the cyclic variations are observed. Therefore we consider that a model which best explains the behaviour of the envelope of ζ Tau, would be represented by the coexistence of and interaction between an elliptical disk in slow precessional motion and highly variable streams of matter.

Reference:

1. Delplace, A. M. 1970, A. Ap. 7, 68
2. Delplace, A. M. and Chambon, M. Th., 1976, in IAU symp. No.70, Be and shell stars ed A. Slettebak (Dordrecht:Reide), P.79
3. Guo Yulian and Gao Weishi, 1987, Be Newsletter 16, 9
4. Harmanec, P., 1984, Bull. Astron. Inst. Czech. 35, 164
5. Hubert-Delplace, A. M., Jaschek, M., Hubert, H. and Chanbon, M. Th.,1982,in IAU symp. 98, eds. M. Jaschek and M. G. Groth, P.125
6. Hubert-Delplace, A. M., Mon. M., Ungerer, V., Hirata, R., Paterson- Beeckmans, F., Hubert, H., and Baade,D., 1983, A. Ap. 121, 174
7. Jarad, M. M., 1987, Ap. S. S. 139,83
8. Mon, M., Kogure, T., Suzuki, M., and Singh, M., 1992, Publ. Astron. Soc. Japan,44,73

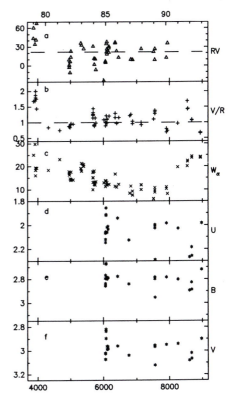

PERIOD DEPENDENT VARIATIONS OF THE SHELL SPECTRUM OF V923 AQL

Lubomir Iliev
Bulgarian Academy of Sciences
Institut of Astronomy
72 Tsarigradsko Shosse Blvd.
BG-1784 Sofia, Bulgaria
tel./fax (+359)-2-758-927
e_mail liliev@bgearn.bitnet

key words: Be and shell stars, binary stars

V923 Aql is a well known Be star with strong shell spectrum. It was included in the list of shell stars of Merrill and Burwell (1949). A detailed study of the radial velocity variations in the spectrum of the star based on wide collection of spectral observations was presented in the work of Koubsky et al. (1988). In this work an orbital period of 214.756 days was suggested for the binary system consisting of a B5-7 primary and low mass (0.5 Msol).

Our study of the variations of the shell spectrum of the star is based on 10spec-trograms obtained at coude-spectrograph of the 2m RCC telescope at BNAO "Rozhen" in three different observing seasons. All spectrograms were digitized with Joice Loebl microdensitometer and the radial velocities were measured with the oscilloscopic comparator at BNAO "Rozhen".

Our results reveal that changes of the Balmer progression correlate with the or-bital phase. Slightly positive at phase 0.0 the progression decreased gradually and at phase 0.4 is already negative and quickly decreased afterwards (fig. 1). The lack of well dispersed over the period observations can not allow us to trace the behaviour of the decrement with more details.

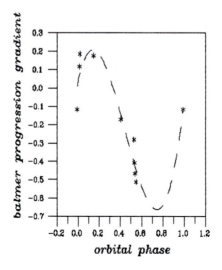

Fig. 1. Period dependent changes of the Balmer progression decrement.

L. A. Balona et al. (eds.), Pulsation, Rotation and Mass Loss in Early-Type Stars, 374–375.
© 1994 *IAU. Printed in the Netherlands.*

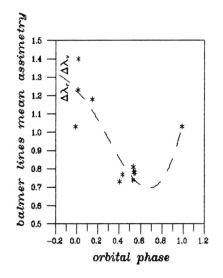

Fig. 2. Period dependent changes of the asymmetry in Balmer shell lines.

Our results are in good agreement with the conclusions of Merrill (1952) that in phases with positive radial velocities the Balmer progression is positive too. It must be specialy noted that the asimmetry of the Balmer shell lines also changes with orbital phase (fig. 2).

Our radial velocity measurements of FeII lines of multipletes 27, 37 and 38 proved the observed from Ringuelet and Sahade (1984) existence of progression in the phase of negative velocities. In all of the multipletes we observed changes of the progression connected with the orbital period following different from that of the Balmer lines behaviour. All variations observed are above the limit of accuracy. The changes of the central depths of the metal shell lines are well pronounced and also show dependance with the orbital phase.

Our results support the conclusion that shell lines of ionized metals and that of the hidrogen not only origin in different layers of the shell but this layers underwent different changes during the orbital motion.

References:
Horn J., Iliev I., Lions R.; 1989, Bull.Astron.Inst.Czechosl., vol.40, p.31.
Koubsky P., Gulliver A.F., Harmanec P., Ballereau D., Chauville J., Graf T.,
Merrill B. and Burwell C.; 1949, Astrophys.J., vol.110, p.387.
Merrill B.W.; 1952, Astrophys.J. vol.116, p.501.
Ringuelet A. and Sahade J.; 1984, Rev.Mex.Astron.Astrof., vol.6, p.208.

LONG-TERM PERIODIC VARIABILITY IN UV

ABSORPTION LINES OF THE Be STAR γ Cas: ON THE

RELATION WITH V/R VARIATIONS IN THE Hβ LINE

JOHN H. TELTING and LEX KAPER

Astronomical Institute Anton Pannekoek and Center for High Energy Astrophysics,
University of Amsterdam, Kruislaan 403, 1098 SJ Amsterdam, Netherlands

Abstract. We present a quantitative study of the variability in ultraviolet resonance lines of N V, Si IV and C IV of the Be star γ Cas, following up on the work of Henrichs et al. (1983). For this purpose we used 133 IUE spectra obtained over a period of eleven years. Variability occurs in the form of discrete absorption components (DACs), which are formed in the fast-outflowing radiatively driven part of the stellar wind. We constructed a template spectrum from spectra containing no or minor extra absorption due to DACs and modelled the isolated DACs in the obtained quotient spectra. Besides the frequently observed narrow components (v_t typically $\leq$ 250 km/s) at high velocity, we found several broad components occurring at low and intermediate wind velocities.

We confirm the finding of Doazan et al. (1987) who reported that the number of observed DACs is associated with the cyclic V/R variability of the Balmer-emission lines. This V/R variability most probably originates in the slowly outflowing high-density equatorial disc-like wind of the star (see e.g. Telting et al. 1993 for the case of γ Cas). We show that when V/R<1 the central optical depth of DACs is significantly lower than when V/R>1. In our interpretation this is due to a correlation between the column density associated with the DACs and the phase of the V/R cycle.

We find that the Hβ observations of Doazan et al. are consistent with a model in which the cyclic V/R variability is due to a global, one-armed oscillation moving through an equatorial disc (Okazaki 1991, Papaloizou et al. 1992, Savonije and Heemskerk 1993). We suggest that the higher column density of DACs in phases of V/R>1 is the result of the higher density in the region of their origin, namely close to or in the part of the equatorial disc which is rotating towards the observer.

For a thorough description of the analysis and our results we refer to our article Telting and Kaper (1994) which is currently in press for publication in A&A main journal. In Fig. 1 we present a concise version of the discussion of that article.

References

Doazan, V., et al.: 1987, Astronomy and Astrophysics 182, L25
Henrichs, H.F., et al.: 1983, Astrophysical Journal 268, 807
Okazaki, A.T.: 1991, Publications of the ASJ 43, 75
Papaloizou, J.C., et al.: 1992, Astronomy and Astrophysics 265, L45
Savonije, G.J., Heemskerk, M.H.M.: 1993, Astronomy and Astrophysics 276, 409
Telting, J.H., et al.: 1993, Astronomy and Astrophysics 270, 355
Telting, J.H., Kaper, L: 1994, Astronomy and Astrophysics 284, 515

L. A. Balona et al. (eds.), Pulsation, Rotation and Mass Loss in Early-Type Stars, 376–377.
© 1994 IAU. Printed in the Netherlands.

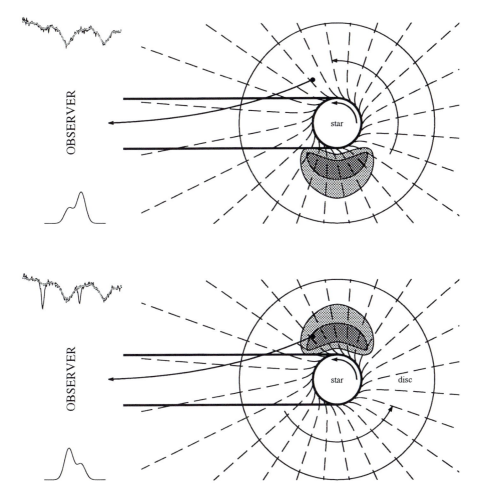

Fig. 1.   Schematic model for V/R variations and DACs in spectra of γ Cas. We show
a projection of the equatorial and polar winds onto the equatorial plane. The grey areas
in the disc represent the high-density part of a one-armed oscillation of the disc (Okazaki 1991, Papaloizou et al. 1992, Savonije and Heemskerk 1993). This high-density part
revolves around the star on the time scale of the V/R variations of the Balmer lines. The
dashed lines are trajectories of individual free stellar wind particles, based on a beta law
for the radial velocity (i.e. the polar wind). The solid curve represents the path of a density
enhancement of the polar wind, as is suggested by comparison of V/R variations and DAC
variability. The structure starts off close to the star, near the equatorial plane, and gradually moves away from the star. Its trajectory will be ruled by the motions in the wind:
outflow and rotation. Only for a particular azimuthal start-off position the structure ends
up in the line of sight, where it can cause a DAC if its density is high enough.   **Top)** The
high-density part of the equatorial disc moves away from the observer; the emission lines
have V<R. Structures of the fast polar wind that end up in the line of sight are formed
near the low-density part of the equatorial disc: DACs are weak (or too weak to be detected).   **Bottom)** The high-density part of the equatorial disc moves towards the observer;
the emission lines have V>R. Now the structures of the polar wind travelling to the line
of sight are formed near the high-density part of the equatorial disc: the observer finds
numerous (strong) DACs

ON THE FORMATION OF BALMER EMISSION LINES
IN THE MODEL ENVELOPES OF Be STARS

T. KOGURE

Bisei Astronomical Observatry, Miyama, Bisei, Okayama

S. SUZUKI

Kanazawa Institute of Technology, Kanazawa, Ishikawa

and

M. MON

Department of Astronomy, Kyoto University, Kyoto

Abstract. We consider the formation of the Balmer emission lines and the decrement $H\alpha/H\beta/H\gamma$, by solving the non-LTE problems in elementary regions of the envelope divided by equal line-of-sight velocities. These envelope-elements are characterized by different optical depths in the Balmer lines and by different dilution factors for the incident stellar radiation. It is shown that the decrements sensitively depend on these parameters of the envelope-elements. We show that the observed spectral-type dependence and large scatter of the decrements among Be stars can be explained in terms of the variation of these physical parameters.

Recently Dachs et al. (1990) and Slettebak et al. (1992) have made spectroscopic observations of Be stars and derived the Balmer decrements. They used the homogeneous plane-parallel slab model of Drake and Ulrich (1980) to discuss the physical properties of the envelope of Be stars. In this model the free parameter was only the electron density, so that it was not sufficient to explain the spectral-type dependence and large scatter of the Balmer decrements among Be stars. This shows the importance of full consideration for the radiation field of the envelopes in order to discuss the properties of Be star envelopes.

In this paper, we consider the basic problems of the formation of emission lines, based on the non-LTE treatment for the radiation field of the envelope of Be stars. Our approach is as follows:

1) The envelope is simplified by a hollow cylinder with finite vertical thickness and outwardly decreasing electron density.

2) The envelope is divided into the elements characterized by the line-of-sight velocity and dilution factor.

3) The radiation field of the envelope is solved for the Balmer lines based on the non-LTE treatment of Kogure (1959). Stellar parameters are taken from Kurucz (1979) model atmospheres.

4) The Balmer decrements $H\alpha/H\beta$, $H\gamma/H\beta$ are derived as a function of the optical depth $\tau(H\alpha)$, dilution factor W of the envelope-element and of the stellar effective temperature.

L. A. Balona et al. (eds.), Pulsation, Rotation and Mass Loss in Early-Type Stars, 378–379.

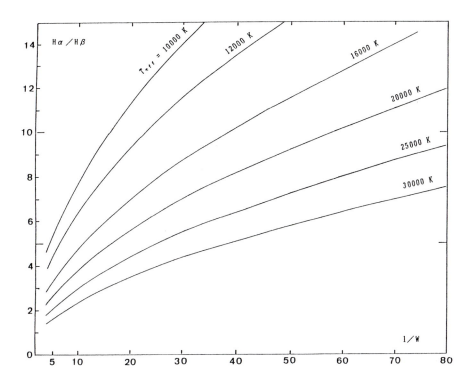

Fig. 1.   Balmer decrement Hα/Hβ formed in envelope-elements

The decrements $H\alpha/H\beta$ thus derived are shown in Fig. 1, for an optically thick envelope-element with $\tau(H\alpha) = 100$. Here we can see the specral-type dependence of the decrement. The construction of the model envelope as an integration of the envelope-elements will be given in future.

References

Dachs, J., Rohe, D., and Loose, A.S.: 1990, *Astronomy and Astrophysics* **238**, 227

Kogure, T.: 1959, *Publications of the ASJ* **11**, 127 and 278

Kurucz, R.L.: 1979, *Astrophysical Journal, Supplement Series* **40**, 1

Slettebak, A., Collins II, G.W., and Truax, R.: 1992, *Astrophysical Journal, Supplement Series* **81**, 335

GLOBAL DISK OSCILLATIONS: THEORETICAL LINE PROFILES

W. HUMMEL* and R.W. HANUSCHIK

Astronomisches Institut, Ruhr-Universität Bochum, D-44780 Bochum, Germany

Abstract. Based on 3D radiative line transfer calculations we present Hα emission line profiles of Be star circumstellar envelopes undergoing one-armed global disk oscillations. The results are in agreement with the observed line profile variability.

1. The model

Kato (1983) and Okazaki (1991) proposed one-armed global disk oscillations (GDO) as an explanation for the long-term variability of optical Be star emission lines. The disk perturbation pattern is rotating retrograde around the star ($P \simeq 10$ years) giving rise to the observed long-term line profile variability (Hanuschik et al., 1993). The perturbed density distribution consists of a significant maximum region located nearby the central star at azimuthal angle $\phi = 0^0$. The particle trajectories are similar to ellipses. Based on Okazaki's perturbation pattern and on a 3D radiative line transfer code (Hummel, 1993) we calculated Hα emission line profiles of circumstellar disks extended up to $R_d = 5R_*$ with a radial density run of $N \sim N_0(\frac{r}{R_*})^{-2}$.

2. Results and discussion

In Fig.1 and in Fig.2 the line profile sequences of different values of ϕ simulate the observed long-term profile variability. At $\phi = 0^0$ the line of sight is parallel to the symmetry axis of the GDO pattern and the maximum density region is in front of the star. At $\phi = 90^0$ the maximum density region is on the left hand side of the star moving towards the observer.

At nearly pole-on view ($i = 10^0$) variations of asymmetric winebottle-type profiles are presented, at $i = 30^0$, typical V/R-ratio variabilities of double-peak profiles are shown and for an edge-on view ($i = 90^0$) the calculated shell profiles exhibit variations of the radial velocity (V_{cd}) and of the flux (F_{cd}) at the central depression of the profile. The observable sequence of $[V/R > 1] \Rightarrow [F_{cd}^{min}] \Rightarrow [V/R < 1] \Rightarrow [F_{cd}^{max}]$ (Cowley & Gugula, 1973) is in agreement with a retrograde pattern rotation (Fig.2; $i = 60^0$).

Acknowledgements: Many thanks go to J. Dachs and D. Baade for helpful discussions. Financial support by the Deutsche Forschungsgemeinschaft under grant Da 75/12-2 is gratefully acknowledged. Plots were executed by means of ESO-MIDAS/91MAY.

* now at: Astrofysisch Instituut, Vrije Universiteit Brussel, Pleinlaan 2, B-1050 Brussel, Belgium

L. A. Balona et al. (eds.), Pulsation, Rotation and Mass Loss in Early-Type Stars, 382–383.
© 1994 IAU. Printed in the Netherlands.

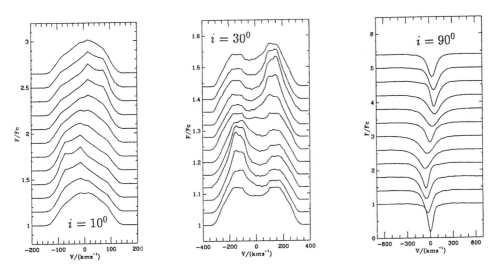

Fig. 1.   Three sequences of calculated emission line profiles of a nearly Keplerian disk $(N_0 = 10^{13} \mathrm{cm}^{-3})$ at $i = 10^0, i = 30^0$ and $i = 90^0$. In each sequence the azimutal angle ϕ varies from bottom to top: $\phi = 0^0, 30^0, 60^0, \ldots 330^0$.

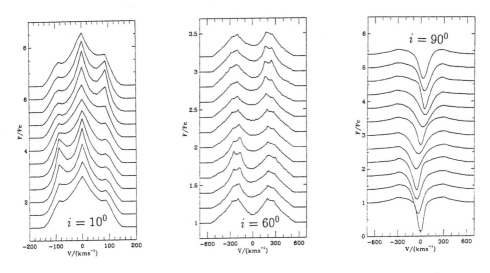

Fig. 2.   As Fig.1, except $N_0 = 10^{14} \mathrm{cm}^{-3}$ and $i = 10^0$, $i = 60^0$ and $i = 90^0$.

References

Cowley, A., Gugula, E.: 1973, Astron. Astrophys. **22**, 203
Hanuschik, R.W., Hummel, W., Dietle, O., Dachs, J., Sutorius, E.: 1993, *these proceedings*
Hummel, W.: 1993, *PhD.*, Bochum
Kato, S.: 1983, *Publ. Astron. Soc. Japan* **35**, 24
Okazaki, A.T.: *Publ. Astron. Soc. Japan* **43**, 75

GLOBAL OSCILLATIONS IN Be STAR ENVELOPES: OBSERVATIONAL EVIDENCE

R. W. HANUSCHIK*

Astronomical Institute, University of Tübingen, D-72076 Tübingen, Germany

W. HUMMEL*

Astrofysisch Instituut, Vrije Universiteit Brussel, Pleinlaan 2, B-1050 Brussel, Belgium

and

O. DIETLE,** J. DACHS and E. SUTORIUS

Astronomical Institute, Ruhr-Universität Bochum, D-44780 Bochum, Germany

1. The observations

Since 1982, we are performing a long-term spectroscopic observing programme of emission-lines in Be stars (Hanuschik 1987, Hanuschik et al. 1988, Dachs et al. 1992, Sutorius 1992, Dietle 1993). We are using ESO's 1.4m CAT, at resolution $R \geq 50\,000$ and $S/N = 100$–1000. Spectral lines chosen are the optically thick $H\alpha$, $H\beta$ lines and the optically thin Fe II $\lambda 5317$ line. The latter line is an extremely sharp tracer ($\Delta v_{\text{th}} = 2$ km s^{-1}) for the kinematics in the disks. We believe that our atlas shows the full range of intrinsic structure of these emission lines.

We find that the apparent wealth of line profile shapes can be ordered into a simple two-class scheme: *symmetric* lines, which are often double-peaked, in case of $H\alpha$ often of winebottle-type shape (so-called class 1; see Fig. 1); and *asymmetric* lines, in case of Fe II often with a typical single-peaked, "steeple"-type profile shape (= class 2; Fig. 2). Shell lines are considered as special cases within the two main classes. The shape of class 1 profiles usually remains approximately constant, while class 2 profiles often show a typical $V/R < 1 \Longrightarrow = 1 \Longrightarrow > 1$ variability pattern with cycles of the order of 7–10 years.

2. The interpretation

Taken together with other evidence (see Hanuschik, these proceedings), class 1 profiles are likely to occur in an axisymmetric disk-like envelope in steady state. The typical shape and variability pattern of class 2 profiles, however,

---

* On leave from: Astronomical Institute Bochum
** Affiliated to: Fachbereich Physik, Universität-GH Essen, D-45141 Essen, Germany

L. A. Balona et al. (eds.), Pulsation, Rotation and Mass Loss in Early-Type Stars, 384–385.
© 1994 *IAU. Printed in the Netherlands.*

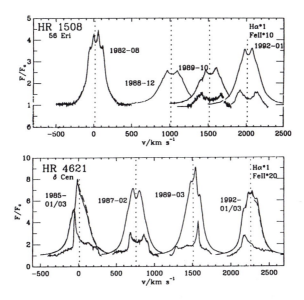

Fig. 1. Typical *class 1* (top) and *class 2* (bottom) line profiles. Broken lines indicate $\pm v \sin i$ and 0 in the stellar frame. Flux scale for Fe II line is enlarged by the factors given.

must be caused by a large-scale density inhomogeneity in the disk which is not co-rotating, but slowly precessing. We believe that such density perturbation is caused by a one-armed global oscillation as recently proposed by Okazaki (1991) and Papaloizou et al. (1992). As Hummel & Hanuschik (these proceedings) show, model line profiles for such perturbation show similarity with the observed steeple-type class 2 profiles. As a consequence of the rotational flattening of the central star, the perturbation is expected to slowly precess, causing the long-term V/R variations known since long (e.g., Doazan et al. 1987).

The origin of the density perturbation is unknown, but possibly related to outbursts on the star and thereby related to the mass ejection processes ultimately giving birth to the whole circumstellar disk.

References

Dachs J., Hummel W., Hanuschik R.W.: 1992, *A&AS* **95**, 437.
Dietle O.: 1993, Diploma thesis (Bochum).
Doazan V., Rusoni L., Sedmak G., Thomas R.N., Bourdonneau B.: 1987, *A&A* **182**, L25.
Hanuschik R.W.: 1987, *A&A* **173**, 299.
Hanuschik R.W., Kozok J., Kaiser D.: 1988, *A&A* **189**, 147.
Hummel W., Dachs J.: 1992, *A&A* **262**, L17.
Okazaki T.: 1991, *PASJ* **43**, 75.
Papaloizou J.C., Savonije G.J., Henrichs H.F.: 1992, *A&A* **265**, L45.
Sutorius E.: 1992, Diploma thesis (Bochum).

RAPID LINE PROFILE VARIABILITY OF Hα

IN TWO A0E HERBIG STARS OF THE P CYG-SUBGROUP

AND A MODEL FOR THEIR CIRCUMSTELLAR ENVELOPE

M.A.POGODIN

Central Astronomical Observatory of the Russian Academy of Sciences at Pulkovo,
196140 Saint–Petersburg, Russia, e-mail:pogodin@gaoran.spb.su

Abstract. The present report gives the results on the Hα line profile investigation of two the most well-known A0e Herbig stars of the so-called P Cyg -subgroup: AB Aur and HD 163296 . 35 high-resolution CCD spectra ($R \sim 50\,000$) of HD 163296 (ESO, CAT+CES, July 1991 and 1992) and 43 CCD spectra ($R \sim 30\,000$) of AB Aur (Crimean Observatory, 2.6-meter telescope, January 1993) were obtained. A striking profile variability is discovered in both objects on the timescale from one our to a few days. The most dominant part of rapid variations ($\tau \sim$ hours) is the monotonous flux drift of different profile components. Positional shift of sharp spectral bumps, found in HD 163296 is probably connected with local inhomogeneities moving in the envelope. Shape variability of the absorption P Cyg -component of the Hα -line in AB Aur is suspected to be periodic with $P = 35$ or 70 hours. A model for a circumstellar envelope is proposed to explain the observed variability in this type of objects. It supposes the existence of an active non-stable region near the star, formed by equatorially concentrated stellar wind and an outer cool shell.

All Hα line profiles of HD 163296 correspond to the Beals's III P Cyg type with a main emission peak and a blueshifted secondary one divided by absorption (Fig. 1, left). Additional absorption components, corresponding to greater negative velocities appeared on some dates. The main P Cyg -absorption remained constant both in strength and position during the entire observing run in 1991 and was slightly variable in 1992.

All these variations can be qualitatively explained in the frame of the model of non-stable equatorially concentrated stellar wind with addition of a cool outer shell, where $r > 10\,R_*$ (Fig. 1, right).

Now AB Aur is not so active, as earlier or in comparison with HD 163296 in 1991–1992. The main type of variability of the II P Cyg -type Hα profile is the change of the P Cyg -absorption in shape (Fig. 2, the upper panel). Periodicity of these variations can be suspected with $P = 70$ or 35 hours.

Residual spectra with respect to the nightly mean Hα spectrum show obvious signs of a rapid line profile variability ($\tau \sim$ hours) in both HD 163296 and AB Aur. One can see standing waves on the residuals, reflecting the monotonous change of different profile components during a night (see Fig. 3–4).

Extremely sharp strong features can be easily identified in the red region of the profile in HD 163296 (Fig. 3, left). The theoretical traces of moving point inhomogeneities were constructed for different rotation phases and for a number of parameters, describing the kinematics of the stellar wind.

Acknowledgements

I'd like to thank Dr. Dietrich Baade for useful discussion of all the results on HD 163296. I am grateful to Drs. Stephen Warren, Cristian Gouiffes and Stanislav Štefl (ESO) and Drs. N.Rostopchin, S.Berdugina, O.Kozlova and V.Scherbakov (CrAO) for their assistance in observations and data reductions.

L. A. Balona et al. (eds.), Pulsation, Rotation and Mass Loss in Early-Type Stars, 386–388.
© 1994 *IAU. Printed in the Netherlands.*

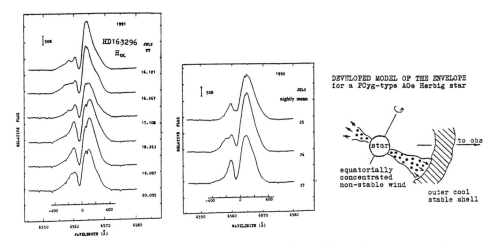

Fig. 1. Night-to-night variations of the Hα-line profile in HD 163296 and the qualitative envelope model.

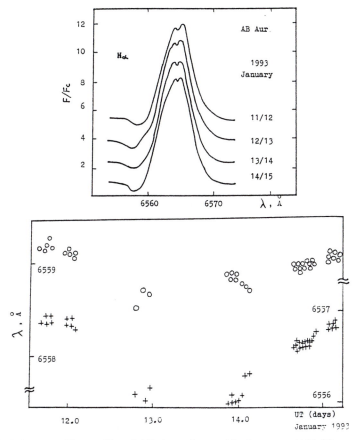

Fig. 2. Nightly mean Hα-profiles of AB Aur, observed in January, 1993 (the upper panel). Variations of the blue (+) and red (○) edges of the P Cyg-absorption on the 0.7 F_C-level (the lower panel).

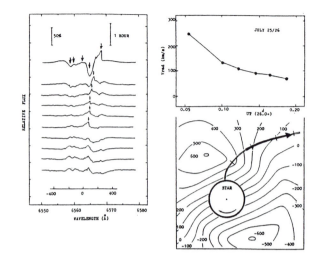

Fig. 3. Left: Residuals from the nightly mean profiles of HD 163296, observed on July, 25 1992. The offset of the residuals from the respective uppermost one is propotional to the time elapsed (time increases downwards). The radial velocity scale is given in km/s. Vertical bars provide the scales for flux (in the units of continuum F_c) and time. Remarkable moving and standing features are shown by arrows (at the top and at the bottom, respectively).

Right: Observed velocity changes of the moving bump (top) and theoretical trace of a point inhomogeneity, which is in the best agreement with observations (bottom). The trace (thick arrow) and surfaces of equal radial velocities are plotted in the non-moving star rest frame. Velocities are in km/s. The 1.5 h intervals are marked on the trace. Calculations were performed for the following kinematic parameters sample: $R_m = 2\,R_*$, $R_a = 3\,R_*$. The observer is at the bottom.

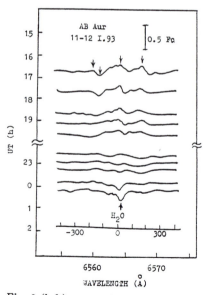

Fig. 4. The same as Fig. 3 (left) except for AB Aur (January, 11/12 1993).

PROFILE VARIABILITY OF Hα IN THE B3E HERBIG STAR
HD 200775
AS INDICATION OF MATTER INFALL AND INHOMOGENEOUS
STELLAR WIND

N.G. BESKROVNAYA and M.A.POGODIN

Central Astronomical Observatory of the Russian Academy of Sciences at Pulkovo,
196140 Saint-Petersburg, Russia, e-mail:beskr@gaoran.spb.su

and

A.G.SCHERBAKOV and A.E.TARASOV

Crimean Astrophysical Observatory, 334413 Nauchnyi, Crimea, Ukraine

Abstract. Significant profile variability of Hα on the timescale of months is investigated on the basis of high-dispersion CCD spectra of the B3e Herbig star HD 200775 . Variations in position and shape of the central absorption component are considered to indicate complex structure of circumstellar environment, containing infalling streams and inhomogeneous stellar wind.

The well-known B3e Herbig star HD 200775 is situated in the centre of a compact region of star formation and associated with the reflection nebular NGC 7023. The circumstellar matter around the star has a rather complex structure. A rotation gaseous disc and stellar wind are likely to exist in the envelope. Distinct indications of their outer parts can be seen on the radio map of a molecular cloud in the vicinity of the star (Watt *et al.* 1986).

52 high-resolution ($R \sim 30\,000$) CCD spectra were obtained at the coude focus of the 2.6 m telescope of the Crimean Astrophysical observatory in 1986–1990. The normalization of all the spectra was performed relative to the level of quasicontinuum F_{qc} (a straight line drawn across narrow spectral bands, centered on two velocity values $-600\,\mathrm{km\,s^{-1}}$ and $+600\,\mathrm{km\,s^{-1}}$).

The results of our observations confirm the presence of strong Hα profile variations on the timescale of a month (Fig. 1–8). Positional shift of the central absorption was ranged from $-9 \pm 0.5\,\mathrm{km\,s^{-1}}$ (May 1986) upto $\sim 70\,\mathrm{km\,s^{-1}}$ (September 1990). This gives evidence of the existence of the radial motions in the circumstellar envelope in both directions—towards and outwards the star.

The variability, observed in July, 1986 (appearence of a number of discrete details, moving from $-50\,\mathrm{km\,s^{-1}}$ upto $+40\,\mathrm{km\,s^{-1}}$ in Fig. 3, 4) can be qualitatively interpreted as a portional matter infall to the star as a result of a stellar wind interaction with a cool outer envelope.

References

Watt,G.D., Burton, W.B.,Choe, S.-U., List, H.S.: 1986, *Astron.Astrophys.* **163**, 194

L. A. Balona et al. (eds.), Pulsation, Rotation and Mass Loss in Early-Type Stars, 389–391.

390

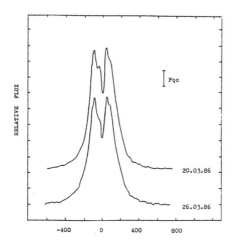

Fig. 1. Line profiles of Hα observed in HD 200775 in March, 1986. The vertical bar provides the flux scale in the units of quasicontinuum F_{qc} (see in the text). No attempts have been made to remove telluric water vapour lines.

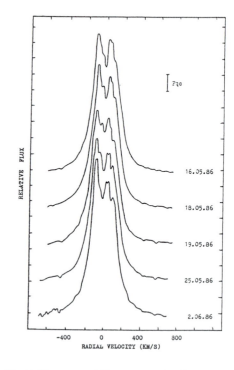

Fig. 2. The same as Fig. 1 except for May–June, 1986

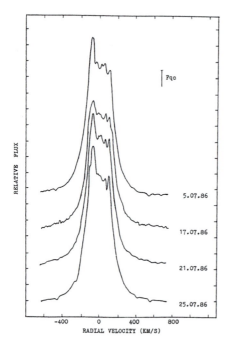

Fig. 3. The same as Fig. 1 except for July, 1986

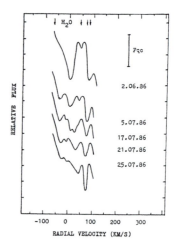

Fig. 4. The large-scale fragments of the line profiles, observed in July, 1986 in comparison with the spectrum of June,2 1986. Discrete emission and absorption features display variations with characteristic time less than a day. Water vapour lines are marked by arrows.

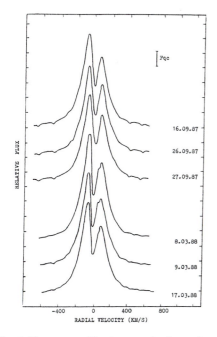

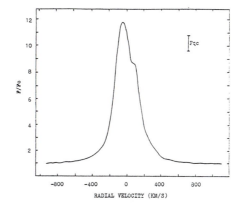

Fig. 6. Line profile of Hα, observed on September,1 1990 in the units of real local continuum F_C. The vertical bar shows the level of quasicontinuum F_{qc}.

Fig. 5. The same as Fig. 1 except for September, 1987 and March, 1988.

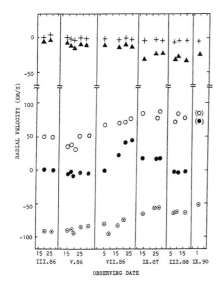

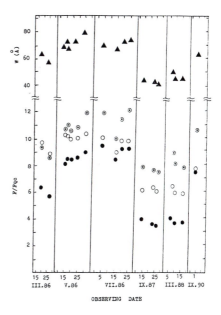

Fig. 7. Time dependence of radial velocities for different profile components, denoted by:
(+)—for the centre of gravity (V_c),
(△)—for the bisector at the $4F_{qc}$ level (V_{bis}),
(⊙)—for the blue emission peak (V_{be}),
(◯)—for the red emission peak (V_{re}) and
(◯—for the central absorption (V_a).

Fig. 8. Time dependence of emission equivalent width W (△) and of relative flux for the blue emission peak (⊙), the absorption component (◯), and the red emission peak (◯). All the values are calculated using the quasicontinuum level.

DUST AND GAS IN SHELLS AROUND HERBIG Ae/Be STARS

N.V. VOSHCHINNIKOV

MPG "Dust in Star-Forming Regions", Jena, Germany and
Astronomical Observatory, St. Petersburg University, St. Petersburg, Russia

and

V.B. IL'IN, A.F. KHOLTYGIN

Astronomical Observatory, St. Petersburg University, St. Petersburg, Russia

Herbig Ae/Be (HAEBE) stars are young objects with strong infrared excesses, variable brightness, intrinsic polarization and complex spectral line profiles. Almost all of these features are due to the presence of circumstellar (CS) shells. We consider the group of HAEBE stars with non-periodic Algol-like minima (UX Ori, WW Vul, etc.). They have irregular brightness drops by 2–3 mag in the visual, accompanied by an increase of linear polarization.

Model. We use the model with variable CS extinction (Grinin, 1988; Grinin *et al.*, 1991). The unusual photometric behaviour of these stars is explained by the presence of very young protoplanetary systems most likely seen edge-on. Many CS dust clumps of different size and mass rotate around such a star and screen off its radiation from an observer from time to time. The shape of the CS shell is assumed to be spheroidal with semiaxes ratio A/B. We consider various silicate-graphite mixtures of dust grains with a power-law size distribution.

Single Scattering Modelling. Using the observations of WW Vul in two deep minima, a model of the dust shell around this star has been constructed (see Voshchinnikov and Grinin, 1991 for details). In both cases, the CS shell has similar characteristics (see Table I), but the properties of CS clumps screening the star are different. The absence of small grains may be explained by their sweeping out of the shell by radiation pressure (Il'in and Voshchinnikov, 1993). If the outer shell radius is at $A = 140$ AU, the mass of dust grains is $\sim 10^{25} g$. Apparently, only this part of the shell gives the main contribution to the scattered radiation and polarization. Taking into account the total duration of minima, the radius of the clouds ($\approx$ 0.2–0.7 AU) has been estimated. The clouds have orbits with large eccentricity and major semiaxes $\geq$ 10 AU. In order to investigate the cloud dynamics in detail, it is necessary to know their radial velocities which can be obtained from observations of the gas absorption lines originating in the clumps.

Multiple Scattering Modelling. On the basis of a new modification of the Monte Carlo approach, a numerical code for polarized radiation transfer in non-spherical CS shells has been developed (see Voshchinnikov *et al.*, 1993

L. A. Balona et al. (eds.), Pulsation, Rotation and Mass Loss in Early-Type Stars, 392–393.

TABLE I

Parameters of dust shell around WW Vul

Parameter	a_-, μm	a_+, μm	q	n_{Si}/n_C	A/B	$\tau_{90}^{ext}(0.36\ \mu m)$
Model	0.055	0.25	5.0	0.25	3	0.55
Standard MRN mixture	0.005	0.25	3.5	1.07		

and Voshchinnikov and Karjukin, 1993 for details).

Our calculations for the model of WW Vul show that there are polarization reversals at $\lambda \approx 1600$ Å and 1900 Å. They are caused by the peculiarity of the dust mixture. The radiation scattered by dust grains may be up to 20–30 % of the total radiation from the system: star + envelope and is the most significant at $\lambda \approx 2000$–3000 Å. The fraction of scattered radiation becomes smaller in the red part of the spectrum. Since the thermal infrared radiation of grains begins to distort the spectrum of HAEBE stars only at $\lambda > 1\ \mu$m, we can conclude that the influence of a CS dust shell on the observed radiation is minimal around $\lambda \approx 1\ \mu$m.

Ionization Structure Modelling. The ionization balance in the outer layers of CS shells has been considered assuming that the sources of gas ionization are: stellar photosphere, stellar chromosphere, and H II region (Kholtygin *et al.*, 1993). We made calculations of column densities (N) of various ions as well as equivalent widths (W_λ) of the D_1, D_2 lines of Na I and H, K lines of Ca II which are most suitable to be observed during the minima of HAEBE stars. It was shown that the values of N and W_λ depend only weakly on the characteristics of the H II region and the CS clumps. The strongest effect is produced by the variations of the gas to dust ratio, element depletion, and velocity of the large scale motion in the clumps. The calculated values of W_λ for Na I and Ca II lines definitely indicate that they may be observed with current big telescopes during minima of the bright HAEBE stars.

Thus, the theoretical basis for the investigations of protoplanetary disks around Herbig Ae/Be stars is created. The photometric, polarimetric, and spectral observations of these stars at various brightness levels can be used to obtain more information on the process of planet formation.

References

Grinin, V.P.: 1988, *Sov. Astron. Lett.* **14**, 65.

Grinin, V.P., Kiselev, N.N., Minikulov, N.Kh., Chernova, G.P., and Voshchinnikov, N.V.: 1991, *Astrop. Sp. Sci.* **186**, 283.

Il'in, V.B. and Voshchinnikov, N.V.: 1993, *Astron. Zh.* **70**, 721.

Kholtygin, A.F., Il'in, V.B., and Voshchinnikov, N.V.: 1993, *Astrop. Sp. Sci.*, in prep.

Voshchinnikov, N.V. and Grinin, V.P.: 1991, *Astrofizika* **34**, 181.

Voshchinnikov, N.V., Grinin, V.P., and Karjukin, V.V.: 1993, *A&A*, submitted.

Voshchinnikov, N.V. and Karjukin, V.V.: 1993, *A&A Supp.*, submitted.

CIRCUMSTELLAR ENVIRONMENT OF THE B[e] STAR
MWC 349 (V 1478 CYG)

R.V. YUDIN

Central Astronomical Observatory of the Russian Academy of Sciences at Pulkovo,
196140 Saint-Petersburg, Russia

Quasi-simultaneous optical and IR photometric (BVRIJHK-bands) and optical polarimetric (RI-bands) observations were made at the 1m telescope of the Assy Observatory, Institute of Astrophysics of Kazakhstan between October 1985 and August 1992. We used the photopolarimeter of the Pulkovo Observatory described by Bergner et al. (1988). Most of our measurements were done with an aperture of 26″. We obtained 26 polarimetric measurements of MWC 349, mainly in the R band. Our data show significant variations in the polarization which has not been reported previously. The amplitude of the polarization parameters is: $\Delta P = 8\%$ and $\Delta P.A. = 15°$. During 12–13.09.90 five polarimetric observations have been obtained and $\Delta P = 3\%$ on a time scale of 7 hours was detected. We found strong photometric variability in all photometric bands. There is however no correlation between brightness and polarization in any photometric band.

1. Interstellar polarization and reddening

The interstellar polarization of the Cyg OB2 association was investigated by Whittet et al. (1992). For 12 out of 19 stars in about 15′ around MWC 349 we obtained P.A. $= 100° \pm 20°$ with an average of interstellar polarization degree of about 3–4%. If we adopt O6.5–B0 ZAMS for MWC349, $(B-V)_0 = -0^m.32$, R $= 3.1$ in the Cygnus region and use the observed color index $2^m.41 < (B - V)_{obs} < 3^m.31$, the expected range of total extinction is $8^m.5 < A_V < 10^m.7$. The average interstellar reddening component would therefore be $A_V(is) < 8^m.5$ and the additional variable extinction $(A_V(cs) > 2^m.0.)$ arises in the circumstellar envelope.

After removing the interstellar component the position angle of intrinsic polarization in MWC 349 is about P.A.(int.) $= 172° \pm 3°$ and $5\% < P(int.) < 13\%$. If scattering in dust grains in the circumstellar disk of MWC 349 is the dominant mechanism of the observed polarization, the dust disk must be oriented approximately in the East-West direction (P.A. $= 82° \pm 3°$).

2. The circumstellar environment of MWC 349

We have evidence for two circumstellar disks in MWC 349: an inner dust disk (P.A. $= 82°$, this paper) and a large-scale outer disk (P.A. $= 100°$,

L. A. Balona et al. (eds.), Pulsation, Rotation and Mass Loss in Early-Type Stars, 394–395.

Cohen et al. 1985), which make an angle of 15° – 20°. Using the formulae from Dolginov et al. (1979), from the alignment of the geometrical sizes of the dust disk and its declination one expects a polarization degree of 10% – 15%, which is in good agreement with what is observed. However, the main question concerns the explanation of the large and rapid variations in the polarization degree. One could explain these variations in terms of episodic eclipses of direct (unpolarized) stellar light by dust clouds moving through the line of sight (Grinin et al. 1988). The increase of polarization degree during the eclipse can then be described by: $p(\lambda) = p_0(\lambda) \, 10^{0.4\Delta m_\lambda}$, where $p_0(\lambda)$ is the intrinsic polarization of the star without eclipse and m_λ is the decrease of star's brightness at the given λ. In our case the observed ΔP must correspond to a decrease in brightness of $> 2^m.0$, which contradicts the observations. This means that the polarimetric changes must be due to variable physical conditions in the gas disk. The observed time scale of variability allows a rough estimate for the size of the circumstellar region where the polarization variability occurs. The average size derived from the several observed polarimetric variations is about 20 – 40 AU. Note that such sizes of a circumstellar shell are comparable to regions of masers. Maser features originate in two spots which arise from clumps of high-density ionized gas located in the near part of the gas disk surface (Planesas et al. 1992). Martin-Pintado et al. (1989) have calculated an upper limit for the size of the maser region of $< 4 \, 10^{14} cm^{-3}$ ($< 27 AU$) which is close to what is observed. One can suppose that the variability of the polarization degree is associated with the presence of these clumps. Using the size and electron density of this region we estimate the polarization degree to be about 7%–8%.

One can propose the following model for the explanation of the polarimetric behavior of MWC 349: during the eclipse of direct starlight by inhomogeneous dust condensations in the disk plane, the electron density in the dense clumps has changed. The speed by which the Strömgren sphere can respond to a change in the Lyman-continuum luminosity is set by the recombination time scale, $t_{rec} = 10^5 (yrs.)/Ne$ (Thum et al. 1992).

For the densities of interest here ($Ne > 10^8 cm^{-3}$, Hamann and Simon, 1986, 1988), t_{rec} is 8 hours or less, which is close to the observed time scale.

Bergner, Yu.K., et.al.: 1988, Izv.GAO AN SSSR, **205**, 142 (in Russian)

Cohen, M., Biegning, J.H., Dreher, J.W., Welch, W.J.: 1985, *Astrophys. J.* **292**, 249

Dolginov, A.Z., Gnedin, Yu.N., Silant'ev, N.A.: *Propagation and polarization of radiation in Space*, Moskow, Nauka, 1979 (in russian)

Grinin, V.P., Kiselev, N.N., Minikulov, N., Chernova, G.: 1988, *Sov. Astr. Lett.* **14**, 514

Hamann, F., Simon, M.: 1986, *Astrophys. J.* **311**, 909

Hamann, F., Simon, M.: 1988, *Astrophys. J.* **327**, 876

Martin-Pintado, J., Thum, C., Bachiller, R.: 1989, *Astron. Astrophys.* , L9

Planesas, P., Martin-Pintado, J., Serabyn, E.: 1992, *Astrophys. J.* **386**, L23

Thum, C., Martin-Pintado, J., Bachiller, R.: 1992, *Astron. Astrophys.* **256**, 507

Whittet, D.C.B., et.al: 1992, *Astrophys. J.* **386**, 562

MWC 314 – A NEW GALACTIC B[e] SUPERGIANT

ANATOLY S. MIROSHNICHENKO

*Central Astronomical Observatory of the Russian Academy of Sciences at Pulkovo,
196140 Saint-Petersburg, Russia, e-mail:anat@gaoran.spb.su*

We present a new study of MWC 314 = BD +14°3887 – a poorly investigated object with strong emission lines and IR excesses. Merrill (1927) payed attention to it because of the presence of hydrogen and Fe II emissions in its spectrum. Swensson (1942) also detected interstellar lines H and K Ca II and 4430 Å band, Balmer emissions from H_α to H_8, Na I 5890 and 5896 Å emissions and estimated its spectral type as gG2-3 or dG4-5 from the SED in continuum, and B2 from the excitation degree. Photospheric lines and spectral features of late-type stars were not observed. Allen (1973) noted that the object's SED corresponds to that of a late-type star but it might be a symbiotic system or a reddened normal star. The IRAS fluxes were obtained only at 12 and 25 μm. The object is unknown as a radio source. From this we can conclude that this system consists of, at least, a hot star surrounded by a gaseous envelope.

We obtained 7 UBVRIJHK and 25 UBVRI observations at the 1-m telescope of the Astrophysical Institute of Kazakhstan Academy of Sciences with the two-channel photometer – polarimeter (Bergner et al., 1988) in 1989-1993 (diaphragm 26″). We have detected $\sim 0\overset{m}{.}3$ variability in all photometric bands. The changes in the optical are synchronous in general. A Fourier analysis of the optical light curves, using Deeming's (1975) technique, show that we can discriminate a regular variability with a period of $4\overset{d}{.}16$. Phase diagrams are most clear in the B and V bands (the amplitudes are $0\overset{m}{.}25$ and $0\overset{m}{.}15$ respectively). The spectra were obtained in July 1991 at the 6-m telescope of the Special Astrophysical Observatory at the Northern Caucasus with a photoelectric scanner in the range of 4000–7400 Å and a dispersion of 50 Å/mm.

The profiles of hydrogen emissions are symmetric without P Cyg-type absorption. The Balmer decrement is $I_{H_\alpha} : I_{H_\beta} : I_{H_\gamma} = 1 : 0.060 : 0.0011$. We have also detected some absorptions which we suppose to be photospheric. These lines together with the weak He I emissions imply that the underlying star is not later than B2. There is no evidence for circumstellar dust radiation between 1 and 25 μm. We have used Lamers and Waters'(1984) method to fit the SED and have found the best fit with the following parameter values: $A_V = 5\overset{m}{.}5$, spectral type - B1 ($T_* = 23\,500$ K), wind temperature $T = T_*$. A comparison of the dereddened Balmer decrement with numerical results for spherical decelerating stellar winds obtained by Pogodin's(1986) method have given the following values of electron density number near the

L. A. Balona et al. (eds.), Pulsation, Rotation and Mass Loss in Early-Type Stars, 396–397.
© 1994 *IAU. Printed in the Netherlands.*

photosphere and volume emission measure: $\log N_{eo} = 11.5$ and $\log \varepsilon \sim 60.5$. It is more difficult to determine the star's mass or radius because we do not know its distance (D). An upper limit of the star's radius can be estimated from the suggestion that the regular variability is connected with the region of the envelope very close to the photosphere, because more clear phase pictures are observed in the bands where the envelope's influence is smallest. Hence we can use the Kepler's third law. Together with the expression for bolometric luminosity it gives the relation between L and M_*. Comparing it with the evolutionary tracks (Maeder and Meynet, 1987) we obtain the best agreement between them for $\log(L/L_\odot) = 5.5$, $M \sim 30\,M_\odot$, and an upper limit of $R_* = 35\,R_\odot$. The characteristics of H_α, ε, and the parameters of the SED fitting give values of the star's radius near $30\,R_\odot$ and a mass-loss rate of about $10^{-6}\,M_\odot/\text{yr}$. Thus we can conclude that MWC 314 is a supergiant. Supposing $R_* = 30\,R_\odot$, the distance towards the object is about 1.7 kpc, and the radio flux at 5 GHz $\sim$ 6 mJy.

It is difficult to determine the age of MWC 314 because the tracks for stars in this part of HR diagram are almost parallel. Its characteristics are close to that of LBVs and B[e] supergiants, but the photometric behaviour has not been studied long enough to make a comparison. We know that most of B[e] supergiants have circumstellar dust, and it is suggested that they are transition objects between Of and WR stars (Zickgraf et al., 1986). It has been unclear in which evolutionary stage massive stars can have dust shells. So, MWC 314 may be an object in a stage before or after the dust formation process.

References

Allen D.A., MNRAS, 1973, **161**, 145

Bergner Yu.K., Bondarenko S.L., Miroshnichenko A.S. et al., Izvestia Glavn. Astron. Obs. v Pulkove, 1988, **205**, 142, (in Russian)

Deeming T.J., Ap&SS., 1975, **36**, 137

Lamers H.J.G.L.M. and Waters L.B.F.M., A&A, 1984, **136**, 37

Maeder A. and Meynet G., A&A, 1987, **182**, 243

Merrill P.W., ApJ, 1927, **65**, 286

Pogodin M.A., Astrofizika, 1986, **24**, 491 (in Russian)

Swennson J.W., ApJ, 1942, **97**, 226

Zickgraf F.-J., Wolf B., Stahl O., Leitherer C., Appenzeller I., 1986, A&A, **163**, 119

THE STRUCTURE OF THE CIRCUMSTELLAR MATERIAL IN BE STARS

L.B.F.M. WATERS

Astronomical Institute Anton Pannekoek
University of Amsterdam, Kruislaan 403, 1098 SJ Amsterdam

SRON Laboratory for Space Research Groningen, P.O. Box 800, 9700 AV Groningen

and

J.M. MARLBOROUGH

Astronomy Department, University of Western Ontario
London, Ontario N6A 3K7, Canada

Abstract. We review observations that are relevant for the determination of the structure and geometry of the circumstellar envelopes of Be stars. Evidence is summarized suggesting that Be stars have rotating discs. Infrared and radio continuum and line measurements and their interpretation are discussed.

1. Introduction

One of the controversial issues in trying to understand the nature of Be stars has been the question of the geometry of their circumstellar envelopes. Extensive discussions in the literature can be found (e.g. Slettebak & Snow 1987). In this review we discuss recent observations which we believe are relevant to the question of the structure of Be star envelopes. The conclusion that can be drawn from these studies is that the envelopes of Be stars are very likely rotationally flattened rather than being (roughly) spherically symmetric, and hence the term *disc-like* would be appropriate. However the detailed structure (velocity field, both rotational and radial; density, radial and in the z direction) remains quite uncertain despite considerable efforts.

2. Optical Emission Lines

The strength and shape of emission lines in the optical spectra of Be stars are important tools to constrain the geometry of their envelopes. This has been investigated in some detail by Dachs and coworkers (Dachs *et al.* 1986; 1992; Hanuschik 1987, 1989) using high resolution and high S/N spectra of Balmer lines and of Fe II emission lines. These studies show that there is a correlation between the FWHM of Hα, its strength and $v \sin i$ (Dachs *et al.* 1986). Such a correlation strongly points to rotation as being the cause of the line broadening in Be star discs and also suggests that the geometry should be highly non-spherical. More recently, Hanuschik (these

399

L. A. Balona et al. (eds.), Pulsation, Rotation and Mass Loss in Early-Type Stars, 399–411.
© 1994 IAU. Printed in the Netherlands.

Proceedings) showed that a correlation between the FWHM and $v \sin i$ also exists for the Fe II lines.

Both the Hα line and the Fe II lines are found to be (much) wider than the rotation of the underlying star would permit. For Hα, part of the line width can be explained by electron scattering, which produces broad, weak emission wings that can easily extend to $\pm$ 1000 km s^{-1}. However even when electron scattering is included the Hα line width in many Be stars exceeds the value expected on the basis of $v \sin i$. This additional broadening was explained by Hummel & Dachs (1992) as a radiative transfer effect in very optically thick lines, where the line photons tend to escape from the line by slightly shifting their frequency (non-coherent scattering) thus broadening the observed line. However such an explanation cannot account for the observed widths in the optical Fe II lines, that also are systematically wider (by about 20 per cent) than expected on the basis of $v \sin i$ (e.g. Hanuschik these Proceedings). Therefore the inner regions of the discs of Be stars may rotate more rapidly than the underlying star, perhaps being in Keplerian orbits. If this is the case, some angular momentum transfer is necessary, and it might solve the problem of the dynamical support of the gas in the disc. Alternately, the $v \sin i$ values of *all* Be stars may be underestimated by some 20 per cent. However this would imply that several stars actually rotate at their Keplerian speed.

3. Hydrogen IR Recombination Lines

The recent improvement in IR detector technology allows us to observe the H I and He I recombination lines of Be stars the near-IR and mid-IR with high spectral resolution and high S/N. These lines are important diagnostic tools because they cover a large range in wavelength and line strength, and thus probe different layers of the envelope. Furthermore the underlying photospheric absorption lines are weak and we get a better view of the disc emission. In Fig. 1 we show the Brγ lines of ψ Per and 59 Cyg, taken in July of 1992 with the CGS4 spectrograph, UKIRT, Hawaii. Both profiles look different from what would be expected from a rotating, roughly Keplerian disc which is optically thin.

In ψ Per the Brγ profile shows peculiar wings that suggest that the bulk of the emission is superimposed on a plateau with a width of $2 \times v \sin i$. The shape of the wings rules out an origin in terms of electron scattering, since that will only produce smooth features. The wings disappear in Brα, probably because the optical depth of the line and the free-free continuum increase to the point that the layers in which this emission is formed are blocked from view. Because the wings are symmetric they are probably located near the surface of the star in the disc. Interestingly, the Hα line of ψ Per has a width which is comparable to the Brγ total width, but does not

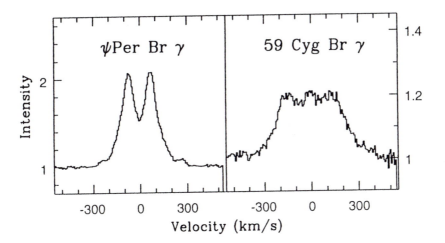

Fig. 1. High-resolution, high S/N Brγ lines for ψ Per and 59 Cyg. Notice the plateau in ψ Per and the broad emission in 59 Cyg.

show the plateau.

The Brγ profile of 59 Cyg is most peculiar. The emission is weak, i.e. the star has a weakly developed disc. The width of the line is about 810 km s^{-1} which considerably exceeds $2 \times v \sin i$ ($v \sin i = 260$ km s^{-1}, Slettebak 1982). Since the emission is weak it is unlikely that the broadening is due to electron scattering and must be due to doppler broadening. The symmetry of the line then points to rotational broadening. This implies that either the star rotates much faster than thought up to now, or that the disc rotates faster than the star. If this is the case then angular momentum must be transferred to the gas in the disc. The line profile also strongly deviates in shape from what is expected for a low-density disc. The flat top, with even a hint for *three* peaks, cannot be explained in terms of a rotating Keplerian disc. Flat-topped profiles are expected for optically thin spherical winds with constant outflow velocity, but that situation probably does not apply in this case.

4. The Infrared and Radio Continuum

The increase in sensitivity of (sub-)mm and radio telescopes in the past five years has resulted in the detection of a few IRAS-bright Be stars at those wavelengths. The first detection was of the B5 Ve star ψ Per at 6 cm using the *VLA* (Taylor *et al.* 1987). This detection demonstrates that the extent of Be star envelopes can be very large indeed, and that the term 'wind' for the high-density part of the envelope seems appropriate at least in some cases. The spectral slope of the continuum was found to be significantly steeper than that found from IRAS data.

A subsequent more extensive *VLA* survey of 22 IRAS-selected Be stars at 2 cm resulted in the detection of another 5 Be stars (Taylor *et al.* 1990) covering the spectral range from B0.5 to B8. In almost all cases the spectral index found from the radio measurements was significantly steeper than that from IRAS 12 to 60 μm data, typically $S_\nu \propto \nu^\alpha$ with $\alpha \approx$ 0.6-1 in the IRAS data and >1 in the radio. Therefore a *turnover* in the spectrum must occur somewhere between the IR and radio. A *VLA* radio survey of Be stars by Apparao *et al.* (1991) did not yield detections.

The position of the turnover was investigated by Waters *et al.* (1991) who reported the detection of the 6 Be stars, previously detected with the *VLA*, at wavelengths of 0.8 and 1.1 mm using the *JCMT* Bolometer UKT14. The turnover in the spectral index occurs in the sub-mm wavelength range for most of the stars, except perhaps for γ Cas whose turnover was found roughly at 60 μm.

It is important to recognise that the steepening of the radio spectrum is quite contrary to the behaviour expected for a wind flowing out and reaching a terminal velocity. In the case of a spherically symmetric outflow or an outflow in a disc with constant opening angle a spectral index of 0.6 is expected (e.g. Wright & Barlow 1975). The origin of the steepening of the spectral index is as yet not well understood. Taylor *et al.* (1990) suggest several possibilities, such as recombination at large distances from the star, re-acceleration or a change in geometry.

Model calculations of the ionisation/excitation equilibrium in circumstellar discs around B stars indicate that the degree of ionisation stays roughly constant or increases slightly outwards (Waters *et al.* 1991), suggesting that recombination cannot explain the observations. Similar conclusions were also found by Poeckert & Marlborough (1982) using their model. However all calculations so far assumed a constant temperature, which may not be a very realistic approximation.

Chen *et al.* (1992) have interpreted the shape of the IR to radio continuum in terms of an additional acceleration of the gas at large distance from the star, and found that the driving force, which they called F_x, should be very small near the star but should dominate all other forces at distances beyond 10-100 R_*. It is not clear however what the nature of this force could be. Chen *et al.* (1992) suggest that stellar UV continuum radiation, unable to penetrate deep into the disc near the star where densities are high, could succeed in doing so at larger distance, especially if the disc near the star is very thin in the z direction. Marlborough *et al.* (1993) suggest that in Be stars the force F_x dominates all other forces at a smaller distance from the star than in shell stars.

Could the turnover be the result of a change in geometry? Indeed slab models have steeper continuum slopes than models with a diverging geome-

TABLE I

Effect of geometry on the slope of the excess flux
ratio Z_ν

n	spherical	pole-on slab	edge-on slab
2	0.67	0.50	0.33
2.5	0.50	0.40	0.25
3.0	0.40	0.33	0.20
3.5	0.33	0.29	0.17
4.0	0.29	0.25	0.14

try with the same density gradient (e.g. Cassinelli & Hartmann 1979; Waters 1986a). In Table 1 we compare the expected slope of the excess flux ratio $Z_\nu = F_\nu^{tot}/F_\nu^*$ as a function of the optical depth parameter E_ν (Waters 1986a), where $E_\nu \propto \lambda^2(g+b)$, for a simple disc model and for a slab model using a power-law density distribution $\rho(r) \propto r^{-n}$. The slab models are capable of reproducing the observed mm-cm slope for n = 2, i.e. a constant outflow velocity model.

The consequence of this interpretation would be that the discs of Be stars have an inner region which has a diverging geometry (in order to explain the flat IRAS continuum slope) and an outer region which is more slab-like. The question then arises what causes such a change in geometry, at a relatively large distance from the star. We will come back to this point in Sect. 7.

The near-IR energy distribution of a large sample of Be stars was studied by Dougherty et al. (1994) using the simple disc model of Waters (1986a). The IR excesses were derived by constructing optical-IR colour-colour diagrams for normal stars using the Geneva system and the ESO JHKL system (Dougherty et al. 1993), and by comparing the observed colours of Be stars to these intrinsic colour-colour relationships. We show the excess colour-colour diagrams in Fig. 2. Panel (a) shows the effect of different density gradients ($\rho(r) \propto r^{-n}$, n=2,2.5,..,5) for pole-on discs (solid lines) and edge-on discs (n = 3,4; dotted lines). Panels (b) to (d) show the effect of disc radius, opening angle and disc temperature for edge-on models. For small excesses or at short wavelengths the emission is optically thin and so no information on the density structure can be obtained. However for a significant fraction of stars the optical depth in the disc at the L-band is already significant (large K-L colour excess). Comparison of models and observations shows that the simple model can describe the near-IR excess of Be stars in most cases, but that the value of the density gradient parameter n is larger than about 3. This is steeper than the value found by Waters et al. (1987) from IRAS data but agrees reasonably well with the range of slopes found from

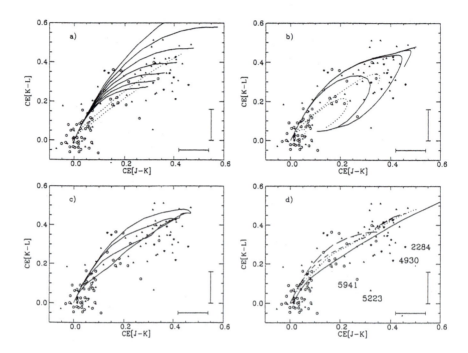

Fig. 2.   Near-IR excess colour-colour diagram for Be stars (taken from Dougherty *et al.* (1994)). See text for details.

the mm-radio continuum. Dougherty *et al.* (1994) show that part of this effect may be due to the assumption that the disc is viewed pole-on. Some stars fall significantly below the model curves for any reasonable choice of the density distribution. These stars may have a different disc geometry (e.g. slab-like), or a small outer disc radius.

In summary the near-IR slope of the continuum energy distribution *and* the mm-radio continuum may point to a steeper spectral index than the IRAS data. As discussed above, this may be related to changes in geometry in combination with changes in the density/velocity gradient.

5. Mass Loss Rates

The mass loss rates of Be stars are quite uncertain because of our poor knowledge of the wind geometry and ionisation structure. The latter uncertainty especially affects the UV mass loss rates derived from e.g. Si IV and C IV (for a review see Snow 1987). The equatorial mass loss rates are mainly based on the IR excess (Waters 1986a; Waters *et al.* 1987) and are uncertain because of the unknown opening angle of the disc and the initial outflow velocity $v(r = R_*)$, which were fixed to 15 degrees and 5 km s^{-1} respectively. This initial velocity was based on estimates derived from detailed fitting of the Hα line of a limited number of Be stars (in particular γ Cas and ϕ Per) by Poeckert & Marlborough (1978a; 1979).

In the case of γ Cas the initial velocity of 7 km s^{-1} probably is overestimated considerably. This is because the Hα line profile that was used by Poeckert & Marlborough was taken when the star was in a phase when the Violet peak of the emission line was weaker than the Red one. Such a line profile can be explained with a rotating and expanding disc. However the shape of the Balmer emission lines in γ Cas is affected by the cyclic V/R variability, in which phases with V>R and V<R alternate on a timescale of several years (e.g. Doazan et al. 1987). The V/R variability probably results from one-armed spiral density waves that propagate in the disc (Okasaki 1991; Papaloizou et al. 1992; Savonije & Heemskerk 1993). Indeed, Telting et al. (1993) have shown that the V/R variations in γ Cas do not affect the radial density structure of the disc strongly. Therefore the velocity law in γ Cas cannot be derived from Hα unless the effects of such density waves on the line profiles are taken into account.

The effect of a high initial outflow velocity on the Hα line was already pointed out by Poeckert & Marlborough (1978b) who used their γ Cas model but assumed an inclination angle of 90 degrees (for γ Cas the inclination angle was taken at 45 degrees). The resulting line shows a P Cygni profile, which is hardly ever observed in Be stars that are assumed to be viewed edge-on, suggesting that the radial outflow velocity in the γ Cas model is too high.

An analysis of the infrared recombination lines of ψ Per by Zijlstra et al. (in preparation) showed that in order to reproduce the shape of the line profiles, an expansion velocity of less than about 1 km s^{-1} at r = R$_*$ is required, i.e. more than a factor 5 lower than assumed by Waters et al. (1987). This would result in IR mass loss rates that are a factor 5 lower; for ψ Per it would reduce to 2.5 10^{-9} M$_\odot$/yr, compared to an UV mass loss rate of 8 10^{-11} M$_\odot$/yr (Snow 1981). Note however that in the analysis of Waters et al. (1987) a total opening angle of the disc of 30 degrees was used. Reducing this opening angle would result in an increase of the equatorial mass loss rate.

Can the mass loss rates from the UV and from the IR be roughly equal? Such a question is difficult to answer given the uncertainties noted above. The X-ray luminosities observed in Be/X-ray binaries indicate that the *mass flux* near the neutron star must be considerably higher than expected on the basis of the UV mass loss rates (Waters et al. 1988).

6. The Wind Compressed Disc Model

The wind compressed disc (WCD) model developed by Bjorkman & Cassinelli (1993; see also J. Bjorkman these Proceedings) predicts the formation of a very thin disc in the equatorial regions of rapidly rotating stars that have radiation-driven winds. This idea is very attractive because for the first time a model is able to explain the existence of discs quantitatively by using properties of hot stars that are well understood. The numerical calculations of Owocki *et al.* (1993) confirm the analytical calculations by Bjorkman & Cassinelli (1993). These theoretical developments as well as advances in observational techniques strongly suggest that Be stars indeed have discs.

However, the WCD model does have some serious difficulties in explaining the observed properties of Be stars (some of these were already pointed out by Bjorkman & Cassinelli). The main difficulty is the predicted density in the disc, which is a factor 100 lower than observed. Also the infall of material inside the stagnation point, with outflow beyond the stagnation point, does not agree with observations of line profile shapes of optical and IR HI recombination lines. These line profiles clearly point to a rotating disc without strong radial motions. A third difficulty is the opening angle of the disc, which is of the order of 1 degree. Such thin, dense discs would show a very strong dependence of IR excess on inclination, which is not observed (e.g. Dougherty *et al.* 1991; Waters 1986b). It would also be difficult to explain the statistics of shell stars, which require opening angles of the order of 10 degrees (Hanuschik these Proceedings).

The steady-state WCD model by definition is unable to explain the strong variability in Be stars. A critical test of the WCD model will be a determination of the rate of increase of the emission measure of the disc during an outburst of a Be star such as the ones observed in μ Cen by Baade *et al.* (1988), since in the WCD model the disc is fed by the radiation-driven wind (we thank L. Kaper and J. Telting for pointing this out). If the rate of increase of the mass in the disc cannot be accounted for, another mass loss mechanism must be effective in Be stars.

7. The Geometry of Be-star Discs

Be stars have now been imaged directly in the Hα line (see Vakili these Proceedings) and in the radio (Dougherty & Taylor 1992). In the case of ψ Per the alignment between the optical and radio images is very good, and also the position angle of the linear polarisation is perpendicular to the semi-major axis of the intensity distribution. These observations strongly support the disc-like geometry of Be star envelopes.

Useful information about the geometry of Be star discs can also be obtained from a comparison of the emission in e.g. Hα and the near-IR.

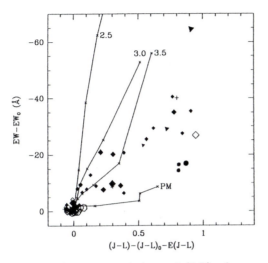

Fig. 3.   Correlation between Hα excess emission and (J-L) colour excess (taken from van Kerkwijk *et al.* 1994). The solid lines are the model curves for the disc model with different radial density gradients, and the PM model. Both models have difficulty explaining the observations

Many studies have shown that a correlation between IR excess and Hα line strength exists (e.g. Ashok *et al.* 1984; Dachs *et al.* 1988) but only a few attempts have been made to reproduce quantitatively the observed correlations.

Kastner & Mazzali (1989) used a pole-on slab model to calculate Hα line flux and near-IR excess, and found that they can reproduce the observed correlations quite well. It is not clear however what the effects of a different inclination angle would be on their results. More recently, van Kerkwijk *et al.* (1994) have used two models, the disc model introduced by Waters (1986a) and the Poeckert & Marlborough (1978a; PM) model, to calculate theoretical correlations between IR excess and Hα emission (Fig. 3), and compared them to simultaneously obtained optical and near-IR data. They find that the disc model produces *much too strong* Hα emission for reasonable choices of the density gradient parameter n, and that the PM model gives *much too weak* Hα emission. Since the near-IR and Hα probe quite different regions in the disc it is obvious that in both models the density/velocity structure is wrong. The disc model has too much material at large distance, the PM model too little. The fact that the disc model produces too much Hα is consistent with the fact that it also predicts too much mm and cm radiation.

The WCD model may have a strong effect on the geometry of an (existing) disc. The collimating effect of the fast radiation-driven wind near the star, which in the WCD model is responsible for *producing* the disc, may, in the

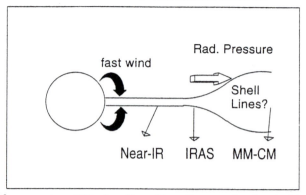

Fig. 4. A possible geometry of Be star discs. Near the star the disc is thin and confined because of the collimating effect of the polar wind. At some distance the disc flares (not necessarily to the same thickness as the stellar radius) and the gas is accelerated radially by the strongly forward peaked radiation field.

case of an *existing* disc produced by some other mechanism, cause this disc to be effectively confined to regions very close to the equatorial plane. At larger radial distance from the star this confinement is much less efficient and therefore the disc may flare. When this happens, the radiation force from the central star, which is strongly forward peaked, can accelerate the gas which is flaring out. The resulting geometry is sketched in Fig. 4, and we call this model the wine-bottle model (not to be confused with so-called wine-bottle Hα line profiles).

Such a model would produce a continuum energy distribution which is rather steep in the near-IR, would flatten somewhat when the disc opens up, and would steepen again when the gas is accelerated to a terminal velocity. As discussed in Sect. 4, such an energy distribution would not be inconsistent with observations. However detailed model calculations are required to verify this. A flaring of the disc would also circumvent the difficulty of the statistics of shell stars, since the shell absorption lines could be formed in the region where the disc begins to flare out. The model is consistent with the observations of Be/X-ray binaries, since it can have both an inner region dominated by rotation (where the Hα line is formed) and an outer region dominated by expansion (where the neutron star is located).

Acknowledgements

We thank Tom Geballe for assistance in obtaining the IR line profiles. Many thanks also to John Telting, Lex Kaper, Sean Dougherty, Jacqueline Coté, Jan-Willem Zijlstra and Peter Zaal for stimulating discussions. This research is supported financially by the Royal Dutch Academy of Arts and Sciences and the Natural Sciences and Engineering Council of Canada.

References

Apparao, K.M.V., Rengarajan, T.N., Tarafdar, S.P. and Ghosh, K.K.: 1991, *Astron. Astrophys.* **229**, 141.

Ashok, N.M., Bhatt, H.C., Kuhlkarni, P.V. and Joshi, S.C.: 1984, *Mon. Not. Roy. Astr. Soc.* **211**, 471.

Baade, D., Dachs, J., can de Weygaert, R. and Steeman, F.: 1988, *Astron. Astrophys.* **198**, 211.

Bjorkman, J.E. and Cassinelli, J.P.: 1993, *Astrophys. J.* **409**, 429.

Cassinelli, J.P. and Hartmann, L.: 1979, *Astrophys. J.* **212**, 488.

Chen, H., Marlborough, J.M. and Waters, L.B.F.M.: 1992, *Astrophys. J.* **384**, 604.

Dachs, J., Hanuschik, R., Kaiser, D. and Rohe, D.: 1986, *Astron. Astrophys.* **159**, 276.

Dachs, J., Engels, D. and Kiehling, R.: 1988, *Astron. Astrophys.* **194**, 167.

Dachs, J., Hummel, W. and Hanuschik, R.W.: 1992, *Astron. Astrophys. Suppl.* **95**, 437.

Doazan, V., Rusconi, L., Sedmak, G., Thomas, R.N. and Bourdonneau, B.: 1987, *Astron. Astrophys.* **182**, L25.

Dougherty, S.M., Taylor, A.R. and Clark, T.A.: 1991, *Astron. J.* **102**, 1753.

Dougherty, S.M. and Taylor, A.R.: 1992, *Nature* **359**, 808.

Dougherty, S.M., Cramer, N., van Kerkwijk, M.H., Taylor, A.R. and Waters, L.B.F.M.: 1993, *Astron. Astrophys.* **273**, 503.

Dougherty, S.M., Waters, L.B.F.M., Burki, G., Cramer, N., van Kerkwijk, M.H. and Taylor, A.R.: 1994, *Astron. Astrophys.* in press.

Hanuschik, R.W.: 1987, *Astron. Astrophys.* **173**, 299.

Hanuschik, R.W.: 1989, *Astrophys. Space Sci.* **161**, 61.

Hummel, W. and Dachs, J.: 1992, *Astron. Astrophys.* **262**, L17.

Kastner, J.H. and Mazzali, P.A.: 1989, *Astron. Astrophys.* **210**, 295.

Kogure, T.: 1990, *Astrophys. Space Sci.* **163**, 7.

Marlborough, J.M., Chen, H. and Waters, L.B.F.M.: 1993, *Astrophys. J.* **408**, 646.

Okasaki, A.T.: 1991, *Publ. Astr. Soc. Japan* **43**, 75.

Owocki, S.P., Cranmer, S.R. and Blondin, J.M.: 1993, *Astrophys. J.* in press.

Papaloizou, J.C., Savonije, G.J. and Henrichs, H.F.: 1992, *Astron. Astrophys.* **265**, L45.

Poeckert, R. and Marlborough, J.M.: 1978a, *Astrophys. J.* **220**, 940.

Poeckert, R. and Marlborough, J.M.: 1978b, *Astrophys. J. Suppl.* **38**, 229.

Poeckert, R. and Marlborough, J.M.: 1979, *Astrophys. J.* **233**, 259.

Poeckert, R. and Marlborough, J.M.: 1982, *Astrophys. J.* **252**, 196.

Savonije, G.J. and Heemskerk, M.H.M.: 1993, *Astron. Astrophys.* **276**, 409.

Slettebak, A.: 1982, *Astrophys. J. Suppl.* **50**, 55.

Slettebak, A. and Snow, T.P.: 1987, *Physics of Be Stars: IAU Colloquium 98*, Cambridge Univ. Press: Cambridge.

Snow, T.P.: 1981, *Astrophys. J.* **251**, 139.

Snow, T.P.: 1987, in Slettebak, A. and Snow, T.P., eds., *Physics of Be Stars: IAU Colloquium 98*, Cambridge Univ. Press: Cambridge, 250.

Taylor, A.R., Waters, L.B.F.M., Lamers, H.J.G.L.M., Persi, P. and Bjorkman, K.S.: 1987, *Mon. Not. Roy. Astr. Soc.* **228**, 811.

Taylor, A.R., Waters, L.B.F.M., Bjorkman, K.S. and Dougherty, S.M.: 1990, *Astron. Astrophys.* **231**, 453.

Telting, J.H., Waters, L.B.F.M., Persi, P. and Dunlop, S.R.: 1993, *Astron. Astrophys.* **270**, 355.

Van Kerkwijk, M.H., Waters, L.B.F.M. and Marlborough, J.M.: 1994, *Astron. Astrophys.* submitted.

Waters, L.B.F.M.: 1986a, *Astron. Astrophys.* **162**, 121.

Waters, L.B.F.M.: 1986b, *Astron. Astrophys.* **159**, L1.

Waters, L.B.F.M., Coté, J. and Lamers, H.J.G.L.M.: 1987, *Astron. Astrophys.* **185**, 206.

Waters, L.B.F.M., Taylor, A.R., van den Heuvel, E.P.J., Habets, G.M.H.J. and Persi, P.: 1988, *Astron. Astrophys.* **198**, 200.

Waters, L.B.F.M., van der Veen, W.E.C.J., Taylor, A.R., Marlborough, J.M. and Dougherty, S.M.: 1991, *Astron. Astrophys.* **244**, 120.
Wright, A.E. and Barlow, M.J.: 1975, *Mon. Not. Roy. Astr. Soc.* **170**, 41.

Discussion

Friedjung: As a non-member of the Be community, I would like to be clear about your assumptions when you determine disc densities and mass loss rates. Are you sure that all the hydrogen of the disc is ionized? If you have enough matter, all will not be ionized by radiation from the star.

Waters: The observed IR excess gives information about the ionized part of the envelope only. The fact that some Be stars, even of late B spectral type, are detected in the radio, suggests that a significant fraction of the envelope remains ionized out to large distances from the star. If a large amount of neutral material were present, one would expect to see emission from neutral metals (Fe I, Ni I, etc.) and also the formation of dust grains. Such things are not generally observed.

Kogure: What is the essential difference between ordinary Be and shell stars? It is evident that the discs of shell stars are optically thicker than those of Be stars, but what determines the structure of these two types of stars?

Waters: The shell stars may simply be Be stars with a very dense disc seen at inclination angles close to 90°. Possibly shell stars have a different disc structure in that the disc flares out relatively close to the star so that a large fraction of the stellar disc is covered by material from the disc even at moderate inclination angles, resulting in deep absorption lines.

Kogure: I would like to stress the importance of analysing the lines of the higher members of the Balmer series. These lines provide useful information on the envelope structure of shell stars (Kogure 1990).

Polosukhina: What do polarization observations tell us?

Waters: The linear polarization observed in Be stars indicates that the scattering electrons are distributed in a non-spherically symmetric manner. As such, polarization places important constraints on the geometry, but in many cases the solution is not unique (model dependant). See K. Bjorkman (these Proceedings) for more detail.

Hanuschik: I would like to bring to your attention observations of the Fe II $\lambda5317$ shell line in *o* Aqr in 1989 October. Shell lines are sensitive tracers of the *radial* component of the velocity field in the disc. Now, the profile of this line is the narrowest feature we ever observed in a Be/Be-shell star. It is unresolved at the 6 km s^{-1} instrumental resolution. The full Doppler width of iron at 10^4K being about 4 km s^{-1}, the residual range for any radial motion (inflow/outflow) in front of the star is only 1 or 2 km s^{-1}! This is a very tough upper limit for any outflow model.

Waters: The Hα and IR hydrogen recombination lines of ψ Per also indicate that there is only a very small radial motion, although the IR continuum suggests that $v(r)$ increases roughly as $\sqrt{r}$.

Owocki: How can one reconcile the evidence you mention (e.g. in X-ray binaries) for outflow of the order of ~ 100 km s^{-1} in the outer disc with observations of narrow circumstellar Fe lines with very low outflow velocity (< 10 km s^{-1})?

Waters: Shell lines clearly indicate much lower radial outflow velocities than the Be/X-ray binaries suggest. The Be/X-ray binaries known so far, however, have early spectral type (in the range O9–B2III) and radiation pressure exerted by the central star may have a significant effect on the radial outflow of these stars. This is consistent with the fact that γ Cas has the steepest IR to radio continuum of the objects detected so far. I would expect, if radiation forces play a role, that later B-type stars should have significantly lower outflow velocities than earlier types and also that it depends on the density in the disc.

Saraswat: I have two comments regarding Be/X-ray binaries:

(1) We have used the "short" period Be/X-ray binary 4U1907+09 to test the predictions of the "ellipsoidal" model proposed by Doazan & Thomas. We find this model is unable to predict the X-ray luminosity of this binary. The neutron star lies in the coronal/wind region where the density is low and the wind velocity is high. Hence the X-ray luminosity predicted by this model is lower by several orders of magnitude.

(2) The disc model can explain the X-ray light curve of this source. However, we find that the disc is *not* continuous with a uniform density, but the Be star ejects multiple *rings* with different initial density values and gradients. Also these rings have different outflow velocities (in the range 100–250 km s^{-1}). One could say that the Be star ejects "puffs" of matter.

Waters: The nature of 4U1907+09 is not clear. It may be a supergiant system rather than a Be/X-ray binary.

INFRARED SPECTROSCOPY OF Be STARS[*]

R.M. TORRES, A. DAMINELI-NETO and
J.A. DE FREITAS PACHECO
Instituto Astronômico e Geofísico – Universidade de São Paulo
CP 9638, 01065-970 São Paulo, SP, Brasil.

1. Introduction

FeII emission lines are present in a variety of astrophysical objects and, in particular, in Be stars, where in some situations they can also be seen in absorption. Selvelli & Araujo (1984) studied a sample of classical Be stars that have FeII emission lines in the optical region. The analysis of IUE spectra of those stars revealed that, for the majority of the objects, neither absorption nor emission FeII features were present in the UV. The conclusion was that their data could not support excitation of FeII by continuum fluorescence. On the other hand, FeIII of circumstellar origin is often seen in absorption in the UV spectra of Be stars (Snow & Stalio 1987 and references therein). This could be an indication that the optical FeII emission lines are originated from recombination and cascade. However, Selvelli & Araujo (1984) argued that, since the multiplet UV 191 of FeII does not appear in emission, that mechanism is probably not relevant. In the present work we report new spectroscopic observations in the near infrared of a sample of 60 Be stars, including the prominent FeII 999.7 nm emission line. This line is also present in the spectra of superluminous B stars for which mass loss rates have recently been estimated (Lopes, Damineli-Neto & Freitas Pacheco 1992). We derived mass loss rates from the infrared line luminosities, in agreement with those derived by other methods. We also found a new evidence of the Be envelope flattening through the FeII/Paδ line ratio.

2. Observations and Results

Most of the observations were carried out at the National Laboratory for Astrophysics (Brasópolis – Brazil), using the 1.6-m telescope. Observations cover the period from 1988 May to 1992 December. The Coudé spectra covered a range of 23 nm in our CCD, encompassing the FeII 999.7 nm and Paδ (HI 1004.9 nm) lines with a resolution of $\lambda/\Delta\lambda \sim 10000$. Some data for northern objects were obtained at the Bologna Astronomical Observatory

---

[*] Based on observations made at the Laboratório Nacional de Astrofísica (Brasópolis, Brazil) and at the Osservatorio Astronomico di Bologna (Loiano, Italy)

L. A. Balona et al. (eds.), Pulsation, Rotation and Mass Loss in Early-Type Stars, 412–413.
© 1994 *IAU. Printed in the Netherlands.*

(Italy) from 1989 September until 1990 January, using the 1.52-m telescope. These spectra have $\lambda/\Delta\lambda \sim 2000$ resolution.

The most important result from our data is the strong correlation between the line luminosity of FeII 999.7 nm and that of Paschen-delta. The best fit equation represents also quite well the superluminous B stars data. The FeII 999.7 nm line is probably excited by $Ly\alpha$ photons (Johansson 1984) and our data suggest that such a mechanism may be operative in both classes of stars. We found, from our previous work (Lopes, Damineli-Neto & Freitas Pacheco 1992), that the mass loss rates from superluminous B stars are strongly correlated with the luminosity in the FeII 999.7 nm (or, equivalently, in the Paδ) line. From our data we obtained the relation

$$\log (dM/dt) = -5.57 + 0.71 \log [L(\text{FeII})/L_\odot], \tag{1}$$

where mass loss rates are in solar masses per year. If such a relation can be extrapolated to lower luminosities, where classical Be stars are found, then the mass loss rate interval for the objects in this class is

$$-7.7 < \log (dM/dt) < -6.3. \tag{2}$$

This range is in quite good agreement with previous analyses (see, e.g., Damineli-Neto & Freitas Pacheco 1982) and with the theoretical predictions of asymmetric wind models of Araujo & Freitas Pacheco (1989).

We have also distributed the stars of our sample into bins of 50 km s^{-1} projected rotational velocity and averaged the corresponding FeII 999.7/Paδ ratio in each bin. There is a net correlation between this line ratio and $V \sin i$ (i.e., and the aspect of the equatorial region), indicating a polar flattening of the classical Be envelopes, as suggested by other methods. The observed correlation can probably be explained in terms of optical depth effects in the Paδ line, as the envelope seems to be optically thin to the FeII 999.7 line. An increasing line opacity from the pole towards the equator by a factor of $2.0 - 2.5$ may explain quite well the observed behaviour.

Acknowledgements

This research was partially supported by FAPESP.

References

Araujo, F.X. and Freitas Pacheco, J.A. de: 1989, *MNRAS* **241**, 543.
Damineli-Neto, A. and Freitas Pacheco, J.A. de: 1982, *MNRAS* **198**, 659.
Johansson, S.: 1984, *Phys. Scr.* **T8**, 63.
Lopes, D.F., Damineli-Neto, A. and Freitas Pacheco, J.A. de: 1992, *A & A* **261**, 482.
Selvelli, P.L. and Araujo, F.X.: 1984, *Proc. 4th European IUE Conf.*, ESA SP-218, p. 301.
Snow, T.P. and Stalio, R.: 1987, *Exploring the Universe with the IUE Satellite*, ed. Y. Kondo, p. 183.

Infrared emission lines of Mg II in B stars

T. A. AARON SIGUT and J. B. LESTER
Department of Astronomy, University of Toronto
60 St. George Street, Toronto, Ontario Canada M5S 1A7

1. Introduction

Recently, Chang et. al. (1992) and Carlsson, Rutten and Shchukina (1992) (CRS) demonstrated the non-LTE formation mechanism behind the 12 μm Mg I emission lines $(6g-7h, 6h-7i)$ observed in the solar spectrum (Murcray et. al., 1981). CRS stress the generality of this mechanism showing that it is a natural consequence of the recombination flow from the large Mg II reservoir through the Rydberg levels of Mg I. We have noted the close parallel between Mg I in the solar atmosphere and Mg II in the atmospheres of B stars (where Mg III plays the role of the reservoir) and investigated the operation of this mechanism in high-ℓ infrared transitions of Mg II. We have employed a 58 level Mg II atom including all energy levels through $n = 25$ and a total of 491 linearized radiative transitions. The coupled equations of radiative transfer and statistical equilibrium were solved with the MULTI code in its local operator form (Carlsson, 1992).

2. Results

Figures 1(a) and 1(b) show the $5g-6h$ and $6h-7i$ transitions of Mg II near their maximum strengths in $T_{\rm eff}$. The emission results from a population divergence $b_l < b_u$ which causes the monochromatic source function to rise with height. This also leads to strong limb brightening of the emergent intensity as shown in Figure 1(c) for $5g-6h$. This sensitivity to the variation of viewing angle over the surface, coupled with a strong pressure dependence, suggests that non-spherical disk integrations should be investigated. We have incorporated the effect of rapid uniform rotation in the Roche approximation following Collins (1963). An example is shown for $5g-6h$ in Figure 1(d) for the case of critical rotation, $\omega_f = 1.0$. The non-spherical profile is noticeably weaker than the best fit spherical profile computed with the same M and L but $R = R_{\rm p}$. The main difference is that for a star seen nearly pole on, the average value of μ over the surface will increase with ω_f. For a spherical model, $< \mu >= 2/3$ while for $\omega_f = 1.0$, $< \mu >= 0.746$ due to the absence of viewing angles $\mu < 0.5$.

414

L. A. Balona et al. (eds.), Pulsation, Rotation and Mass Loss in Early-Type Stars, 414–415.
© 1994 IAU. Printed in the Netherlands.

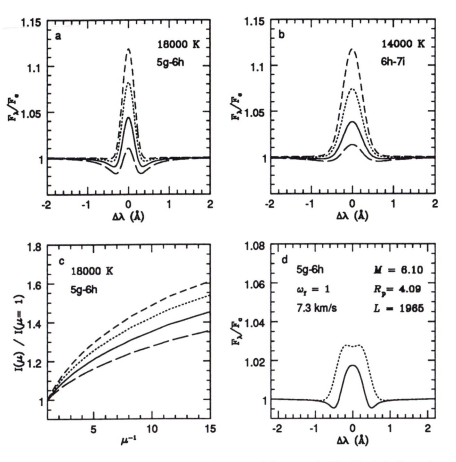

Fig. 1. (a) Relative flux for the transition $5g-6h$ (1.86 μm). The T_{eff} is indicated and the model gravities are identified by $\log(g) = 4.5$ (long dash), 4.0 (solid), 3.5 (dotted), and 3.0 (short dash). (b) same for $6h-7i$ (3.09 μm). (c) Line center limb brightening of $5g-6h$. (d) Non-spherical profile (solid) compared to the best fit spherical profile (dotted). Model parameters are given in solar units; R_{p} refers to the polar radius and both R_{p} and L were assumed unaffected by rotation. The $v \sin i$ of the spherical profile is also given.

References

Carlsson M., Rutten R.J., Shchukina N.G.: 1991, *Astr. Ap.* **253**, 567

Carlsson, M.: 1992, 'The MULTI non-LTE program' in M. S. Giampapa and J. A. Bookbinder, ed(s)., *Cool Stars, Stellar Systems and the Sun*, A.S.P. Conf. Ser. Vol. **26**, 499

Chang E.S., Avrett E.H., Mauas P.J., Noyes R.W., Loeser R.: 1991, *Ap. J.(Letters)* **379**, L79

Collins G.W.: 1963, *Ap. J.* **138**, 1134

Murcray F.J., Goldman A., Murcray F.H., Bradford C.M., Murcray D.G., Coffey M.T., Mankin W.G.: 1981, *Ap. J.(Letters)* **247**, L97

IDENTIFYING NEAR-IR VARIABLE Be STARS

S.M. DOUGHERTY
John Moores University, Liverpool, England

and

A.R. TAYLOR
University of Calgary, Calgary, Canada

Abstract. Near-IR variable Be stars are identified from multi-epoch observations spanning 20 years using a statistical technique. In this manner, observations from different observing sites can be meaningfully combined and compared. A more thorough investigation of the incidence and properties of IR variability in Be stars as a class of stars is then possible.

Although IR variations on time scales ranging from months to years have been observed in Be stars, the statistics and properties of the IR variations in the few studies undertaken differ dramatically. To date there has been no systematic search for variations at near-IR wavelengths from a large sample of Be stars. A large sample of IR observations over a number of epochs allows better statistics of the long term variations of Be stars to be determined than in previous analyses. Furthermore, with a large sample of stars it is possible to use the sample itself to correct for systematic differences between epochs.

The data set consists of magnitudes from several epochs of observation, both new observations and those taken from the literature (Dougherty and Taylor, 1994). Observation of the i th star at the j th epoch yields a magnitude $m_{ij} \pm \sigma_{ij}$, σ_{ij} being the 1σ uncertainty in the magnitude. Consider the distribution of deviations Δ_{ij}, of the magnitudes m_{ij} from the mean magnitude of the i th star, $\overline{m}_i$ *i.e.*

$$\Delta_{ij} = m_{ij} - \overline{m}_i.$$

To identify stars that have variable magnitudes, a sample of non-variable stars was first determined. The deviations for the non-variables will be mainly confined within $\pm 3\sigma$ of the mean of the distributions of Δ_{ij}. An initial sample of non-variables was defined as those stars for which the deviations satisfy $|\Delta_{ij}| \leq 2.5\sigma_{\Delta_{ij}}$.

Once the non-variable stars were identified, the distribution of deviations for non-variables at each individual epoch were corrected so that a non-variable star has, on average, the same magnitude at all epochs, and that the width of the distributions Δ_{ij} of non-variable stars are consistent with the quoted rms uncertainties in the magnitudes m_{ij}. Assuming the uncertainties are uniform at epoch j, then the distribution of Δ_{ij}/σ_{ij} for non-variable stars at epoch j will have a standard deviation of unity.

Stars for which $|\Delta_{ij}| > 4\sigma_{ij}$ at any passband were identified as IR variables. The possibly variable stars were defined using the deviation that

416

L. A. Balona et al. (eds.), Pulsation, Rotation and Mass Loss in Early-Type Stars, 416–417.

TABLE I

Summary of number of variable stars

	J	H	K	L
# of stars with $\Delta/\sigma > 4$	10	15	21	13
Total # of stars	112	116	117	89
%	8.9	12.9	17.9	14.6

defines a sample in which there will be, on average, one non-variable event. This corresponds to a deviation of 3.2σ for this data sample. Hence, stars with $3.2\sigma_{ij} < |\Delta_{ij}| < 4\sigma_{ij}$ were identified as possible IR variables.

In this manner, the variable Be stars are: o Cas, BK Cam, 120 Tau, ω Ori, HR 2309, FV CMa, ω CMa, FW CMa, HR 2921, HR 3858, HR 4930, μ Cen, V767 Cen, χ Oph, 66 Oph, V3031 Sgr, V4024 Sgr, V923 Aql, HR 7983 59 Cyg, 31 Peg, π Aqr, 8 Lac, ϵ PsA, EW Lac, β Psc. The following stars are possible IR variables: γ Cas, ψ Ori, ζ Tau, HR 3135, κ Dra, η Cen, 48 Lib.

A more complete version of this analysis can be found in Dougherty and Taylor (1994) along with a discussion of the properties of the variations.

References

Coté, J. and Waters, L.B.F.M., 1987, *Astronomy and Astrophysics*, **176** , 93

Dougherty, S.M., Waters, L.B.F.M., Burki, G., Coté, J., Cramer, N., van Kerkwijk, M.H., and Taylor, A.R., 1994, *Astronomy and Astrophysics,submitted.*

Dougherty, S.M., and Taylor, A.R., 1994,*Monthly Notices of the RAS, submitted.*

STRUCTURE OF THE CIRCUMSTELLAR ENVELOPE OF
ψ PERSEI AT LARGE RADII

S.M. DOUGHERTY
John Moores University, Liverpool, England

and

A.R. TAYLOR
University of Calgary, Calgary, Canada

Abstract. The radiative transfer equation is solved for a generalised disc model of the circumstellar envelope around ψ Persei to investigate the nature of the long wavelength turndown in the continuum spectrum of Be stars. The flux density and source dimensions at 15 GHz rule out truncation of the circumstellar envelope around ψ Persei as the cause of the turndown.

The excess thermal continuum spectra of radio detected Be stars exhibit a turndown longward of the far-IR wavelength regime (Taylor *et al.*, 1990). This feature has not been observed in the continuum spectra of other hot stars. Several mechanisms have been proposed to account for this turndown, including truncation of the circumstellar envelope.

The far-IR region of Be star spectra have a power-law form which can be interpreted as arising from a plasma with a radial density distribution of the form $\rho \propto r^{-\beta}$, where β is the density index. For such a circumstellar plasma the optical depth along a line of sight z is given by

$$\tau_\nu(z) \propto \int_\zeta^\xi r^{-2\beta}\, dz,$$

where ζ and ξ are limits of integration determined by the geometry of the circumstellar plasma envelope, namely the radius of the disc, the inclination of the rotation axis of the star to the line of sight (i), and the half-opening angle of the disc (θ). Constraints on the model parameters θ and i can be determined from this model in conjunction with the observed minimum aspect ratio of 1.6 for the 15 GHz emission region (Dougherty and Taylor, 1992). In addition, the presence of shell lines *e.g.* Oegerle and Polidan (1984) implies $\theta + i \geq 90°$. It is found that $i > 67°$ for ψ Persei (Dougherty, 1993).

The best fit model (Figure 1) was determined by using a weighted minimum χ^2. The most striking feature of the fit is that to account for the spectral turndown requires that the disc model is truncated at a radius of $560 R_\odot$. This is much smaller than the observed lower limit to the radius of the emitting region of $1850 R_\odot$ (Dougherty and Taylor, 1992). A disc model with a radius of $1850 R_\odot$ gives radio fluxes that are too high. Clearly, a truncated disc model cannot account for both the radio flux densities and the

L. A. Balona et al. (eds.), Pulsation, Rotation and Mass Loss in Early-Type Stars, 418–419.
© 1994 *IAU. Printed in the Netherlands.*

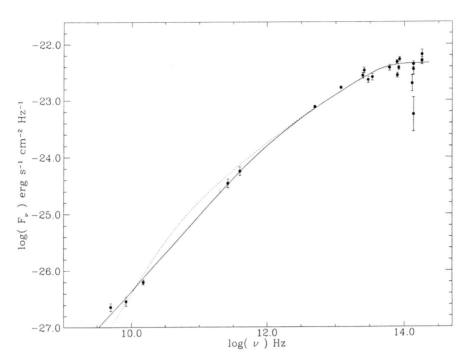

Fig. 1. Optimal fit of the truncated (dashed) and double density index (solid) models to the observed excess spectrum of ψ Persei. Clearly, the truncated model gives an unsatisfactory fit to the radio and millimeter observations, whereas the solid line represents an excellent fit at *all* wavelengths.

observed size of the radio emitting region. This rules out disc truncation as the mechanism causing the long wavelength turndown observed in ψ Persei.

The spectrum is best fit by an infinite disc model that has a change in the density index at a radius of $166 \pm 57 R_\odot$ (Figure 1). Clearly, a change in the density index at $\sim 170 R_\odot$ can account for both the radius of the 15 GHz emitting region and the long wavelength turndown in the excess spectrum. Re-combination of electrons and ions, a disc opening angle that is a function of radius, and acceleration at large distances from the underlying star are all mechanisms that could give rise to such a change in density index (Taylor *et al.*, 1990; Waters *et al.*, 1991)

References

Dougherty, S.M., and Taylor, A.R. 1992, *Nature*, **359**, 808
Dougherty, S.M. 1993, PhD thesis, University of Calgary.
Oegerle, W.R., and Polidan, R.S. 1984, *Astrophysical Journal*, **285**, 648
Taylor, A.R., Waters, L.B.F.M., Bjorkman, K.S., Dougherty, S.M. 1990, *Astronomy and Astrophysics*, **231**, 453.
Waters, L.B.F.M., van der Veen, W.E.C.J., Taylor, A.R., Marlborough, J.M., Dougherty, S.M. 1991, *Astronomy and Astrophysics*, **244** , 120

A HIDDEN CLASS OF BE STARS?

P.A. ZAAL
Kapteyn Astronomical Institute, Groningen
L.B.F.M. WATERS
Astronomical Institute 'Anton Pannekoek', Amsterdam
SRON Laboratory for Space Research, Groningen
J.M. MARLBOROUGH
Astronomy Department, University of Western Ontario, London, Canada

and

T.R. GEBALLE
Joint Astronomy Centre, Hilo, Hawaii, USA

1. Introduction

Recently, Brα and Brγ emission was detected in the infrared spectrum of the B0.2V star τ Scorpii, without noticeable emission in Hα (Waters et al. A&A **272**, L9-L12, 1993). Here we present simple HI recombination line calculations in the infrared and in the optical that demonstrate that there could be a class of B stars with low-density discs, a factor of 100 lower in density compared to normal Be stars, which may have escaped detection so far.

2. The disc model

We carried out a parameter study for 4 B stars of different T_{eff} to find out under which conditions a hot star with a circumstellar disc can have IR emission lines without obvious Hα emission. For the underlying photospheric continuum (from Kurucz, 1979) we used T_{eff}= 30000, 23000, 15000 and 12000 K and $R_* =$ 6, 4.3, 3 and 2.5 R$_\odot$ for resp. spectral type B0, B2, B5 and B8. We used a simple disc model which consists of a disc with a opening angle θ, with $\theta = 5°$, and a density distribution, $\rho(r) = \rho_0(r/R_*)^{-2.5}$ inside the disc. For mass continuity this corresponds to a velocity , $v(r) = v_0\sqrt{r/R_*}$. We adopted a Keplerian rotation, $v_\phi(r) = v_{\phi,0}\sqrt{R_*/r}$, for the disc where $v_{\phi,0} = 0.7 \cdot v_{br}$, with a breakup velocity, v_{br} of about 800 km/s for a B0V star.

The Hα line was calculated solving the equations of statistical equilibrium for the levels 1 to 4 for a gas consisting of pure H (Marlborough J.M., ApJ **156**, 135, 1969), taking into account the underlying photospheric absorption line. We further assume that the disc is isothermal at a temperature of $0.6 \cdot T_{eff}$. The IR recom-

420

L. A. Balona et al. (eds.), Pulsation, Rotation and Mass Loss in Early-Type Stars, 420–421.
© 1994 *IAU. Printed in the Netherlands.*

bination lines were calculated with an optically thin approximation. It calculates the line intensity of a volume of gas with density $\rho(r)$ and temperature T:

$$I_{approx} = EM \cdot \alpha_{line,eff} \cdot h\nu_{line}$$

With EM the emission measure of the emitting gas, $h\nu_{line}$ the energy of the line and $\alpha_{line,eff}$ the effective recombination coefficient for case B from Hummer and Storey (MNRAS, **224**, 801, 1987).

The approximate line intensity can be compared to numerical calculations. The difference between the two is mostly due to optical depth effects. The boundaries in density for our possible new class of Be stars are due to two facts. The upper limit is due to the emission feature which appears in Hα. One gets the already known Be stars. The lower limit is due to the fact that no line emission in the infrared is visible any more below that density. To enlarge the density range for detection of the disc in the IR we have to find the HI line in the infrared with the strongest line over continuum ratio. Between 2 μm and 50μm we find a maximum for the 10-09 HI line transition at 38.9 μm.

The line over continuum ratio of the IR line depends strongly on the inclination angle of the disc. A disc viewed edge on will give fewer detectable emission lines. The dashed lower limit and solid lower in figure 1 shows the difference between resp. pole on and edge on.

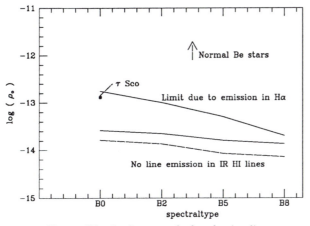

Fig. 1.   The density range for low-density discs.

3. Conclusions

We propose that τ Sco is a member of a new class of B stars with low-density circumstellar gas, possibly a disc. The HI (10-09) line at 38.9 μm is most sensitive to low density circumstellar gas. These stars will be great targets for ISO.

ENERGY DISTRIBUTION OF Be STARS

GOPAL C. KILAMBI

Department of Astronomy, Osmania University, Hyderabad - 500 007, India

1. Introduction

Be stars are defined to be non-supergiant early-type stars of spectral type B showing at times Balmer emission lines in their spectra. These stars often develop strong stellar winds considered to be variable in nature (Slettebak 1988) and have high rotational velocities compared to normal stars of similar spectral types. They also tend to show an excess amount of energy in the near- and far-infrared region compared to normal stars which is presumed to be due the surrounding material around the central star. Thus, the observed energy is a combination of that due to the stellar source and the surrounding material. Various attempts have been made to disentangle the stellar energy component from that of the circumstellar component in order to understand the nature, size and temperature of the envelope. These include:

a) Radius determination based on IR excess (Gehrz et al. 1974, Dachs and Hanuschik 1984; Waters et al. 1987),

b) Radius estimates from polarization and spectrophotometric data (Jones 1979),

c) Envelope dimensions derived from the width of shell absorption cores (Kogure 1969; Hirata and Kogure 1977),

d) Dachs et al. (1992) attempted to understand the physical properties, flow patterns and density distribution of the gas by a comparison of synthetic emission line profiles and empirical profiles measured for real Be stars.

Thus, the estimated disk radii are different for different authors due to the way the optical depth is defined. In this analysis, an attempt is made to estimate the photospheric temperatures, distances, the envelope size and temperature without any recourse to the definition of optical depth from the published energies in various passbands with a simple assumption that both the central star and the surrounding disk or envelope will be radiating like a blackbody and the decrement in the observed radiation strictly follows the inverse square of the distance.

2. Data and reduction

Published broadband magnitudes for 23 selected Be stars have been corrected for interstellar absorption and the corrected magnitudes have been converted into absolute flux units through the calibrations given by Johnson (1966) and in the IRAS Supplementary Catalogue. The observed energy distribution of each star is fitted with a representative blackbody tempera-

L. A. Balona et al. (eds.), Pulsation, Rotation and Mass Loss in Early-Type Stars, 422–424.
© 1994 IAU. Printed in the Netherlands.

ture by covering maximum number of observed energies. This temperature of the photospheric flux is thus estimated on the basis of visual inspection of the best fit between the observed and the predicted, covering as many passbands as possible while keeping in view the spectral type and luminosity classification of the star. The zero-point shift between the observed and the predicted allows us to estimate the distance to the star. Once we have established the flux emitted by the photosphere of the star in each passband through a blackbody analysis, the difference between the observed and the predicted flux could be attributed to the contribution by the circumstellar material. The distribution of this residual flux with wavelength allows us to estimate the temperature of the material. The relevant shift factor would lead to an estimation of the extension or spread of this material in terms of stellar radii. The parameters thus derived are given in Table I for all 23 stars and compared with other earlier estimates.

3. Results

The present analysis suggest that HD 5394, HD 22192, HD 50013, HD 63462 and HD 148184 have two component shell structure, while, HD 6811, HD 10516, HD 30076, HD 372023, HD 41335, HD 50138, HD 142983 and HD 217891 showed a single component structure. In the case of HD 20336, HD 24534, HD 35439, HD 83953, HD 86612, HD 91120, HD 120324, HD 138749, HD 142926 and HD 217675 no predominant shell structure has been noticed except some slight excess emission at longer wavelengths in spite of that they are classified as Be stars. The present analysis suggests that our estimated shell temperatures are lower than those of Waters et al. (1987). In addition, both Waters et al. and Andrillat et al. (1990) had suggested a predominant shell structure for HD 20336 and HD 35439, while the present analysis could not confirm this. Further investigations are necessary.

This research has been carried out under grant No. SP/S2/023/89 from the Department of Science and Technology, Government of India, which is gratefully acknowledged.

References

Andrillat, A., Jaschek, M, Jaschek, C. (**AJJ**): 1990, *Astronomy and Astrophysics* **84**, 11

Dachs, J., Hanuschik, R.W.: 1984, *Astronomy and Astrophysics* **138**, 140

Dachs, J., Hummel, R.W. (**DH**): 1992, *Astronomy and Astrophysics* **95**, 437

Gehrz, R.D., Hackwell, J.A., Jones, T.W.: 1974, *Astrophysical Journal* **191**, 675

Hirita, R., Kogure, JT.L.: 1977, *Publications of the ASJ* **29**, 477

Johnson, H.LO.: 1966, *ARA& A* **4**, 193

Jones, T.W.: 1979, *Astrophysical Journal* **228**, 787

Kastner, J.A., Mazzali, P.A. (**KM**): 1989, *Astronomy and Astrophysics* **210**, 295

Kogure, T.: 1969, *Publications of the ASJ* **27**, 165

Slettebak, A.: 1988, *Publications of the ASP* **100**, 770

Waters, L.B.F.M., Coté, J., Lamers, H.J.G.L.M. (**WCL**): 1987, *Astronomy and Astrophysics* **185**, 206

TABLE I

Estimation of derived stellar temperatures and distances, and temperature and radii of the shells

HD	Sp.T.	Temperature (K)				Distance (pc)			Temperature (K)			Radius(R_*)			
		Sp.T.	Pres. An.	DH[a]	KM[a]	M_V	Flux	KM[a]	Sh. 1	Sh. 2	WCL[a]	Sh. 1	Sh. 2	WCL[a]	AJJ[a]
24534	O9.5 V	31500	30600			462	417								
63462	B0 V	30000	29550	30000	28000	402	389		4000	1000		3.96	16.03	>3.0	
5394	B0 IV	29500	30200			245	223		3350	750	25000:	5.69	21.14	>5.6	
35439	B1 V	25400	25850			316	342	490				9.77		>3.6	1.31; 3.66
37202	B1 IV	24700	20000			174	183	135	1000			1.86		>4.8	1.01; 1.77
50013	B1.5 V	23700	23700	22500	25000	218	238		10000	400	11000		54.95	>5.3	
148184	B1.5 V	23700	24000	22500		149	140		3000	300	19000	5.37	47.86	>7.4	
10516	B2 V	22000	23000			108	139	205	1000		16000	15.14		>6.8	
41335	B2 V	22000	22000	22500	21000	239	277		2500			4.17		>4.5	1.66; 3.43
30076	B2 V	22000	22000		20000	378	482		3500			3.63			1.37; 6.00
120324	B2 IV-V	22000	25850	22500		111	185							>1.6	
20336	B2.5 V	20350	20500			214	274	275						>2.4	1.09; 1.72
142983	B3 IV	17800	15000			246	300	230	1250			9.44		>3.0	
86612	B4 V	17000	15000			297	321								
22192	B5 V	15400	20500			104	142	115	3350	750	12000	2.45	18.19	>6.2	1.70; 4.48
83953	B6 V	14000	13500	15500	16600	130	121								
138749	B6 V	14000	13350			99	96								
217891	B6 V	14000	13600		14800	117	111		600			10.23		>2.4	
217675	B6 III	14100	15000			111	153								
6811	B7 V	13000	13000			85	80		700			6.17		>1.8	
50138	B7.5 V	12450	13000			202	224		1100			64.56			
91120	B9 V	10500	10500			114	145								
142926	B9 V	10500	11000			129	120								

[a]See references

CONTINUOUS SPECTRA OF CIRCUMSTELLAR
ENVELOPES OF Be STARS

D. ROHE-KOTHS and J. DACHS

Astronomisches Institut, Ruhr-Universität Bochum, D-44780 Bochum, Federal Republic of Germany

Line emission in Be star spectra is accompanied by continuous emission both in the Balmer continuum and in the infrared spectral region, due to the same process that is responsible for Balmer line emission, i.e. to recombination radiation from ionized hydrogen in the extended circumstellar disks surrounding the hot central stars.

In the present study a grid of continuous spectra for model Be stars has been calculated. Computed model continua are then fitted to measured spectra, covering the wavelength range between 320 nm and 60 μm, for a total of 20 bright Be-type stars with spectral types between B0 and B7. For these stars, optical spectrophotometry in the wavelength range from $\lambda = 320$ nm to $\lambda = 850$ nm performed in 1981 or 1982 (Kaiser (1987), Dachs et al. (1989)) was combined with infrared broadband photometry in the J, H, K, L, M and N bands (λ 1.25...10.3 μm) obtained in 1982 (Dachs et al. (1988)) and with space-borne photometry collected in 1983 by the IRAS satellite at wavelengths 12 μm, 25 μm, and 60 μm (Beichman et al. (1988)). Measured Be star spectra were dereddened by applying the interstellar extinction curve taken from Cardelli, Clayton and Mathis (1989) and from Mathis (1990), and using appropriate values of the colour excess E^{is}_{B-V} of the interstellar reddening.

Model spectra were computed by adding continuous free-free and free-bound hydrogen recombination radiation from a circumstellar disk to photospheric fluxes from the central stars. Starting values for the effective temperatures, T_{eff}, and gravities g were assigned to program stars according to their MK spectral types as determined by Slettebak (1982), using the calibrations of Schmidt-Kaler (1982). Photospheric continuous fluxes for the central B stars were derived from the tables of Kurucz (1979) for wavelengths up to 20 μm. For $\lambda > 20$ μm, the central B star is assumed to radiate like a black body at temperature T_{eff}. The circumstellar disk is taken to be a rotating, isothermal plane-parallel slab with inner radius R_i equal to the stellar radius R_*, and with outer radius R_a, and to consist of pure hydrogen. Surface density N in the disk is assumed to decrease as N $\sim$ $r^{-\alpha}$ with distance r from the central star.

Theoretical models were fitted to measured Be star spectra by varying the following fit parameters: T_{eff}, E^{is}_{B-V}, the fraction x_e of disk radiation

L. A. Balona et al. (eds.), Pulsation, Rotation and Mass Loss in Early-Type Stars, 425–426.
© 1994 IAU. Printed in the Netherlands.

to total (photospheric + disk) radiation at the normalization wavelength of $\lambda = 656.3$ nm, the electron temperature in the disk, T_e, using $T_e = 10000$ K as a starting value, the outer radius of the disk, R_a, in units of the stellar radius R_*, the optical depth τ_o of the emitting disk at the inner radius R_i for the wavelength of $\lambda = 656.3$ nm and for electron temperature T_e, and the exponent α of the radial density law in the disk.

Values of the stellar radius R_* are obtained for the program stars from the compilation by Schmidt-Kaler (1982) according to their MK spectral types. The inclination angle i was derived from the values of v sin i (Slettebak (1982)) under the assumption that Be stars rotate at a uniform velocity taken to be $\sim$ 450 km/s for B0...B1e stars and $\sim$ 350 km/s for B2...B7 stars. For shell-type spectra, partial absorption of continuous photospheric radiation by the circumstellar disk is taken into account. Important criteria for the goodness of fits are the height of the Balmer discontinuity and the slopes of continuous spectra in the ultraviolet, optical and infrared regions of the spectra.

Preliminary model fits obtained for the 20 program stars show that the fraction of continuous envelope radiation at λ 656.3 nm typically ranges between 1% and 24% of the total fluxes of the Be star systems. This causes circumstellar reddening of $E_{B-V}^{cs} = 0^m00...0^m09$. Electron temperatures of the models are found between 8000 K and 12000 K. Outer envelope radii range from 6 to 40 stellar radii and optical depths τ_o at λ 656.3 nm from 0.05 to 4.5, being about unity on the average. Stars with weak line emission from their envelopes (μ Cen, η Cen) seem to have extended disks with small optical depths τ_o. Exponents α of the power law of surface density in the disk range from 1.4 to 2.6, most often lying between 1.4 and 1.8. Volume emission measures EM defined by

$$EM = \int_{V_{envelope}} N_i(r)N_e(r)dV$$

(with V = volume of the circumstellar envelope) were also calculated for the best-fit models for program stars and found to range between $6.3 \cdot 10^{58}$ cm^{-3} and $1.3 \cdot 10^{61}$ cm^{-3}.

Close correlations exist for program stars between the several characteristic measures of the strength of continuous emission in Be star spectra, viz. the fraction x_e of disk to total continuous radiation at λ 656.3 nm, the colour excess E_{B-V}^{cs} caused by circumstellar excess radiation, the logarithm of volume emission measure, log EM, and the intensity of Balmer emission as given by Hα emission line equivalent widths, $W_e'(H\alpha)$.

This work has been supported by the Deutsche Forschungsgemeinschaft under grants Da 75/13-1 and Da 75/13-2.

HIGH–PRECISION CONTINUUM RECTIFICATION *

Towards an Abundance Analysis of Be and Bn Stars

M. KOLB and D. BAADE

European Southern Observatory
Karl-Schwarzschild-Str. 2, D-85748 Garching, Germany

Abstract. We sketch a new method for the accurate flux calibration and normalization of stellar spectra. This is of particular importance for the analysis of rapidly rotating early–type stars. Some preliminary log g determinations by profile fitting of $H\gamma$ are presented.

1. Scientific Goals and Observations

Apart from the presence of emission lines in Be stars, the overall appeerence of optical spectra of Be and Bn stars seems to be indistinguishable. We have therefore started a model atmosphere analysis of 42 bright stars (20 Be, 18 Bn, and 4 O(e) stars). The observations were carried with the ESO 1.52m telescope and two different instruments: The Boller & Chivens spectrograph at a resolving power of 4,000 between 380 and 510 nm and the Echelec spectrograph at a resolving power of 30,000 between 400 and 480 nm.

2. The Rectification Problem and the Solution Adopted

The limiting factor of measurements in line profiles is not the signal-to-noise ratio or the spectral resolution but residual curvature of the continuum. This especially affects gravity determinations, which depend sensitively on the wings of Balmer lines, and abundance analyses of rapidly rotating stars. Furthermore the crowding of circumstellar features in Be stars makes a reliable continuum rectification very difficult. The often used method of interactive interpolation of continuum regions is not objective or physical and especially problematic for echelle spectra, if the orders with their relatively short wavelength coverage shall be merged. Therefore we have developed a new technique, which iterates on spectra of the same objects but different resolutions, starting with low resolution spectra of flux standard stars.

The key to overcoming the problems arising from the presence of spectral lines is to find a good model for the line spread function (LSF) of the tabulated flux standard spectra. Then, the observed standard spectra are convolved with that LSF and devided by the tabulated data, yielding a smooth instrumental response curve. Residual spikes originating from spectral features can be further reduced by an advanced filtering procedure. Finally the spectra of

---

* Based on observations at the European Southern Observatory, La Silla, Chile

L. A. Balona et al. (eds.), Pulsation, Rotation and Mass Loss in Early-Type Stars, 427–428.

the program stars are divided by the filtered curve, which yields spectra with a flux distribution closely resembling the tabulated data. This procedure can be iterated to flux–calibrate the spectra with the next higher resolution.

In order to approximate the instrumental profile we chose VOIGT–functions. Our experience shows that it is, in fact, *essential* to take the LORENTZ–component into account, since it approximates the instrumental straylight quite well, which appeared to be significant in all spectra analyzed by us (incl. the standards of HAMUY et al.: 1992, PASP 104, 533). The actual LSFs are computed by means of χ^2–fitting. In the general case it is possible to minimize the residuals of spectral and other features in the instrumental response curve by optimizing the VOIGT–parameters Gaussian σ and Lorentzian δ. Alternatively, the lines in comparison arc spectra can be replaced with "DIRAC-δ" functions and subsequently convolved with trial VOIGT profiles until an optimum match of the original spectrum is obtained.

3. Results

We have derived tertiary flux standard spectra of three bright southern stars (HR 718, HR 3454, HR 9087) with a resolution of 0.1 nm between 380 and 510 nm. They are available on request. Determinations of *log* g have been attempted for a subsample of our program stars by fitting NLTE, line–blanketed stellar atmosphere models of $H\gamma$ (K.BUTLER: 1993, priv. comm.). A simultaneous spectroscopic determination of T_{eff} from the same line woud be possible with our data, if model spectra would predict line cores more accurately. Some preliminary results are compiled in the following table:

HR	Name	MK	v sin i [km/s][1]	T_{eff}[K][2]	$log\ g_{ph}$ [3]	$log\ g_{sp}$ [4]
779	δ CET	B2IV	22	21,000	3.5	3.70
1679	λ ERI	B2IVne	345	23,000		3.70
1899	ι ORI	O9III	130	30,500		3.80
7121	σ SGR	B2.5V	230	18,000	4.1	4.00
8858	ψ^2 AQR	B5V	341	15,000	3.6	3.65

[1] Source: P.L. BERNACCA, M. PERINOTTO (1973), *A Catalogue of Stellar Rotational Velocities*, Contr. Oss. Astrof. Padova, 239, 250.

[2] Calculated from Strömgen [u–b] indices (B. HAUCK, M. MERMILLIOD: 1980, A&AS 40, 1) and the temperature calibration of R. NAPIWOTZKI et al.(1993, A&A 268, 653). We adopt a standard error of ±2000 K, because the calibration is established only for slowly rotating main sequence stars. In Be stars, [u–b] is furthermore affected by the disk.

[3] Derived from Strömgen β, the calibration of T.T.MOON, M.M.DWORETZKY (1985, MNRAS 217, 305) – established for $T_{eff} \leq 20,000$ K – and the correction term introduced by NAPIWOTZKI et al.(1993).

[4] This paper. The errors of our method intrinsically are ±0.05 dex but together with the uncertain temperatures amount to ±0.25 dex.

IR-RADIO SPECTRAL INDICES FOR Be STARS

H. CHEN and A. R. TAYLOR
University of Calgary, Canada

S.M. DOUGHERTY
John Moores University of Liverpool, UK

and

L.B.F.M. WATERS
SRON Laboratory for Space Research, The Netherlands

Abstract. Using published data and new data from mm observations, we calculate spectral indices $\alpha(12\mu\text{-}25\mu)$, $\alpha(25\mu\text{-}1.1\text{mm})$ and $\alpha(1.1\text{mm-2cm})$ for Be stars. The index $\alpha(25\mu\text{-}1.1\text{mm})$, obtained for 8 stars, shows two characteristics: 1) for "normal" Be stars the index decreases from earlier types towards later types, i.e., later type Be stars tend to have shallower spectra in the IR-radio region than earlier types; 2) the index for shell stars appears to have smaller values than that for normal Be stars with same spectral types.

1. Introduction

In an initial VLA survey carried out by Taylor et al. (1987), five Be stars were observed at λ=6 cm, resulting in the detection of one star, ψ Per, the only shell star in the program stars. In a second radio survey (Taylor et al. 1990, Taylor survey, hereafter), 21 Be stars were observed at λ=2 cm, among which five are shell stars and sixteen others are "normal" Be stars. Six stars were detected. The detection rate at radio is 60% and 19% for shell stars and "normal" Be stars, respectively. This may suggest that shell stars are more likely to be radio emitters than normal Be stars.

In Taylor survey, the distribution of program stars is similar to the frequency of Be star, i.e., it peaks at B2 and decreases towards both directions. However the detected stars do not show any preference to B2. In fact, the detection rate for the stars later than B5 is 3/7=43% relative to 3/14=21% for the stars earlier than B5. This suggest that stars of later types might have relatively stronger emission at radio wavelengths than stars of earlier types.

In this investigation, we re-examine IR, mm and radio spectral indices for Be stars to see if there is any difference between "normal" Be stars and shell stars and between Be stars of earlier and later spectral types.

2. Results

The infrared spectral indices, $\alpha(12\mu\text{-}25\mu)$, for 58 stars show no dependence on spectral types, although the values for earlier types tend to be more scattered than those for later types. Neither does it show dependence on

L. A. Balona et al. (eds.), Pulsation, Rotation and Mass Loss in Early-Type Stars, 429–430.
© 1994 *IAU. Printed in the Netherlands.*

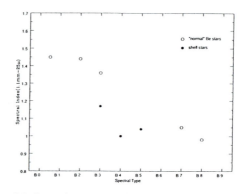

Fig. 1. Spectral index $\alpha(25\mu\text{-}1.1\text{mm})$ versus spectral types for eight stars.

whether the star is a shell star or a normal Be star. This is consistent with the fact that the distribution of progam stars in Taylor survey is similar to the relative frequency of Be stars, since the program stars in Taylor survey were chosen according to their flux in the IRAS data.

In Fig. 1, we plot the infrared-millimeter spectral index, $\alpha(25\mu\text{-}1.1\text{mm})$, for eight stars. The index shows two characteristics: 1) for "normal" Be stars the index decreases towards later types, i.e., later type Be stars tend to have shallower spectra in the IR-radio region than earlier types; 2) the index for shell stars appears to have smaller values than that for normal Be stars with same spectral types. The first characteristic suggests that the density in the envelopes of later type Be stars may drop less fast with the increasing distance from the underlying star than that for earlier types. Therefore, later type stars tend to emit more towards longer wavelengths relative to earlier type stars. This would explain why the detection ratio appears higher for later type stars in the Taylor survey. The second characteristic suggests that shell stars have more emitting material at larger distance from the underlying star than "normal" Be stars. As the result, one would expect that shell stars are more likely to be detected at radio, which is consistent with the result in Taylor survey.

This result is supported by the cm-mm spectral index $\alpha(1.1\text{mm}\text{-}2\text{cm})$ calculated for six stars. For "normal" Be stars, the index tends to have about the same value of 1.4, while the index for shell stars tends to be smaller and more scattered.

3. The References

Taylor, A.R., Waters, L.B.F.M., Lamers, H.G.L.M., Persi, P., & Bjorkman, K.S., *Mon. Not. R. ast. Soc.* **288**, 811 (1987).

Taylor, A.R., Waters, L.B.F.M., Bjorkman, K.S. & Dougherty, S.M., *Astr. Astrophys.* **231**, 453 (1990). **172**, 225 (1987).

PROBLEMS OF MODELIZATION OF CIRCUMSTELLAR MATTER OUT OF EQUILIBRIUM

J-P. J. LAFON and E. HUGUET
Observatoire de Paris-Meudon (DASGAL, URA CNRS D0335)
92195 Meudon Cedex (France)

1. Introduction

Circumstellar envelopes of young and evolved stars are responsible for many important phenomena concerning the exchange of matter, angular momentum, energy and maybe magnetic field between the core structure of stars and the interstellar medium. In particular, it is through them that matter enriched in heavy elements flows from evolved stars towards the interstellar gas, submitted to complex ordinary chemistry or photochemistry and condensation into solid particles.

Circumstellar envelopes are not a quiet medium : they are sites of complicated dynamics, chemistry, thermodynamics and energy exchanges, with typical values of basic parameters ranging in very large domains.

In envelopes of evolved stars, densities run from 10^{15} particles per cm^{-3} close to the star to a few particles per cm^{-3} far from it, with temperature of the orders of 2,000-3,000K close to the photosphere, 20,000K in the chromosphere, if any, and a few 10K in the outer parts.

Very different phases out of equilibrium are mixed in circumstellar envelopes; at least in their inner part, there is no simple symmetry : there may even be mixtures of phases in "partial" equilibrium (i.e. not in equilibrium with the other phases), cold and hot, optically thick and optically thin, dense and tenuous; moreover the phases can be distributed in layers with some (spherical for instance) symmetry or maybe in mixed clumps (Lafon, 1993).

According to likely theories, within 100 stellar radii the winds are highly structured at many different scales. However this is not obvious depending on the angular resolution. Indeed, due to the distance of closest stars, 100 stellar radii correspond to something like 0.1" whereas for early type stars this more likely corresponds to the milliarcsecond. This means that a resolution of a milliarcsecond could provide precise informations on the phenomena close to or on the surface of an evolved stars, but rather on the global structure of the envelope for an early type star (Lafon, 1993).

Finally constraints on the models describing the phenomena occuring within a few stellar radii from the star surface are fundamental for under-

431

L. A. Balona et al. (eds.), Pulsation, Rotation and Mass Loss in Early-Type Stars, 431–433.
© 1994 IAU. Printed in the Netherlands.

standing the physics and the structure of a zone close to the star, to which the farther parts of the envelope is very sensitive.

2. Two examples of phenomena out of equilibrium

2.1. DUST CONDENSATION

There is a controversy concerning the compatibility of dust with chromospheres observable around evolved stars : are chromospheres quenched by dust or is dust destroyed by chromospheric radiation? Or finally are dust and chromosphere simultaneously present?

Another point in discussion is the interaction of dust and shock waves (if any) in star envelopes : is dust nucleation and growth inhibited or stimulated by propagation of shock waves? In a similar way does dust make easier the dissipation of Alfvén waves?...and so on...

In any case, dust condensation and evolution occur obviously out of equilibrium and one should use cautiously quantities well defined only in case of equilibrium such as condensation temperatures or thermodynamical parameters of bulk material...(Lafon, 1991; Lafon and Berruyer, 1991).

2.2. RADIATIVE SHOCK WAVES : a summary of self-consistent non linear phenomena in media out of equilibrium

Strong radiative shock waves are an important phenomenon in atomic or/and molecular circumstellar atmospheres. Their structure is a summary of various media out of equilibrium in various ways.

The waves we talk about are propagating in highly collisional dense hydrogen atmospheres; investigations and modelization of shock waves have lead to amazing conclusions emphasized by recently available models (Huguet, Gillet and Lafon, 1992; Huguet, Lafon and Gillet, 1993a, b), as follows :

- The hydrodynamic discontinuity is included in a much larger consistent structure covering large distances on both sides.
- Behind the shock front there is no thermal equilibrium between electrons and heavy particles : the temperature can differ by orders of magnitude.
- The models are very sensitive to the presence or the absence of molecules which can still be present in gases at 100,000K.
- The medium is optically thin behind the front for the Lyman-α radiation, but at larger distances it is optically thick; before the front the medium absorbs the radiation in a "radiative precursor".
- The same structure exists for the Lyman continuum radiation, but over much larger distances, so that the Lyman-α dominated structure is included within the Lyman continuum dominated structure.

- The same structure exist also for the Balmer continuum, but the wake is optically thick at very much larger distances and the radiation crosses practically freely before the discontinuity.
- Specific unusual phenomena such as "three body recombinations" can compete with radiative recombinations in gases far from equilibrium of the populations of the energy levels, under very unusual and not well investigated ranges of parameters.
- Sites of line formation can be located in various place of such structures, in particular in places where the velocity deduced from usual spectroscopic analysis is much smaller than that of the shock front; accurate spectroscopic analysis would require good models (not still available!).

Thus, radiative shocks are an interesting "laboratory" of strange phenomena which can also be observed under other circumstances and may be the source of unexpected or misunderstood observations (in other atmospheric structures for instance).

3. The References

Huguet E., Gillet D., Lafon J-P. J. "Radiative shocks in atomic and molecular stellar-like atmospheres, V Influence of the excited level of the hydrogen atom : the precursor structure", Astron. Astrophys., **255**, 233, 1992

Huguet E., Lafon J-P. J., Gillet D. "Radiative shocks in atomic and molecular stellar-like atmospheres, VI Influence of the Lyman-α flux on the precursor structure", Astron. Astrophys., 1993a, in press

Huguet E., Lafon J-P. J., Gillet D. "Radiative shocks in atomic and molecular stellar-like atmospheres, VII Influence of the excited level of the hydrogen atom : the wake structure", Astron. Astrophys., 1993b, in press

Lafon J-P. J., Dust formation and evolution in circumstellar media, in "The infrared spectral region of stars", Proc. int. coll. held in Montpellier (France), 16-19 Oct. 1990, C. Jaschek and Y. Andrillat eds, 1991, Cambridge University Press.

Lafon J-P. J., Prospects for stellar Physics, Proc. ESA Coll. on "Target for space-based interferometry", Beaulieu (France), ESA SP-354, 265-269, 1993

Lafon J-P. J. and Berruyer N. " Mass loss mechanisms in evolved stars", Astron. Astrophys. Review, **2**, 249-289, 1991.

OPTICAL RESOLUTION OF Be STAR ENVELOPES

F. VAKILI, D. MOURARD AND P. STEE
OCA-Fresnel-GI2T, 06460 Caussols, France

Abstract. Probing the atmospheres of Be stars by means of Optical Long Baseline Interferometry (OLBI) is a method still in its infancy. Published observational results in this field can be counted on one hand. In this paper we introduce the principles of OLBI as they apply to observations of Be stars and review results from existing long baseline interferometers. Regular operation of OLBI is most effective when carried out in conjunction with standard techniques such as photometry and spectroscopy. With the next generation of synthetic arrays combining sub-milliarcsecond resolution and a spectral resolution better than 10000, OLBI should be able to address the direct detection of pulsation and magnetic fields.

1. Introduction

Optical resolution of early-type stars was pioneered by Hanbury Brown and collaborators in the 1960s with the Narrabri Stellar Intensity Interferometer in Australia (Hanbury Brown *et al.* 1974a). In an intensive study of the spectroscopic binary γ^2 Vel (WC8+O7), the Narrabri team measured the angular diameter of the C III-IV emission region surrounding the Wolf-Rayet component, which they found to be 5 times larger than the central star (Hanbury Brown *et al.* 1970). In another survey which combined interferometry and polarimetry, Hanbury Brown and co-workers set an upper limit of 6.5 percent on the variation of the angular size of β Ori in two orthogonal polarizations (Hanbury Brown *et al.* 1974b). From a simple model based on electron scattering in a hot corona, they were able to estimate an upper limit of $2 \times 10^{-5} M_\odot$ y^{-1} on the mass loss of this blue supergiant. More appropriate models for comparing predicted interferometric measurements to actual data were used to interpret the observations of the O star ζ Pup (Cassinelli & Hoffman 1975). The possibility of resolving the envelope of Be stars in the Balmer lines had already been formulated by Hanbury Brown *et al.* (1974b), who concluded that such observations demanded a much more sensitive instrument. By the time their observing programme stopped, the Narrabri group had already paved the way for future studies of extended atmospheres in early-type stars.

The invention of speckle interferometry by Labeyrie in 1970, and his successful operation of a prototype optical amplitude interferometer with physically separated telescopes in 1974, opened new perspectives for measuring the geometry of Be star envelopes (Labeyrie 1978). Since Be stars are generally quite distant, it is not surprising that speckle observations have only set upper limits on the apparent diameters of their envelopes in the Balmer lines (Blazit *et al.* 1977). Interestingly enough, a significant number

L. A. Balona et al. (eds.), *Pulsation, Rotation and Mass Loss in Early-Type Stars*, 435–447.

of spectroscopic binaries which include a B or Be component were resolved during systematic speckle observations (McAlister & Hartkopf 1988). β Cep and o And are examples of such multiple objects resolved with the Palomar 5-m telescope (Bonneau et al. 1980, Labeyrie et al. 1974). Besides the detection of multiplicity, patrol observations of B or Be stars using true imaging techniques up to the diffraction limit of large telescopes, enable us to predict episodes of close contact between the components. Optical long baseline interferometers can be used to study the nature and evolution of accretion disks, or the disks formed by tidal interactions in close binaries, as these disks usually give rise to emission-lines in the visible region.

2. Optical Long Baseline Interferometry and Be Stars

2.1. AN INTRODUCTION TO OLBI

OLBI can be defined as the art of synthesizing a giant optical telescope, of say 100-m diameter, from an array of much smaller apertures placed at comparable distances. In principle, the angular resolution of such an array attains the diffraction limit of a giant telescope, but the smaller light-collecting surface lowers the limiting magnitude. The theory and implementation of OLBI is similar to that of radio interferometry. The main difference lies in the nature of the noise (which is mostly photon noise), the dramatic effect of atmospheric turbulence, and the absence of phase-tracking amplifiers at visible wavelengths. Besides the difficult task of coherently combining primary beams in the presence of mechanical instabilities, atmospheric turbulence drastically complicates the operation of optical interferometers. At present, only about five interferometers are operational. A comparable number of aperture synthesis projects should start operating in the coming decade. Further information is available in review papers in more specialized colloquia (see for instance Robertson & Tango 1994).

An interferometer can be thought of as a spatial filter which samples the two-dimensional Fourier transform of the brightness distribution of the object of interest. Each sample is a complex quantity called the visibility which is a function of spatial frequency (u,v). The spatial frequency is defined as the ratio of the baseline to the mean wavelength at which the interferometer is operated. When N telescopes are operated together, $N(N-1)/2$ such spatial frequencies are measured simultaneously. This is often referred to as the instantaneous uv-plane coverage. If N is large enough, say $N \geq 20$ as in the case of the radio Very Large Array, a snapshot can almost be obtained from a straightforward Fourier transform of visibility data. If N is small, the domain of spatial sampling can be extended by using earth-rotation synthesis. With good uv-coverage, true images can be reconstructed using numerical inversion techniques. The dynamic range and detail in the reconstructed image depends both on the signal-to-noise ratio of measured visibilities and

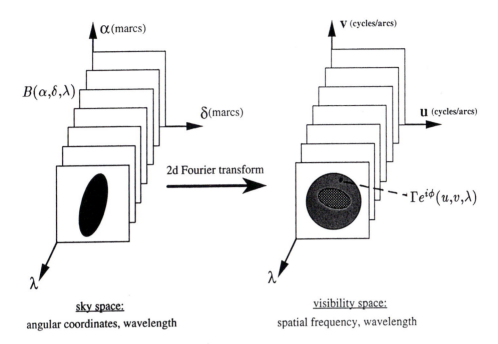

Fig. 1. Schematic representation of the brightness of a Be star as projected on the sky, $B(\alpha,\delta,\lambda)$, as a function of wavelength (left). For each λ-slice of B there corresponds a complex visibility map $\Gamma e^{i\phi}(u,v,\lambda)$ which is the two-dimensional Fourier transform of B.

the uv-plane filling factor achieved during a synthesis run. Until now, only interferometers with two telescopes have produced data on Be stars.

2.2. APPLICATION OF OLBI TO BE STARS

Most Be stars are notoriously variable with a morphology that depends on the wavelength across various scales of angular and spectral resolutions. Therefore, if OLBI is to be used for deducing morphological structures, it would be most efficiently operated in spectrally-dispersed light. By adding wavelength information to high angular resolution, the image of a Be star can be envisaged as a hypercube of brightness information which depends on the spatial coordinates, wavelength, and eventually polarization and time. A subset of such a representation is shown in Fig. 1.

What OLBI measures is the visibility function $\Gamma e^{i\phi}(u,v,\lambda)$ (right panel), and what we want to determine is the brightness distribution of the source $B(\alpha,\delta,\lambda)$ (left panel). The problem is that the sampling of the right cube is often sparse due to the limited spectral bandwidth and number of interferometric sub-apertures. Therefore, inversion from the measured right cube to the left cube is not straightforward. A natural way to interpret OLBI observations is to build the left cube from a physical model, to compute $\Gamma e^{i\phi}(u,v,\lambda)$ from it, and to iterate on the input parameters of B until Γ fits the observations. Since OLBI is a newcomer to the field of Be stars, this method of modelling is not familiar to most theoreticians, but hopefully the situation will change in the coming years.

Traditionally, a two-telescope interferometer can measure only the modulus of the Fourier transform of sky brightness and only provides dimensional information concerning the star of interest. If the interference pattern is dispersed by a spectrograph, one can simultaneously observe in as many spectral channels as can be resolved across the total bandwidth. In practice, the only limitation to this technique comes from the detector size and the signal-to-noise ratio. Thus we can compare the size of the program star at different wavelengths and detect chromatic effects. In addition, we can measure the relative phase of the interferometric signal at different Doppler shifts along an emission line compared with the continuum. To first order, the detected phase offsets, if any, correspond to differences of position on the sky. They tell us about the spatial distribution of iso-radial velocity regions in the envelope.

However, these methods should be used with caution. For instance, if the continuum object is partially resolved it does not provide a safe signal for calibrating the visibility across the line. Be-star models (Poeckert & Marlborough 1978; PM hereafter) estimate a significant contribution of the envelope to continuum flux due to free-free emission and to scattering of stellar photons by the envelope. Nevertheless, the extent of the envelope region producing this emission must lie close to the star. The corresponding visibility contribution of the envelope weakly contaminates the continuum visibility. Therefore differential calibration techniques between the line and continuum apply for the present range of baselines of operational interferometers (Stee & Araujo 1994).

3. Expected Angular Diameter of Be Star Envelopes from Spectroscopy

Operational optical interferometers are unlikely to resolve the photospheric diameter of even the brightest Be stars (Table 1). This becomes evident from a study of stellar angular diameters calculated from colour indices (Ochsenbein & Halbwachs 1982). Accordingly, all Be stars are smaller than 1 mas. Thus a factor 10 improvement in angular resolution, corresponding to kilometric baselines, is necessary for measuring stars up to the 6th visual magnitude. Though the construction and operation of such an instrument remains a real challenge, it would be highly desirable. For instance, the measurement of the angular diameters of both the Be photosphere and the envelope will provide a check on the effectiveness of spectroscopic methods which estimate the extent of the envelope from the separation between the peaks of emission profiles (Huang 1972). This method is widely used for deriving the inner and outer extent of Balmer and Fe II emitting envelopes (Hanuschik *et al.* 1988, Slettebak *et al.* 1992) with an uncertainty of about 20 percent (Jaschek & Jaschek 1993). The interaction between OLBI observations and spectroscopy would refine this method and might provide us with an independent means of inferring the distance of Be stars.

TABLE I

Expected angular diameters for photospheric and circumstellar envelopes of 8 northern hemisphere Be stars which have so far been observed by means of OLBI. The envelope extends in Hα and Hβ are given in mas where possible, and otherwise in stellar radii (r_*).

Star	Spec.type	m_v	Φ_*	$\Phi_{H\alpha}$	$\Phi_{H\beta}$
γ Cas	B0.5IV	2.47	0.45	8.6	4.2
ϕ Per	B1V	4.0	binary	$23.8(r_*)$	$13.8(r_*)$
ψ Per	B5III	4.2	?	$18.1(r_*)$	$10.5(r_*)$
η Tau	B7III	2.9	0.63	12.0	2.6
48 Per	B3V	4.0	?	?	$34.8(r_*)$
ζ Tau	B1IV	3.1	0.40	6.5	5.5
β CMi	B8V	2.9	0.63	10	4.2
o And	B6III	3.6	quadruple	$1.4(r_*)$	$0.9(r_*)$

4. Measured Angular Diameters of Be-Star Envelopes by OLBI

4.1. γ CAS

Attempts to measure Be stars can be traced back to the early 1980s with the observations of the I2T. Using a maximum baseline of 48 m and a visual method with 10–20 percent uncertainties, the I2T was able to set an upper limit of 0.9 mas on the continuum angular diameter of γ Cas (Vakili *et al.* 1984). In 1985, photoelectric observations of γ Cas which combined spatial and spectral resolutions were able to resolve the Hα emitting envelope with intermediate baselines of the order of 20 m (Thom *et al.* 1986). The use of the relative visibility phase between the red and blue wings of Hα enabled the I2T to find that the receding regions of the envelope were to the north of the star and the approaching regions were to the south. Furthermore, the I2T found that the envelope asymmetry must be smaller than 0.7 stellar radii at the time of observations.

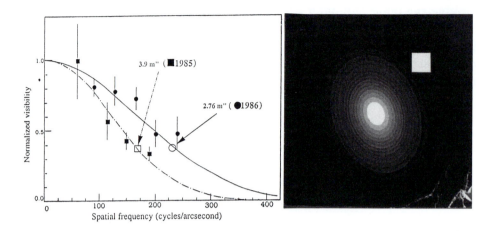

Fig. 2.   Results from OLBI observations of γ Cas. Left: the angular diameter of the Hα
emitting envelope measured by I2T. The size of the envelope, assumed to have a gaussian
limb-darkening law, was found to be about eight times larger than the photosphere and
most probably shrank in size within a one year interval from 1985–1986. Right: a major
contribution by Mark III observations to the understanding of γ Cas using earth-rotation
synthesis. The angular scale is given by the white square on the top right which corresponds
to 1 mas. North is to the top, east to the left. This Hα image was computed by the author
using recent results published by Quirrenbach *et al.* (1993).

There is a long-standing controversy concerning the mechanism which
produces Balmer line V/R variations. Such observations, which are of particular interest, were conducted on γ Cas in 1986. They indicated that the
envelope had shrunk by about 10 percent since 1985 (Percheron *et al.* 1994),
while V/R in Hα increased during the same period (Doazan *et al.* 1987).
If confirmed, this isolated observation (Fig. 2) is more consistent with the
pulsating envelope scenario than with the more popular precessing elliptical-
disk model (Tetling *et. al.* 1993). Unpublished observations of γ Cas in 1986
have also set an upper limit of 4 mas to the envelope in the Hβ line (Vakili
1990).

More recent observations of γ Cas by the GI2T (Mourard *et al.* 1989)
not only confirmed the 1985 results of the I2T, but opened new perspectives
for studying the kinematics of Be-star envelopes. With an improved spectral
resolution of as much as 0.2 nm, and with twice the angular resolution of
the I2T, Mourard and co-workers discovered morphological structures of the
envelope through the study of the modulus and phase variations of the visibility across Hα. As many as 7 contiguous spectral channels were obtained
across the Hα line (Fig. 3). The interpretation of these observations were
based on the PM model of γ Cas. The objection could be made that at the

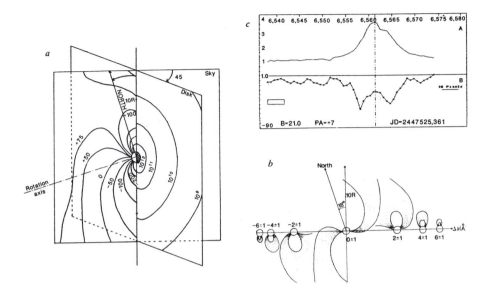

Fig. 3.   Interpretation of spectrally resolved OLBI observations of γ Cas according to Mourard *et al.* (1989). a) Iso-radial velocity contours in the sky plane obtained from the projection of the velocity law computed by Poeckert & Marlborough (1978). b) Inferred intensity maps corresponding to different iso-radial velocity areas of the envelope. c) Observed Hα profile (upper curve) and corresponding visibilities as a function of doppler-shift across the line (below).

time of the observations the emission profile of Hα was in a V>R regime, so that the PM model could not apply. However, this model was used merely as a guideline and the conclusion was that the measurements agreed well with a rotating envelope. The GI2T results showed that it is not easy to compare OLBI observations with any presently existing Be model. Unless model builders modify their code to produce emission line intensity maps from which visibilities can be predicted, OLBI observations have no straightforward interpretation. The situation will be different once sub-milliarcsecond maps are derived directly from multi-aperture OLBI.

A major step towards this target has been taken by the Mark III interferometer. Using earth-rotation synthesis exceeding 100° for some baselines, Quirrenbach *et al.* (1993) have clearly shown that the envelope of γ Cas is oblate. Their observations, made through a 1 nm Hα filter, were fitted to an elliptical Gaussian with an axial ratio of 0.74 and a FWHM major axis diameter of 3.2 mas (Fig. 2). This agrees well with the I2T and GI2T results and, besides establishing the disklike geometry of the envelope, also provides

Fig. 4. The Hα map of ζ Tau modelled as an elliptical Gaussian. The major axis is tilted by 59° NW. North is at the top and east to the left. The white square gives the angular scale of the map and corresponds to 1.2 mas on the sky.

the orientation of the rotational axis. The 21° NE tilt is in excellent agreement with the 20° positional angle obtained by polarimetry. The oblateness of γ Cas is indisputable: even raw visibilities bear a clear signature of the asymmetry (see Fig. 2 of Quirrenbach *et al.* 1993). However, the question of asymmetry with respect to the rotation axis remains open until continuous observations, which include the phase information, can be obtained.

4.2. ζ TAU

The first resolution of the Be shell star ζ Tau is another illustration of OLBI findings with the Mark III interferometer (Quirrenbach *et al.* 1994). From a Maximum Entropy map reconstructed in Hα, Quirrenbach and co-workers deduce a maximum elongation of (3.55 ± 0.33) mas, tilted by 59° NW (Fig. 4). By considering the perspective, they also determine the inclination angle of the rotational axis and estimate the rotation velocity at 280 km s^{-1}, which is 85 percent of the breakup limit.

5. Other Results

Besides the resolution of γ Cas and ζ Tau in Hα, observations of 5 other Be stars by Mark III are reported by Quirrenbach (these Proceedings). These results confirm the asymmetry of two other objects, ψ Per and φ Per, and indicate that 48 Per and η Tau are seen almost pole on. It also appears that envelope oblateness is clearly correlated with $v \sin i$ and that the orientations of the major axes are in excellent agreement with polarimetric observations. This supports the hypothesis of electron scattering in a disklike envelope.

Fig. 5. Intensity maps at various Doppler-shifts across the Hα emission line for a B1 Ve star and computed from a rotating expanding-envelope model. From left to right: envelope maps computed from $\delta v=-1000$ km s^{-1} to $\delta v=+1000$ km s^{-1} in steps of 400 km s^{-1} and averaged on spectral channels each 500 km s^{-1} wide. The rotation is 45° along a vertical axis. Such maps predicted by Be-models are highly desirable for direct interpretation of OLBI observations (Courtesy of Stee & Araujo).

Nevertheless, we must bear in mind that Mark III observations are based on visibility modulus information only: they cannot detect departures from axisymmetry.

Mourard *et al.* (1992) reported spectrally resolved OLBI observations of β Lyr at various orbital phases. Their data include the recording of the Hα, Hβ as well as the He I λ6678 emission lines obtained with interferometric baselines ranging from 16–51 m. There is no clear indication that the binary system is resolved. There is, however, a partial resolution of the Hα region which is centered on the continuum object, whilst the He I region remains unresolved.

Time-resolved spectroscopy of Pleione during a lunar occultation has been reported by Gies *et al.* (1990), who detect a time lag between the occultation of the continuum and the emission line. This is interpreted as an asymmetry of the envelope extending to 34–55 stellar radii which would be formed by gravitational influence of the binary companion discovered recently by speckle-interferometry.

6. Targets for the Future

6.1. THE TEMPORAL VARIATION OF BE STARS

The temporal variability of Be stars has different time scales and amplitudes. On one extreme, variations of a few minutes across the Hα profiles have been reported by Anandarao (these Proceedings). If these variations result from photospheric activities which propagate through the envelope, local structures of the latter should be modified and presumably become detectable by means of OLBI. However, the characteristic cycles are very short and the interferometric signal can only be integrated over a few minutes. Most probably such observations can only be achieved with synthesis arrays of the GI2T class which use sub-apertures of the 1–2 m class. On the other extreme, variations of a few years in V/R are generally attributed to one-armed density waves propagating in nearly Keplerian disks (Okazaki, these Proceedings) as opposed to pulsating envelopes (Doazan 1987). Spectrally resolved OLBI can address this question directly by tracing the variations of the envelope size and correlating it with V/R.

6.2. POLARIZATION AND CHROMATIC EFFECTS

OLBI also allows high spatial resolution observations in polarized light (Vakili 1981). Simultaneous observations in two different polarizations permit the detection of magnetic structures and phenomena such as electron scattering in the envelope.

Observations with high spectral resolution over a large bandwidth can be used to probe the circumstellar structure. Observing in two bands centered on Hα and Hβ enables us to disentangle the flux contribution of the star, the free-free emission of the envelope, and the photospheric flux scattered by the envelope. Other phenomena such as starspots or differential rotation could potentially be detected with spectrally resolved OLBI. However, these observations demand an order of magnitude improvement in angular resolution and sensitivity.

6.3. DIRECT DETECTION OF NRP

The direct detection of nonradial pulsations (NRP) can be envisaged using two different methods. Using differential speckle or long baseline interferometry we can locate the two-dimensional spatial distribution of iso-radial velocity regions on the stellar disk. This would allow the detection of tesseral as well as sectorial modes (Vakili & Percheron 1991). Rough estimates of signal to noise ratio (Petrov, private communication), indicate that high degree modes, say $l=-m=12$, are observable with interferometers using 2-m class telescopes. Another approach consists of placing the array of telescopes to suit the spatial frequency of the pulsations that we wish to detect. As argued by Balona (1990), NRP must modulate at some level the photospheric disk. The spatial filter which is the interferometer can be tuned to the scales of the NRP which we want to detect. The presence of a visibility signal would signify the detection of an NRP mode. Although such observations are not forseeable in the immediate future, it is important to bear them in mind.

6.4. TOWARDS INTERFEROMETRIC MODELLING OF BE-STAR ENVELOPES

The pioneering work at OCA and the achievements of Mark III clearly demonstrate that direct constraints provided by OLBI are applicable. Several optical aperture synthesis arrays will start operating in the coming decade (Robertson & Tango 1994). Many ad hoc or physical models have been reported during this symposium. Theoreticians take note: as with photometry, polarimetry, and spectrometry, models for Be-star envelopes should provide observable parameters for interferometry!

Acknowledgements

The authors are indebted to P. Lawson for his careful reading of the manuscript and to L. Balona for his generous help during the editing. We acknowledge A. Quirrenbach and D. Bonneau for having kindly provided us with data from Mark III and speckle-interferometric observations.

References

Balona, L.: 1990, *Mon. Not. Roy. Astr. Soc.* **245**, 92.
Blazit A., Bonneau, D., Koechlin, L. and Labeyrie, A.: 1977, *Astrophys. J. Let.* **214**, 79.
Bonneau, D., Blazit D., Foy, R. and Labeyrie, A.: 1980, *Astron. Astrophys. Suppl.* **42**, 185.
Cassinelli, J. and Hoffman, J.P.: 1975, *Mon. Not. Roy. Astr. Soc.* **173**, 789.
Doazan, V., Rusconi, L., Sedmak, G., Thomas, R.N. and Bourdonneau, B.: 1987, *Astron. Astrophys.* **182**, L25.
Doazan, V.: 1987, in Slettebak, A. and Snow, T.P., eds., *IAU Colloq. 92: Physics of Be stars*, Cambridge Univ. Press: Cambridge, 384.
Gies, D.R., McKibben, W.P., Kelton, P.W., Opal, C.B. and Sawyer. S.: 1990, *Publ. Astr. Soc. Pacific* **100**, 1601.
Hanbury Brown, R., Davis, J., Herbison-Evans D. and Allen, L.R.: 1970, *Mon. Not. Roy. Astr. Soc.* **148**, 103.
Hanbury Brown, R., Davis, J. and Allen, L.R.: 1974a, *Mon. Not. Roy. Astr. Soc.* **167**, 121.
Hanbury Brown, R., Davis, J. and Allen, L.R.: 1974b, *Mon. Not. Roy. Astr. Soc.* **168**, 93.
Hanuschik, R.W., Kozok, J.R. and Kaiser, D.: 1988, *Astron. Astrophys.* **189**, 147
Huang S.S: 1972, *Astrophys. J.* **171**, 549
Jaschek, C and Jaschek, M.: 1993, *Astron. Astrophys. Suppl.* **97**, 807.
Labeyrie, A.: 1978, *Ann. Rev. Astron. Astrophys.* **16**, 77.
Labeyrie, A., Bonneau, D., Stachnik, R.V. and Gezari. D.Y.: 1974, *Astrophys. J. Let.* **214**, 79.
McAlister, H. and Hartkopf, W.I.: 1988, *Second Catalog of Interferometric Measurements of Binary Stars*, CHARA Contribution No. 2.
Mourard, D., Bosc, I., Labeyrie, A., Koechlin, L. and Saha, S.: 1989, *Nature* **342**, 520.
Mourard, D., Bonneau, D., Blazit, A., Labeyrie, A., Morand, F., Percheron, I., Tallon-Bosc, I., Vakili, F.: 1992, in McAlister, H.A. & Hartkopf, W.I., eds., *IAU Colloq. 135, Complementary Approaches to Double and Multiple Star Research*, ASP Conf. Series, 32
Ochsenbein, F and Halbwachs, J.L.: 1982, *Astron. Astrophys. Suppl.* **47**, 523.
Percheron, I., Rabbia, Y. and Vakili, F.: 1994, *Astron. Astrophys.* submitted.
Poeckert, R. and Marlborough, J.M.: 1976, *Astrophys. J.* **206**, 182.
Poeckert, R. and Marlborough, J.M.: 1978, *Astrophys. J.* **220**, 940 (PM).
Quirrenbach, A., Hummel., C.A., Buscher, D.F., Armstrong, J.T., Mazurkewich, D. and Elias, N.M.: 1993, *Astrophys. J. Let.* **416**, 25.
Quirrenbach, A., Buscher, D.F., Mazurkewich, D., Hummel, C.A. and Armstrong, J.T.: 1994, *Astron. Astrophys.* submitted.
Robertson J.G. and Tango, W.J.: 1994, *IAU Colloq. 158, Very High Angular Resolution Imaging*, Kluwer: Dordrecht.
Slettebak, A., Collins, G.W. and Truax, R.: 1992, *Astrophys. J. Suppl.* **81**, 335.

Stee, P. and Araujo, X.F.: 1994, *Astron. Astrophys.* submitted.

Tetling T.J.H., Waters L.B.F.M., Persi P. and Dunlop S.R.: 1993, *Astron. Astrophys.* **270**, 355.

Thom, C., Granes, P. and Vakili, F.: 1986, *Astron. Astrophys.* **165**, L13.

Vakili, F.: 1981, *Astron. Astrophys.* **101**, 352.

Vakili, F.: 1990, *PhD. Thesis*, University of Nice

Vakili, F., Granes, P., Bonneau, D., Noguchi, M. and Hirata, R.: 1984, *Publ. Astr. Soc. Japan* **36**, 231.

Vakili, F. and Percheron, I.: 1991, in Baade, D., ed., *Rapid Variability of OB-Stars: Nature and Diagnostic Value*, ESO: Garching, 77.

Discussion

Gies: In ϕ Per has the binary companion been resolved?

Quirrenbach: The Mark III observations have not seen the signature of the companion in the continuum data and we concluded that probably $\Delta m \geq 2.5$.

Vakili: ϕ Per's companion was once resolved by speckle-interferometric observations on the Palomar 5-m telescope in 1975. On JD=2442643.5 Bonneau (private communication) estimated a separation of $\rho = 20$ mas, $\theta = 330°$ and $2 < \Delta m < 3$.

Smith: We are looking forward to the day when you can resolve the inner radius of the disk to see whether it is distinct from the stellar disk.

Vakili: Even the brightest Be stars are smaller than 1 mas and will require interferometric baselines exceeding a few hundred meters. One might address these questions by using Differential Speckle Interferometry introduced by J. Beckers. In principle DSI attains sub-milliarcsecond resolutions even with monolithic telescopes of the 10-m class.

Sareyan: Is it true that with existing techniques any asymmetry in the Hα envelope would appear as symmetrical in the image reconstruction?

Quirrenbach: Reconstructed images from Mark III observations are based on visibility modulus data from earth-rotation synthesis. To get a final answer on asymmetries, phase data are clearly needed.

Friedjung: Do stars with elliptical images tend to have shell episodes and do they tend to have a higher $v \sin i$ than those with circular images?

Vakili: Ellipicities determined for 6 bright northern Be stars by Quirrenbach and co-workers correlate with the degree of polarization that one would expect from equatorially-flattened disks. The correlation between the degree of polarization and $v \sin i$ is now well established (e.g., Poeckert & Marlborough 1976). Shell features are not permanent in Be stars and interferometric observations should be repeated as often as possible to track any variation in the shape of the envelope.

Hirata: Can spectrally-resolved interferometry provide any information on the circular velocity field, i.e., Keplerian or otherwise?

Vakili: Mourard *et al.* (1989) have actually done this on γ Cas some 4 years ago with the GI2T. Other interferometers under construction such as the Big Optical Array will also observe in dispersed light. In order to obtain the exact dynamics of the envelope, models should provide the intensity map of iso-radial velocity areas at a given Doppler shift along the emission profile. Predicted visibilities could thus be directly compared to interferometric measurements and vice-versa.

Harmanec: Would an unrecognized radial velocity of say, the Hα line, fool your image reconstruction?

Quirrenbach: In addition to the 1 nm-wide filter, we have also used a 10 nm-wide filter centered on Hα. The results of the two filters agree with each other in all cases.

Vakili: If the Hα extends beyond the spectral width of the narrow filter, the corresponding iso-radial velocity areas are rejected and the reconstructed image is incomplete. This is obviously true for Be stars with large projected rotational velocities, especially when the emission is strongly asymmetric as in ζ Tau.

Arsenijevic: Could you please make some comments on the necessity to have additional information, i.e. to use different techniques simultaneously (photometry, spectroscopy, polarimetry, etc.)?

Vakili: In principle, all these techniques can be combined on an optical interferometer by observing in polarized and spectrally dispersed light. Nevertheless, simultaneous observations by more well-developed techniques are required for calibrating the data from long baseline interferometry. Most of all, they can monitor the observational strategies of interferometry, for instance when active episodes occur on target Be stars.

CANDIDATES OF BE STARS FOR GI2T
INTERFEROMETRY

R. HIRATA

Department of Astronomy, Kyoto University, Kyoto, 606-01, Japan

Abstract. Eighteen bright Be stars have been observed polarimetrically and line- photometrically for searching for candidates for a direct measurement of the envelope diameters of Be stars by GI2T. I found that o Cas and 28 Tau are suitable in addition to γ Cas.

Mourard et al.(1989) have succeeded in the direct measurement of the envelope diameter of γ Cas at Hα. The GI2T has a potentiality which brings us direct information on the envelope structure around Be stars on a milliarcsecond scale. The candidates for GI2T must satisfy the following two conditions, in addition to a large stellar diameter: 1) the envelope projected on the sky is located near the north-south direction, or pole-on stars, 2) the Hα emission is strong.

I listed up 18 Be stars which are accessible by GI2T and whose stellar diameters are expected to be larger than about 0.3 milliarcsec(mas), assuming the envelope radius is at least three times larger than its stellar radius, i.e., 1 mas. Table 1 shows the list of the candidates.

Observations have been done, using the 8-channel photo-polarimeter attached to the 91-cm telescope at the Dodaira station of National Astronomical Observatory, since the winter season of 1988. The broad-band filters were employed with the effective wavelengths of 0.36, 0.42, 0.455, 0.53, 0.64, 0.69, 0.76 and 0.88μm. The accuracy is better than 0.05% in general. In November 1989, I introduced the Hα filters with FWHF of 260Å and 56Å, centered on 6563Å. Direct calibration of Dodaira Hα index was done for common Be stars through quasi-simultaneous spectroscopic observation with the Thomson CCD attached to the 60-cm telescope at the Ouda station, and from Kyoto University in 1989 October-December season. Both polarimetry and Hα photometry were repeated to examine their variability since then.

Very fortunately, the intrinsic polarization angles θ_* for eleven stars in Table 1 have been determined by Poeckert et al.(1979), and Poeckert and Marlborough (1976, 1979). For these stars, I made a consistency check, using data from us and from the literature, and found no contradiction except κ Dra. The time variation was examined for a direct estimation of θ_*. New values of θ_* were determined for three stars, including κ Dra. Regarding the residuals, the polarization degree is too small for further analysis. The last column in Table 1 shows the mean equivalent width of Hα (negative in emission). The mark 'v' indicates a gradual variation or large variation with time.

L. A. Balona et al. (eds.), Pulsation, Rotation and Mass Loss in Early-Type Stars, 448–449.
© 1994 *IAU. Printed in the Netherlands.*

As for candidates for GI2T observation, I pose the following two criteria: 1) the envelope is directed within 30° from north ($60° < \theta_* < 120°$) or pole-on ($Ve \sin i <200$km s^{-1}, and 2) the Hα emission is strong, say, < -20Å.

Then, o Cas and 28 Tau are the best candidates in addition to γ Cas. β CMi and η Tau are possible candidates from their large stellar angular diameters, though their directions on the sky are unclear. o And and ζ Oph could be candidates when they show emission, though the former is a quadruple system.

References

Mourard, D., Bosc, I., Labeyrie, A., Kochelin, L., and Saha, S.: 1989, *Nature* **342**, 520

Poeckert, R., Bastien, P., and Landstreet, J. D.: 1979, *Astronomical Journal* **84**, 812

Poeckert, R., and Marlborough, J. M.: 1976, *Astrophysical Journal* **206**, 182

Poeckert, R., and Marlborough, J. M.: 1979, *Astrophysical Journal* **220**, 940

TABLE I

Object stars and Results

name	sp.type	V	$Ve \sin i$	distance	angular diameter	θ_*	source	$<W\alpha>$
		mag	km s^{-1}	parsec	mas	degree		A
o Cas	B5IIIe	4.5	255	210	.27	78	present	−24.4v
γ Cas	B0IVe	2.2	260	230	.39	105	PM79	−25.0v
ϕ And	B7Ve	4.3	75	140	.36	80	present	+1.8
ϕ Per	B2Vep	4.1	505	210	.26	26	PBL79	−38.3v
ψ Per	B5Ve	4.2	375	170	.34	44	PBL79	−30.3
17 Tau	B6IIIe	3.7	215	125	.43	?		+7.5
23 Tau	B6IVe	4.2	285	125	.35	?		−6.4
η Tau	B7IIIe	2.9	210	125	.72	?		−1.7
28 Tau	B8Ve	5.1	345	125	.28	73	PBL79	−25.9v
48 Per	B3Ve	4.0	230	110	.39	137	PBL79	−19.0
ζ Tau	B4IIIpe	3.0	320	160	.43	35	PM76	−14.2
ν Gem	B6IIIe	4.2	225	190	.34	34	PBL79	+1.8
β CMi	B8Ve	2.9	285	40	.73	?		+0.4
κ Dra	B6IIIpe	3.9	230	160	.41	15	present	−16.0
θ CrB	B6Vnne	4.1	385	100	.34	?		+9.2
ζ Oph	O9.5Vn	2.6	385	170	.49	133	PBL79	+9.4
o Aqr	B7IVe	4.7	305	130	.30	27	PBL79	−12.0
o And	B6IIIpe	3.6	320	115	.48	111	PBL79	+6.5v

SEVEN Be STARS RESOLVED BY OPTICAL
LONG-BASELINE INTERFEROMETRY

A. QUIRRENBACH

NRL/USNO Optical Interferometer Project and USRA

1. Introduction

The recent progress in optical interferometry has made the direct study of the circumstellar matter around Be stars possible. The Hα emission region around γ Cas has been resolved with the I2T (Thom et al. 1986), GI2T (Mourard et al. 1989) and MkIII (Quirrenbach et al. 1993) instruments; the results support the basic picture of a rotating disk-shaped envelope (e.g. Poeckert and Marlborough 1978). Nearly spherical geometries can be ruled out. In this paper, first results from MkIII observations of a small sample of Be stars (see Table I) will be presented.

2. MkIII Observations of Be Stars

The observations were carried out on 13 nights between September 27 and November 27, 1992, using the variable north-south baseline of the MkIII interferometer, operated by the Remote Sensing Division of the Naval Research Laboratory (NRL). The instrument has been described by Shao et al. (1988); details of the data reduction and calibration procedures are given by Mozurkewich et al. (1991). For the Be star observations, baseline lengths ranging from 4.0 to 31.5 m were used. Two filters, 1 and 10 nm wide, were centered on the Hα emission line, while a 25 nm wide filter centered at 550 nm was used to measure the continuum for comparison.

None of the observed stars was strongly resolved at 550 nm, as expected from estimates of the photospheric diameters (0.35 to 0.75 mas). In the Hα line, however, all objects were clearly resolved, and deviations from circular symmetry are evident for some stars even in the raw data (e.g. γ Cas, see Quirrenbach et al. 1993). Model fits to the data in the narrow Hα channel with an elliptical Gaussian are presented in Table I; for β CMi the amount of data was insufficient to constrain the fitting parameters well, so a circular Gaussian was used. It should be noted that the contribution of photospheric emission to the flux in the Hα channel will systematically bias the fitted major axis a towards smaller values. This bias decreases with increasing E_α. The values of r derived for φ Per, ψ Per, and ζ Tau may actually be upper limits, since the resolution is not sufficient along the minor axes.

3. Discussion

The disk diameters (FWHM of the Gaussians) a correspond to 2 – 10 photospheric diameters; they increase to 3 – 12 photospheric diameters, if the bias to a mentioned above is taken into account. The two stars with the smallest values of E_α, which also have the latest spectral types, have the smallest disks.

The sky brightness distributions for γ Cas, φ Per, ψ Per, and ζ Tau are clearly not circularly symmetric; this proves that near-spherical models for the geometry

450

L. A. Balona et al. (eds.), Pulsation, Rotation and Mass Loss in Early-Type Stars, 450–451.

TABLE I

The MkIII observing list. The $H\alpha$ excesses E_α are from Coté and Waters (1987), $v \cdot \sin i$ from Uesugi and Fukuda (1982), and polarization position angles χ from Clarke (1990) and Poeckert et al. (1979). The right part of the table gives Gaussian model fits to the interferometer data, with formal uncertainties. The models have three free parameters: major axis a, axial ratio r, and position angle of the major axis ϕ.

Star	HR No.	V	E_α [Å]	$v \cdot \sin i$ [km s^{-1}]	χ [°]	a [mas]	r	ϕ [°]
γ Cas	264	2.47	22.7	260	-68	2.89 ± 0.02	0.77 ± 0.02	24 ± 2
ϕ Per	496	4.07	57.6	505	26	2.36 ± 0.17	0.47 ± 0.05	-64 ± 5
ψ Per	1087	4.23	43.2	375	44	2.78 ± 0.21	0.54 ± 0.07	-37 ± 11
η Tau	1165	2.87	11.9	210		1.57 ± 0.05	0.98 ± 0.06	(48)
48 Per	1273	4.04	24.2	230	-43	2.40 ± 0.52	0.86 ± 0.18	81 ± 43
ζ Tau	1910	3.00	26.9	320	33	3.55 ± 0.33	0.30 ± 0.02	-59 ± 4
β CMi	2845	2.85	11.5	285		1.54 ± 0.03	$-$	$-$

of Be star shells are not adequate. Under the assumption of axial symmetry, lower limits for the inclinations i can be derived from $i \geq \arccos r$. The values obtained for ϕ Per, ψ Per, and ζ Tau suggest that these stars are seen almost edge-on, while γ Cas has an intermediate inclination $i \approx 45°$. η Tau and 48 Per are either intrinsically more spherical, or viewed more face-on; the latter possibility seems more plausible and is consistent with the relatively low values of $v \cdot \sin i$.

The polarization of Be stars is generally attributed to Thomson scattering in the circumstellar disk; this explanation predicts that the polarization position angle χ is perpendicular to the equatorial plane. In agreement with this expectation, we find that for the four stars which are significantly elongated the position angle ϕ of the major axis is indeed perpendicular to χ. For ψ Per, our value of ϕ agrees with the one found by Dougherty and Taylor (1992) in the radio regime on a $\sim$ 40 times larger scale.

Acknowledgements

This work has been done in collaboration with Drs. J.T. Armstrong, D.F. Buscher, N.M. Elias, C.A. Hummel, and D. Mozurkewich. I thank the Alexander von Humboldt foundation for support through a Feodor Lynen fellowship.

References

Clarke, D.: 1990, A&A 227, 151
Coté, J., and Waters, L.B.F.M.: 1987, A&A 176, 93
Dougherty, S.M., and Taylor, A.R.: 1992, Nature 359, 808
Mourard, D., Bosc, I., Labeyrie, A., Koechlin, L., and Saha, S.: 1989, Nature 342, 520
Mozurkewich, D., et al.: 1991, AJ 101, 2207
Poeckert, R., Bastien, P., and Landstreet, J.D.: 1979, AJ 84, 812
Poeckert, R., and Marlborough, J.M.: 1978, ApJ 220, 940
Quirrenbach, A., et al.: 1993, ApJ, in press
Thom, C., Granes, P., and Vakili, F.: 1986, A&A 165, L13
Uesugi, A., and Fukuda, I.: 1982, Revised Catalogue of Stellar Rotational Velocities, Kyoto

7. OB STELLAR WINDS

THE WIND-COMPRESSED DISK MODEL

J.E. BJORKMAN
Univ. of Wisconsin, Dept. of Astronomy
Madison, WI, 53706, USA

Abstract. We discuss the effects of rotation on the structure of radiatively-driven winds. When the centrifugal support is large, there is a region, at low latitudes near the surface of the star, where the acceleration of gravity is larger than the radiative acceleration. Within this region, the fluid streamlines "fall" toward the equator. If the rotation rate is large, this region is big enough that the fluid from the northern hemisphere collides with that from the southern hemisphere. This produces standing shocks above and below the equator. Between the shocks, there is a dense equatorial disk that is confined by the ram pressure of the wind. A portion of the flow that enters the disk proceeds outward along the equator, but the inner portion accretes onto the stellar surface. Thus there is *simultaneous* outflow and infall in the equatorial disk. The wind-compressed disk forms only if the star is rotating faster than a threshold value, which depends on the ratio of wind terminal speed to stellar escape speed. The spectral type dependence of the disk formation threshold may explain the frequency distribution of Be stars. Observational tests of the wind-compressed disk model indicate that, although the geometry of the disk agrees with observations of Be stars, the density is a factor of 100 too small to produce the IR excess, Hα emission, and optical polarization, if current estimates of the mass-loss rates are used. However, recent calculations of the ionization balance in the wind indicate that the mass-loss rates of Be stars may be significantly underestimated.

1. Introduction

Rapid rotation has an important effect on the structure and lives of many stars. There is gravity darkening at the equator according to the von Zeipel theorem. Rotation affects the internal structure, which changes the non-radial pulsation modes, as well as the subsequent evolution of the star. If there is a magnetic field, angular momentum is transported by the stellar wind, causing the star to "spin down". The most extreme effects of rotation are seen in the Be stars. These stars, which are among the most rapidly rotating stars, are thought to have a dense equatorial disk in their circumstellar envelope (see the review by Waters elsewhere in these Proceedings). However, the connection between rotation and the formation of this equatorial disk is poorly understood. Since rotation and mass-loss affect the structure and evolution of massive stars, and since it leads to the formation of equatorial disks, it is important to understand the effects of rotation on the structure of the stellar wind.

2. Structure of Rotating Winds

In this section we review recent results concerning the two-dimensional structure of the winds from rotating stars. In particular, we discuss the Wind-Compressed Disk (WCD) model of Bjorkman & Cassinelli (1992, 1993) as

L. A. Balona et al. (eds.), *Pulsation, Rotation and Mass Loss in Early-Type Stars*, 455–468.

well as the time-dependent hydrodynamics simulations of Owocki, Cranmer & Blondin (1993).

In a rotating 2-D axisymmetric model, the streamlines are not radial, but instead bend toward the equator due to the centrifugal and Coriolis forces. To determine the structure of the wind, we must solve the fluid equations and find the location of the resulting streamlines.

For simplicity we assume a steady-state isothermal wind. With these assumptions, the continuity and momentum equations are

$$\nabla_j \left(\rho v^j \right) = 0 \,, \tag{1}$$

$$\nabla_j \left(\rho v^j v_i \right) = -a^2 \partial_i \rho + \rho F_i \,, \tag{2}$$

where ∇ is the covariant derivative, ρ is the fluid density, v is the fluid velocity, a is the isothermal sound speed, and F is the external force per unit mass. These equations are quite difficult to solve; however, an enormous simplification occurs in the supersonic portion of the flow.

2.1. FLUID EQUATIONS IN THE SUPERSONIC LIMIT

Consider the forces acting on the fluid. For an axisymmetric geometry, the pressure gradient only has r- and θ-components. Although the θ-component is large at the stellar surface (to enforce hydrostatic equilibrium), it drops rapidly beyond the sonic radius, r_s. The other forces are gravity and radiation, which are central forces. Thus, beyond the sonic point, there are no external torques, so both the θ- and ϕ-components of the velocity are determined by angular momentum conservation. This implies that v_θ and v_ϕ are $O(V_{rot}R/r)$, where R is the stellar radius. Typically for an early-type star, the rotation speed, V_{rot}, is highly supersonic. So, as long as $r \gg r_s$ and $r \not\gg R$, all three velocity components are highly supersonic.

Note that the left hand side of the momentum equation (2) is $O(v^2)$, but the pressure gradient on the right hand side is $O(a^2)$. As long as all three velocity components $v_r, v_\theta, v_\phi \gg a$, then we may completely ignore the pressure gradient. If there are no pressure forces, there are no interactions between the individual fluid particles. This implies that the streamlines are free particle trajectories corresponding to the external forces. To find the location of the streamlines, we simply integrate Newton's equations of motion using gravity and radiation for the forces.

2.2. MOTION IN THE ORBITAL PLANE

Much can be learned about the location of the streamline by recalling that gravity and radiation are central forces. Therefore, the *total* angular momentum is constant along a streamline. Consequently, the streamline lies in a plane (shown in Fig. 1) that contains the initial radius and velocity, V_0.

Fig. 1 shows two trajectories labeled (a) and (b), that correspond to different initial conditions. Trajectory (a) has a slow initial acceleration and

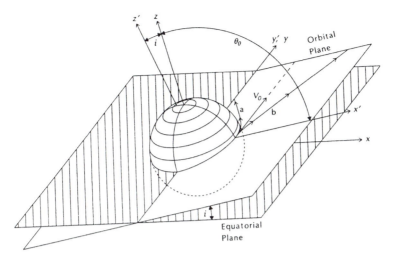

Fig. 1. Orientation of the orbital plane for a streamline originating at a polar angle θ_0. The streamline labeled (a) is a case with a high rotation rate and the streamline labeled (b) denotes a low rotation rate. (Figure from Bjorkman & Cassinelli 1993).

occurs when there is a large rotation rate. Trajectory (b) has a fast initial acceleration and occurs when there is a slow rotation rate. Note that as trajectory (a) wraps around the star, it has a decreasing altitude, z, and eventually crosses the equator. Conversely, trajectory (b) deflects outward and has an increasing altitude.

The curvature of the streamline depends on the forces and is most easily understood in the non-rotating reference frame. In this frame, there are only two forces, and each is in the radial direction; gravity points inward, and the radiation force points outward. To produce a net force with a negative z-component, we must have $F_{\mathrm{grav}} > F_{\mathrm{rad}}$. Thus, the equator-crossing trajectory (a) corresponds to initial conditions where the force of gravity exceeds the radiation force, and trajectory (b) occurs when the radiation force is larger than gravity.

2.3. FORCES IN ROTATING WINDS

The location where the radiation force exceeds gravity depends on the subtle interaction of the radiation force with the velocity gradient, dv_r/dr. In the orbital plane, the r-component of the momentum equation is

$$v_r \frac{\partial v_r}{\partial r} = -\frac{a^2}{\rho} \frac{\partial \rho}{\partial r} + F_{\mathrm{grav}} + F_{\mathrm{rad}} + \frac{v_\phi{}^2}{r} . \tag{3}$$

The last term on the right hand side is the centrifugal force, so the velocity gradient, dv_r/dr, is determined in the *rotating* reference frame. To maintain an outward flow, a line-driven wind must constantly accelerate to higher velocities, so that there is always a supply of unattenuated stellar photons.

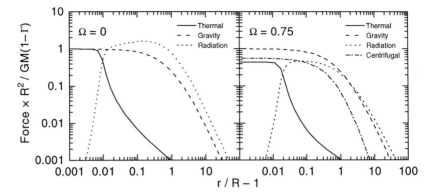

Fig. 2. Forces vs. radius in a Friend & Abbott (1986) 1-D equatorial rotating wind model that has $v_\infty/v_{esc} = 1.4$. Shown are a non-rotating case, $\Omega \equiv V_{rot}/V_{crit} = 0$, and a rapidly rotating case with $\Omega = 0.75$. (Figure from Bjorkman & Cassinelli 1993).

Since the radiation force depends on dv_r/dr, the velocity adjusts until the radiation force maintains a positive dv_r/dr, and the velocity increases monotonically.

Fig. 2 compares the forces for rotating and non-rotating winds. In the non-rotating case, $\Omega \equiv V_{rot}/V_{crit} = 0$, thermal pressure supports the flow out to the sonic radius, $r_s \approx 1.01R$. Beyond the sonic point, the thermal pressure support is negligible; therefore, the radiation force must increase until it is larger than gravity (so that dv_r/dr is positive). Note that the radiation force exceeds gravity at all locations beyond the sonic point. In the rapidly rotating case, $\Omega = 0.75$, most of the support is instead from the centrifugal force. When the thermal pressure support is lost at the sonic point, the radiation force again increases to supply the missing force, but because of the large centrifugal support, the required amount is smaller than gravity. The centrifugal support falls as $1/r^3$ (much slower than the thermal pressure) and it is not until the centrifugal support is lost that the radiation force finally exceeds gravity at about $3R$. Thus there is a region between the sonic point and about $3R$ where gravity is larger than the radiation force. Within this region, the streamlines fall toward the equator, and if the outer radius of this region is large enough, the streamlines attempt to cross the equator.

2.4. STREAMLINES

To build a model for the entire 2-D structure of the wind, we calculate the shape of the streamlines as a function of initial latitude on the surface of the star. If the star is not rotating, then the streamlines are entirely in the radial direction. On the other hand, if the star is rotating, then the streamlines fall toward the equator between the sonic point and the location where the radiation force exceeds gravity (see Fig. 3). Near the pole, the

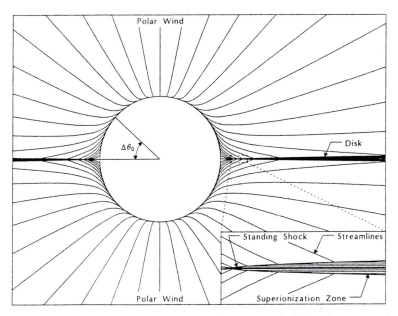

Fig. 3.   Diagram of the stellar wind and wind-compressed disk. Shown are the wind streamlines, which fall toward the equator. The expanded view shows the standing shocks that form above and below the disk. (Figure from Bjorkman & Cassinelli 1993.)

rotation velocity is small, so the streamlines are radial. But for streamlines closer to the equator, the rotation velocity is higher, the region where gravity exceeds the radiation force is larger, and the streamlines fall farther before turning in the radial direction. If the equatorial rotation rate of the star, V_{rot}, is above a threshold value, V_{th}, then for latitudes less than $\Delta\theta_0$, the streamlines attempt to cross the equator (see Fig. 3). This equator-crossing latitude, $\Delta\theta_0$, is approximately given by $V_{rot} \sin \Delta\theta_0 \sim V_{th}$.

When the streamlines cross the equator, they collide with the streamlines from the opposite hemisphere of the star. Streamlines cannot cross because the density diverges. Instead, the increase in density causes a large pressure gradient, and since the flow velocity (perpendicular to the equator) is supersonic, a pair of shocks form above and below the equator. The pressure at the equator must balance the ram pressure of the wind, so between the shocks there is a dense equatorial disk. Additionally, the shock temperature is large enough (a few 10^5 K) that the disk is bounded by a thin superionization zone (shown in Fig. 3). Note that this disk forms only when the star is rotating faster than the equator-crossing threshold.

The disk formation (equator-crossing) threshold depends on the ratio of the terminal speed of the wind to escape speed of the star, v_∞/v_{esc}. This is because a faster wind implies a larger radiative acceleration, which decreases the size of the region where the streamlines fall toward the equator. To compensate, the stellar rotation rate must be increased. Thus the disk formation threshold increases with increasing v_∞/v_{esc}.

2.5. TIME-DEPENDENT HYDRODYNAMICS

The Wind-Compressed Disk (WCD) model developed by Bjorkman & Cassinelli (1993, hereafter BC) is only valid in the supersonic region of the flow, and thus requires initial conditions at the sonic point. To obtain these initial conditions, BC assume that the subsonic expansion is in the radial direction, i.e., θ is constant and $v_\theta = 0$ for $r < r_s$. Another approximation BC employ is to assume a shape for the WCD shock surface, because the actual shape depends on the detailed dynamics of the disk.

To assess the validity of the WCD approximations and to examine in detail the dynamics of the disk, Owocki, Cranmer & Blondin (1993, hereafter OCB) developed a 2-D time-dependent numerical simulation of the wind from a rotating star. Aside from properly including shocks and gas pressure, OCB also included an oblate lower boundary condition that accounts for the rotational distortion of the star. Starting with a wind that is initially spherically symmetric, OCB find that, after about 50000 s, the time-dependent solution relaxes to a steady-state solution with a thin equatorial disk (see Fig. 4).

The qualitative appearance of the disk agrees quite well with that predicted by BC. The thickness of the disk is about 3° in latitude (BC predicted $0°\!\!.5$), and the disk density is about two orders of magnitude higher than the density at the pole, which is somewhat lower than predicted by BC. Interestingly, a weak disk persists even at rotation rates below the rotation threshold predicted by BC.

At first, quantitative comparison of the velocity of the wind (before entering the disk) did not agree with the analytic results of BC. The discrepancy arose because BC integrated their streamline trajectories starting at the stellar surface, $r = R$, instead of the sonic radius, $r = r_s \approx 1.01R$. Surprisingly, this small correction lowers the θ-component of the velocity by about a factor of two. After correcting this, BC's analytic approximations are in good agreement with OCB's numerical results (see Fig. 12 of Owocki *et al.* 1993); however, many details (mostly concerning the properties of the disk) are somewhat different in the numerical simulations.

There are two fundamental differences between OCB's results and the predictions by BC. Firstly, the disk is not detached from the stellar surface (compare Figs. 3 and 4). Secondly, there is a stagnation point in the disk. Exterior to the stagnation point, the disk material flows outward. Interior to the stagnation point, the material falls back onto the stellar surface. Thus there is simultaneous outflow and infall in the disk.

In the previous sections, we examined the structure of a radiatively-driven rotating wind and discussed the conditions whereby an equatorial disk forms.

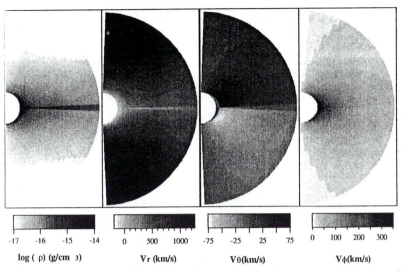

Fig. 4. Time-dependent numerical simulation of the wind-compressed disk for a B2.5 V star with $V_{rot} = 350$ km s^{-1}. Shown (from left to right) are the density and the r-, θ-, and ϕ-velocity components. (Figure from Owocki *et al.* 1993).

We have seen that the formation of a thin disk is inevitable, *if* the terminal speed of the wind is slow enough, *and if* the star is rotating fast enough. We suspect that disk formation in rotating winds is a general feature that occurs whenever there is the combination of rapid rotation and slow initial acceleration. In the next section, we explore the possibility that wind-compressed disks may be responsible for the Be star phenomena.

3. Application to Be Stars

To establish that Be stars might have wind-compressed disks, we must first determine if the stars are rotating faster than the disk formation threshold.

3.1. ROTATION THRESHOLD

The disk formation threshold depends on the ratio of the wind terminal speed to stellar escape speed. Fig. 5 shows both the observed and theoretical values of the wind terminal speed ratio as a function of main-sequence spectral type. The observed and theoretical values mostly agree for O stars; however, for late B stars, the observed terminal speeds are significantly lower than the theoretical terminal speeds. The observed terminal speeds of B stars are estimated from the edge-velocities of the C IV and Si IV line profiles (K.S. Bjorkman 1989). Typically these profiles are quite weak, so it is quite likely that the observed B star terminal speeds are systematically underestimated. On the other hand, there are many uncertainties in the the-

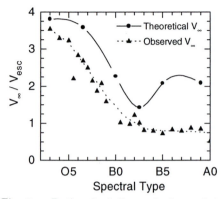
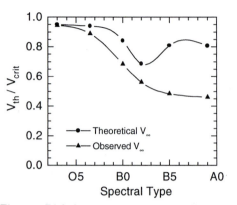

Fig. 5. Ratio of wind terminal speed to stellar escape speed vs. spectral type. The solid curve is obtained from theoretical calculations of the terminal speed, and the dashed curve is from observations. (Figure from Bjorkman & Cassinelli 1993.)

Fig. 6. Disk formation threshold vs. spectral type. The curve with filled circles is the threshold obtained using the theoretical values of the terminal speed, and the curve with triangles employs the observed terminal speeds.

oretical calculations. At this time, it is unclear which terminal speeds are more reliable for B stars.

Using the terminal speeds shown in Fig. 5, we find the disk formation threshold, shown in Fig. 6, as a function of spectral type. Note that these (and all subsequent) results contain the correction to the streamline locations found by OCB that is discussed in section 2.5. Assuming that the winds of B stars are radiatively-driven, Fig. 6 indicates that any B star that rotates faster than the rotation threshold will have a thin equatorial disk. Since observations of Be stars seem to indicate that they possess a dense equatorial disk (see the review by Waters in these Proceedings), one is tempted to conclude that wind-compressed disks may be responsible for the Be phenomena. Note that the rotation threshold has a minimum at B2, if one uses the theoretical terminal speeds. This minimum may qualitatively explain the frequency distribution of Be stars, i.e., why Be stars are most common at a spectral type of B2.

To investigate more quantitatively whether or not the WCD model can explain the Be phenomena, we must determine the properties of the disk. For a concrete example, we present results for a B2 V star with a mass-loss rate of 10^{-9} $M_\odot$ yr^{-1} and a terminal speed ratio $v_\infty/v_{esc} = 1$.

3.2. SHOCK TEMPERATURE AND DISK DENSITY

The WCD shock temperature is determined by the shock velocity, which is approximately the θ-component of the velocity of the wind when it enters the disk. The shock temperature, shown in Fig. 7, is typically a few 10^5 K, which is large enough to produce C IV and Si IV by collisional ionization.

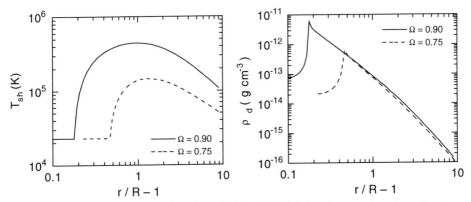

Fig. 7. Shock temperature as a function of radius for two values of the rotation rate, Ω. Note the increase in the maximum shock temperature as the rotation rate increases.

Fig. 8. Disk density as a function of radius for two values of the rotation rate, Ω. Note that the slope of the curve indicates that the density falls as $(r/R - 1)^{-3}$.

Thus the disk is bounded above and below by a thin superionization layer. Note that the maximum shock temperature depends on the stellar rotation rate, so more rapidly rotating stars will produce higher ionization states in the WCD shocks. In addition to the shocks that produce the disk, there is also an accretion shock (due to the disk inflow) in the equator at the stellar surface. The disk infall velocities are typically a few hundred km s^{-1}, so the accretion shock can have temperatures of order 10^6 K and may produce soft X-rays. Note that the temperature due to the accretion shock is not shown in Fig. 7. Since all of these shocks can produce superionization, they may be responsible for the excess superionization of Be stars compared to normal B stars (Grady, Bjorkman & and Snow 1987; Grady et al. 1989).

The disk density may be estimated by equating the gas pressure in the disk with the ram pressure of the wind entering the disk. Assuming that the shocked material that enters the disk cools to the same radiative equilibrium temperature as the stellar wind (OCB's numerical simulations indicate that the shock cooling length is thinner than the disk), we find $\rho_{\text{disk}} \approx \rho_{\text{wind}}[1 + (v_\theta/a)^2]$, where the right hand side is evaluated in the wind just prior to the shock that forms the disk.

The disk density is shown in Fig. 8. Note that the disk density is $\rho = \rho_0(r/R - 1)^{-n}$, where $\rho_0 \approx 10^{-13}$ g cm^{-3} and $n \approx 3$. Using the slope of the IR free-free continuum excess, Waters, Coté & Lamers (1987) estimated values of n in the range of 2–3.5; however, the required disk density is $\rho_0 \approx 10^{-11}$ g cm^{-3}. Thus the WCD model predicts the correct radial distribution of material, but if a polar mass-loss rate of 10^{-9} $M_\odot$ yr^{-1} is assumed, it produces a density that is about two orders of magnitude too small.

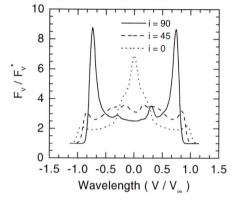 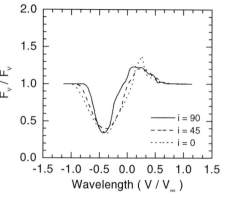

Fig. 9. Hα line profile produced by a wind-compressed disk for three different inclination angles, i. The polar mass-loss rate of 2×10^{-7} $M_\odot\,\mathrm{yr}^{-1}$ was chosen to produce a peak intensity (relative to the continuum) that is typical of a strong Hα emission line produced by a Be star.

Fig. 10.   C IV line profile produced by a wind-compressed disk for three different inclination angles, i. The polar mass-loss rate of $\dot{M}q = 5 \times 10^{-10}$ $M_\odot\,\mathrm{yr}^{-1}$ was chosen to produce an absorption depth (relative to the continuum) that is typical for a Be star.

3.3. OBSERVATIONAL PREDICTIONS

Similarly, the optical linear polarization and Hα emission produced by the disk are too small. The optical polarization of a typical Be star is a few tenths of a per cent, but the wind-compressed disk for our example star produces only about 0.004 per cent. Since the polarization is linearly proportional to the disk mass, the disk must contain about 100 times more material. Preliminary calculations of the Hα emission are shown in Fig. 9 (the small bumps are numerical artifacts). The edge-on ($i = 90°$) profile shows more or less symmetric, double-peaked structure (characteristics that are often observed in Be stars); however, to produce as much emission as is shown requires a polar mass-loss rate of 10^{-7} $M_\odot\,\mathrm{yr}^{-1}$, which is again a factor of 100 times larger than our example star.

The IR excess, Hα, and optical polarization observations all indicate that the disk density must be larger. To find out whether there is too little shock compression or instead too little mass-loss from the star itself, we have examined synthetic UV line profiles for the WCD model. The UV lines are sensitive to the material in the polar portion of the flow, so they provide a measure of the polar mass-loss rate. Fig. 10 shows a WCD model C IV line profile. The polar mass-loss rate required is only $\dot{M}q = 5 \times 10^{-10}$ $M_\odot\,\mathrm{yr}^{-1}$, where q is the ionization fraction. This agrees with Snow's (1981) mass-loss rates as well as the mass-loss rate used for our example star, *if* C IV and Si IV are the dominant ionization states in the wind (so that $q \sim 1$).

Recently, MacFarlane (elsewhere in these Proceedings) has calculated NLTE ionization fractions for B stars. He includes an X-ray source that matches the emission measure inferred from *ROSAT* observations. Assuming an X-ray luminosity, $L_X = 10^{29}$ erg s^{-1}, and temperature, $T_X = 10^6$ K, for our example star, he finds that Si V and C III are the dominant ionization states in the polar regions and that the ionization fractions of Si IV and C IV are of order 0.1. Additionally, the ionization fractions are sensitive enough to the density that it is quite likely that there are ionization gradients from the pole to equator. With these considerations, it appears that the mass-loss rates of Be stars could be underestimated by an order of magnitude or more, depending on the X-ray luminosity and temperature.

This is still one order of magnitude less than required for producing the IR excess, Hα emission, and optical polarization observations with a wind-compressed disk. However, increasing the mass-loss rates would reduce the radiative line-driving force, and might decrease the initial acceleration of the wind. If this is the case, then the streamlines would fall closer to the stellar surface, the θ-component of the velocity would be larger, and there would be more compression of the disk. Furthermore, the higher disk density would decrease the expansion velocity of the disk, which in turn increases the total mass of material in the disk.

4. Concluding Remarks

We have seen that rapid rotation of a B star leads naturally to the formation of a wind-compressed disk. Whether or not this disk causes many of the Be phenomena remains to be seen. It can only explain Be stars if there is a mechanism that increases the mass of the disk. Some possibilities are: 1) the mass-loss rate of the underlying star might be larger than previously thought, 2) the velocity in the disk might be smaller if the radiation-driving forces have been over estimated, and 3) a weak magnetic field might transfer angular momentum to the disk, stop the infall of material, and instead store the material in a Keplerian disk.

References

Bjorkman, J.E. and Cassinelli, J.P.: 1992, in L. Drissen, C. Leitherer and A. Nota, eds., *A.S.P. Conf. Series, Vol. 22, Nonisotropic and Variable Outflows from Stars*, A.S.P.: San Francisco, 88.

Bjorkman, J.E. and Cassinelli, J.P.: 1993, *Astrophys. J.* **409**, 429 (BC).

Bjorkman, K.S.: 1989, Ph.D. Thesis, University of Colorado.

Friend, D.B. and Abbott, D.C.: 1986, *Astrophys. J.* **311**, 701.

Grady, C.A., Bjorkman, K.S. and Snow, T.P.: 1987, *Astrophys. J.* **320**, 376.

Grady, C.A., *et al.*: 1989, *Astrophys. J.* **339**, 403.

Owocki, S.P., Cranmer, S.R. and Blondin, J.M.: 1993, *Astrophys. J.* submitted (OCB).

Snow, T.P.: 1981, *Astrophys. J.* **251**, 139.

Waters, L., Coté, J. and Lamers, H.: 1987, *Astron. Astrophys.* **185**, 206.

Discussion

Hummel: How can you get symmetric Hα line profiles from an expanding disk?

Bjorkman: There is infall as well as outflow in the disk, and the maximum infall velocity is about the same as the maximum outflow velocity. This produces more or less symmetric absorption.

Hubert: Can you explain the presence of shell lines in Be stars with moderate $v \sin i$?

Bjorkman: I agree that it is difficult to explain shell stars with a thin disk for two reasons. Firstly, unless the inclination angle of the star places the observer within the opening angle of the disk, a thin disk can cover at most half of the solid angle of the star. So, even if the disk has infinite optical depth, it cannot produce profiles that are more than 50 per cent black. If the observer is instead within the opening angle of the disk, then only the disk material at large radii completely covers the stellar disk. Secondly, the shell lines are narrow and at low velocity. This implies that they would have to be formed only near the stagnation point in the disk—not at large radii. Perhaps shell stars, which are only a small fraction of Be stars, are formed by a completely different mechanism, or, since some Be stars change between B, Be, and Be-shell phases, perhaps the mechanism that creates the variability forces the disk into different configurations at different times.

Percy: Given that there is now a mechanism for producing a disk, what is the role of non-radial pulsations (NRP's), which are found in most Be stars? Do the NRP's enhance the Be phenomena, could the NRP's be caused by the disk infall, or are the two unrelated?

Bjorkman: If NRP's increase the mass-loss rate from the star, they will increase the disk density and enhance the Be phenomena. Whether or not they could supply the extra mass-loss needed by the WCD model is unclear. Looking at the disk connection to NRP's, I suppose the disk infall could act as a small noise source which might excite NRP's, but the mass involved in infall is tiny compared to the mass involved in the pulsations and it is doubtful that the disk infall would be "tuned" to any particular frequency.

Smith: Of course the devil lies in the details. There are a few problems with the actual numbers. Firstly, Ω for Be stars is less than about 0.5; very few Be stars show $\Omega > 0.7$. Secondly, as you point out, there is a problem with the densities of the disk. Third, the episodic appearance of Be outbursts and disks is even more important. I would like to suggest that all these problems together do not negate the WCD. They merely argue strongly that a time-dependent "something else" is needed to make the WCD model work.

Bjorkman: First of all, Owocki's simulations show a weak disk even when $\Omega = 0.5$, which is less than the equator-crossing threshold of the WCD model. This indicates that the disk formation thresholds are somewhat too large. If you believe that Owocki's results support the basic WCD scenario, then many (if not all) Be stars will have a wind-compressed disk, whether we like it or not. The real question is whether the disk densities are large enough to explain the Be phenomena. I would suggest that your "something else" is the mechanism that increases the mass of the disk. Finally, as far as time variability is concerned, there are many possibilities including: time-dependent mass-loss from the star, changes in the ionization balance that affect the driving forces in the wind, and stability of the disk itself.

Harmanec: Firstly, I found your model quite convincing and acceptable; however, since one can find B absorption stars (usually classified as Bn) with $v \sin i > 300$ km s^{-1}, why are all rapidly rotating stars not Be stars? Secondly, since you find infall in the inner part of the disk, it becomes even more difficult to observationally distinguish accretion disks from disks created by outflow.

Bjorkman: Remember that the disk formation threshold depends on the spectral type of the star, but more fundamentally on the ratio v_∞ / v_{esc}. Before concluding that any given star is rotating faster than the disk formation threshold, one must also determine the terminal speed of its wind. If it turns out that some of these Bn stars are rotating faster than their threshold, then a disk will be present. This would indicate that something beyond the mere presence of the disk is required for making a Be star. As far as distinguishing between accretion disks and wind-compressed disks, I agree that it would be useful to find an observational indicator that differentiates the two. Certainly there is more than one way to make a disk, and some Be binaries must have accretion disks.

Lafon: One should be very cautious when trying to observationally distinguish between disks with or without infall, because the shock is probably subject to instabilities—both sides of the strong shock have opposing flow at large velocities, which probably produces two-stream instabilities.

Bjorkman: The question of the stability of the disk is very interesting. Firstly, if the disk is unstable, then the instability itself could cause the notorious variability of Be stars. Secondly, the instability would likely impose a very complicated multiple shock structure that propagates outward through the disk and thus alters the geometry and density of the disk. This in turn affects the radiation driving forces in the disk, which changes the disk velocity and indirectly changes the disk thickness. Thus the disk could be "clumpy" and geometrically thicker, which would help with the shell star problem.

Waters: The WCD model predicts infall near the star with high velocities.

If this is the case, then the optical and IR emission lines should have very different shapes, but this is not observed. So probably infall does not occur.

Bjorkman: We need to calculate the IR hydrogen line profiles for the WCD model. If the difference between the optical and IR hydrogen lines can prove that there is no disk infall, it would be a very important constraint, because one suggestion for increasing the disk density is to stop the inner disk inflow.

Dougherty: Waters mentioned in his talk that the far IR observations may indicate a flaring (increase in opening angle) of the disk at about 5–$10R_*$. Likewise the radio spectrum can be explained by a flaring of the disk at somewhat larger radii. At what radius does the WCD begin to "flare"?

Bjorkman: The wind streamlines enter the disk at very oblique angles when r becomes large, so the disk will begin to "flare" when the ram pressure of the wind becomes small. Flaring can be seen as soon as few stellar radii ($\sim 3R$) in the highest rotation rate cases in Owocki's simulations.

Lafon: The expression for the CAK radiation force has been determined assuming spherical symmetry and includes many lines. It is probably highly perturbed close to the shock and stagnation point.

Bjorkman: There are many simplifications that have been employed in the radiation-driving force. An example arising from the lack of spherical symmetry is that the azimuthal velocity gradient produces an asymmetry in the shape of Sobolev surface. The resulting asymmetry in photon escape probabilities produces a radiative torque acting on the fluid. As you point out, the ionization balance in the wind could play an even more important role. Our calculations so far have assumed that the CAK force multiplier parameters (k, α, δ) are constants independent of position. I think one of the potentially most important effects that has been neglected is how changes in the ionization balance change the CAK parameters as a function of position.

Moss: Can you make any qualitative comments (e.g., disk thickening) about the possible effects of a modest (unobservable) magnetic field in your model?

Bjorkman: A weak magnetic field might be the missing ingredient needed to increase the disk mass. One problem with the disk is that it leaks— material falls back onto the star. To plug the leak requires adding angular momentum to the disk material so that it is rotationally supported. A magnetic field is one of the few ways to exert a torque on the fluid. There are many advantages to producing a Keplerian disk, namely the long timescales associated with the variability. One interesting scenario involves wrapping up the magnetic field lines in the rotating disk. Eventually the magnetic stresses might disrupt the disk causing large scale changes.

TWO-DIMENSIONAL HYDRODYNAMICAL SIMULATIONS
OF WIND-COMPRESSED DISKS
AROUND RAPIDLY ROTATING B-STARS

S. P. OWOCKI and S. R. CRANMER
*Bartol Research Institute, University of Delaware,
Newark, DE 19716, U.S.A.*

and

J. M. BLONDIN
*Department of Physics, North Carolina State University,
Raleigh, NC 27695, U.S.A.*

We use a 2-D PPM code to simulate numerically the hydrodynamics of a radiation-driven stellar wind from a rapidly rotating Be-star. The results generally confirm predictions of the semi-analytic "Wind Compressed Disk" model recently proposed by Bjorkman and Cassinelli to explain the circumstellar disks inferred observationally to exist around such rapidly rotating stars. However, this numerical simulation is able to incorporate several important effects not accounted for in the simple model, including a dynamical treatment of the outward radiative driving and gas pressure, as well as a rotationally oblated stellar surface. This enables us to model quantitatively the compressed wind and shock that forms the equatorial disk. The simulation results thus do differ in several important details from the simple model, showing, for example, cases of inner disk *inflow* not possible in the heuristic approach of assuming a fixed outward velocity law. In addition, the disk opening typically has a half-angle of 2-4 degrees, somewhat larger than the $\sim 0.5^0$ predicted from the analytic model, and there is no evidence for the predicted detachment of the disk that arises in the fixed outflow picture.

Despite these differences, if the radial velocity in the analytic model is fixed to be the same as that found in the dynamical simulation, then the analytically derived equatorward flow that characterizes the equatorial focusing effect is in excellent agreement with that in the fully dynamical model. Hence, the general predicted effect of disk formation by wind focussing toward the equator is substantially confirmed.

This paper has been accepted for publication in the April 1, 1994 issue of the Astrophysical Journal.

L. A. Balona et al. (eds.), Pulsation, Rotation and Mass Loss in Early-Type Stars, 469.
© 1994 *IAU. Printed in the Netherlands.*

LATITUDE DEPENDENT RADIATIVE WIND MODEL FOR
Be STARS: LINE PROFILES AND INTENSITY MAPS

PH. STEE AND F.X. DE ARAUJO

Observatoire de la Côte d'Azur, CNRS URA 1361,
B.P. 229, F-06304 Nice-Cedex 04, France.

1. The hydrodynamical code

We present theoretical line profiles and intensity maps from an axi-symmetric radiative wind model from a fast rotating Be star (Araújo & Freitas Pacheco, 1989; Araújo *et al*, 1993). The introduction of a viscosity parameter in the latitude dependent hydrodynamic code enables us to consider the effects of the viscous force in the azimuthal component of momentum equations. The line force is the same as Friend and Abbott (1986), but it is introduced a varying contribution of thin and thick lines from pole to equator by adopting latitude-dependent radiative parameters. The numerical calculation for parameters characteristic of early Be stars gives a density contrast between equator and pole of the order of 100. The total mass loss rates range from 10^{-8} to 10^{-7} solar mass per year. Furthermore, the "opening-angle" usually adopted in had-hoc models (Lamers and Waters, 1987; Waters et al.,1991) arises naturally as a result of our model. The velocity fields and density laws derived from the hydrodynamic equations have been used for solving the statistical equilibrium equations. By adopting the Sobolev approximation, we could easily obtain a good estimate of both electronic density and hydrogen level populations throughout the envelope (Stee and Araújo, 1994).

2. Results

We obtain double-peaked (asymmetric) Hα and Hβ emission profiles from our rotating and expanding wind model. Our computation takes into account the portion of the envelope occulted by the stellar disc and the absorption photospheric line. However the full width at half maximum of these profiles are larger than those usually seen in Be stars. This large FWHM reflects the strong radial flow present at all stellar latitudes in addition to the rotational field due to the fact that the star rotates at 70% of its breakup velocity.

On the other hand we have shown intensity maps of the circumstellar envelope in the lines and continuum. From them we could also estimate that the Hα emission region is wider then the Hβ one. Moreover, our results show clearly that the envelope morphology depends not only on the inclination angle but also strongly on the central observational wavelength and

L. A. Balona et al. (eds.), Pulsation, Rotation and Mass Loss in Early-Type Stars, 470–472.

bandwidth. In addition the $\lambda = 0.65$ μm map supports the usual interferometric calibrations based on a "point-like" continuum source which is used as a reference signal (Mourard et al., 1989).

The fast acceleration to high terminal velocities is characteristic of line-driven wind models (Poe and Friend, 1986; Koninx and Hearn, 1992), making them inadequate for equatorial regions of Be stars. In order to improve our line profiles for reproducing realistic FWHM we are currently considering several mechanisms for decelerating the wind. For instance Iglesias and Ringuelet (1993) have presented hydrodynamical codes including a weak magnetic field and an arbitrary force term of the form $1/r^2$. They obtain terminal velocities ranging from 24 to 740 km s^{-1}. We remember that the force term has a dependence of the form $\frac{1}{r^2}(\frac{dv}{dr})^\alpha$ with $\alpha = 1$ for optically thick and $\alpha = 0$ for optically thin lines. In addition Shimada *et al* presented in this meeting new calculations of the radiative force including 520000 lines. Their results indicate that the value of the α parameter is smaller than those used up to now, what would decrease the wind velocity. Another way to slow down our wind may consist on including a radial viscosity which however could introduce a meridional velocity gradient in contradiction to our initial assumption of no meridional flow. On the other hand, a supersonic wind produce shock waves which may perhaps slow down the expansion motion. These mechanisms will be investigated within the framework of our model to develop an improved physical model in the future and hopefully used to interpret directly the observations from optical long baseline interferometry.

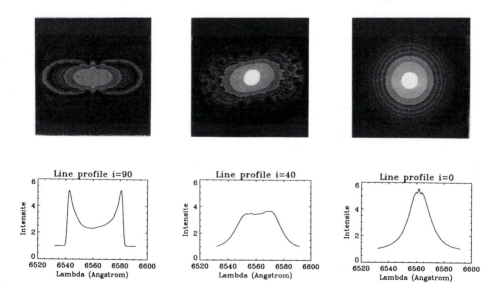

Fig. 1.   Hα intensity maps at various inclinations, centered on 6562Å with a bandwidth of 1Å. Left $i = 90°$, center $i = 45°$, right $i = 0°$ (pole on)

472

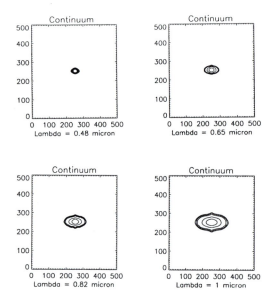

References

Araújo, F.X., Freitas Pacheco, J.A.: 1989, *Monthly Notices of the RAS* **241**, 543
Araújo, F.X., Freitas Pacheco, J.A., Petrini: 1993, *Monthly Notices of the RAS* **267**, 501
Friend, D.B., Abbott, D.C.: 1986, *Astronomy and Astrophysics* **311**, 701
Koninx, J.P.M., and Hearn, A.G.: 1992, *Astronomy and Astrophysics* **263**, 208
Iglesias, M.E., Ringuelet, A.E.: 1993, *Astrophysical Journal* **411**, 342
Lamers, H.J.G.L.M., Waters, L.B.F.M.: 1987, *Astronomy and Astrophysics* **182**, 80
Mourard, D., Bosc, I., Labeyrie, A., Koechlin, L. and Saha, S.: 1989, *Nature* **342**, 520
Poe, C.H., Friend, D.: 1986, *Astrophysical Journal* **311**, 317
Stee, Ph. and Araújo, F.X.: 1994, *Astronomy and Astrophysics* **311**, submitted
Waters, L.B.F.M., van den Veen, W.E.C.J., Taylor, A.R., Marlborough, J.M., Dougherty,
    S.M.: 1991, *Astronomy and Astrophysics* **244**, 120

CONTINUUM FLUX AND POLARIZATION IN B[e] SUPERGIANTS

F.X. DE ARAÚJO, PH. STEE and J. LEFÈVRE
Observatoire de la Côte D'Azur, B.P. 229 Nice Cedex 4, 06304 - France

B[e] supergiants constitute one group of stars in the upper left part of the HR diagram. On the basis of their hybrid spectrum and their intrinsic polarization a two components envelope was suggested (Zickgraf & Schulte-Ladbeck, 1989): a "polar wind" and a denser "equatorial disc".

Recently Araújo *et al* (1993) have presented an axi-symmetric model in which the radiative line force varies from pole to equator. Some typical density profiles (arising from the application to B[e] supergiants) may be seen in fig. 1. In this particular model it was adopted $M = 30M_{\odot}$, $R = 75R_{\odot}$, $T_{eff} = 20000^{\circ}K$ and a rotational rate of 70% of the critical speed. There is a density enhancement towards the equatorial plane which is due to a greater mass flux and a lower expansion.

Assuming a fully ionized, pure hydrogen envelope the density laws give directly the electronic distribution. (A numerical code is being developed in order to overcome this drastic hypothesis, see Stee & Araújo, this Symposium). Fig. 2 shows the optical thickness due to electronic absorption as a function of wavelength (lines) and that due to electronic scattering (circles), which is λ independent. It is clearly seen as opacity increases from pole to equator.

Subsequently the propagation of stellar radiation in the envelope was followed by employing a Monte Carlo type scheme (Lefèvre & Daniel, 1988). Figure 3 shows the flux that escapes from the envelope as a function of λ at some inclination angles: $i = 0^{\circ}$, $i = 45^{\circ}$ and $i = 90^{\circ}$. We see that i) emission decreases with λ (the star radiates as a blackbody with $T = 2.10^{4}K$); ii) the flux is dependent on the line of sight due to different opacities; and iii) the wavelength dependence of absorption is found in the increasing separation of lines. In fig. 4 we plot the global flux that escapes in all directions as a function of the number of scatterings. We may conclude that the single scattering approximation is not reasonable since at least 50% of radiation is scattered twice or more.

Concerning polarisation degrees we stress they are preliminary ones since we have not been able yet to perform computations with a number of interactions great enough to prevent statistical fluctuations. Fig. 5 shows that we have obtained degrees between about 0.5% and 4.5%, increasing with inclination angles. On the other hand we can see in fig. 6 that no dependence on wavelength was obtained. This result however must be kept with great caution since we have not included the envelope emission. It will most likely

L. A. Balona et al. (eds.), Pulsation, Rotation and Mass Loss in Early-Type Stars, 473–474.

add up a no (or weakly) polarised flux which is strongly increasing with λ. In order to compare the results with observational data we must taking into account ionisation equilibrium, including envelope emission and performing more accurate numerical simulations.

References

Araújo, F.X., Freitas Pacheco, J.A. & Petrini, D.: 1993, *Monthly Notices of the RAS* , in press

Lefévre, J. & Daniel, J.-Y.: 1988, *in "Polarized radiation of circumstellar origin", (G. Coyne et al., eds.), Vatican Observatory* , 523

Zickgraf, F.-J., & Schulte-Ladbeck, R.: 1989, *Astronomy and Astrophysics* **214**, 274

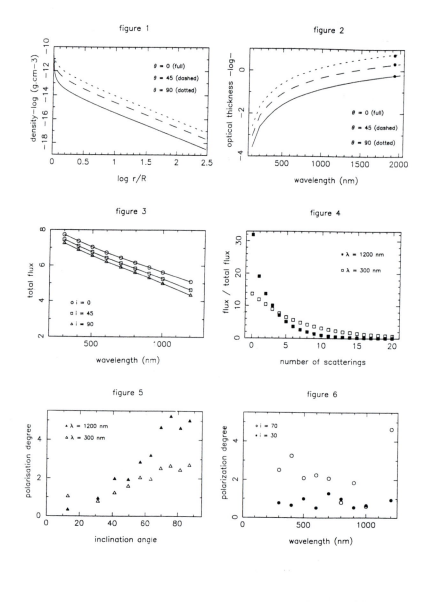

THE BASIC PHYSICS OF HOT-STAR WINDS

STANLEY P. OWOCKI

Bartol Research Institute, University of Delaware, Newark, DE 19716 USA

Abstract. Hot-star winds are believed to be driven by line-scattering of the star's continuum radiation. This review summarizes the physics of this line-driving process and the resulting basic wind properties. I also briefly discuss the generation of wind structure, both at small scale due to inherent instabilities in the line-driving, and at large scales due to various rotational effects.

1. Introduction

Stellar winds from hot stars (i.e., spectral types O, B, and WR) have been known and studied for more than 20 years now. During this time, there has developed an extensive literature, including numerous reviews of both theory (e.g. Cassinelli 1979; Lucy 1985; Abbott 1988; Owocki 1990; Drew 1994) and observations (Conti 1978; Cassinelli & Lamers 1987). (See also the monograph edited by Conti & Underhill 1988.) Given this already rich collection of reviews, I will not attempt here to give any kind of exhaustive list of the literature. Instead, this review will provide a tutorial on the fundamental driving mechanism—line-scattering in the wind of the star's continuum radiation field—and what it implies for basic wind properties (e.g., mass-loss rate, $\dot{M}$, and terminal flow speed, v_∞). I will also briefly summarize processes that can lead to wind structure and variability.

2. The Line Force

Hot-star winds are thought to be driven by line-scattering of the star's continuum radiation flux (Lucy & Solomon 1970). Most modern models are based on the formalism developed by Castor, Abbott, & Klein (1975; hereafter CAK) for including the ensemble effect of many lines. For simplicity let us assume (as did CAK) a point source of radially streaming radiation, but for now consider only a *single, isolated line*. The radiative acceleration associated with absorption of radiation in this line is then

$$g_{\rm rad}(r) = g_{\rm abs}(r) = g_{\rm thin} \int_{-\infty}^{\infty} dx \; \phi(x - v(r)/v_{\rm th}) \, e^{-t(x,r)}, \qquad (1)$$

where v and $v_{\rm th}$ are the flow speed and ion thermal speed, ϕ is the line-profile function, and $x \equiv (\nu - \nu_{\rm L})/\Delta \nu_{\rm D}$ is the frequency displacement from line center, measured in Doppler-units in the star's rest frame. The frequency integration corrects the optically thin acceleration $g_{\rm thin} \equiv \kappa_{\rm L} L_\nu \Delta \nu_{\rm D}/4\pi c r^2$ for

L. A. Balona et al. (eds.), Pulsation, Rotation and Mass Loss in Early-Type Stars, 475–486.
© 1994 IAU. Printed in the Netherlands.

self-absorption, with the optical depth $t(x,r)$ at radius r given by

$$t(x,r) = \int_{R_*}^{r} \kappa_L \rho(r') \phi(x - v(r')/v_{th}) dr'. \tag{2}$$

Here κ_L is the line mass-absorption coefficient, and L_ν is the stellar luminosity at the line frequency ν.

For a scattering line, there is, in principle, an additional force component associated with the diffuse, scattered radiation field; however, in a smooth, supersonic flow such scattering is nearly fore-aft symmetric (Sobolev 1960), which implies that the diffuse force is quite small, of order v_{th}/v times g_{abs} (Castor 1974). Another implication of this so-called 'Sobolev approximation' for supersonic flow is that the variation in the integrand of Eq. (2) is dominated by the Doppler-shift associated with changes in the velocity $v(r')$; by switching the variable of integration to the comoving frame frequency $x' \equiv x - v(r')/v_{th}$, we thus find

$$t(x,r) \approx \tau \int_{x-v(r)/v_{th}}^{\infty} \phi(x') dx', \tag{3}$$

where the Sobolev optical depth $\tau \equiv \kappa_L \rho v_{th}/(dv/dr)$ represents a collection of spatial variables that are assumed to be approximately constant over a Sobolev length $L \equiv v_{th}/(dv/dr)$. For example, in the case of a smooth, steady flow this means that the density scale length

$$H \equiv \frac{\rho}{d\rho/dr} \approx \frac{v}{dv/dr} \gg \frac{v_{th}}{dv/dr} \equiv L, \tag{4}$$

and hence that $v \gg v_{th}$. Since the ion thermal speed is on the order of the sound speed a, we can expect that the approximation will be well satisfied in the supersonic portions of a smooth flow. Applying Eq. (3) to Eq. (1) we see that *both* required frequency integrations can now be done *analytically*, yielding

$$g_{rad}(r) = g_{thin} \frac{1 - e^{-\tau}}{\tau}. \tag{5}$$

Note that for a weak line $\tau \ll 1$ we recover the optically thin expression, while for a strong line $\tau \gg 1$ we obtain the optically thick form

$$g_{rad}(r) = \frac{g_{thin}}{\tau} = \frac{\nu_L L_\nu}{L_*} \frac{L_*}{4\pi \rho r^2 c^2} \frac{dv}{dr}. \tag{6}$$

The latter is independent of the opacity κ_L, depending instead on the velocity gradient dv/dr. Once a line is optically thick, all photons shifted into resonance with it are scattered, and further increases in the opacity only push the first scattering further into the blue wing. The number of photons scattered per unit length is thus independent of the opacity, and depends instead on the slope of the photon red-shift, which, in turn, depends on the velocity gradient. Normally, such a velocity gradient is part of an inertial term that represents the acceleration that *results* from a driving force, but here we see that it helps *determine* the driving force. In general terms, just as the optically thin line force helps reduce the effect of gravity on the fluid, so does the optically thick line force reduce the effect of *inertia*.

A simple consideration of just the inertial force balance enables us to estimate the mass-loss rate that results from such line-driving. The acceleration is just

$$v\frac{dv}{dr} = g_{\mathrm{rad}} = \frac{\nu_L L_\nu}{L_*} \frac{L_*}{4\pi \rho r^2 c^2} \frac{dv}{dr}, \tag{7}$$

which implies that the mass-loss rate $\dot{M}_l$ associated with driving by a single, isolated line is

$$\dot{M}_l = 4\pi \rho v r^2 = \frac{\nu_L L_\nu}{L_*} \frac{L_*}{c^2}. \tag{8}$$

Since $\nu_L L_\nu / L_* \approx 1$, we see that the *wind* mass-loss rate driven by a single line is only about the *photon* mass-loss rate! It is essentially for this reason that the original Lucy & Solomon (1970) model, which considered the effect of driving in only a few strong lines, yielded such low mass-loss rates.

3. The CAK Model: Driving by a Line Ensemble

A major advance of the CAK model was to develop a formalism for efficiently including the cumulative effect of a large ensemble of lines, which they effectively assumed to have a flux-weighted number distribution that was a power law in opacity, i.e., $N(\kappa) \sim \kappa^{\alpha-2}$, where $0 < \alpha < 1$. Assuming the lines do not significantly overlap, the cumulative force can then be computed by integrating the expression (1) over this number distribution. Applying, as before, the Sobolev approximation, one obtains the CAK line-force

$$g_{\mathrm{CAK}} = \frac{KL_*}{r^2} \left(\frac{1}{\rho} \frac{dv}{dr} \right)^\alpha, \tag{9}$$

where K is a normalization constant for the line-distribution (related to the CAK constant k by $K = k(\kappa_e^{1-\alpha}/4\pi c v_{\mathrm{th}}{}^\alpha)$, where κ_e is the electron scattering opacity.)

Gas pressure forces play a relatively minor role in driving these winds, and so we can derive the basic wind characteristics by considering just the balance among inertial, gravitational, and radiative forces, ignoring gas pressure altogether. The requirement of momentum balance thus reduces to the condition,

$$F \equiv w' + 1 - C w'^\alpha = 0, \tag{10}$$

where $w' \equiv r^2 vv'/GM_*$ and

$$C \equiv \frac{K}{GM_*} \left(\frac{4\pi GM_*}{\dot{M}c} \right)^\alpha. \tag{11}$$

Depending on the value of $\dot{M}$, the constant C will be such that Eq. (10) will admit either 0, 1, or 2 solutions. The critical case with one solution has the property that $F = \partial F/\partial w' = 0$, which implies the critical values $w'_c = \alpha/(1-\alpha)$ and $C_c = (1-\alpha)^{\alpha-1}/\alpha^\alpha$. Since Eq. (10) has no explicit spatial dependence, this critical value w'_c must hold throughout the flow. Integrating w'_c from the stellar radius R_*, we thereby obtain the usual "CAK velocity law",

$$v_{\mathrm{CAK}}(r) = v_\infty \sqrt{1 - \frac{R_*}{r}}, \tag{12}$$

where the asymptotic wind speed is given by

$$v_\infty \equiv \sqrt{\frac{2\alpha GM_*}{(1-\alpha)R_*}} = \sqrt{\frac{\alpha}{1-\alpha}} v_{\mathrm{esc}}. \tag{13}$$

Likewise, if we solve the definition (11) for the mass-loss rate and apply the critical value C_c, we obtain the CAK mass-loss rate formula,

$$\dot{M}_{\mathrm{CAK}} = 4\pi\alpha \left(\frac{GM_*}{1-\alpha} \right)^{1-1/\alpha} (KL_*)^{1/\alpha}. \tag{14}$$

These are just the relations derived by CAK for the case of a point-star; taking account of the finite angular size of the stellar disk tends to make the mass-loss rate lower, the velocity law shallower, and the terminal speed higher (Friend & Abbott 1986; Pauldrach, Puls, & Kudritzki 1986). The effect of retaining a finite gas pressure is comparatively small, as is the effect of using a full comoving frame transfer solution instead of the CAK/Sobolev line force (9) (Pauldrach et al. 1986). Recent further efforts (e.g. Gabler et al. 1989) have focussed on developing unified wind-atmosphere models that abandon the artificial distinction between "core" and "halo", and that solve for the basic wind parameters k and α from a more complete treatment of the NLTE statistical equilibrium (Pauldrach 1987). A major overall goal of these efforts is to carry out "quantitative spectroscopy", i.e., to compare quantitatively theoretical and observed spectra in order to infer more precisely and reliably the basic parameters of both wind and star. Further details of these efforts are given in the review by Kudritzki & Hummer (1990) and references therein.

4. Multi-Line Scattering in Overlapping Lines

For a typical OB supergiant model (e.g., ζ Pup), the CAK mass-loss rate is $\sim$100 times the single line value, implying that there are effectively about 100 optically thick lines spread throughout the spectrum. On the other hand, note from Eq. (13) that the terminal flow speed v_∞ scales with surface escape speed v_{esc}, and is independent of the number of strong lines. Typically v_∞ is on the order of 1% of the speed of light, so that, quite coincidentally, in the CAK model

$$\dot{M}v_\infty \approx \frac{L_*}{c}\left(N_{thick}\frac{v_\infty}{c}\right) \approx \frac{L_*}{c}, \tag{15}$$

which is often referred to as the "single scattering limit". This means that, fortuitously, the CAK model is just barely self-consistent in ignoring line overlap, if one assumes the lines are smoothly spread throughout the stellar spectrum.

In reality, of course, there is always some line overlap, arising both from chance frequency coincidences of lines as well as from the tendency for lines to occur in multiplets. An early but quite elegant examination of the dynamical effects of such overlap was carried out by Friend & Castor (1983), who derived an extension of the CAK model to include overlap under the assumption that the line frequency spacing is Poisson distributed. Although this ignored the tendency of lines to occur in multiplets, it demonstrated quite clearly the major effect that overlapping lines can drive mass-loss in excess of the single scattering limit, i.e., with $\dot{M}v_\infty > L_*/c$. Similar results were obtained for more realistic line lists by Abbott & Lucy (1985) from a Monte-Carlo calculation, and by Puls (1987), who obtained approximate solutions of the coupled system of radiative transfer equations.

For O and B stars such overlap effects are typically minor, but they probably play a much more fundamental role in the Wolf-Rayet stars, which are inferred observationally to have $\dot{M}v_\infty/(L_*/c) \approx 5--50$ (Abbott & Conti 1987). Lucy & Abbott (1993) have recently applied their Monte-Carlo code to derive mass-loss rates for a multi-line scattering wind model assuming a fixed wind velocity law. For typical WR star parameters they were able to achieve $\dot{M}v_\infty \approx 10L_*/c$. These results make it clear that there is nothing fundamental about the single scattering limit, and that this limit can be exceeded by a large factor, *if* photons can be sufficiently *trapped* in the envelope. Thus, although the difficulty in explaining the large inferred mass-loss rates for Wolf-Rayet stars is often characterized as the "momentum" problem, it really is more of an "opacity" problem. In the Lucy & Abbott (1993) model, there is a tendency for the ionization balance to adjust so that there is always lots of line opacity near the peak of the photon spectrum, even as this spectrum shifts to longer wavelengths at larger radial distances in the cooling envelope. A further physical discussion of how such photon trapping enhances momentum deposition in the wind is given by Gayley & Owocki (1994).

5. Generation of Wind Structure

There is much evidence that hot-star winds are not the smooth, steady, spherically symmetric outflows envisaged in the above models, but rather are highly variable and structured on both relatively large and small scales. Such structure could arise in several ways, including perturbations from stellar pulsation, rotational modulation associated with underlying atmospheric structure (possibly magnetic), or simply from the intrinsic instabilities in the wind itself.

The most direct evidence for relatively large-scale structure comes from the variable "discrete absorption components"(DAC) of unsaturated UV lines, which are observed in IUE monitoring campaigns to form at $v \sim v_\infty/3$ and then to migrate to higher frequencies. Such DAC's typically recur quasi-regularly, with both a recurrence and evolution time scaling roughly with the inverse of the projected rotation velocity, typically a day or two (Prinja 1988; Kaper 1993; see also contributions by Henrichs and by Prinja in these Proceedings). They are thought to arise from accelerating wind density enhancements that are large enough (i.e., a stellar radius or so) to cover a substantial fraction of the stellar disk. Mullan (1984; 1986) has proposed that these could arise from "co-rotating interaction regions" (CIR's) between fast and slow speed wind streams. In the solar wind, such wind speed differences can arise from the differences in flow divergencies associated with magnetic field structure in the lower solar corona. MacGregor (1988) has shown that changes in flow divergence can have a similarly large influence on the wind speeds in a line-driven flow. For normal OB stars, observational upper limits on the global magnetic field are typically about $B < 100G$, but even fields below this could have an important effect. In general a field can divert a flow up to the Alfven critical point, which for these winds occurs at a velocity $v \approx 1500$ km s$^{-1}(B_{100}^2 r_{12}^2/\dot{M}_{-6})$, where $B_{100} \equiv B/100$ G, $r_{12} \equiv r/10^{12}$ cm, and $\dot{M}_{-6} \equiv \dot{M}/10^{-6} M_\odot$ y^{-1}. This implies that undetected fields, e.g. with $B \gtrsim 10$ G, could have quite a significant dynamical influence on the acceleration of OB winds.

The relatively low signal to noise ($\sim$20) in IUE spectra has severely limited our ability to directly detect wind fluctuations on a smaller scale in UV lines for OB stars. However, such small-scale fluctuations are directly measured in the much higher signal to noise ($\sim$500), ground-based observations of optical emission lines formed in WR star winds (McCandliss 1991; Robert 1991), and these have been interpreted as implying the existence of a "turbulent" spectrum of relatively small-scale structures in these winds (Moffat et al. 1988, 1993). Furthermore, even in OB winds, there is *indi-*

rect evidence for such small-scale structure, in the form of the nearly black absorption troughs of saturated UV lines, which seem to require a multiply nonmonotonic velocity field (Lucy 1982).

It seems likely that such small-scale, "stochastic" structure could arise naturally from the strong instability of the line-driving mechanism itself (Rybicki 1987). A small-scale increase in radial flow speed Doppler-shifts the local line-frequency out of the absorption shadow of intervening material, leading to an increased line-force which then tends to further increase the flow speed. Formal stability analyses (MacGregor, Hartmann, & Raymond 1979; Carlberg 1980; Owocki & Rybicki 1984, 1985) yield a growth rate nearly a hundred times the wind expansion rate, implying that even very small amplitude perturbations should quickly become nonlinear.

Since the instability occurs primarily for perturbations at a scale near and below the Sobolev length L (cf. Eq. (4)), it cannot be analyzed in terms of the simple, *local* force expressions (5) or (9), which are based on the Sobolev approximation that all variations occur on a scale much larger than L. Rather it requires expensive computation of numerical integrals occurring in *nonlocal* force expressions, either based on a *pure-absorption* approximation (Owocki, Castor, & Rybicki 1988; hereafter OCR), or the "Smooth Source Function" approximation (Owocki 1991) for line-scattering. Numerical simulations of the nonlinear evolution (OCR; Puls, Owocki, & Fullerton 1994; Feldmeier 1993, 1994) show a wind embedded with multiple reverse shocks and dense shells. Recent models with greater resolution to resolve relatively high-frequency structure have synthetic spectra that reproduce quite well the broad, black absorption troughs observed in saturated lines (Owocki, Puls, & Fullerton 1994; Puls *et al.* 1994). The resulting wind shocks also seem to provide a potential explanation for the soft X-ray emission typically observed from OB stars (Cooper & Owocki 1991, 1993). Work is also underway to try to generalize such simulations to WR winds, with the aim to model the small-scale fluctuations seen in optical wind lines (Gayley & Owocki 1994).

In unsaturated UV lines, the dense shells that form in these instability simulations also give rise to multiple narrow absorption features which superficially resemble DAC's, but which have acceleration time scales that are more characteristic of the mean wind (i.e., with velocity law exponent $\beta \approx 1$) than of the much more slowly evolving DAC's (for which typically $\beta \approx 3 - -10$). Considering also that their recurrence is nearly stochastic, it seems unlikely that such intrinsic, small-scale wind instabilities could be the direct cause of the semi-regularly recurring DAC's. Emphasis has thus shifted to trying to simulate such DAC's as resulting from a relatively large scale disturbance at the wind base. For example, a recent simulation (Owocki, Puls, & Fullerton 1994) shows that, if the density in the lower wind (at $v \approx 100$ km s^{-1}) is artificially increased by just a factor of two, then the outer wind develops a large, slow-moving dense shell with roughly the same slow acceleration time scale seen in DAC's.

Such time-dependent instability models have so far been restricted to 1-D, owing to the great computational expense of computing the nonlocal radiative force. In a more realistic 3-D model, the relatively narrow shells that form in such 1-D models would almost certainly be broken into small-scale clumps which, due to global averaging, would have much weaker spectral line signatures. However, even though the 3-D wind structure would thus not have the lateral phase coherence of a strictly 1-D model, it still seems likely that velocity fluctuations would be predominately 1-D radially polarized, since 3-D linear stability analyses (Rybicki, Owocki, & Castor 1990) indicate that lateral velocity fluctuations are strongly damped by the line-drag effect of the diffuse radiation field (Lucy 1984). There appears to be some evidence for such a predominance of radial velocity fluctuations in WR winds (McCandliss 1991; Robert 1991).

If one is interested in only relatively large-scale wind structure, i.e., larger than a Sobolev length L, then it is possible to carry 2-D or 3-D simulations using the simple, *local* CAK/Sobolev force expression (9), which is much cheaper to compute than the nonlocal expressions needed to study small-scale instabilities. Owocki, Cranmer, & Blondin (1994) have recently carried out 2-D hydrodynamical simulations of CAK winds from rapidly rotating stars, in order to test the Bjorkman & Cassinelli (1993) "Wind Compressed Disk" (WCD) model. With some modifications, these simulations basically confirm the original WCD idea, but do show some interesting new features, e.g., inner disk *infall*. Such WCD models have a natural appeal for explaining the Hα emission from the relatively rapid rotating Be stars, which have long been associated with disks. Unfortunately, as discussed in the paper by Bjorkman in these Proceedings, it now appears that, the simple wind compression effect results in only a relatively low-density, optically thin disk that seems inadequate to explain either the observed Hα emission or the observed optical continuum polarization from such stars. To overcome these limitations, it may be worthwhile for future models to consider the possible role of magnetic fields in further guiding the wind toward the equator and in spinning up the disk.

Acknowledgements

This work was supported in part by NSF grant AST 91-15136 and NASA grants NAGW-2624 and NAG5-1567, and by an allocation of supercomputer time from the San Diego Supercomputer Center. I thank A. Fullerton and S. Cranmer for helpful comments.

References

Abbott, D.C.: 1988, in Pizzo, V.J., Holzer, T.E., Sime, D.G., eds., *Proceedings of the Sixth International Solar Wind Conference, Vol. 1*, NCAR: Boulder, 149.

Abbott, D.C. and Conti, P.S.: 1987, *Ann. Rev. Astron. Astrophys.* **25**, 113.

Abbott, D.C. and Lucy, L.B.: 1985, *Astrophys. J.* **288**, 679.

Bjorkman, J.E. and Cassinelli, J.P.: 1993, *Astrophys. J.* **409**, 429.

Carlberg, R.G.: 1980, *Astrophys. J.* **241**, 1131.

Cassinelli, J.P.: 1979, *Ann. Rev. Astron. Astrophys.* **17**, 275.

Cassinelli, J.P. and Lamers, H.G.J.L.M.: 1987, in Kondo, Y. *et al.* eds., *Exploring the Universe with the IUE Satellite*, Reidel: Dordrecht, 140.

Castor, J.I.: 1974, *Mon. Not. Roy. Astr. Soc.* **169**, 279.

Castor, J.I., Abbott, D.C. and Klein, R.I.: 1975, *Astrophys. J.* **195**, 157 (CAK).

Conti, P.S.: 1978, *Ann. Rev. Astron. Astrophys.* **16**, 371.

Conti, P.S. and Underhill, A.B.: 1988, *O Stars and Wolf-Rayet Stars*, NASA: Washington.

Cooper, R.G. and Owocki, S.P.: 1991, in Drissen, L., Leitherer, C. and Nota, A., eds., *Nonisotropic and Variable Outflows from Stars*, Astron. Soc. Pacific: San Francisco, 281.

Cooper, R.G. and Owocki, S.P.: 1993, in Moffat, A. *et al.* eds., *Instability and Variability in Hot Star Winds*, Kluwer: Dordrecht, in press.

Drew, J.E.: 1994, in Clegg, R., Meikle, P. and Stevens, I., eds., *Circumstellar Media in the Late Stages of Stellar Evolution*, Cambridge Univ. Press: Cambridge, in press.

Feldmeier, A.: 1993, *Ph. D. Thesis*, Univ. Munich: Germany.

Feldmeier, A.: 1994, in Moffat, A. *et al.* eds., *Instability and Variability in Hot Star Winds*, Kluwer: Dordrecht, in press.

Friend, D.B. and Abbott, D.C.: 1986, *Astrophys. J.* **311**, 701.

Friend, D.B. and Castor, J.I.: 1983, *Astrophys. J.* **272**, 259.

Gabler, R., Gabler, A., Kudritzki, R.P., Puls, J. and Pauldrach, A.W.A.: 1989, *Astron. Astrophys.* **226**, 162.

Gayley, K. and Owocki, S.P.: 1994, *Astrophys. J.* , submitted.

Kaper, L.: 1993, *Ph.D. Thesis*, Univ. Amsterdam: The Netherlands.

Kudritzki, R.P. and Hummer, D.G.: 1990, *Ann. Rev. Astron. Astrophys.* **28**, 303.

Lucy, L.B.: 1982, *Astrophys. J.* **255**, 286.

Lucy, L.B.: 1984, *Astrophys. J.* **284**, 351.

Lucy, L.B.: 1985, in Mihalas, D. and Winkler, K-H.A., eds., *Radiation Hydrodynamics in Stars and Compact Objects, IAU Colloq. No. 89*, Springer: Berlin, 75.

Lucy, L.B. and Abbott, D.C.: 1993, *Astrophys. J.* **405**, 738.

Lucy, L.B. and Solomon, P.M.: 1970, *Astrophys. J.* **159**, 879.

MacGregor, K.B.: 1988, *Astrophys. J.* **327**, 794.

MacGregor, K.B., Hartmann, L. and Raymond, J.C.: 1979, *Astrophys. J.* **231**, 514.

McCandliss, S.R.: 1991, in Drissen, L., Leitherer, C. and Nota, A., eds., *Nonisotropic and Variable Outflows from Stars*, Astron. Soc. Pacific: San Francisco, 214.

Moffat, A.F.J., Drissen, L., Lamontagne, R. and Robert, C.: 1988, *Astrophys. J.* **334**, 1038.

Moffat, A.F.J., Lepine, S., Henriksen, R.N. and Robert, C.: 1993, *Kinematics and Dynamics of Diffuse Astrophysical Media*, Astophys. Space Sci., in press.

Mullan, D.J.: 1984, *Astrophys. J.* **283**, 303.

Mullan, D.J.: 1986, *Astron. Astrophys.* **165**, 157.

Owocki, S.P.: 1990, *Rev. Modern Astron.* **3**, Springer: Berlin, 98.

Owocki, S.P.: 1991, in Crivellari, L., Hubeny, I. and Hummer, D.G., eds., *Stellar Atmospheres: Beyond Classical Models*, Kluwer: Dordrecht, 235.

Owocki, S.P. and Rybicki, G.B.: 1984, *Astrophys. J.* **284**, 337.

Owocki, S.P. and Rybicki, G.B.: 1985, *Astrophys. J.* **299**, 265.

Owocki, S.P., Castor, J.I. and Rybicki, G.B.: 1988, *Astrophys. J.* **335**, 914 (OCR).

Owocki, S.P., Cranmer, S.R. and Blondin, J.B.: 1994, *Astrophys. J.* , in press.

Owocki, S.P., Puls, J. and Fullerton, A.W.: 1994, in Moffat, A. *et al.* eds., *Instability and

Variability in Hot Star Winds, Kluwer: Dordrecht, in press.

Pauldrach, A.: 1987, *Astron. Astrophys.* **183**, 285.

Pauldrach, A. and Puls, J.: 1990, *Astron. Astrophys.* **237**, 409.

Pauldrach, A., Puls, J. and Kudritzki, R.P.: 1986, *Astron. Astrophys.* **164**, 86.

Prinja, R.K.: 1988, *Mon. Not. Roy. Astr. Soc.* **231**, 21P.

Puls, J.: 1987, *Astron. Astrophys.* **184**, 227.

Puls, J., Owocki, S.P. and Fullerton, A.W.: 1994, *Astron. Astrophys.* , in press.

Puls, J., Feldmeier, A., Springmann, U., Owocki, S.P. and Fullerton, A.W.: 1994, in Moffat, A. *et al.* eds., *Instability and Variability in Hot Star Winds*, Kluwer: Dordrecht, in press.

Robert, C.: 1991, *Ph.D. Thesis*, Univ. de Montreal: Canada.

Rybicki, G.B.: 1987, in Lamers, H.J.G.L.M. and de Loore, C.W.H., eds., *Instabilities in Luminous Early Type Stars*, Reidel: Dordrecht, 175.

Rybicki, G.B., Owocki, S.P. and Castor, J.I.: 1990, *Astrophys. J.* **349**, 274.

Sobolev, V.V.: 1960, *Moving Envelopes of Stars*, Harvard Univ. Press: Cambridge.

Discussion

Lafon: (1) The problem of comparing L/c and $\dot{M}v_\infty$ is not well posed! Indeed momentum is a vectorial quantity and should be added as a vector so that the net flux of momentum from the star is always zero in azimuthal and plane (equatorial symmetry), or in spherical symmetry. Thus there is no problem with finding ratios $\dot{M}v_\infty c/L$ of the order 50 or larger.

(2) The "mass-loss rate", $\dot{M}$, is only defined in steady state. Otherwise some amount of matter elevated from the surface of the star can fall back to the star later. (This is a well known phenomenon for shock perturbed atmospheres of evolved stars.) One should not talk of "mass-loss variability" where observing line variation for lines formed or changed in the wind.

(3) Instabilities, even in spherically or azimuthally symmetric models, should be investigated in three dimensions, even if the global structure of the wind is not perturbed on large scales. First the instabilities may produce blobs or clumps instead of shells. Then one should be cautious with comparisons with the solar wind in which large scale structures are maintained by freezing of magnetic field, contrary to what occurs in hot star winds.

Owocki: (1) As noted in my talk, I basically agree that, in principle, $\dot{M}v_\infty c/L$ can become very large. It requires, however, a large opacity distributed throughout the spectrum to keep photons trapped in the wind, and, in practice, this can be quite difficult to achieve. This is why I prefer to characterize WR wind models as suffering an "opacity" rather than a "momentum" problem.

(2) If one simply chooses to define $\dot{M} \equiv 4\pi\rho v r^2$ at each location in the wind, then this quantity can vary in time or space in a time-dependent, or a non-spherically symmetric wind model. Such variations in local mass flux often have quite distinct, detectable consequences in observed spectra, and so I believe it is quite appropriate to speak in this way of "mass-loss variability".

(3) Rybicki, Owocki & Castor (1990) investigated the linear instability of winds in 3-D, but so far nonlinear simulations have had to be limited to 1-D, simply because of the great computational expense of computing the nonlocal line force. As for the role of magnetic fields, this in my opinion is still an open question, since large scale fields with $B < 100$ Gauss, roughly the observational upper limit for most OB winds, could have dynamical effects as strong or stronger as those seen in the solar wind.

Harmanec: You mentioned that the α coefficient of proportionality is basically given by the specific mixture of absorbing lines. Now, many Be stars exhibit long-term spectral variations, developing strong shell lines at certain epochs. Would you agree that the α vary correspondingly and that this could be a way to further test the theory?

Owocki: The basic line-driving parameters α and k can switch abruptly with stellar conditions, such as in the Pauldrach & Puls (1990) "bi-stability" mechanism. Determining the role, if any, such switching may play in the epochal variations you mention requires further investigation.

Kaper: Which conditions have to fulfilled to get fast and slow moving streams in a radiatively driven wind?

Owocki: Within the context of a steady, CAK/Sobolev type line-driving, MacGregor (1988) showed that a faster-than-radial flow divergence—such as can occur in region of open, diverging magnetic field—can lead to more than a factor of two increase in asymptotic flow speed. Within the context of time-dependent models, inducing base variations in, e.g., mass flux, can lead to even larger variations in wind speed, owing to the nonlocal, nonlinear character of the radiative driving. My own expectation is that achieving flow speed variations is, in principle, much easier in hot-star winds than in the solar wind, where factor two fluctuations in speed are quite common.

Smith: Your wind-velocity diagram for a B2.5e star showed low wind velocities in the equatorial zone. Why, then, do we observe high velocity DAC's in spectra of Be-stars viewed nearly edge-on?

Owocki: The model I showed (see Fig. 2 of Owocki, Cranmer & Blondin 1994) is just a simple simulation meant primarily to investigate the Bjorkman/Cassinelli Wind Compressed Disk effect, but it would still allow for relatively high-velocity DAC's arising in the flow just outside the very thin disk, which would still be seen projected against the star.

Prinja: Based on your latest calculations, do you predict that the intrinsic wind instabilities (resulting in small-scale structure) would be observable as time-dependent variations in the wind profiles of OB stars?

Owocki: The small-scale structure would necessarily produce direct *variability* that could only be detected at sufficiently high signal to noise, probably more than 100, and in any case much higher than typical for IUE. The presence of small scale structure is nonetheless implied by the overall profile shape, viz. the width and blackness of absorption troughs in saturated lines, which I do believe can only be produced by nonmonotonic velocity variations with a high spatial frequency.

Mourand: Have you plans to compute brightness distributions in your model in order to make direct comparisons with interferometric data?

Owocki: We do hope in general to develop 2-D and 3-D radiative transfer codes to synthesize from these models a wide range of observational diagnostics, including line-spectra, polarization, and, eventually, interferometry.

RADIATIVELY DRIVEN WINDS OF OB STARS

MICHIHIRO R. SHIMADA, MASAKI ITO*, RYUKO HIRATA
Department of Astronomy, Kyoto University, Sakyo-ku, Kyoto, 606-01, Japan

and

TOSHIHIRO HORAGUCHI
National Science Museum, Ueno Park, Taito-ku, Tokyo 110, Japan

Abstract. We newly calculated the line radiative force with 520,000 atomic lines, which is twice as many as those of Abbott (1982), for OB supergiants. Our results are as follows. (1) The mass loss rates for O stars with $T_{\text{eff}} = 50,000$K are seven times as large as Abbott's (1982) because of contribution from Fe IV lines. (2) Contribution from many weak lines increases the mass loss rates and decreases the wind velocities of OB stars within a temperature range of $10,000$K $\leq T_{\text{eff}} \leq 30,000$K. This result is qualitatively in accordance with the results from the recent observations of O stars. (3) The mass loss rates of OB stars depend on metallicity with $\dot{M} \sim Z$.

OB stars lose their mass through stellar winds. Line radiative force is thought to be the dominant mechanism for driving winds. Analytical formulation for line driven wind was developed by Castor, Abbott & Klein (1975). Abbott (1982) calculated the line radiative force with a list of 250,000 atomic lines and investigated the properties of the force in detail. The numerical values of the line radiative force of Abbott (1982) have been widely used in the study of mass loss phenomena of OB stars so far.

We calculated the line radiative force with ATMLINE database, which is more complete database of opacity and excitation level of atomic lines. We took into account the first to sixth stages of ionization of the elements H–Ni and the first to forth stages of ionization of the elements Cu–U. Total number of absorption lines amounts to 520,000, which is twice as many as those of Abbott (1982).

The line radiative force was calculated with Sobolev approximation, as in Castor, Abbott & Klein (1975) and Abbott (1982). Following Pauldrach et al. (1986), the line radiative force per unit mass is expressed as

$$\frac{\kappa_e F}{c} f_{\text{rad}} F_{\text{corr.}},$$

where $\kappa_e F/c$ is the acceleration due to continuum radiation pressure and $F_{\text{corr.}}$ is a correction factor for a finite core angle. The force multiplier, f_{rad}, can be approximated by

$$f_{\text{rad}} \sim k t^{-\alpha} s^{\delta}$$

* Deceased on 27 March, 1989

L. A. Balona et al. (eds.), Pulsation, Rotation and Mass Loss in Early-Type Stars, 487–489.
© 1994 IAU. Printed in the Netherlands.

as a function of t and s, where the value t and s are defined as

$$t = \kappa_e \rho v_{th} \left| \frac{dv}{dr} \right|^{-1}, \quad s = \frac{\rho}{m_H W} \times 0.85 \times 10^{-11}.$$

All quantities have the same meaning as in Abbott (1982).

The force multiplier and the values k, α and δ were calculated for OB supergiants within the temperature range $10,000K \leq T_{eff} \leq 50,000K$ and within the metallicity range $0.002 \leq Z \leq 0.165$. The results for the case with $Z = 0.02$ are listed in Table 1 and Table 2.

We found that the line radiative force for the star with $T_{eff} = 50,000$ K is larger by 0.3–0.5 dex than Abbott's (1982) because our new atomic line database includes much more highly ionized ions, such as Fe IV. Accordingly the mass loss rate for this star is seven times as large as Abbott's (1982).

Our new atomic line database also includes much more weak lines than before. Therefore, the force multiplier parameter k becomes larger by a factor of 1.6–3.8 and α becomes smaller by 0.05 than Abbott's (1982) within a temperature range of $10,000K \leq T_{eff} \leq 30,000K$. Groenewegen et al. (1989) and Lamers & Leitherer (1993) recently pointed out that the predicted terminal velocities for O stars are systematically larger than the observed values by about a factor of 1.4, and that the predicted mass loss rates are systematically smaller than the observed values by about -2.5 dex. Since our force multiplier parameters tend to increase mass loss rates and decrease wind velocities, it is expected to dissolve the discrepancies between predictions and observation. This result shall be refined by future works.

We found that the mass loss rates for metal deficient stars depend on metallicity roughly with $\dot{M} \sim Z$. This result is close to the result of Abbott (1982) rather than that of Kudritzki et al. (1987). We all have succeeded in interpretation of the wind velocities of Magellanic O stars but failed to interpret the observational mass loss rates of Magellanic O stars. This is one of the most challenging problems of the stellar wind theory.

References

Abbott, D.C. 1982, *Ap. J.*, **259**, 282.

Castor, J., Abbott, D.C. and Klein, R. 1975, *Ap. J.*, **195**, 157.

Groenewegen, M.A.T., Lamers, H.J.G.L.M. and Pauldrach, A.W.A. 1989, *A. Ap.*, **221**, 78

Kudritzki, R.P., Pauldrach, A.W.A. and Puls, J. 1987, *A. Ap.*, **173**, 293.

Lamers, H.J.G.L.M and Leitherer, C. 1993, *Ap. J.*, **412**, 771.

Pauldrach, A.W.A., Puls, J. and Kudritzki, R.P. 1986, *A. Ap.*, **164**, 86.

Table 1: The force multiplier for OB supergiants

T_{eff}	$\log g$	s	$\log f_{\text{rad}}$						
			$t = 10^{-7}$	10^{-6}	10^{-5}	10^{-4}	10^{-3}	10^{-2}	10^{-1}
50,000	4.5	3.1×10^{-3}	3.307	3.026	2.531	1.979	1.430	0.842	0.192
		3.1	3.314	3.039	2.566	2.079	1.637	1.087	0.420
		3.1×10^{3}	3.465	3.250	2.788	2.245	1.729	1.115	0.407
40,000	4.0	1.8×10^{-3}	3.186	2.838	2.274	1.683	1.083	0.493	-0.140
		1.8	3.227	2.858	2.333	1.788	1.297	0.799	0.180
		1.8×10^{3}	3.321	3.059	2.589	2.086	1.595	1.007	0.335
30,000	3.5	1.0×10^{-3}	2.931	2.437	1.863	1.336	0.864	0.377	-0.189
		1.0	3.260	2.894	2.243	1.565	1.031	0.522	-0.050
		1.0×10^{3}	3.336	3.001	2.469	1.955	1.513	1.039	0.475
20,000	2.5	3.0×10^{-4}	2.861	2.495	1.903	1.348	0.810	0.375	-0.091
		3.0×10^{-1}	2.952	2.646	2.283	1.811	1.357	0.863	0.275
		3.0×10^{2}	3.196	2.857	2.414	1.948	1.505	1.031	0.476
15,000	2.0	1.3×10^{-4}	2.499	2.231	1.900	1.377	0.891	0.430	-0.125
		1.3×10^{-1}	2.841	2.411	1.909	1.442	0.987	0.544	0.026
		1.3×10^{2}	3.254	2.938	2.599	2.283	1.831	1.279	0.660
10,000	1.5	3.2×10^{-5}	2.560	2.300	1.981	1.633	1.163	0.603	-0.010
		3.2×10^{-2}	2.919	2.623	2.304	1.902	1.391	0.808	0.185
		3.2×10	3.099	2.757	2.379	1.942	1.417	0.835	0.214

Table 2: The parameters of the force multiplier for OB supergiants

T_{eff}	$\log g$	this work			Abbott (1982)		
		k	α	δ	k	α	δ
50,000	4.5	0.917	0.510	0.040	0.240	0.561	0.083
40,000	4.0	0.483	0.526	0.061	0.205	0.578	0.098
30,000	3.5	0.375	0.522	0.099	0.222	0.561	0.107
20,000	2.5	0.709	0.470	0.089	0.429	0.510	0.084
15,000	2.0	0.922	0.446	0.134	0.524	0.489	0.126
10,000	1.5	0.866	0.454	0.058	0.494	0.490	0.047

RADIATION-ACCELLERATED IONS IN HOT STAR WINDS:
HEATING AND "TURBULENCE" EFFECTS

E.YA. VILKOVISKIJ

V.G. Fessenkov Astrophysical Institute of the National Academy of Sciences, 480068
Almaty, Observatory AFI, Kazakhstan

In the papers by Vilkoviskij (1981), Vilkoviskij and Tambovtzeva (1988) and Springmann and Pauldrach (1992) it was shown that ions can reach velocities $v_i \geq \overline{v}_{Tp}$ ($\overline{v}_{Tp} = (2kT/m_p)^{1/2}$) relative to the wind plasma when the radiation pressure force exceeds the protons frictional force. The condition for this transition can be estimated as

$$v_7 \geq 0.7 \dot{M}_6 z_i^2 / (R_1^2 T_4 \phi(\lambda_i T_*)) \quad , \tag{1}$$

where v_7 is the wind velocity in units of 10^7 cm/s, $\dot{M}_6 = dM/dt$ the stellar mass-loss rate in units of $10^{-6} M_\odot/$yr, z_i the ion charge, $R_1 = R_*/(10R_\odot)$, $\phi(\lambda_i T_*) = f_i \lambda_5^{-3} / e^{14.4/(\lambda_5 T_4)-1}$, $\lambda_5 = \lambda_i /10^3$Å and $T_4 = T_e/10^4$K. The ion distribution function consists of two parts: the part DF_1 approximates the "shifted Maxwell" distribution, and the non-Maxwellian part DF_2 is for "runaway" particles:

$$DF(v_i) = (1 - q_2)DF_{1i} + q_2 DF_{2i} \quad , \tag{2}$$

where $q_2 > 0$ when condition (1) is fulfilled. The specific heat power, divided by N_p^2 is

$$H = N_i/N_p^2 \int_0^\infty v_i DF(v_i) F_{ip} dv_i \quad , \tag{3}$$

where N_i and N_p are the ion and proton densities and F_{ip} is the frictional force. If the "runaway condition" (1) is not fulfilled, the mean velocity of ions relative to the wind plasma $\overline{v}_i$ is less then $\overline{v}_{Tp}$, but it rises to about the electron thermal velocity $\overline{v}_{ep} \sim 43\overline{v}_{Tp}$ in the opposite case. So we can estimate

$$H = H^m(\overline{v}_i/v_{Tp}(1 - q_2) + 43q_2) \quad , \tag{4}$$

where $H^m \cong \pi e^4 z_i^2 N_i (N_p k T_p)^{-1} \ln \Lambda \ v_{Tp} = (N_i)$
which is $\sim 2 \cdot 10^{-22} z_i^2 T_{p4}^{-1/2} (n_i/10^{-4})$ erg cm^3/s, where $n_i = N_i/N_p$.
Model calculations show that for the O- and early B-type stars this "kinetic heat" is sufficient for heating of the wind to a temperature $T_e \sim 10^5$ K at $R \geq 2R_*$ and to $T_e \geq 10^6$ K at $R \geq 100R_*$, and the ion's velocity distribution can manifest itself as "turbulence" in spectral lines.

L. A. Balona et al. (eds.), Pulsation, Rotation and Mass Loss in Early-Type Stars, 490–491.

So the physical picture including kinetic heat resembles the empirical "warm wind" model, and moreover, it predicts an outer hot corona at $R \geq R_k \sim 100 R_*$. The X-ray luminosity of the corona is $L_x = 4\pi \int_{Rk}^{\infty} \epsilon_X R^2 dR$. With $\dot{M} = 4\pi \rho v r^2$, $\epsilon_X \cong 2 \cdot 10^{-27} N_e^2 T_e^{1/2}$ erg/cm^{3}s, we have

$$L_x \cong 5 \cdot 10^{32} T_{e7}^{1/2} \dot{M}_6^2 / (R_1 v_8^2 (r_k/100)) \ \text{erg/s} \ , \tag{5}$$

where R_1 is the stellar radius in units of $10 R_\odot$, $r_k = R_k/R_*$, $T_{e7} = T_e/10^7$K and v_8 is the wind velocity in units of 10^3km/s at R_k.

We can predict the wide dispersion of X-ray luminosities as due to the dispersion in stellar parameters, and the X-ray variability (with a characteristic time scale of several hours to days) as due to stellar wind variability.

References

Springmann U.V.E., Pauldrach A.W.A.: 1992, *Astronomy and Astrophysics* **262**, 515
Vilkoviskij E.Ya.: 1981, *Astrofisica (Sov)* **17**, 310
Vilkoviskij E.Ya., Tambovtzeva L.V.: 1988, *in: "Mass Outflows"*, Bianchi L., Gilmozzi R. (eds.) , p. 195

A simple model for a radiatively driven metallic wind in A0 stars

J. BABEL

Service d'Astrophysique, Centre d'Etudes de Saclay, 91191 Gif-sur-Yvette, France

Abstract. Litle is known about the wind of main sequence A7 to B4 stars. We here investigate the case of a radiatively driven wind for an A0 star and obtain stringent limits on the mass loss rate. We also show that frictional heating lead to the presence of a chromospheric region very near the photosphere.

1. A radiatively driven metallic wind in A stars?

For main sequence stars with $T_{eff} \lesssim 14000$ K, the maximum radiative acceleration in expanding atmospheres, $g_{rad,tot}^{max}$, is always smaller than gravity (Abbott 1982) and there is no possibility of homogeneous radiatively driven winds. Individually however, metals receive a large radiative acceleration which can overcome gravity even in the photosphere (Babel & Michaud 1991). Approximating the radiative acceleration transmitted to metals (as a whole), $g_{rad,Z}^{max}$, by $g_{rad,tot}^{max}/Z$, with Z the mass fractions of metals, we obtain that $g_{rad,Z}^{max} > g$ and that a *metallic* wind can potentially be present, but cannot drag H and He in the subsonic domain.

The individual momentum equation for a stationary multicomponent flow is:

$$\frac{\partial_r V_i}{V_i}(a_i^2 - V_i^2) + 2\frac{a_i^2}{r} - \frac{g}{r^2} + g_{rad,i} + \frac{z_i E}{m_i} - \sum_{j \neq i} n_j \frac{R}{kT}\frac{z_i z_j}{m_i} g(X_{ij})\frac{V_i - V_j}{|V_i - V_j|} \equiv 0 \quad (1)$$

and the global energy equation is (Braginskii 1965), for $T_i = T \; \forall i$,

$$\sum_{i \neq e} n_i V_i kT \left(\frac{3}{2}\frac{\partial ln(T)}{\partial r}(1 + z_i) + \frac{5}{2}\frac{\partial z_i}{\partial r} - z_i \frac{\partial ln(n_e)}{\partial r} - \frac{1}{n_i}\frac{\partial n_i}{\partial r}\right) =$$

$$- \sum_i div q_i + Q_{rad} + Q_{fric} \quad (2)$$

We make here the approximation of Maxwellian distribution for all components, nebular approximation for the ionisation equilibrium and assume that $g_{rad,i}$ is not a function of $\partial_r V_i$. This last approximation is justified by the very small optical depth of the medium. In the case of A0 stars, H and He can be considered as being in hydrostatic equilibrium. In this case and if we consider all metals as one average population (label Z) there is only one critical point for $V_Z^2 = kT/m_Z$. It does not fix the mass loss rate. The mass loss rate is fixed instead by the populations of each species at the lower boundary (photosphere).

L. A. Balona et al. (eds.), Pulsation, Rotation and Mass Loss in Early-Type Stars, 492–493.
© 1994 IAU. Printed in the Netherlands.

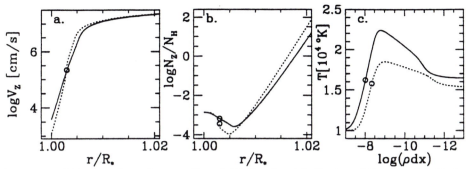

Fig. 1. a. Velocity of metals *vs* radius. Solid line (model A), dotted line (model B). b. Same as in (a) but for the concentration of metals. c. Temperature as a function of the column density. A circle indicates the position of the critical point in the two models.

2. Results and discussion

Equations 1 and 2 were solved simultaneously for a ZAMS star with $T_{eff} = 10000$ K, $\log g = 4.37$, for a force multiplier $M_{max} = 10^{2.7}$ (Abbott 1982) and for lower boundary conditions:$T^\circ = T_{eff}$, $N_H^o = 10^9$ cm^{-3}, $Y^\circ = Y_\odot$ and $Z^\circ = Z_\odot$ (we consider the mean metallic atom as an iron atom). Model A is for a case of constant radiative acceleration $g_{rad,Z} = g_{rad,Z}^{max}(R_*/r)^2 = 1.7 \ 10^5(R_*/r)^2$ and model B is for a case case where $g_{rad,Z}$ varies with V_m in a way which simulates the effect of Doppler shift on the radiative acceleration.

Results are shown in Fig. 1. The *metallic* mass loss rate corresponding to the 2 models are: (A) 6.5 10^{-16} M$_\odot$yr^{-1}, (B) 1.9 10^{-16} M$_\odot$yr^{-1}. Any additional decrease, relative to the cases considered here, of the radiative acceleration or any additional coupling of the gases below (or at) the critical point decreases the mass loss rate. The estimates from our simple model give thus an approximate upper limit on the mass loss rate due to radiative pressure. They are of similar order as obtained from the diffusion theory (Babel 1992).

Even if $\dot M$ is very small, the frictional heating, Q_{fric}, plays a major role, as in τ Sco (Springmann & Pauldrach 1992), and leads to a large increase of the gas temperature near the critical point. It could thus lead, for A and B stars, to much larger *EUV* radiation than expected from hydrostatic (*LTE* or *NLTE*) models in radiative equilibrium. This could help to solve the problem caused by the *EUV* flux of β Cma (Cassinelli et al., this volume) .

References

Abbott, D. C.:1982, ApJ 259, 282
Babel, J.:1992, A&A 258, 449
Babel, J., Michaud, G.:1991, ApJ 366, 560
Braginskii, S. I.:1965, in Reviews of Plasma Physics, New York, p. 205
Springmann, U. W. E, Pauldrach, A. W. A.: 1992, A&A 262, 515

SMOOTH WINDS AND CLUMPED WINDS FOR EARLY-TYPE STARS

R. BLOMME

Royal Observatory of Belgium, Ringlaan 3, B-1180 Brussels, Belgium

and

M. RUNACRES

Vrije Universiteit Brussel, Pleinlaan 2, B-1050 Brussels, Belgium

Abstract. The use of the IR and radio continuum as a clumping indicator for the stellar wind of early-type stars is investigated.

1. Introduction

The instability of hot star winds was already pointed out in 1970 (Lucy and Solomon) and has been confirmed in recent years by various observations. This instability causes the wind to be highly structured: it is clumped rather than homogeneous (Owocki et al. 1988). Free-free and bound-free scattering are the most important opacity sources at infrared and radio wavelengths and they depend on the square of the density. Hence the flux at these wavelengths can be used as a clumping indicator.

Furthermore, the infrared continuum is formed close to the surface of the star, whereas the radio continuum is formed far out in the wind. Hence by looking at different wavelengths, we can scan through different geometrical regions of the stellar wind. In this study we shall compare both the smooth models (density monotonously decreasing with radius) and clumped models ('wild' density variations) with observations.

2. Results and Conclusions

We applied our own NLTE model to a sample of stars (for details we refer to Runacres and Blomme 1994). We find good agreement between our model predictions and the observations of the few stars we already studied. As an example we show ϵ Ori (figure 1).

However, for the extreme O star ζ Pup we find an infrared excess which is slightly too low (figure 2). This probably points to clumping due to instability in part of the stellar wind. To test this hypothesis we ran a model with a clumped density structure damping out before 10 stellar radii. It is clear that this produces an effect in the right direction: even though we have taken a rather arbitrary damping scale length, the clumped model agrees with the observations better than the smooth one does.

L. A. Balona et al. (eds.), Pulsation, Rotation and Mass Loss in Early-Type Stars, 494–495.

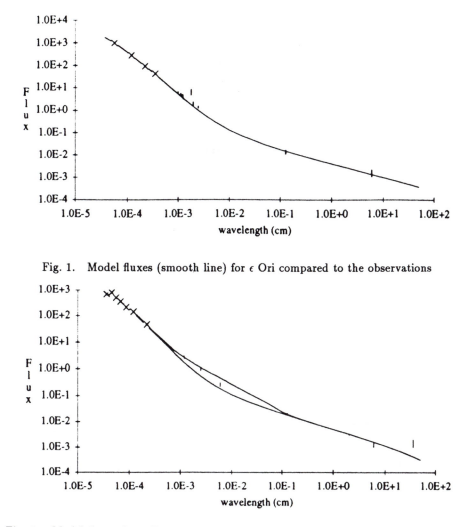

Fig. 1. Model fluxes (smooth line) for ϵ Ori compared to the observations

Fig. 2. Model fluxes for ζ Pup compared to the observations. The lower line shows the results for the smooth model, the upper one for the clumped model

It appears that clumping might not be all that severe or frequent among O and B stars. However, more definite conclusions will be drawn once the study is completed.

References

Lucy, L.B., Solomon, P.M.: 1970, *ApJ* **159**, 879
Owocki, S.P., Castor, J.I., Rybicki, G.B.: 1988, *ApJ* **335**, 914
Runacres, M.C., Blomme, R.: 1994, in *Instability and Variability of Hot Star Winds*, Kluwer, in press

LINE PROFILES IN EXPANDING ATMOSPHERES

L. CIDALE

Observatorio de La Plata - Paseo del Bosque S/N - 1900 La Plata - Argentina

1. Introduction

The analysis of the *UV* resonance transitions of superionized species observed in early-type stars suggests the presence of a high temperature region
and an outward flow of matter (Snow & Marlborough 1976).

With the aim of contributing to the knowledge of the atmospheric structure in Be stars we computed line profiles of Mg II and C IV assuming the
existence of a chromosphere in an expanding medium with spherical symmetry. In particular, we select the atmospheric models that were previously
used for the computation of the Hα line profiles (Cidale & Ringuelet 1993).

2. The Atmospheric Model

Taking into account the different ionization degrees and Doppler shifts observed in the spectrum of Be stars, we assume the existence of a sequence of
spherically symmetric layers with different thermodynamical and kinematical properties; that is, *i*) a classical photosphere in radiative and hydrostatic
equilibrium, *ii*) an expanding high temperature region with $T > T_{\text{eff}}$ (the
chromosphere), and *iii*) a cool envelope with $T < 10^4$ K.

For the construction of the atmospheric model we follow, basically, the
model adopted by Catala & Kunasz (1987). The velocity law increases monotonically with radius throughout the extended atmosphere until it reaches
a constant value. We try with different velocity curves considering high and
low velocity gradients at the base of the chromosphere (see figure 1A).

3. Results

Once the atmospheric model and the velocity field have been selected, the
radiative transfer equation is treated for a spherically symmetric medium
applying the comoving-frame method (Mihalas & Kunasz 1978). The statistical equilibrium equations are solved, simultaneously, for a Mg II atomic
model consisting of 14 energy levels plus continuum. The solution is obtained
by means of the equivalent two-level atom approach (ETLA).

We compute Mg II resonance lines for a wide range of effective temperature. Since the results related to the velocity law are valid for all spectral
types, we present results corresponding to a late Be-type star ($T_{\text{eff}} = 14000$

L. A. Balona et al. (eds.), Pulsation, Rotation and Mass Loss in Early-Type Stars, 496–497.

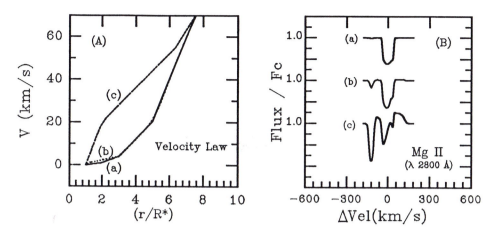

Fig. 1. A) Different velocity laws, B) A sequence of Mg II profiles computed with the velocity laws shown in figure 1A

K) obtained with different velocity laws (cases a, b, c). We observe that, 1) The shape of the resonance line profiles (multiplet UV 1) depend on the velocity gradients in the region next to the photosphere. Low-velocity gradients, at the base of the wind, yield a single absorption line profile; however, high velocity gradients in the same layers predict multiple components (either in absorption or in emission) as the ones showed in figure 1B. The blue shifted absorption component, when it appears, may give information about the velocity field in the most outer layers of the atmosphere. 2) In all our models multiplet UV 3, as well as $\lambda 4481$, appear in absorption and they are not sensitive to the radial velocity distribution. These lines present, mainly, a photospheric line-forming region. 3) Mg II lines depend slightly on the chromospheric temperature law.

Finally, we compute P Cyg profiles of C IV considering a two-level atom and a continuum. We observe that increasing the distance of the temperature maximum the emission component becomes stronger while the absorption component shifts to the blue.

All our results are consistent with observations and with the previously computed Hα profiles.

References

Catala, C., and Kunasz, P. B.: 1987, *A&A* **174**, 158
Cidale, L. S., and Ringuelet, A. E.: 1993, *ApJ* **411**, 874
Mihalas, D., and Kunasz, P. B.: 1978, *ApJ* **219**, 635
Snow, T. P., and Marlborough, J. M.: 1976, *ApJ* **203**, L87

NON-THERMAL VELOCITIES IN THE STELLAR WIND OF EARLY-TYPE STARS

R. DE VOS

IH-CTL-Gent, Voskenslaan 270, B-9000 Gent, Belgium

and

R. BLOMME

Royal Observatory of Belgium, Ringlaan 3, B-1180 Brussels, Belgium

Abstract. Turbulence is known to be important in the stellar wind of early-type stars. We explore the influence of turbulence that depends on the distance to the star.

1. Introduction

The P-Cygni profiles seen in UV resonance lines are due to the presence of a stellar wind. The SEI (Sobolev with Exact Integration) method as developed by Lamers et al. (1987) provides a semi-empirical model for fitting these profiles. Contrary to the more classical Sobolev approach, this method takes into account the intrinsic broadening of the spectral line (due to thermal and turbulent velocities).

Groenewegen and Lamers (1989) found that a best fit could only be obtained by assuming a large intrinsic broadening, of the order of 100-300 km/s, which they called "turbulence". It is suspected that this "turbulence" is somehow related to the instabilities found in the time-dependent models of stellar winds (Owocki et al., 1988).

In their work Groenewegen and Lamers assumed a "turbulence" that was constant in the wind. We investigated the effect of a "turbulence" that varies with distance to see whether a better fit could be obtained. Knowledge of the radial dependence of this "turbulence" could put constraints on the ab-initio calculations of the time-dependent radiatively driven wind. In our exploratory study we specifically tried the following law for the turbulent velocity: $v_{turb}/v_\infty = 0.01 + 0.09(v(r)/v_\infty)^3$ and applied it to ζ Pup.

2. Results and Conclusions

Fig. 1 shows the best fit we obtained with this variable turbulence law compared to a constant law $(v_{turb}/v_\infty = 0.1)$ for a saturated profile (C IV). Fig. 2 shows a similar comparison for an unsaturated profile (N IV). The turbulence law we studied does not result in an obvious improvement over the constant turbulence. It should be stressed however that in future we shall explore the large parameterspace possible with these variable turbulence

L. A. Balona et al. (eds.), Pulsation, Rotation and Mass Loss in Early-Type Stars, 498–499.

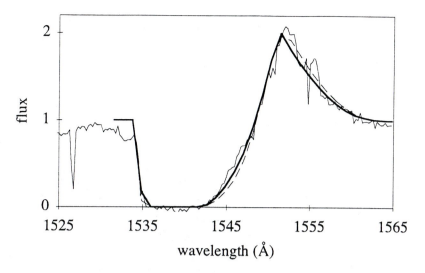

Fig. 1. The variable turbulence (full line) compared to the constant turbulence (dashed line) for C IV. The observed profile is also shown

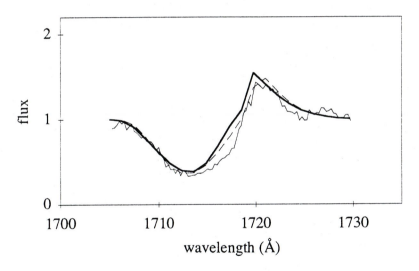

Fig. 2. The same as Fig. 1 but for N IV

laws and that in some cases we might find an improvement over constant turbulence.

References

Groenewegen, M.A.T., Lamers, H.J.G.L.M.: 1989, *A&AS* **79**, 359
Lamers, H.J.G.L.M., Cerruti-Sola, M., Perinotto, M.: 1987, *ApJ* **314**, 726
Owocki, S.P., Castor, J.I., Rybicki, G.B.: 1988, *ApJ* **335**, 914

UNDERSTANDING THE P CYGNI PROFILE

N. RONS
I.W.O.N.L. bursary holder
Astrofysisch Instituut, Vrije Universiteit Brussel, Pleinlaan 2, B-1050 Brussel, Belgium

1. Introduction

Narrow Absorption Components (NACs) have been detected in essentially all unsaturated P Cygni profiles of O-type stars (Howarth & Prinja, 1989). Time-series of observations have revealed their migration through the profile's absorption trough (Prinja *et al.*, 1987, Henrichs *et al.*, 1988). The interpretation of the NACs is essential for the understanding of the dynamic structure of the winds of O stars and of the driving mechanisms behind it. The P Cygni profile substructures caused by isolated reverse and forward shocks were calculated in a 'line only' approximation for a spherically symmetric, isothermal stellar wind. The Comoving Frame method was used for an accurate treatment of the complex radiation transfer in the shock environment. The underlying velocity structure follows a β-law (except in the innermost layers where a polynomial law was used) while the ionization was taken to be constant throughout the entire wind.

2. Calculations and Interpretation

Migration behaviour. Both reverse and forward shock calculations result in NACs which 'appear' at roughly half of the edge velocity, gain in strength as they migrate to gradually faint away as they approach the edge velocity (Fig. 1.). Shocks moving through the lower velocity regions of the wind cause some variation of the P Cygni profile, but do not produce features recognizable as NACs in individual observations. Almost no influence of shocks is seen in the emission peak of the P Cygni profile. The main difference between NACs caused by reverse and forward shocks is in the order in which the increase and decrease in absorption occur in the profile with respect to the 'unshocked' model. It is not easy to distinguish between these two cases in the observed individual profiles and even in time-series this may not be straightforward as the NACs often are organized in a closely spaced recurrent scheme. The high velocity stage was found to be the best phase in which to identify the 'reverse' or 'forward' nature of the shocks in observed P Cygni profiles. After their migration the observed NACs remain at their highest velocity during a considerable lapse of time (Prinja, 1992). This behaviour is reproduced by the reverse shock models (Fig. 2), while the

500

L. A. Balona et al. (eds.), Pulsation, Rotation and Mass Loss in Early-Type Stars, 500–501.
© 1994 *IAU. Printed in the Netherlands.*

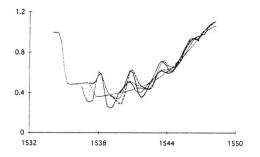

Fig. 1. Migration of theoretical NACs in the C IV λλ 1548.19 1550.76 doublet, caused by reverse (full line) and forward (dashed line) shocks. The dotted line is the 'unshocked' profile.

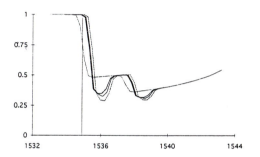

Fig. 2. Theoretical high velocity NACs in the C IV λλ 1548.19 1550.76 doublet, caused by reverse shocks. The dotted line is the 'unshocked' profile.

forward shock models result in features above the edge velocity disappearing more rapidly as the shocks move through the outer wind layers.

The NACs as a Source of Information. The NACs contain a wealth of information about the wind structure. In the case of reverse shocks, an NAC with a faint decreased absorption component points towards a shock with a sharp low velocity edge. A 'sharp edged' NAC points in the direction of relatively low thermal (and microturbulent) velocities. The width of the NAC is linked to the depth of the shock in the wind.

References

Henrichs, H.F., Kaper, L., Zwarthoed, G.A.A.: 1988, *Ap. J.* **317**, 389

Henrichs, H.F., Kaper, L., Zwarthoed, G.A.A.: 1988, 'Rapid Variability in O Star Winds' in E.J. Rolfe, ed(s)., *A Decade of UV Astronomy with IUE*, Proc. Celebratory Symposium, GSFC, Greenbelt, USA, ESA SP-281, Vol. 2, 145

Howarth, I.D., Prinja, R.K.: 1989, *Ap. J. Supp. Ser.* **69**, 527

Prinja, R.K., Howarth, I.D., Henrichs, H.F.: 1987, *Ap. J.* **317**, 389

Prinja, R.K.: 1992, 'UV P Cygni Profile variability in O Stars' in L. Drissen, C. Leitherer, A. Nota, ed(s)., *Nonisotropic and Variable Outflows from Stars*, ASP Conference Series, Vol. 22, 167

IMPROVED MODELLING FOR SPATIALLY RESOLVED
SPECTROSCOPY OF A P CYG ENVELOPE

M.BOURGUINE[1,2] AND A. CHALABAEV[1]

[1] CNRS, Observatoire de Grenoble, UJF-B.P.53X,38041 Grenoble CEDEX France

[2] FIAN Astro-space Center, Moscow, Russia

1. General remarks

Our study, of which we give here a progress report, addresses two problems. The first is to develop methods and software permitting to compare a wide range of theoretical models of stars with envelopes with observational data produced or expected to be produced by high angular resolution optical interferometry combined with spectroscopy. The second problem is to find out the modes of interferometric observations (base length, spectral resolution etc) that are most informative for determining the physical parameters of stellar envelopes.

We started with the relatively simple case of a spherically symmetric stellar envelope of the Bep star P Cyg. Because of the diversity of theoretical models, the number of parameter sets to be considered in comparing observational data with theoretical predictions is rather large even in the case of spherical symmetry. In order to be able to interpret the observational data using a small- or medium-scale computer, it is practical, at least at the first stage, to employ a simplified model permitting a rapid performance of the calculations for many parameter sets. The more refined and time consuming codes like those used by Pauldrach and Puls (1990), and de Koter et al. (1993) may be employed at the final stage for the analysis of a limited number of models.

Furthermore, in non-spherically symmetric cases (e.g. Be stars) both the number of parameters to be determined and the complexity of theoretical models are even larger, and computational efficiency becomes one of the critical points.

For time being, we use a code that computes the emergent monochromatic intensity distributions for any reasonable stationary spherically symmetric model defined by radial velocity $V(R)$ and envelope temperature $T(R)$ as functions of radial distance R, and mass-loss rate dM/dt.

Calculations of the line source function are performed for a 3-level + continuum model of the H atom and are limited to the Hα line. Following Drew (1985), the envelope is taken to be completely opaque in the Lyman continuum, so that direct recombinations to and photoionizations from the ground level cancel out.

In calculating statistical equilibrium, the line radiation transfer was treated in the Sobolev approximation. Computation of the emergent intensity is

L. A. Balona et al. (eds.), Pulsation, Rotation and Mass Loss in Early-Type Stars, 502–504.

based on the code developed by Bertout (1984).

The simplicity of the 3-level atomic model used so far allows to solve the statistical equilibrium equations very rapidly, although with poor accuracy. It should be improved at the next stage, when we are planning to take into account the influence of higher levels. Preliminary analysis shows that except for the innermost part of the envelope, the influence of levels $n > 3$ can be accounted for accurately enough without noticeable decrease in efficiency.

2. Interferometry vs. spectroscopy

At which extent can the interferometric spectro-imaging provide interesting physical information that cannot be obtained with much more simple spatially unresolved spectroscopy? In order to answer this question, it is necessary to compute a grid of multiparametric models of the object under study that covers sufficiently well the parametric space domain of interest, and to search for pairs of models, which yield similar spectral line profiles, while differing significantly in monochromatic intensity distributions.

This was done for the case of P Cyg. We considered a grid of 100 3-parameter models of the envelope. The varied parameters were mass-loss rate dM/dt ($5\,10^{-6} - 5\,10^{-5} M_\odot/y$), parameter α (3.0 - 4.5) of the velocity law $V(R) = V_{inf}(1 - R_*/R)^\alpha$, and envelope temperature T (10000 K - 14000 K). The terminal velocity was taken to be 300 km/s.

For each model we calculated the Hα profile I(λ) and the effective radius $R_{eff}(\lambda)$. The usefulness of $R_{eff}(\lambda)$ for a condensed representation of the results of interferometric observations is discussed in Burgin and Chalabaev (1992). The preliminary result is that on the considered grid the interferometric observations are not much more informative than conventional spectroscopy. This conclusion disagrees with our earlier result (Burgin and Chalabaev 1992). The reason is that in that paper the calculation for envelopes with a different temperature structure were based on a single model taken from Drew (1985), and the dependence of excited levels population on the temperature was estimated very approximately. It was supposed that the Menzel departure coefficients are nearly constant, and that the dependence on temperature may be sufficiently well represented by the dependence of Boltzmann factors, which is not the case for photoionization-recombination controlled levels.

A qualitative assessment of the comparative merits of interferometric spectro-imaging and classical spectroscopy for differentiating between various theoretical model would require an analysis of the influence of observational errors and depends on the instrumentation considered. Application to realistic situations would also require the extension of the analysis to the case of observations in several lines. Both these problems will be addressed in further work.

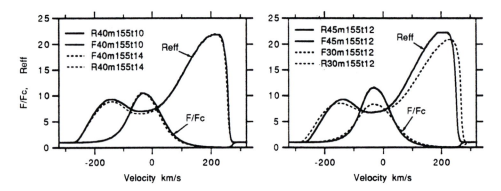

Fig. 1. Left: comparison of the results for models with the same velocity law (α=4.0) and various envelope temperatures (T=10000 K and T=14000 K). Right: the same for fixed T=12000 K and various velocity laws (α=3.0 and α=4.5). In both cases $dM/dt = 1.5\,10^{-5}\,M_\odot/y$. Effective radius R_{eff} is measured in units of stellar radius, positive velocities correspond to blue-shifted emission.

References

Bertout C.: 1984, *Astrophysical Journal* **285**, 269

Burgin M.S, and Chalabaev A.A.: 1992, *Proceedings of the ESA Colloquium on Targets for Space-based Interferometry, Beaulieu, France, 13-16 October 1992* **ESA SP-354**,

Drew J.E.: 1979, *Monthly Notices of the RAS* **217**, 867

de Koter A., Schmutz W., & Lamers H.J.G.L.M.: 1993, *Astron. Astroph.* **277**, 561

Pauldrach A.W.A. & Puls J.: 1990, *Astronomy and Astrophysics* **237**, 409

THE HARD LIFE OF CLUMPS IN ATMOSPHERES OF EARLY-TYPE STARS

A.F. KHOLTYGIN

Astronomical Institute, St. Petersburg University,
198904, St. Petersburg, Petrodvoretz, Russia
E-Mail: afx@astro.lgu.spb.su, vayak@astro.lgu.spb.su

Abstract. As has been recently established, the atmospheres of early-type stars are strongly disheveled. These atmospheres consist of dense clumps and a more rarefied inter-clump medium. We consider the forces that determine the motion of the clumps. We show that the main force is gravitational attraction of the clumps to the star. For stars with a high effective temperature ($T_{\mathrm{eff}} \geq 6 \cdot 10^4$ K) it appears that the clump/interclump gas number density ratio is ≥ 30. In this case we can conclude that the clumps are formed by shock waves. Some evidence as to the existence of gas jets in the atmospheres of Wolf-Rayet stars have recently been presented. We propose that such jets may be a common feature of all early-type stars. Profiles of HeI lines in spectra of clumped atmospheres with jets have been calculated.

There is now much evidence (both theoretical and observational) that inhomogeneities exist in the atmospheres (envelopes) of early-type stars (Antokhin et al. 1988, hereafter AKC; Andrillat and Vreux 1988; Floquet et al. 1992, Lamers et al. 1988). These inhomogeneities can be described in the framework of the Clump model. In the Clump model (see, for example, AKC, Cherepaschuck 1990) stellar atmospheres are presented as consisting of numerous dense clumps embedded in a rarefied interclump medium. Ions of low and moderate ionization stages are found only in clumps while the interclump medium is strongly ionized.

In the case of early-type stars the prime reason for the formation of clumps seems to be the instability of the radiation driven stellar wind (see, for example, Owocki et al. 1990). After the initial perturbations are formed they can be compressed by shock waves to provide a ratio of the clump/interclump number densities as large as $30 - 100$, which is needed to keep the ions of the low and intermediate ionization stages in the deeper layers of the clumps (Kholtygin 1988). A special case is presented by the clumps in shells of Herbig Ae/Be stars. They have huge, dense and extreme-ly cold clumps with comet like orbits. The formation of clumps in Herbig Ae/Be star shells are probably connected with the gravitational collapse of primordial molecular clouds (Grinin 1988, Voschinnikov and Grinin 1991).

We have considered the forces that determine the motion of clumps in an atmosphere and concluded that the main force is gravitational attraction to the core and that the clumps should therefore be decelerated in the atmo-sphere. However, the cluster of clumps seems to be accelerated (e.g., Lamers et al. 1988). So we must conclude that either the clumps are formed in the whole atmosphere, or that the cluster of clumps, which form the profile

L. A. Balona et al. (eds.), Pulsation, Rotation and Mass Loss in Early-Type Stars, 505–506.

bump, include clumps with very different velocities.

The Clump model has been used as the base for line profiles calculations. We have considered lines of ions in lower ionization stages (HeI, CII, NII, etc.) which are disposed only in clumps.

In spectra of some WR stars the profiles of the HeI $\lambda10830$ lines have bumps with a constant shape and velocity displacement (Eenens 1991). Similar bumps have been observed in optical HeI lines (Robert et al. 1992). For this reason we have proposed that there are only one or two jets or streams consisting of both rarefied gas and clumps. To evaluate the line profiles we proposed that the ratio of the clump number densities in the jets and in the remaining atmosphere is constant. The calculated profiles of the HeI $\lambda10830$ line in the spectra of the star WR 134 are in a good accordance with the observed ones (Eenens 1991). The jet-like structures also exist in the atmospheres of Herbig Ae/Be stars. They seems to be connected with the existence of a strong magnetic field (Pogodin 1990). We propose that jets are a common feature in all early-type stars.

Acknowledgements

We are grateful to Dr. P.R.J. Eenens for presenting a copy of his Ph.D. Thesis. This work was partly supported by ESO grant A-01-093 within the C&EE programme of East European Astronomers support and by grant G 01.3.5.10 of the Russian Ministry of Education and Science.

References

Andrillat, Y., Vreux, J.-M.: 1988, *Astr. Ap.* **253**, L37

Antokhin, I.I., Kholtygin, A.F., Cherepaschuck, A.M.: 1988, *Sov. Ast.Zh.* **32**, 285

Cherepaschuck, A.M.: 1990, *Astron. Zh.* **67**, 955

Eenens, P.R.J.: 1991, *Ph.D. Thesis, Univ. of Edinburg, 1991* ,

Floquet, M., Hubert, A.M., Janot-Pacheco, E., Mekkas, A., Hubert, H., Leister, N.V.: 1992, *Astr. Ap.* **254**, 177

Grinin, V.P.: 1988, *Pis'ma v Astron Zh.* **14**, 65

Kholtygin, A.F.: 1988, *"Wolf-Rayet stars and related objects", Tallinn* , 115

Lamers, H.J.G.L.M., Snow, T.P., de Jager, C., Langerwerf, A.: 1988, *Ap. J* **325**, 342

Owocki, S.P., Castor, J.I., Rybicki, G.B.: 1988, *Ap. J* **335**, 914

Pogodin, M.A.: 1990, *Astrofizika* **32**, 371

Robert, C., Moffat, A.F.J., Drissen, L., Lamontagne, R., Seggewiss, W., Niemela, V.S., Cerruti, M.A., Barrett, P., Bailey, J., Garcia, J., Tapia, S.: 1992, *Ap. J* **397**, 277

Voschinnikov, N.V., Grinin, V.P.: 1991, *Astrofizika* **34**, 181

TIME-DEPENDENT PHENOMENA IN OB STAR WINDS

RAMAN K. PRINJA

Department of Physics and Astronomy, University College London, Gower Street,
London WC1E 6BT, U.K.

Abstract. Spectroscopic observations are discussed which indicate that the winds of luminous OB stars are highly structured in space and variable in time. Blueward migrating discrete optical depth enhancements in the absorption troughs of P Cygni profiles are the characteristic signatures of the evolving structure. Constraints on physical mechanisms are discussed, provided by the observed accelerations of the migrating features, the evidence for rotationally modulated variability, and the behaviour of the winds at very low velocities.

1. Observations of Evolving Wind Structure

I shall discuss here some of the recent developments in our understanding of the time-variable nature of winds in luminous OB stars. The view I present is mostly as witnessed by the remarkable *IUE* satellite, since UV spectroscopy still provides one of the most sensitive probes of hot, stellar outflows.

One of the important discoveries made with *IUE* data is that the UV resonance line P Cygni profiles of OB star are *systematically* variable on time scales down to less than 1 hour, and that the profile changes are primarily due to variable "Discrete Absorption Components" (DACs). Recent reviews of DAC properties in OB stars have been provided by Henrichs (1988), Prinja (1992) and Howarth (1992). The DACs are initially recognised as localised optical depth enhancements at typically ≥ 0.3 of the wind terminal velocity (v_∞), in the absorption troughs of the UV P Cygni profiles. They subsequently accelerate to the shortward wing of the profile over ~ 1–2 days, and usually recur at intervals of ≥ 1 day. The full-width at half-maximum of the absorption enhancements may vary over $\sim 0.05v_\infty$ to $0.25v_\infty$, with some dependence on the central velocity. Typical Si^{3+} column densities due to DACs range from $\sim 1 \times 10^{13}$ cm^{-2} to $\sim 3 \times 10^{14}$ cm^{-2} (assuming locally plane-parallel geometry). Migrating DACs have now been reported in the UV wind lines of a wide range of hot stars, including early, mid and late O-type dwarfs, giants and supergiants, plus early B supergiants, and a WN7 Wolf-Rayet star. Individual case studies combine with surveys of the time-averaged wind properties (e.g. Howarth & Prinja 1989), to suggest that evolving optical depth enhancements are present all the time in the wind-formed UV lines of most O stars.

The progressive DACs provide direct evidence that the winds of OB stars are highly structured in space. The structures must be substantial and large scale since DACs can account for $\sim 20\%$ to 50% of the wind line profile, with observed central optical depths which often exceed 0.5. Clues that

L. A. Balona et al. (eds.), Pulsation, Rotation and Mass Loss in Early-Type Stars, 507–516.
© 1994 IAU. *Printed in the Netherlands.*

the DAC line-formation region is *not* spherically symmetric come from noting that the observed profile variability is confined (in *IUE* data) to the absorption regions only, with no corresponding fluctuations evident in the emission components of the P Cygni profiles. The structured nature of hot star winds has some bearing on the determination of important parameters such as mass-loss rate, terminal velocity, ionization balance, and on theories concerned with the hydrodynamics of the outflows. An attractive physical process for generating a degree of evolving wind structure is that due to strong line-driven instabilities (see e.g. reviews by Owocki 1991, 1992). The radiation hydrodynamics simulations of Puls *et al.* (1993) show for example that the evolution of dense clumps formed in the wind due to the action of the radiative instability do indeed reproduce some of the observed DAC behaviour.

I discuss below three aspects of the observed time-dependent nature of OB star winds which are relevant to the development of physical models addressing evolving wind structure; i.e. (i) the observed DAC accelerations, (ii) evidence for rotationally modulated variability, and (iii) the behaviour of the wind at very low velocities ($\ll 0.3v_\infty$).

2. The Observed DAC Accelerations

The observations of DACs moving through line profiles provides an important, and rare, opportunity to determine directly the acceleration of material associated with hot stellar winds. The ranges of DAC accelerations for 10 hot stars are listed in Table 1, based on UV and optical spectroscopic time-series studies. Typically, the acceleration of the UV absorption enhancement is largest (~ 2–3×10^{-2} km s^{-2}) close to its initial observed central velocity (~ 0.3–$0.5v_\infty$), where the feature may be more than 600 km s^{-1} broad. The acceleration decreases to $\leq 1 \times 10^{-3}$ km s^{-2} towards the 'end-point' of the DAC evolution, where a narrow absorption component (≤ 100 km s^{-1} wide) becomes 'lost' in the blue edge of the line profile. The trend of acceleration as a function of velocity is variable for different DACs in an individual star. The accelerations of sequential features in a star are not always the same for example at similar velocities. Indeed, there are no clear cases where it may be argued—from the recorded trends of central velocity, width, acceleration, and strength—that a given structure has entered the line-of-sight more than once during time-series observations typically spanning ~ 2–5 days.

All the DAC accelerations listed in Table 1 are slow in comparison with steady-state wind model predictions. The observed accelerations of representative DACs in the UV spectra of ζ Pup, 68 Cygni, and HD 64760 are plotted as a function of central velocity in Fig. 1. The corresponding accelerations

TABLE I

Observed DAC accelerations

Star	Sp. type	DAC velocity range (v_c/v_∞)	DAC acceleration range (10^{-2} km s^{-2})	Reference
HD 93131	WN 7	0.55–0.85	~ 1.0–0.15	Prinja & Smith (1992)
ζ Pup	O4 I(n)f	0.30–0.75	~ 3.2–0.10	Prinja *et al.* (1992)
λ Cep	O6 I(n)fp	0.26–0.75	~ 2.2–0.20	Kaper (1993)
68 Cyg	O7.5 III:n((f))	0.47–0.95	~ 3.0–0.05	Fullerton *et al.* (1991)
ξ Per	O7.5 III(n)((f))	0.70–0.90	~ 3.0–0.50	Prinja *et al.* (1987)
HD 152408	O8:Iafpe	0.05–0.45	~ 0.15–0.05	Prinja & Fullerton (1994)
HD 151804	O8 Iaf	0.12–0.48	~ 0.35–0.10	Fullerton *et al.* (1992)
19 Cep	O9.5 Ib	0.53–0.73	~ 0.10	Prinja (1992)
ζ Oph	O9.5 V	0.80–0.96	~ 0.20–0.05	Howarth *et al.* (1993)
HD 64760	B0.5 Ib	0.17–0.95	~ 2.0–0.10	Massa *et al.* (1994)

expected for a velocity law of the type $w = w_0 + (1 - w_0)(1 - R_*/r)^\beta$ for $\beta = 1$ are also shown (e.g. steady-state wind calculations of Friend & Abbott 1986; Pauldrach, Puls & Kudritzki 1986). The observed DAC accelerations are frequently less than 50% of the acceleration expected for the ambient stellar wind. However, the successes of the steady-state wind models (see e.g. Pauldrach, Puls & Kudritzki 1986), the overall observed morphology of the P Cygni profiles (e.g. Groenewegen & Lamers 1989), and the global O star density structures based on Hα observations (e.g. Leitherer 1988), all combine to indicate persuasively that the mass-flux in the time-averaged hot star wind follows a $\beta \sim 0.8$–1 type velocity law. This apparent discrepancy is resolved by the notion that the migrating DACs are due to perturbations or structures in the outflow, through which stellar wind material flows (i.e. we are not observing the same mass-conserving feature at all times). Fullerton & Owocki (1992), for example, picture this perturbation as a plateau in the flow that varies in velocity and radial extent as a function of time. The wind material flows through the plateau, and the observed acceleration of the optical depth enhancement is then associated with the (slow) motion of the plateau, rather than the acceleration of the global mass-flux.

Ultimately, numerical hydrodynamical simulations of (intrinsic or induced) wind perturbations, yielding DAC-like structures, need to match, (i) the initial broadening (to FWHM $\sim 0.25 v_\infty$) and subsequent narrowing ($\leq 0.1 v_\infty$) of the absorption enhancement as a function of increasing velocity, (ii) the

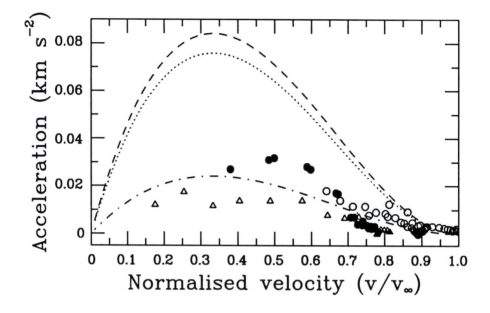

Fig. 1.   The observed accelerations of representative DACs in ζ Pup ($\bullet$), 68 Cyg (O) and HD 64760 (filled triangles), compared with predictions from a standard "$\beta=1$ type" velocity law (ζ Pup, dotted line; 68 Cyg, dashed line; HD 64760, dot-dashed line).

corresponding acceleration range of $\sim 2 \times 10^{-2}$ km s^{-2} to $< 1 \times 10^{-3}$ km s^{-2}, (iii) a separation time between successive episodes of ≥ 1 day, and (iv) changes in the total absorption equivalent width of the spectral line due to DACs that can exceed 50%.

3. Rotationally Modulated Wind Variability

An important finding in recent studies of stellar outflows is that the observed systematic wind variability in OB stars has some relation to the stellar projected rotation velocity ($v_e \sin i$). In particular, the time scales for the development and recurrence of discrete absorption features in O stars is dependent on stellar rotation, with a trend of faster-developing, and more frequent features, with increasing $v_e \sin i$ (see Prinja 1988; Henrichs *et al.* 1988; Prinja 1992; Kaper 1993). Some estimates for DAC recurrence times based on *IUE* spectroscopy, are given in Table 2 for 8 OB stars, together with values of $v_e \sin i$ and the maximum rotation period. Discrete absorption features typically recur every 15 – 30 hours in the fast rotators. For the low $v_e \sin i$ cases, the upper limit for the DAC recurrence time is not constrained by data sets covering ≥ 3 days.

TABLE II

DAC recurrence times versus stellar rotation

Star	Sp. type	$v_e \sin i$ (km s^{-1})	P_{rot} (max.) (days)	DAC recurrence time (days)
ζ Oph	O9.5 V	400	1.1	~ 0.8
68 Cyg	O7.5 III:n((f))	315	2.3	~ 0.7
HD 64760	B0.5 Ib	238	4.9	~ 0.9
ζ Pup	O4 I(n)f	230	4.0	~ 0.6
ξ Per	O7.5 III(n)((f))	215	2.6	~ 1.3
HD 164402	B0 Ib	90	13.5	> 3.0
HD 162978	O7.5 II((f))	80	10.1	> 3.0
19 Cep	O9.5 Ib	40	22.8	> 2.5

This is an intriguing result that highlights the rotation period of a star as playing a key role in determining the characteristic time scale of variability in early-type stars. However, the origin and nature of this connection is not known. Since the majority of currently available *IUE* hot star time-series data sets barely cover 1–2 rotation cycles, the repeatability of the DAC behaviour (and wind structure) has yet to be established or discredited. An important unresolved issue is whether the recurrence of the DACs is *strictly* periodic. Spectroscopic monitoring of wind lines over at least 4 or 5 successive rotation periods is required for a rigorous determination of the DAC recurrence time, and to test whether it is commensurate with reasonable estimates of the stellar rotation period. Detailed DAC properties may be used to establish whether a given structure in the wind enters the line-of-sight on several occasions, in phase with the rotation of the star.

In the absence of such rigorous observational details, physical models attempting to connect wind activity and stellar rotation are poorly constrained, and perhaps premature. The key question of whether stellar rotation is simply a characteristic time scale for the wind variability, or whether it is directly related to the formation of DACs has a major impact on models of wind activity. In the former case the rotation related time scale may result, for example, from increased shear near the stellar surface, with observations probing the interaction between rotation and structure due to wind instabilities. Alternatively, if intensive observations demonstrate that the individual DACs are indeed periodic and tied to stellar rotation, then it is difficult to avoid a direct link to features rooted on the surface of the star. For example, Mullan (1984, 1986)—by analogy to the case of the solar wind—suggested that the entire hot star wind may be pictured in terms of 'corotating inter-action regions', created at the interface of fast and slow wind streams (tied to surface magnetic fields). The interaction regions may then result in the formation of periodic DACs. The notion of surface magnetic fields playing a substantial role in hot star wind activity raises several open issues: (i) the precise physical form expected for O star winds, which are clearly dominated by radiation-pressure driving, (ii) the ubiquity of magnetic fields (and their detection) in luminous hot stars with a wide range of e.g. effective temper-ature, log g, luminosity (i.e. the range over which DACs are observed), (iii) the predicted morphology of the wind-formed line profiles.

4. The Stellar Wind at Low Velocities

The low-velocity, near-star, regions of the outflows deserve special attention since details on the time-dependant behaviour in this regime impact on the growth rate of perturbations, the stability of the inner wind, and the origin of wind variability. I discuss in this section new results concerning low-velocity variability, based on recent optical and UV case studies of HD 152408 (O8: Iafpe; Prinja & Fullerton 1994) and HD 64760 (B0.5 Ib; Massa, Prinja & Fullerton 1994).

4.1. Observations of structure down to $\leq 0.1v_\infty$

Prinja & Fullerton (1994) presented an optical spectroscopic analysis of wind lines in the extreme O supergiant HD 152408. Systematic variability on $\sim$ hourly time scales is particularly apparent in the well developed, recombina-tion formed, He I λ5876 P Cygni profiles. These changes indicate the pres-ence of evolving wind structure, which takes the form of blueward-migrating, discrete optical depth enhancements. Four distinct features may be identi-fied in the full data set over $\sim$ 3 days, progressing slowly bluewards from a velocity of ~ -50 km s^{-1} ($\sim 0.05v_\infty$) at formation, to ~ -500 km s^{-1} at the blue edge of the He I λ5876 profile. The variations in width (FWHM between about $0.08v_\infty$ and $0.25v_\infty$) and the slow accelerations of the DACs (see Table 1) are reminiscent of the behaviour of migrating features gener-ally seen at $\geq 0.3v_\infty$ in UV P Cygni profiles of OB stars. The stellar wind of HD 152408 is therefore structured down to very small velocities, and presumably, very small heights above the photosphere.

Although the unambiguous identification of individual, discrete absorp-tion features in the UV spectra of OB stars is mostly limited to initial velocities $\geq 0.3v_\infty$, Massa, Prinja & Fullerton (1994) have demonstrated the presence of DACs at substantially lower velocities in HD 64760 (B0.5 Ib). A gray scale image showing the time-dependent behaviour of the Si IV $\lambda\lambda$1393.76, 1402.77 resonance line P Cygni profile is shown in Fig. 2. Up to six sequential DACs may be identified in Si IV over $\sim$ 6 days, with formation

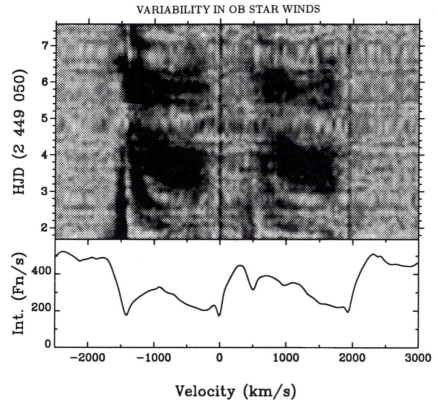

Fig. 2. Gray scale representation of the variable behaviour of the Si IV $\lambda\lambda1393.76$, 1402.77 P Cygni profile in HD 64760 over ~ 6 days. Individual spectra are normalised by a minimum absorption (maximum flux) template (bottom panel). Blueward migrating DACs are evident from very low velocities (dark shades).

velocities spanning $\sim 0.03v_\infty$ to $\sim 0.24v_\infty$. (Similar results are obtained from the Si III $\lambda1206.51$ wind-formed line). HD 64760 is thus another example of a hot star exhibiting extensive stellar wind structure deep in the outflow (see also the optical study of HD 151804 by Fullerton, Gies & Bolton 1992). In the context of the "line-driven instability model" (*e.g.* Owocki 1991, 1992), these results indicate that the growth rate of perturbations must remain large deep in the wind, despite the effect of line drag. Since UV observations show that, in general, the high-velocity portion of OB star wind is also variable, it seems likely that systematic structure occurs at some level throughout the hot stellar outflow.

4.2. A DEEP-SEATED CONNECTION?

Another similarity between the HD 152408 and HD 64760 is that in both cases the recorded appearance of a low-velocity DAC was preceded by velocity disturbances near the photosphere of the stars. Prinja & Fullerton (1994) detected one instance of a blueward shift of ~ 7 km/s in the weak metallic

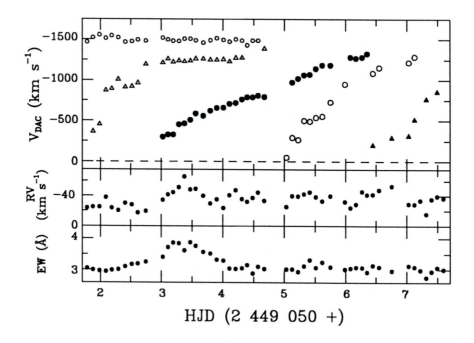

Fig. 3.   The central velocities of progressive DACs in the Si III λ1206.51 profiles of
HD 64760 (top panel). The corresponding changes in the total absorption equivalent width
of the Si III λ1300 (photospheric) triplet and the central velocity of the Si III λ1299 line
are also shown (bottom and middle panels, respectively). Note the substantial change in
the Si III λ1300 strength and velocities during the formation on the strong DAC at ∼
HJD 2 449 053.0.

emission lines of N III λλ5320, 5327 and C III λ5696, ∼ 5 hours prior to
the initial detection of a DAC at ∼ −40 km/s. The metal emission lines are
likely formed very close to the stellar surface.

   The central velocities of DACs in the Si III λ1206.51 line of HD 64760
are shown in Fig. 3 as a function of time. Also plotted here are the total
absorption equivalent width of the Si III λ1300 (photospheric) triplet and
the observed central velocity of the Si III λ1299 component. A substantial
increase in Si III λ1300 strength, plus a velocity shift of ∼ 40 km/s, is noted
in clear association with the formation of the strongest observed DAC at
∼ HJD 2 449 053.0. The Si III λ1300 UV photospheric lines are sensitive
temperature and surface gravity diagnostics in early B stars, and may also
show wind effects (e.g. Massa 1989, Massa et al. 1992). Therefore these lines
likely sample both the photosphere and the densest regions of the stellar
wind. It is an interesting possibility then that the observed disturbances

in these lines are perhaps a precursor to the appearance of DACs at low velocities. These changes may for example diagnose the initial evolution and growth of the line-driven instability.

Further observations are needed to establish whether the perturbations near the photosphere in HD 152408 and HD 64760 are *causally* connected with the development of the DACs, and therefore provide the trigger for the subsequent formation of extended stellar wind structures.

References

Friend, D.B. and Abbott, D.C.: 1986, *Astrophys. J.* **311**, 701.

Fullerton, A.W., Bolton, C.T., Garmany, C.D., Gies, D.R., Henrichs, H.F., Howarth, I.D., Kaper, L., McDavid, D., Prinja, R.K. and Zwarthoed, G.A.A.: 1991, in Baade, D., ed., *ESO Workshop on Rapid Variability of OB-Stars: Nature and Diagnostic Value, ESO Conf. proc. 36*, ESO: Garching, 213.

Fullerton, A.W., Gies, D.R. and Bolton, C.T.: 1992, *Astrophys. J.* **390**, 650.

Fullerton, A.W. and Owocki, S.P.: 1992, in Drissen, L., Leitherer, C. and Nota, A., eds., *Nonisotropic and Variable Outflows from Stars*, Astr. Soc. Pacific Conf. Series 22, 177.

Groenewegen, M.A.T. and Lamers, H.J.G.L.M.: 1989, *Astron. Astrophys. Suppl.* **79**, 539.

Henrichs, H.F.: 1988, in Conti, P.S. and Underhill, A.B., eds., *O Stars and Wolf-Rayet Stars*, NASA SP-497, 199.

Henrichs, H.F., Kaper, L. and Zwarthoed, G.A.A.: 1988, in *A Decade of UV Astronomy with the IUE satellite.* ESA SP-281 **2**, 145.

Howarth, I.D.: 1992, in Drissen L., Leitherer C. and Nota A., eds., *Nonisotropic and Variable Outflows from Stars*, Astr. Soc. Pacific Conf. Series 22, 155.

Howarth, I.D., Bolton, C.T., Crowe, R.A., Ebbets, D.C., Fieldus, M.S., Fullerton, A.W., Gies, D.R., McDavid, D., Prinja, R.K., Reid, A.H.N., Shore, S.N. and Smith, K.C.: 1993, *Astrophys. J.* in press.

Howarth, I.D. and Prinja, R.K.: 1989, *Astrophys. J. Suppl.* **69**, 283.

Kaper, L.: 1993, *Ph.D. Thesis*, Univ. Amsterdam: The Netherlands.

Leitherer, C.: 1988, *Astrophys. J.* **326**, 356.

Massa, D.: 1989, *Astron. Astrophys.* **224**, 131.

Massa, D., Shore, S.N. and Wynne, D.: 1992, *Astron. Astrophys.* **264**, 169.

Massa, D., Prinja, R.K. and Fullerton, A.W.: 1994, in preparation.

Mullan, D.J.: 1984, *Astrophys. J.* **283**, 303.

Mullan, D.J.: 1986, *Astron. Astrophys.* **165**, 157.

Owocki, S.P.: 1991, in Crivellari L., Hubeny I. and Hummer D.G., eds., *Stellar Atmospheres: Beyond Classical Models*, Kluwer: Dordrecht, 235.

Owocki, S.P.: 1992, in Heber U. and Jeffery C.S., eds., *The Atmospheres of Early-Type Stars*, Springer: Berlin, 393.

Pauldrach, A., Puls, J. and Kudritzki, R.P.: 1986, *Astron. Astrophys.* **164**, 86.

Puls, J., Owocki, S.P. and Fullerton, A.W.: 1993, *Astron. Astrophys.* in press.

Prinja, R.K.: 1988, *Mon. Not. Roy. Astr. Soc.* **231**, 21P.

Prinja, R.K.: 1992, in Drissen L., Leitherer C. and Nota A., eds., *Nonisotropic and Variable Outflows from Stars*, Astr. Soc. Pacific Conf. Series 22, 167.

Prinja, R.K. and Fullerton, A.W.: 1994, *Astrophys. J.* in press.

Prinja, R.K., Balona, L.A., Bolton, C.T., Crowe, R.A., Fieldus, M.S., Fullerton, A.W., Gies, D.R., Howarth, I.D., McDavid, D. and Reid, A.H.N.: 1992, *Astrophys. J.* **390**, 266.

Prinja, R.K., Howarth, I.D. and Henrichs, H.F.: 1987, *Astrophys. J.* **317**, 389.

Prinja, R.K. and Smith, L.J.: 1992, *Astron. Astrophys.* **266**, 377.

Discussion

Owocki: You mention that DACs can involve up to a factor of 10 increase in absorption equivalent width. It is very difficult for a theoretical model to do this without producing a comparable increase in emission equivalent width. The IUE spectra (even with their low S/N) don't show clear emission variability. Is this a real puzzle?

Prinja: Matching the observed variable UV P Cygni profiles with a line-synthesis code, for example, would imply changes in the product of mass-loss rate and ionization of up to a factor of 10 (over $\sim$ 1 day!). The fact that the emission components of P Cygni profiles are fairly constant even during large changes in the absorption trough, most likely indicates that the DAC line-formation geometry is not spherically symmetric. Small scale fluctuations are apparent in the emission part of the He I $\lambda5876$ P Cygni profile of HD 152408 (in S/N $\sim$ 200 optical data). However they are substantially less than corresponding absorption changes due to DACs.

Heap: High S/N spectra of ξ Per with HST *do* show variations in the emission component of the Si IV resonance doublet.

Have you looked at changes in the ionization state of the wind at low velocities?

Prinja: A big advantage of studying a B supergiant like HD 64760 is that the evolution of DACs can be monitored in several unsaturated lines, including Si III $\lambda1206$, Si IV $\lambda1400$, and N V $\lambda1240$. Our analysis is still at a preliminary stage, and we are currently addressing the ionization balance issues. There is evidence for a trend of increased ionization with velocity, though in the strongest DAC event a much larger increase of Si III to Si IV is noted at low velocities (with a decrease in N V).

WIND VARIABILITY IN O-TYPE STARS

H.F. HENRICHS, L. KAPER

Astronomical Institute "Anton Pannekoek", University of Amsterdam,
and Center for High-Energy Astrophysics,
Kruislaan 403, 1098 SJ Amsterdam, Netherlands

and

J.S. NICHOLS

Science Programs, Computer Sciences Corporation, 10000-A Aerospace Rd.,
Lanham-Seabrook, MD 20706, U.S.A.

Abstract. The search for the cause of the discrete absorption components (DACs) in P Cygni profiles in UV spectra of O stars continues: pulsation, rotation and magnetic fields are probably all in various degrees important. We highlight some key observations which show that DACs start close to the stellar surface, and emphasize the possible role of localized magnetic fields (which are still to be detected).

1. Introduction

Discrete absorption components (DACs) in P Cygni lines in UV spectra of O and B stars were first detected more than 15 years ago, but their origin is still uncertain. They are readily recognized in individual spectra as 'unexpected' absorption features at high velocities (-1000 to -2000 km/s) in stellar wind lines (N V, C IV, Si IV in O and early B stars, and Si III, N IV, C III, Al III in intermediate to late B stars, e.g. Henrichs 1988). The phenomenon that produces DACs is clearly a fundamental property of the winds of early-type stars. Estimates of the frequency of DACs approach 90% for the O stars and B supergiants (Prinja and Howarth 1986), whereas among the nonsupergiant B stars, DACs uniquely occur in Be stars (Henrichs 1984) with a frequency of at least 65% (Grady *et al.* 1987, Prinja 1989). DACs vary in strength and velocity, with a typical recurrence timescale of a few days in O stars, during which a DAC develops and migrates to an asymptotic velocity, while decaying in strength on a typical timescale of a week. In the meantime a new DAC develops, with a similar (but not identical) pattern. The best examples are the O7.5 III stars ξ Per (Prinja *et al.* 1987, Henrichs *et al.* 1994, hereafter HKN) and 68 Cyg (Prinja and Howarth 1989) and ζ Pup O4I(n)f (Prinja *et al.* 1992). Figure 1 gives examples of rapidly and slowly developing DACs. Many other examples can be found in Kaper (1993).

While significant theoretical progress has been made regarding the properties of radiatively driven flows (Owocki *et al.* 1988, Owocki 1991, Puls *et al.* 1993, Owocki, this volume) there is still no answer to the question why a DAC should develop at the first place, in particular, no prediction can be made when, and under which circumstances a DAC should begin. DACs

L. A. Balona et al. (eds.), Pulsation, Rotation and Mass Loss in Early-Type Stars, 517–529.

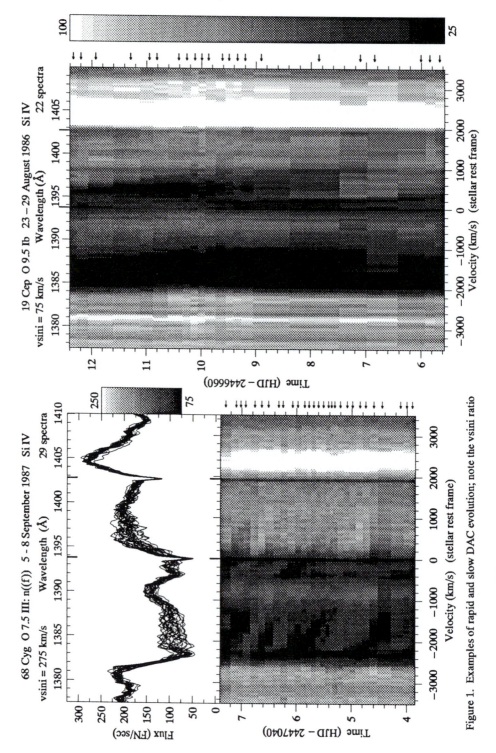

Figure 1. Examples of rapid and slow DAC evolution; note the vsini ratio

may form as a necessary consequence of the intrinsically unstable nature of such winds (Owocki c.s.), or, alternatively, DACs might start to develop as the result of some changing condition at or near the stellar surface (our working hypothesis, e.g. Kaper *et al.* 1992, HKN, Kaper *et al.* 1994). The real question is actually whether the second alternative is needed in addition to the first.

In this review we concentrate on a number of recent observations which all suggest that the outflowing material which is ultimately observed as DACs, can be already identified close to the stellar surface. We first show that the DAC variability is closely associated with variations in the low-velocity and highest-velocity regimes of variable wind lines where no DACs are observed. Secondly, we present some coordinated observations of optical (Hα and He II) and UV wind lines, which suggest that, at least in some O stars, the beginning of a DAC coincides with the appearance of additional blue emission in the optical lines. Finally we discuss a simplified model in which DACs arise from inhomogeneities at the base of the wind, caused by the (hypothesized) presence of weak-magnetic field configuration(s), which temporarily trap some of the wind material on its way out. The observed temporal behavior of the optical lines is in first approximation consistent with such a model.

2. DACs, blue-edge and low-velocity variability

Since the launch of the IUE satellite in 1978, about a dozen O stars have been observed intensively in the UV spectral region with a typical time resolution of one to several hours, resulting in a number of very well planned time series, which show remarkable variety in P Cygni line variability. Three forms of variability are usually distinguished:

(1) the DACs themselves, mostly observed in unsaturated lines, in particular Si IV.

(2) in the blue-edge of saturated P Cygni lines (C IV, N V). The whole edge is observed to significantly shift about 10% in velocity, on a typical timescale of hours to days. The highest velocity of the black bottom of the profile remains approximately at the same position. Figure 2 shows a detailed example for the star ξ Per.

(3) in the low-velocity regime, below 500 km/s, especially in the unsaturated subordinate line N IV λ1718 (see figure 3 (from HKN)).

It is now clear that these three forms are all associated with the behavior of DACs. The precise phase relation, however, between the high-velocity edge and the DACs has not been unraveled for all stars. Figure 2 shows the parallel behavior of the Si IV edge, where DACs dominate the variable behavior, and the strikingly corresponding behavior of the C IV edge, which show that they must be related in origin. Figure 3 shows the parallel behavior

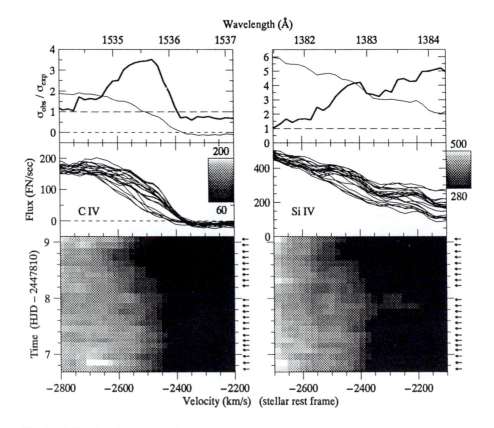

Fig. 2. ξ Per October 1989. The strikingly parallel behavior of the edges of the saturated
C IV line and the unsaturated Si IV line in which DACs dominate the variability, shows
that the blue-edge variability in saturated lines is caused by DACs with a high optical
depth

of the low-velocity N IV region and the DACs in Si IV: again a one-to-one
correspondence of the phenomena.

It is worth noting that a quantification of the significance of variability
in high-resolution IUE spectra, with a S/N < 30, has proven to be crucial
to obtain these results. This problem has been satisfactory solved by HKN:
a two-parameter fit of the form $A \tanh(B/F)$ to the observed signal to noise
ratio as a function of flux level F has been applied to the statistically rigorous
method developed by Fullerton (1990). The upper panels in figures 2 and 4
show the ratio of the observed to the expected variability, which measures
the significance of the variations.

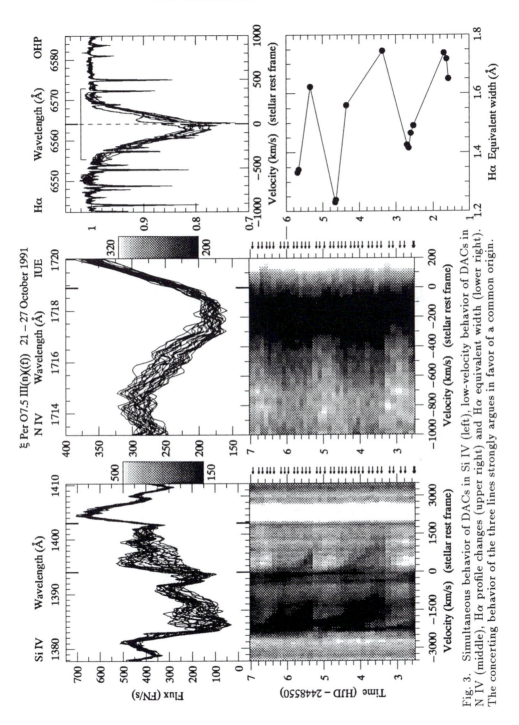

Fig. 3. Simultaneous behavior of DACs in Si IV (left), low-velocity behavior of DACs in N IV (middle), Hα profile changes (upper right) and Hα equivalent width (lower right). The concerting behavior of the three lines strongly argues in favor of a common origin.

3. Simultaneous low- and high-velocity stellar wind observations

A strong argument that wind variability in the form of O-star DACs originates close to the photosphere is that DACs in the Si IV resonance lines
develop at the same time when enhanced absorption appears at low velocity
(< 250 km/s) in the subordinate N IV $\lambda1718$ line (figure 3) (HKN, Kaper
and Henrichs 1994). This N IV line can only be formed close to the star,
because the strength of subordinate lines decreases with (density)^2, which
is unlike the resonance lines, where the strength decreases proportionally to
density and which are therefore observed throughout the wind (see HKN).

Further evidence that DACs have their origin very near the star comes
from the observation that rotation plays a significant role. Starting with
Prinja (1988) and Henrichs et al. (1988), all papers in which more than one
sequence of DACs was described, strongly suggest that DAC patterns repeat
on the rotational timescale of the star (see also Prinja, this volume). Fig. 4
illustrates this over a 4-year period for ξ Per (rotation period 2 to 4 days).

We have observed three cases by means of simultaneous optical and UV
spectroscopy, which strongly suggest that matter which piled up at a certain
confined location close to the photosphere, is carried around by rotation.
This matter is observable as additional Hα or He II 4686 emission as soon
as it appears around the limb of the star. Pulsation cannot cause such an
effect, because the pulsational time scale is too short to account for the
observed changes. The three cases are:

(1) In ξ Per not only the equivalent width of the Hα absorption line varies in
concert with the DACs (figure 3, right), but the additional emission appears
at the blue side of the profile (at about $-v\sin i$) and moves towards the red
(figure 5a). This happened during 3 subsequent DAC events.

(2) Similar behavior was observed in 19 Cep (O9.5Ib) in Hα, which is also
in absorption in this star. Just prior to the epoch that a DAC started,
additional emission appeared on the blue wing of the line and moved to the
red (Kaper et al. 1994).

(3) In the O6I(n)fp star λ Cep simultaneous variations were observed at the
blue side of the He II 4686 line (< 200 km/s) and the onset of DACs (figure
5b, Henrichs 1991).

It looks as if *the beginning of a DAC can be predicted from the rather
sudden appearance of additional blue emission in the optical lines*. This result
has obviously to be confirmed for other stars. The observations above are
consistent with a configuration in which a part of the stellar wind with
enhanced density at a given location just above (because of the emission)
the stellar surface is carried into the line of sight by the rotation of the
star (because of the movement from blue to red). This location is likely
the footpoint of a DAC that will appear somewhat later, depending on the
rotational and flow timescale.

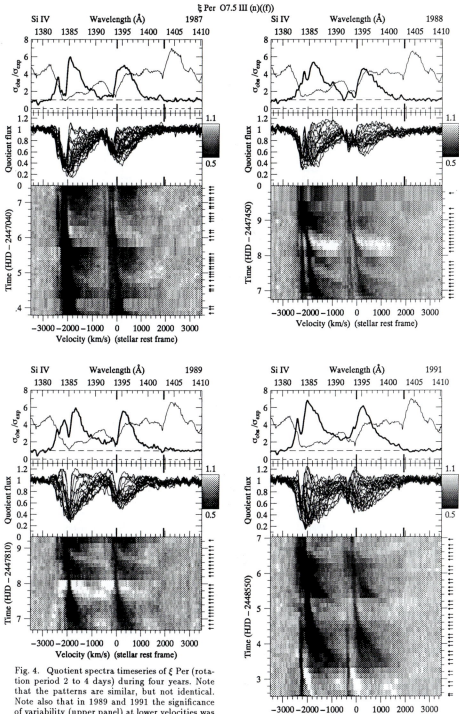

Fig. 4. Quotient spectra timeseries of ξ Per (rotation period 2 to 4 days) during four years. Note that the patterns are similar, but not identical. Note also that in 1989 and 1991 the significance of variability (upper panel) at lower velocities was higher than in 1987 and 1988.

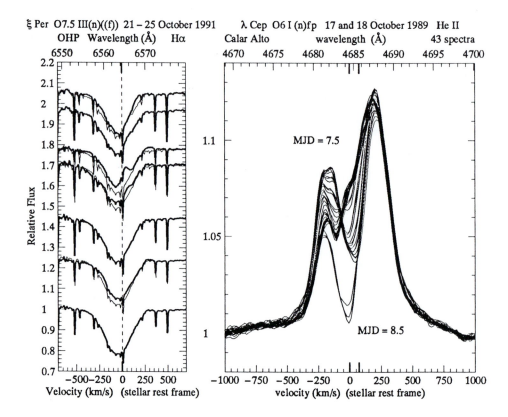

Fig. 5. *a. left:* Representative Hα spectra of ξ Per in October 1991, in the vertical direction shifted proportionally with time. The first spectrum in each of the three episodes of figure 3 is overplotted with a thin line, in order to demonstrate the appearance of additional emission at the blue side of the profile just prior to the beginning of a Si IV DAC.
b. right: Sample profiles He II λ4686 of λ Cep in October 1989, during two consecutive nights. Typical time resolution was 20 min. On top of the asymmetric double-peaked emission line, there is additional emission appearing at −200 km/s, which corresponds to −v sin i, which was rapidly followed by additional central absorption. The next day, the emission was gone.

4. DACs, rotation and magnetic fields in O stars

The considerations above argue in favor of an interpretation that DACs originate at a particular spot on the stellar surface. As put forward by HKN and Kaper *et al.* (1994) the most likely explanation for such behavior is the presence of a localized magnetic field, which traps some fraction of the outflowing wind, and makes a non-axisymmetric wind, which leads in turn to DACs, like in a modified model of corotating interacting regions (CIRs, Mullan 1984). It could also be that simply the fact that the lower boundary condition of the wind at such a magnetic patch is different from elsewhere, is sufficient to create an inhomogeneous wind which, according to its unstable nature, form structures which are observed as DACs. Anyhow, the above given argumentation leads automatically to the next logical step, namely to investigate whether magnetic fields are actually present in O stars.

Till recently, only null detections with 1σ of order 100 G existed for two O stars (ζ Pup, Barker *et al.* 1981, and ζ Oph, Barker *et al.* 1985), but only two or three measurements were obtained. Currently a large long-term survey for magnetic fields in O stars is being carried out by Bohlender (this volume) on a variety of telescopes with the UWO Balmer line polarimeter. To date only upper limits of about 300 G has been established. Also for these objects usually one measurement was obtained, and none of our targets has been included.

With a field configuration as suggested above, it is expected that the longitudinal component of the magnetic field will be modulated with the rotation of the star. This means that in case of a dipolar type field only twice per rotation the maximum field strength can be measured. Single snapshots are therefore unlikely to be obtained during optimum orientation, and longer sampling is needed to establish the presence of a magnetic field.

From the theoretical side, it is difficult to estimate the strength of such a field (see e.g. Moss, this volume), although Maheswaran and Cassinelli (1988, 1992) were able to set constraints. That magnetic fields should exist in O stars is shown by the existence of the very strong fields at the end of their evolution in the form of collapsed neutron stars. An upper limit of about 100 G is set by the requirement that a stronger field would cause a rapid spin down of the star during its main-sequence lifetime (MacGregor *et al.* 1992), which is not the case, as many evolved stars are still moderate rotators.

About the configuration and stability of such weak fields, not much is known. If the wind modulation in O stars, as illustrated in figure 4, indeed reflects the presence of surface fields, it is clear that they are not very stable, with a lifetime of the order of a month, i.e. 10 stellar rotations.

5. Discussion and prospects

Although some progress has been made, we are still far from understanding the different roles played by rotation, non-radial pulsation (undoubtedly present) and magnetic fields (yet to be detected) and their mutual interaction with respect to the behavior of O-star winds. It is important to stress that we have not found direct observational evidence that stellar pulsations affect the wind variability. A tentative picture could be summarized as follows: the non-radial pulsation causes the instabilities in the wind to develop. At places where at the surface a magnetic field "patch" is able to influence the flow behavior, the lower boundary condition of the wind will be different. We observe the less-accelerated, denser material just above the surface in the optical lines, which will be rotationally modulated. The non-axisymmetric flow with contrasting velocity fields (i.e. of the normal wind and above a patch) gives rise to the development of CIR's (like in the solar wind case), which in turn gives rise to density enhancements, which we observe as DACs. It is clear that the crucial test of such a model should have to come from simultaneous UV spectroscopy and magnetic-field measurements. From the theoretical side it would be interesting to investigate the existence and stability of magnetic patches as described above. Finally, much progress is expected from hydrodynamical simulations and line profile calucations by Owocki, Puls and others.

Acknowledgements

The authors are very grateful to their many co-authors and co-observers on various optical telescopes around the globe, especially H. Ando, H. Barwig, K. Bjorkman, A. Fullerton, D. Gies, R. Hirata, E. Kambe, D. McDavid. L. Snoek, H. Volten, R. Zwarthoed and others. Discussions with S. Owocki and D. Bohlender are also gratefully acknowledged. This work was partially supported by NASA grant NAS5-31835 and by the Netherlands Organization for the Advancement of Pure Research (NWO).

References

Barker, P., Landstreet, J., Marlborough, J., Thompson, I., Maza, J.: 1981, *Astrophys. J.* **250**, 300
Barker, P., Landstreet, J., Marlborough, J., Thompson, I.: 1985, *Astrophys. J.* **288**, 741
Fullerton, A.W., 1990, Thesis, University of Toronto
Grady, C.A., Bjorkman, K.S., Snow, T.P.: 1987, *Astrophys. J.* **320**, 376
Henrichs, H.F.: 1984, in ESA, ed., *Fourth European IUE Conference, Roma, Italy*, ESA SP-218, 43.
Henrichs, H.F.: 1988, in P.S. Conti and A.B. Underhill, eds., *NASA/CNRS monograph series*, NASA SP-497, 199.
Henrichs, H.F.: 1991, in Baade, D., ed., *Rapid Variability of OB-stars: Nature and Diagnostic Value*, ESO: Garching, 199.
Henrichs, H.F., Gies, D.R. Kaper, L., Nichols-Bohlin, J.S., et al.: 1990, in ESA, ed., *Evolution in Astrophysics: IUE Astronomy in the era of new space missions*, ESA SP-310, 401.
Henrichs, H.F., Kaper, L. and Zwarthoed, G.A.A.: 1988, in ESA, ed., *Celebratory Symp.: A decade of UV astronomy with IUE*, ESA SP-281, Volume 1, 145.
Henrichs, H.F., Kaper, L. and Nichols, J.S. : 1994, *Astron. Astrophys.* **265**, 685 (HKN)
Howarth, I.D., Prinja, R.K. : 1989, *Astrophys. J. Suppl.* **69**, 527
Kaper, L., 1993, Thesis, University of Amsterdam
Kaper, L., Henrichs, H.F., Zwarthoed, G.A.A. and Nichols-Bohlin, J.: 1990, in L. A. Willson, G. Bowen and R. Stalio, eds., *NATO Workshop on Mass Loss and Angular Momentum of Hot Stars*, Kluwer, 213

Kaper, L., Henrichs, H.F., Nichols-Bohlin, J.S.: 1992, in B. Warner, ed., *Variable Stars and Galaxies, Cape Town, Feb.*, ASP Conference Series, 135.

Kaper, L., Henrichs, H.F.: 1994, in A. Moffat, S. Owocki, A. Fullerton , N. St.-Louis, eds., *Workshop on instability and outflows from stars*, ASP Conference Series,

Kaper, L., Henrichs, H., Ando, H., Bjorkman, K., Fullerton, A., Gies, D., Hirata, R., Kambe, E. McDavid, D., Nichols, J.: 1994, *Astron. Astrophys.* submitted

MacGregor, K.B., Friend, D.B., Gilliland, R.L.: 1992, *Astron. Astrophys.* **256**, 141

Maheswaran, M., Cassinelli, J.P.: 1988, *Astrophys. J.* **335**, 931

Maheswaran, M., Cassinelli, J.P.: 1992, *Astrophys. J.* **386**, 695

Mullan, D.: 1984, *Astrophys. J.* **283**, 303

Owocki, S, Castor, J., Rybicki, G.: 1988, *Astrophys. J.* **335**, 914

Owocki, S.: 1991, in I. Hubeny and L. Crivellari, eds., *NATO Workshop on Stellar Atmospheres: Beyond Classical Models*, Kluwer,

Prinja, R.K.: 1988, *Mon. Not. Roy. Astr. Soc.* **231**, 21P

Prinja, R.K.: 1989, *Mon. Not. Roy. Astr. Soc.* **241**, 721

Prinja, R.K. and Howarth, I.D.: 1986, *Astrophys. J. Suppl.* **61**, 357

Prinja, R.K. and Howarth, I.D.: 1988, *Mon. Not. Roy. Astr. Soc.* **233**, 123

Prinja, R.K., Howarth, I.D. and Henrichs H. F.: 1987, *Astrophys. J.* **317**, 389

Prinja, R.K., et al.: 1992, *Astrophys. J.* **390**, 266

Puls, J, Owocki, S., Fullerton, A.: 1993, *Astron. Astrophys.* **279**, 457

Discussion

Prinja: It is not clear to me in your CIR picture, what the relation is between high-velocity blue-edge changes in the saturated UV P Cygni profiles and the (apparently connected) development of DACs at very low velocity.

Owocki (response to Prinja): The fact that the C IV blue edge is gradual despite the strong opacity of this resonance line has two possible interpretations. First is that it arises from very low-density, high-speed flow, such as seen in the wind instability simulations. A second possibility is that it arises from a high-speed stream forming from a rapidly diverging B field, relatively close to the star, but geometrically narrow so that it occults only a small fraction of the stellar disk. We hope soon to compute a dynamical "CIR" model to see if this idea can be made to match observations.

Lafon: For corotating interacting regions to organize, it is necessary that the magnetic field be frozen in the matter (as in the solar wind). This seems difficult for *weak* fields, in *cool* disks, with *high* density. The validity of the assumption of "quasi-frozen field" (infinite conductivity) must be verified. Maybe this could be applied with better results for hot stars than for Be stars.

Henrichs: I agree, and would like to emphasize that DACs in Be stars are likely to have a different origin then in O stars.

Dachs: Similar DAC development might also be visible in spectra of the Be star HR 2855 (FY CMa) obtained in early 1987, when DACs were visible in Si IV and C IV profiles taken by IUE at about the same time when He I λ 5876Å showed inverse P Cygni structure.

Henrichs: Thank you for mentioning this. It would be interesting to follow the DACs in this star with a high time resolution. As far as I know that has not been done.

Le Contel: How can you predict the appearance of the DACs? I did not understand it. Similar DACs appear in some β CMA stars associated with the phase of the pulsation. It would be interesting to confront the two phenomena.

Henrichs: We found that the beginnings of DACs in O stars can be predicted from the the rather sudden appearance of additional emission at the blue side of Hα in the cooler O stars, and of He II in the hotter O stars. I don't know whether this behavior is also observed in the β CMA stars you mention. It should be looked at.

Smith: You mentioned that there was no observational evidence for your magnetic patch model for O stars, but in fact if you look carefully at the original paper suggesting NRP in an Oe star, Vogt and Penrod (1983) monitored lpv's of two He I profiles, $\lambda6678$ and $\lambda4471$, in ζ Oph. What was left understated in their paper was rather significant differences in the lpv's of these two lines. At that time neither Don Penrod, nor I could understand this. However, the differences in behavior that I described yesterday for λ Eri between $\lambda6678$ and $\lambda4922$ (the triplet analogy of $\lambda4471$) may perhaps be in a rather similar way to what I suggested therefore. Therefore there are some grounds for suspecting that magnetic activity may be present as well in ζ Oph.

Henrichs: I hope that such fields will be searched for soon.

Harmanec: I have three questions:

(1) When I was arguing in favour of local outflow of material from the equator of rapidly rotating stars in my 1991 Garching review, I pointed out that there should be some logical time delay in variations detected in lines formed close to, and far from the photosphere. Can you see evidence of such time delays in your data?

(2) Do you need the local magnetic field to start the outflow? Can it not start from a high wave crest of NRP, where the apparent irregularity could be related to multiperiodicity of NRP waves?

(3) It seems to me that the periodogram for λ Cep you have shown indicates that there may be a systematic variation in the value of the period accross the line (differential rotation??) What do you think about that?

Henrichs: (1) We have not searched for such time delays in different optical lines. We have data for a few stars which possibly carry this information.

(2) The magnetic field is not needed for the outflow to start. Above a magnetic patch, it rather slows down the outlow. If I look at figure 4, the best evidence for long-term behavior we have sofar, it is clear that there is no obvious long-term clock driving the DAC patterns. This suggests that periodic phenomena like (multiperiodic) NRP is unlikely the trigger. This is why we propose short-lived magnetic patches: if they are present, they will be carried around by rotation, but if they disappear and reappear later, the phase is lost.

(3) We have not finished our period analysis for this star, as this is non-trivial. It is an interesting suggestion, which we should investigate.

Baade: Is it possible that the difference in apparent predictability between O stars and Be stars concerning the development of DACs arises from the difference in the ratio of wind flow time to rotation period? If a possible circumstellar pattern gets wound up by rotation, it may be difficult to see correlations.

Henrichs: This is certainly possible, but more factors are likely involved. For instance the observed correlation between V/R variability in Be stars and the strength of the DACs (Doazan et al, Telting et al) suggests that DACs in these stars are related to the conditions in the disk, which is also very different from what we see in O stars.

INTERNATIONAL MULTIWAVELENGTH CAMPAIGNS ON
SHORT-TERM VARIABILITY OF OB STARS:
OPTICAL POLARIMETRY

D. MCDAVID*

Limber Observatory, P.O. Box 63599, Pipe Creek TX 78063-3599

From 1986 through 1992, wide-band optical (B or V filter) linear polarization measurements of eight Be stars and seven O stars were obtained simultaneously with ultraviolet observations from *IUE* and worldwide ground-based optical spectroscopy and photometry in a series of campaigns designed to study the short-term variability of these objects. Each campaign consisted of intensive monitoring of a few carefully chosen stars over a period of several days and nights, with the greatest possible continuity subject to the limitations of instrument scheduling, weather, and the longitudes of the observing sites.

With a typical instrumental uncertainty of about 0.03% for a single observation, no polarization variability was detected at the 3σ level for any of the program stars. Normalized Stokes parameter plots are shown in Fig. 1, where the data points are filled circles, the mean is a cross drawn to the size of the average instrumental uncertainty of a single observation, the standard deviation is represented by a dotted ellipse centered on the mean, and three times the average instrumental uncertainty of a single observation is represented by a solid ellipse centered on the mean.

Since most of the stars showed definite signs of activity in their winds and photospheres during the time intervals covered, it appears that associated changes in polarization are uncommon, or at least too small to measure by current techniques. In some cases, weak periodicities may be present in the polarization at frequencies which match those found in the simultaneous photometric and spectroscopic data sets from the campaigns, but their significance has not yet been thoroughly evaluated.

* Guest Observer, McDonald Observatory, University of Texas at Austin; Visiting Astronomer, Cerro Tololo Inter-American Observatory, National Optical Astronomy Observatories, operated by the Association of Universities for Research in Astronomy, Inc., under contract with the National Science Foundation.

L. A. Balona et al. (eds.), Pulsation, Rotation and Mass Loss in Early-Type Stars, 530–531.
© 1994 *IAU. Printed in the Netherlands.*

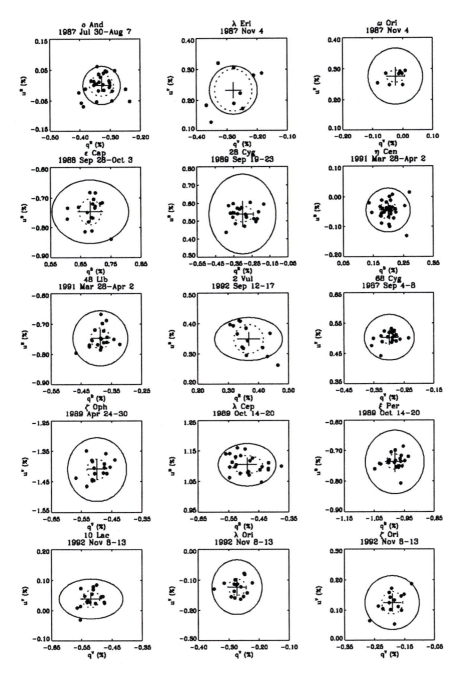

Fig. 1.   Normalized Stokes parameter plots for the program stars. (The symbols are explained in the text.)

HIGH RESOLUTION SPECTROSCOPY AT THE ESO 50 CM TELESCOPE: SPECTROSCOPIC MONITORING OF LUMINOUS HOT STARS

A. KAUFER, H. MANDEL, O. STAHL, B. WOLF and
TH. SZEIFERT, TH. GÄNG, C.A. GUMMERSBACH
Landessternwarte Heidelberg-Königstuhl, D-69117 Heidelberg, Germany

and

J. KOVACS
Gothard Astrophysical Observatory, H-9707 Szombathely, Hungary

Abstract. Galactic Luminous Blue Variables and A- and B-type supergiants were monitored spectroscopically with high resolution in wavelength and time. Line profile variations on different timescales are found.

1. Observations

With the portable fiber-linked echelle spectrograph of the Landessternwarte Heidelberg–Königstuhl FLASH (Mandel, 1988) attached to the ESO 50 cm telescope at La Silla we monitored from February through May 1993 a few galactic Luminous Blue Variables (LBV) and several A- and B-type supergiants (cf. Wolf et al. 1993) at high resolution both in wavelength

TABLE I

List of monitored objects

object	sp	spectra/nights
η Car	pec (LBV)	103/117
θ^1 Ori C	O7 V	78/89
AG Car	Ofpe–A2 (LBV)	91/115
ζ^1 Sco	B1 Ia-O	93/114
β Ori	B8 Ia	86/104
HD 96919	B9 Ia	84/115
HD 92207	A0 Ia	86/115
HD 160529	A2 Ia-O (LBV)	71/101

($R = 20000$) and time (each night one spectrum per object). With our EEV CCD with 1152×770 pixel of $22\,\mu$ we cover 2700 Å in one exposure (standard setup: $4050 - 6780$ Å).

L. A. Balona et al. (eds.), Pulsation, Rotation and Mass Loss in Early-Type Stars, 532–533.
© *1994 IAU. Printed in the Netherlands.*

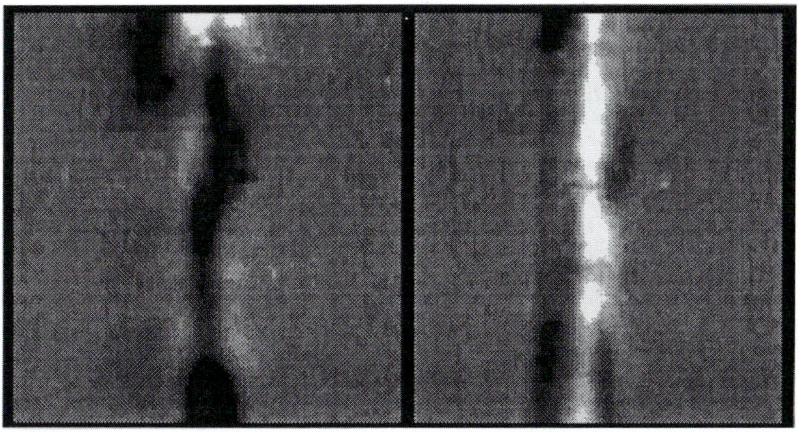

Fig. 1. Line profile variations of He I 6678 (left) and Si II 6347 of AG Car. The lines are centered to the systemic velocity (+20 km/s) of AG Car. The complete width of the abscissa is from −300 to +300 km/s for both lines. The ordinate covers the timespan from JD 2449023(bottom) to JD 2449139(top).

2. Aims and first results

The aim of the campaign is to study the time and depth dependent atmospheric velocity fields of these objects and the mass-loss variations derived from the Hα-profile variations.

All monitored objects show line profile variations from strictly periodic as found for θ^1 Ori C (cf. Stahl et al. 1993) to highly irregular as for β Ori. The observed typical timescales of these variations due to the hydrodynamic processes in the extended atmospheres of these luminous objects range from a few days to a few weeks. Therefore daily observations over months are the right choice to monitor these stars.

References

Mandel, H.: 1988, in IAU Symp. 132, eds. G. Cayrel de Strobel and M. Spite, pg. 9-13, Kluwer.

Stahl, O., Wolf, B., Gäng, Th., Gummersbach, C.A., Kaufer, A., Kovacs, J., Mandel, H., Szeifert, Th.: 1993, A&A, **274**, L29-L32

Wolf, B., Mandel, H., Stahl, O., Kaufer, A., Gäng, Th., Gummersbach, C.A., Szeifert, Th.: 1993, "High resolution spectroscopy at the ESO 50 cm telescope": The ESO Messenger (submitted)

THE STELLAR WIND OF χ^2 ORI AND ITS VARIABILITY

L. SAPAR and A. SAPAR

Tartu Astrophysical Observatory, Tôravere EE2444, Estonia

Abstract. The resonance spectral lines of high-resolution IUE spectra of χ^2 Ori (B2I) have been studied with the aim to determine in more detail the physical status of the stellar wind and its variability due to shell outbursts and clumps superimposed on the isotropic stellar wind. Some general features of resonance spectral line profiles have been studied and explained in physical terms using χ^2 Ori as a specimen.

1. General picture

The spectrum of B2Ia star χ^2 Orionis is rich of spectral lines. In our recent investigation (L. Sapar and A. Sapar 1992) we succeeded to recognize, identify and catalogue by computer codes in high-resolution IUE exposures SWP 3436 and LWR 3021 about 1700 observed spectral features, most of them being multiple blends. The heavily extended stellar wind of χ^2 Ori produces P-Cygni type profiles of resonance lines of SiIV, CII, CIV, AlIII and NV. We study them using the theory of resonance spectral line formation in a spherically symmetric stellar wind with an optically thick and geometrically thin layer, where multiple scattering takes place in the Doppler core of Voigt function (A. Sapar and L. Sapar 1990). We use an idealized model stellar wind with sharp and optically thick stratification boundaries throughout the resonance line profile. From the images SWP 2970, SWP 3436, SWP 4302, SWP 6471 and SWP 40669 we tried to sketch the outflow velocity run, the observed stellar wind extension and their possible variations. We found that the observed moderate variations can be due to both shell outbursts and clumps with deviating radial velocities. The extension and the extreme velocity of the expanding envelope differ for resonance line profiles of each ion, thus showing an ionizational stratification in the stellar wind. However, the nature of the short-term variations of the stellar wind can be studied in detail only by analysing the landscape or darkening picture of the resonance line profile temporal variations. In such a picture a shell outburst is observed as a perturbation which appears at line centre, after which it broadens over the whole line profile. The restoration of the undisturbed line profile takes place in the same way. The small migrating dips and peaks in the smooth spectral line profiles appear due to clumps with perturbed velocities and opacities, and their form mimics their areas and perturbation velocities.

L. A. Balona et al. (eds.), Pulsation, Rotation and Mass Loss in Early-Type Stars, 534–535.

2. Formulae for extension and extreme velocity of the envelope

The ratio of stellar radius r_* and the radius of envelope r in the above mentioned idealized picture can be found by (see L. Sapar and A. Sapar 1993)

$$r_*^2/r^2 = 1 - \mu_L^2 = 1 - 1/(2(\Delta\lambda_1/\Delta\lambda_L)^2 - 1)$$

where $\Delta\lambda_1$ and $\Delta\lambda_L$ are the values of the wavelength shifts towards the blue as measured in the spectral line profile at the blue edge corresponding to relative residual flux values 1/2 and 0, respectively. The latter point can be found by extrapolation of the curve of the steep blue edge of the spectral line profile to its zero-flux value.

The extreme velocity for each stratified spectral line can be found theoretically by $V_e = c\Delta\lambda_L/\mu_L$ or, using a quantity μ_1 which is connected with the above given quantities by $\mu_1^2 - \mu_L^2 = \frac{1}{2}\left(\frac{r_*}{r}\right)^2$, can be obtained from $V_e = c\Delta\lambda_1/\mu_1$.

3. Main results

The study of observed resonance lines of CII, CIV, AlIII and SiIV in χ^2 Ori spectra carried out by us showed that they are variable due to deviating wind velocity both in outbursts and in clumps. It is found that the lower the ion recombination potential, the further off in the stellar wind its resonance line originates. The P-Cyg type line formation with extreme velocity of the stellar wind takes place at a relative distance about 0.2 for CIV and reaches 0.4 stellar radii for the CII resonance lines. An unexpected result is that the outflow velocity is decelerating with distance, evidently due to gravity. The extreme velocity of the stellar wind is about -850 km/s in CIV, -800 km/s in SiIV, -700 km/s in AlIII and -600 km/s in the CII resonance lines. An observational feature in favour of deceleration is the presence of wide zero-flux cores in the absorption components of the resonance lines. However, this feature can also be generated by the presence of large-scale turbulence-like instabilities in the stellar wind.

The results of investigation will be published in more detail in our paper L. Sapar and A. Sapar (1993).

References

Sapar, A. and Sapar, L.: 1990, *Publ. Tartu Astrofüüs. Obs.* **53**, 52
Sapar, L. and Sapar, A.: 1992, *Baltic Astronomy* **4**, 461
Sapar, L. and Sapar, A.: 1993, *Baltic Astronomy* , (in press)

Observed variety of timescales
in variability of P Cygni

I. Kolka, Tartu Astrophysical Observatory, EE2444 Tõravere, Estonia

1. P Cygni is a LBV (B1 Ia+) which exhibits photometric and spectroscopic variability.

2. We present observational data from two periods (1981-82 and 1989-90) pointing to three different variability timescales.

3. Data sources:

a) spectroscopy - our observations at Tartu and results published earlier (Markova, Kolka 1989),

b) photometry - our compilation (Kolka, de Groot, Percy 1993).

4. The features measured:

a) the brightness (m(V)) in V -filter,

b) the residual intensity (R_c) and the radial velocity (V_r in km/s) of the absorption trough in chosen spectral lines.

5. Important quasi-stabil variability timescales (see Fig.1, Fig.2) in our data:

a) cycle-length 230...240 days in the variations of R_c in Balmer-line H9 correlated with the V_r -oscillations in the same line. The m(V) -curve has regularly local maxima near the moment of residual intensity minimum. The deepening of the absorption trough is related to the higher average level of the brightness.

b) cycle-length around 50 days in the behaviour of R_c and V_r in the line HeI3820 (most pronounced in the period 1981 - 82 (Fig.1) when minima in R_c occurred as a rule together with maxima in V_r). This timescale-value is often observable in the lightcurve, too (see also Percy et al. 1988).

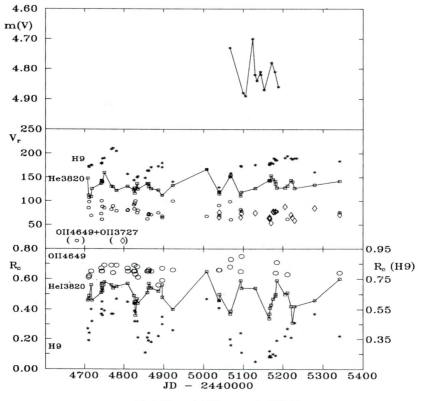

Fig.1 The variability curves in 1981-82

L. A. Balona et al. (eds.), Pulsation, Rotation and Mass Loss in Early-Type Stars, 536–537.
© 1994 IAU. Printed in the Netherlands.

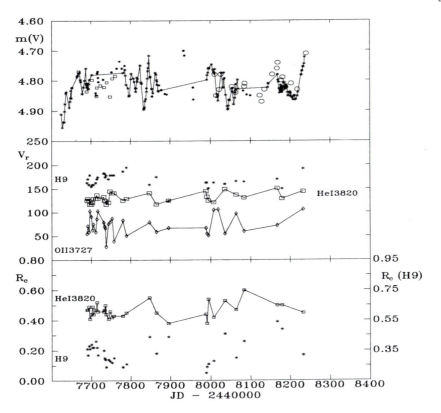

Fig.2 The variability curves in 1989-90

c) cycle-length 15...20 days is common in V_r -variability of weak lines (e.g. OII3727) and in lightcurve data with sufficient time resolution.

6. We believe the time-dependent intensity of matter outflow is responsible for all variability cycles.

The longest cycle (over 200 days) reflects the swinging of the envelope between the states of lower and higher density which may be caused by the bistability of the stellar wind (Pauldrach, Puls 1990).

The shortest timescale (15...20 days) can be in connection with pulsations – having probably the non-radial character (Percy, Welch 1983) – which excite the inhomogeneous non-isotropic mass ejection in the form of " blobs" as it was suggested on the basis of spectropolarimetric measurements (Taylor et al. 1991). The alternative way to get the changing density in the essential volume of the envelope is in this case the changing birthrate and/or the varying dimensions of "blobs" due to the interaction of slightly different periods generated by the pulsational instability.

The cycle with the medium length can be fitted to this scenario by incorporating the rotational modulation of the asymmetrical outflow of "blobs" (the estimated rotation period of P Cygni is ~ 50 days (Percy et al. 1988)).

References

Kolka I., de Groot M., Percy J.R., 1993, (in preparation)

Markova N., Kolka I., 1989, An Atlas of Spectral Line Profiles of P Cygni in 1981-83, Tartu Teated nr.103.

Pauldrach A.W.A., Puls J., 1990, Astron. Astrophys. 237,409-424.

Percy J.R., Welch D.L., 1983, PASP 95,491-505.

Percy J.R. et al., 1988, Astron. Astrophys. 191,248-252.

Taylor M. et al., 1991, Astron. J. 102,1197-1206.

STELLAR WINDS IN A-TYPE SUPERGIANTS

E.VERDUGO and A.TALAVERA*

IUE Observatory, European Space Agency
P.O.Box 50727, E-28080-Madrid, Spain

1. Introduction

A-type supergiants are just at the boundary between the early type super-giants, with strong stellar winds driven by radiation pressure and the cool supergiants in which the origin of mass loss is the dissipation of mechanical energy. The mechanisms involved in the mass loss processes in A supergiants as a whole, are still uncertain.

In a study of A-type supergiants observed with IUE, Talavera and Gomez de Castro (1987) divided them into two groups: Group I contained the less luminous A supergiants which show only weak signs of stellar winds. Group II are the most luminous A supergiants and they show strong evidences of wind and mass loss.

We present here a preliminary study of 33 stars in the visible and in the UV range, whose purpose is the modelling of the envelope-wind complex in A-type supergiants.

2. Hα and Hβ profiles

We have studied 29 stars of our sample in the visible range. The spectral intervals we have selected correspond to the following lines: CaII K line (3933 Å), MgII (4481 Å), Hβ (4861 Å), NaI and HeI (5890 Å) and Hα (6563 Å).

The Hα profiles in A-type supergiants present very different shapes. The less luminous stars show symmetric absorption profiles but when luminosity increases the profile starts to be asymmetric becoming a P Cygni type III profile for the brightest stars (see Fig.1).

The less luminous stars do not show variability in the line profiles. However we have observed variations in some of the stars with asymmetric-emission profiles. In five "peculiar" stars the observed profile shows very strong variations, becoming in two stars a pure emission profile.

In most of the stars the Hβ observed profile does not show evidence of mass loss. However in the five "peculiar" stars with very strong variations in Hα, the Hβ profile is asymmetric and shows emission (see Fig.1).

* Affiliated to the Astrophysics Division, Space Science Department

L. A. Balona et al. (eds.), Pulsation, Rotation and Mass Loss in Early-Type Stars, 538–539.

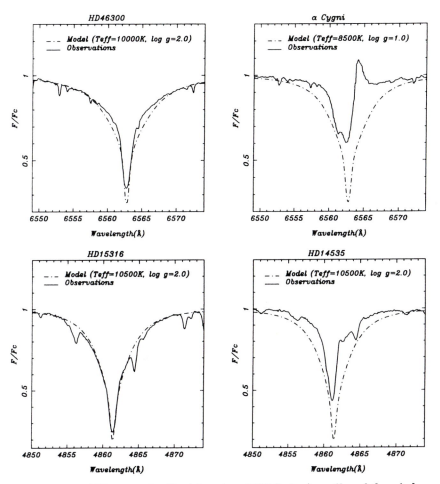

Fig. 1.   Effects on the Hα (above) and Hβ (below) profiles of the wind

3. Model Atmospheres

We have fitted the energy distribution (visible plus uv) using the Kurucz model atmospheres as a preliminary solution. These models assume LTE and radiative equilibrium, and include line blanketing in the opacity.

For stars with mass loss the Kurucz models are inadequate for a detailed explanation of spectral features formed in the wind (see Fig.1).

In the second part of our project we shall try the modelling of these winds taking into account effects of velocity fields in expanding atmospheres.

References

Kurucz, R.L.: 1979a, *Ap. J. Suppl.* **40**, 1
Talavera, A. and Gomez de Castro, A.I.: 1987, *Astron. Astrophys.* **181**, 300

AUTHOR INDEX

546

SUBJECT INDEX